# LITHIUM-SULFUR BATTERIES

# LITHIUM-SULFUR BATTERIES
## Advances in High-Energy Density Batteries

Edited by

**PRASHANT N. KUMTA**
Swanson School of Engineering, University of Pittsburgh, Pittsburgh, PA, United States

**ALOYSIUS F. HEPP**
Nanotech Innovations LLC, Oberlin, OH, United States

**MONI K. DATTA**
Bioengineering Department, University of Pittsburgh, Pittsburgh, PA, United States

**OLEG I. VELIKOKHATNYI**
Bioengineering Department, University of Pittsburgh, Pittsburgh, PA, United States

ELSEVIER

Elsevier
Radarweg 29, PO Box 211, 1000 AE Amsterdam, Netherlands
The Boulevard, Langford Lane, Kidlington, Oxford OX5 1GB, United Kingdom
50 Hampshire Street, 5th Floor, Cambridge, MA 02139, United States

Copyright © 2022 Elsevier Inc. All rights reserved.

No part of this publication may be reproduced or transmitted in any form or by any means, electronic or mechanical, including photocopying, recording, or any information storage and retrieval system, without permission in writing from the publisher. Details on how to seek permission, further information about the Publisher's permissions policies and our arrangements with organizations such as the Copyright Clearance Center and the Copyright Licensing Agency, can be found at our website: www.elsevier.com/permissions.

This book and the individual contributions contained in it are protected under copyright by the Publisher (other than as may be noted herein).

**Notices**
Knowledge and best practice in this field are constantly changing. As new research and experience broaden our understanding, changes in research methods, professional practices, or medical treatment may become necessary.

Practitioners and researchers must always rely on their own experience and knowledge in evaluating and using any information, methods, compounds, or experiments described herein. In using such information or methods they should be mindful of their own safety and the safety of others, including parties for whom they have a professional responsibility.

To the fullest extent of the law, neither the Publisher nor the authors, contributors, or editors, assume any liability for any injury and/or damage to persons or property as a matter of products liability, negligence or otherwise, or from any use or operation of any methods, products, instructions, or ideas contained in the material herein.

ISBN: 978-0-12-819676-2

For information on all Elsevier publications
visit our website at https://www.elsevier.com/books-and-journals

*Publisher:* Matthew Deans
*Acquisitions Editor:* Kayla Dos Santos
*Editorial Project Manager:* Emily Joy Grace Thomson
*Production Project Manager:* Sojan P. Pazhayattil
*Cover Designer:* Greg Harris

Typeset by STRAIVE, India

# Contents

| | | |
|---|---|---|
| *Contributors* | | *xi* |
| *Preface* | | *xv* |

## Part I  Technology background and novel materials

### 1. Introduction to the lithium-sulfur system: Technology and electric vehicle applications   1

Tobias Glossmann, Abhi Raj, Tea Pajan, and Elizaveta Buch

| | | |
|---|---|---|
| **1.1** | Introduction to lithium-sulfur battery | 3 |
| **1.2** | Electric vehicle batteries | 4 |
| **1.3** | Early lithium-sulfur batteries | 5 |
| **1.4** | Lithium-ion and lithium-sulfur batteries | 5 |
| **1.5** | Sulfur | 6 |
| **1.6** | Today's lithium-sulfur batteries | 7 |
| **1.7** | Cathodes | 8 |
| **1.8** | Anode and electrolyte | 9 |
| **1.9** | Fundamental challenge: Low cell voltage | 10 |
| **1.10** | Goal: Commercialized battery | 12 |
| | References | 12 |

### 2. Solid electrolytes for lithium-sulfur batteries   17

Eleni Temeche and Richard M. Laine

| | | |
|---|---|---|
| **2.1** | Introduction to Li-S batteries | 18 |
| **2.2** | Introduction to solid electrolytes | 19 |
| **2.3** | Brief history of solid electrolytes | 19 |
| **2.4** | Introduction to inorganic solid electrolytes | 20 |
| **2.5** | Li-S batteries based on polymer electrolytes | 39 |
| **2.6** | Summary | 43 |
| | Acknowledgments | 44 |
| | References | 45 |

### 3. Applications of metal-organic frameworks for lithium-sulfur batteries   49

Fu-Sheng Ke and Si-Cheng Wan

| | | |
|---|---|---|
| **3.1** | Introduction | 49 |
| **3.2** | MOFs for lithium-sulfur batteries | 54 |

v

vi  Contents

| 3.3 Characterization techniques | 92 |
| 3.4 Summary and outlook | 104 |
| Acknowledgments | 107 |
| References | 108 |

## Part II  Modeling and characterization

### 4. Multiscale modeling of physicochemical interactions in lithium-sulfur battery electrodes  123

Partha P. Mukherjee, Zhixiao Liu, Feng Hao, and Bairav S. Vishnugopi

| 4.1 Introduction | 123 |
| 4.2 The growth of crystalline $Li_2S$ film in cathode | 125 |
| 4.3 Parasitic reactions in anode | 144 |
| 4.4 Summary and outlook | 152 |
| Acknowledgment | 154 |
| References | 154 |

### 5. Reliable HPLC-MS method for the quantitative and qualitative analyses of dissolved polysulfide ions during the operation of Li-S batteries  159

Dong Zheng, Tianyao Ding, and Deyang Qu

| 5.1 Introduction to HPLC-MS | 159 |
| 5.2 Dissolved polysulfide ions and their behaviors in nonaqueous electrolytes | 164 |
| 5.3 Advantages of HPLC-MS vs. other analytical techniques | 166 |
| 5.4 One-step derivatization, separation, and determination of polysulfide ions | 171 |
| 5.5 The mechanism of sulfur redox reaction determined in situ electrochemical-HPLC technique | 175 |
| 5.6 Conclusions | 194 |
| References | 195 |

### 6. Modeling of electrode, electrolyte, and interfaces of lithium-sulfur batteries  201

Venkat Srinivasan and Aashutosh Mistry

| 6.1 Introduction | 202 |
| 6.2 Mathematical description of porous electrode performance | 206 |
| 6.3 Evolution of cathode porous electrode structure | 213 |
| 6.4 Concentrated electrolyte transport effects | 218 |
| 6.5 Dynamics of the polysulfide shuttle effect | 221 |

Contents    vii

6.6 Sources of variability: Mechanisms and properties    225
6.7 Summary and outlook    226
Acknowledgments    227
References    227

# Part III    Performance improvement

## 7.    Recent progress in fundamental understanding of selenium-doped sulfur cathodes during charging and discharging with various electrolytes    235

Chen Zhao, Gui-Liang Xu, Tianshou Zhao, and Khalil Amine

7.1 Introduction    235
7.2 Overview of $Se_xS_y$ cathode composition and electrochemistry    236
7.3 Progress on Li-$Se_xS_y$ batteries with liquid electrolytes    238
7.4 All-solid-state Li-$Se_xS_y$ batteries    253
7.5 Concluding remarks and future design strategies for $Se_xS_y$-based battery systems    256
Acknowledgments    257
References    258

## 8.    Suppression of lithium dendrite growth in lithium-sulfur batteries    261

XiaoLong Xu and Hao Wang

8.1 Introduction    261
8.2 Dendritic growth mechanism    263
8.3 Effect of Li dendrite growth on Li-S batteries    265
8.4 Suppression method    266
8.5 Conclusions    287
References    289

## 9.    The role of advanced host materials and binders for improving lithium-sulfur battery performance    297

Shahid Hussain, Naseem Akhtar, Awais Ahmad, Muhammad Khurram Tufail, Muhammad Kashif Aslam, Muhammad Sufyan Javed, and Xiangzhao Zhang

9.1 Introduction to energy sources and rechargeable batteries    298
9.2 Complex energy storage challenges and solutions    298
9.3 Host materials    300
9.4 Binders    314
9.5 Conclusions and future directions    322
References    322

viii Contents

# Part IV Future directions: Solid-state materials and novel battery architectures

## 10. Future prospects for lithium-sulfur batteries: The criticality of solid electrolytes     327

Patrick Bonnick and John Muldoon

| | |
|---|---|
| 10.1 The advantages of lithium-sulfur batteries | 327 |
| 10.2 The challenges of conventional sulfur electrodes when used with liquid electrolytes | 331 |
| 10.3 Lithium metal electrodes in lithium-sulfur batteries | 339 |
| 10.4 Path forward | 345 |
| Dedication | 346 |
| References | 347 |

## 11. New approaches to high-energy-density cathode and anode architectures for lithium-sulfur batteries     353

Moni K. Datta, Ramalinga Kuruba, T. Prasada Rao, Oleg I. Velikokhatnyi, and Prashant N. Kumta

| | |
|---|---|
| 11.1 Introduction | 354 |
| 11.2 Novel confinement architectures for sulfur cathodes | 359 |
| 11.3 Assembly and testing of pouch cells | 366 |
| 11.4 Coin cells: Preparation of hybrid solid electrolyte-coated battery separators | 373 |
| 11.5 Directly deposited sulfur architectures | 376 |
| 11.6 Computational studies to identify functional electrocatalysts | 393 |
| 11.7 Functional electrocatalysts and related materials for polysulfide decomposition | 399 |
| 11.8 Engineering dendrite-free anodes for Li-S batteries | 421 |
| 11.9 Conclusions | 435 |
| Acknowledgments | 436 |
| References | 436 |

## 12. A solid-state approach to a lithium-sulfur battery     441

Muhammad Khurram Tufail, Syed Shoaib Ahmad Shah, Shahid Hussain, Tayyaba Najam, and Muhammad Kashif Aslam

| | |
|---|---|
| 12.1 Introduction | 441 |
| 12.2 Solid electrolytes | 443 |
| 12.3 Polymer/ceramic hybrid composite electrolytes | 456 |
| 12.4 Stable Li metal anodes for all-solid-state Li-S batteries | 458 |
| 12.5 Sulfur-based cathode composites for all-solid-state Li-S batteries | 467 |

Contents ix

12.6 All-solid-state thin-film batteries 474
12.7 Conclusions 477
References 478

## Part V Applications: System-level issues and challenging environments

### 13. State estimation methodologies for lithium-sulfur battery management systems 491

Faten Ayadi, Daniel J. Auger, Abbas Fotouhi, and Neda Shateri

13.1 Introduction 491
13.2 Lithium-sulfur battery models 493
13.3 Li-S BMS: State estimation methods 497
13.4 Performance of state estimation methods 507
13.5 Conclusions and outlook 525
Acknowledgments 527
References 527

### 14. Batteries for aeronautics and space exploration: Recent developments and future prospects 531

Aloysius F. Hepp, Prashant N. Kumta, Oleg I. Velikokhatnyi, and Moni K. Datta

14.1 Introduction 532
14.2 Energy storage for (solar-) electric aircraft and high-altitude airships 533
14.3 Overview of energy storage for space exploration 552
14.4 Recent NASA missions to Mercury, Mars, and small bodies 553
14.5 Radiation issues and exploration missions to the Jupiter region 565
14.6 Next generation(s) of battery technologies for space exploration 572
14.7 Conclusions 581
References 584

*Index* 597

# Contributors

**Awais Ahmad**
Department of Chemistry, The University of Lahore, Lahore, Pakistan

**Naseem Akhtar**
College of Materials Science and Engineering, Beijing University of Chemical Technology, Beijing, China

**Khalil Amine**
Chemical Science and Engineering Division, Argonne National Laboratory, Lemont, IL; Materials Science and Engineering, Stanford University, Stanford, CA, United States

**Muhammad Kashif Aslam**
Faculty of Materials and Energy, Southwest University, Chongqing, China

**Daniel J. Auger**
School of Aerospace, Transport and Manufacturing, Cranfield University, Bedford, United Kingdom

**Faten Ayadi**
School of Aerospace, Transport and Manufacturing, Cranfield University, Bedford, United Kingdom

**Patrick Bonnick**
Toyota Research Institute of North America, Ann Arbor, MI, United States

**Elizaveta Buch**
Mercedes-Benz AG, Stuttgart, Germany

**Moni K. Datta**
Department of Bioengineering; Center for Complex Engineered Multifunctional Materials (CCEMM), Swanson School of Engineering, University of Pittsburgh, Pittsburgh, PA, United States

**Tianyao Ding**
Department of Mechanical Engineering, College of Engineering and Applied Science, University of Wisconsin Milwaukee, Milwaukee, WI, United States

**Abbas Fotouhi**
School of Aerospace, Transport and Manufacturing, Cranfield University, Bedford, United Kingdom

**Tobias Glossmann**
Mercedes-Benz Research & Development North America, Inc., Redford; Oakland University, Rochester, MI, United States

**Feng Hao**
Department of Engineering Mechanics, Shandong University, Jinan, Shandong, China

## Aloysius F. Hepp
Nanotech Innovations, LLC, Oberlin, OH, United States

## Shahid Hussain
School of Materials Science and Engineering, Jiangsu University, Zhenjiang, China

## Muhammad Sufyan Javed
School of Materials Science and Engineering, Jiangsu University, Zhenjiang, China

## Fu-Sheng Ke
Sauvage Center for Molecular Sciences, College of Chemistry and Molecular Sciences, Wuhan University, Wuhan, China

## Prashant N. Kumta
Department of Bioengineering; Center for Complex Engineered Multifunctional Materials (CCEMM); Department of Chemical and Petroleum Engineering; Department of Mechanical Engineering and Materials Science, Swanson School of Engineering, University of Pittsburgh, Pittsburgh, PA, United States

## Ramalinga Kuruba
Department of Bioengineering; Center for Complex Engineered Multifunctional Materials (CCEMM), Swanson School of Engineering, University of Pittsburgh, Pittsburgh, PA, United States

## Richard M. Laine
Department of Materials Science and Engineering, University of Michigan, Ann Arbor, MI, United States

## Zhixiao Liu
College of Materials Science and Engineering, Hunan University, Changsa, Hunan, China

## Aashutosh Mistry
Chemical Sciences and Engineering Division, Argonne National Laboratory, Lemont, IL, United States

## Partha P. Mukherjee
School of Mechanical Engineering, Purdue University, West Lafayette, IN, United States

## John Muldoon
Toyota Research Institute of North America, Ann Arbor, MI, United States

## Tayyaba Najam
College of Physics and Optoelectronic Engineering, Shenzhen University, Shenzhen, People's Republic of China

## Tea Pajan
Mercedes-Benz AG, Stuttgart, Germany

## T. Prasada Rao
Department of Bioengineering; Center for Complex Engineered Multifunctional Materials (CCEMM), Swanson School of Engineering, University of Pittsburgh, Pittsburgh, PA, United States

**Deyang Qu**
Department of Mechanical Engineering, College of Engineering and Applied Science, University of Wisconsin Milwaukee, Milwaukee, WI, United States

**Abhi Raj**
Mercedes-Benz Research & Development North America, Inc., Redford, MI, United States

**Syed Shoaib Ahmad Shah**
Hefei National Laboratory for Physical Sciences at the Microscale, School of Chemistry and Material Science, University of Science and Technology of China, Hefei, People's Republic of China

**Neda Shateri**
School of Aerospace, Transport and Manufacturing, Cranfield University, Bedford, United Kingdom

**Venkat Srinivasan**
Chemical Sciences and Engineering Division; Argonne Collaborative Center for Energy Storage Science, Argonne National Laboratory, Lemont, IL, United States

**Eleni Temeche**
Department of Materials Science and Engineering, University of Michigan, Ann Arbor, MI, United States

**Muhammad Khurram Tufail**
Key Laboratory of Cluster Science of Ministry of Education Beijing Key Laboratory of Photoelectronic/Electrophotonic Conversion Materials, School of Chemistry and Chemical Engineering, Beijing Institute of Technology, Beijing, China

**Oleg I. Velikokhatnyi**
Department of Bioengineering; Center for Complex Engineered Multifunctional Materials (CCEMM), Swanson School of Engineering, University of Pittsburgh, Pittsburgh, PA, United States

**Bairav S. Vishnugopi**
School of Mechanical Engineering, Purdue University, West Lafayette, IN, United States

**Si-Cheng Wan**
Sauvage Center for Molecular Sciences, College of Chemistry and Molecular Sciences, Wuhan University, Wuhan, China

**Hao Wang**
The College of Materials Science and Engineering, Beijing University of Technology, Beijing, China

**Gui-Liang Xu**
Chemical Science and Engineering Division, Argonne National Laboratory, Lemont, IL, United States

**XiaoLong Xu**
The College of Materials Science and Engineering, Beijing University of Technology, Beijing; School of Materials Science and Engineering, Qilu University of Technology (Shandong Academy of Sciences), Jinan, Shandong Province, China

**Xiangzhao Zhang**
School of Materials Science and Engineering, Jiangsu University, Zhenjiang, China

**Chen Zhao**
Chemical Science and Engineering Division, Argonne National Laboratory, Lemont, IL, United States; Department of Mechanical and Aerospace Engineering, The Hong Kong University of Science and Technology, Kowloon, Hong Kong, China

**Tianshou Zhao**
Department of Mechanical and Aerospace Engineering, The Hong Kong University of Science and Technology, Kowloon, Hong Kong, China

**Dong Zheng**
Department of Mechanical Engineering, College of Engineering and Applied Science, University of Wisconsin Milwaukee, Milwaukee, WI, United States

# Preface

Lithium-sulfur (Li-S) batteries (LSBs), due to their higher theoretical energy densities of $2600\,Wh\,kg^{-1}$ compared to currently used, conventional Li-ion cells of $100\text{--}265\,Wh\,kg^{-1}$, offer significant opportunities for high-energy applications. This technology with a history dating back to the 1960s is considered one of the most promising next-generation energy storage devices. While Li-S battery technology has evolved of late, into systems with significantly improved performance, Li-S batteries have yet to date, meet the performance metrics and energy storage demands for many practical applications. The practical use of Li-S batteries has been hindered by many challenges, the most prominent being the dissolution and shuttling of electrolyte-soluble polysulfides combined with the large volume expansion of conversion of sulfur to $Li_2S$, combined with the low electronic and ionic conductivity of sulfur. Additionally, there is the problem of corrosion and dendrite formation on the Li metal anode which can lead to inferior Coulombic efficiencies, puncture and short-circuiting of the battery rendering them susceptible to fire and explosion. During the past decade, to overcome these barriers preventing widespread adoption, research efforts have progressed and furthered the fundamental understanding of the basic electrochemistry while positing new innovative approaches to materials and cell designs. A simple Scopus search with the terms "Li-S battery" produced a total of 10,300 documents; using two terms, "Li-S battery" and "materials," produced a total of 6030 results. Nearly 80% of all results for both searches have been published from 2018 to 2021, as of the writing of this preface. Recognizing the importance of this topic, the editors of *Lithium-Sulfur Batteries: Advances in High-Energy Density Batteries* selected and organized 14 (and including two coauthored) chapters written by experts in the fields of battery materials science, energy storage, and their applications. The main objective is to present the state-of-the-art and potential future developments in Li-S batteries, including fundamentals, recent advances, electrolyte interface challenges, approaches to achieve high(er) performance, and prospects for current and future energy storage-related applications. This book is divided into five parts, each devoted to an important aspect of Li-S batteries indicated above.

Part I, *Technology background and novel materials*, includes three chapters that provide some technical background and introduce the main topic(s) of the book: Li–S battery materials, technologies, and applications. Chapter 1, authored by a team of researchers from Mercedes Benz led by Dr. Tobias Glossmann (also affiliated with Oakland University, Michigan, United States), provides an overview of the promise of Li–S batteries as an alternative to lithium–ion batteries for electric vehicles and grid storage. This is primarily due to their better theoretical performance, lower cost, and environmental benefits. This introductory chapter provides a high-level perspective of Li–S batteries and their intrinsic challenges. Chapter 2 is coauthored by Dr. Eleni Temeche and Prof. Richard Laine from the University of Michigan; it begins with an introduction to solid electrolytes, including inorganic solid, polymer, and composite electrolytes. The chapter describes the use of solid electrolytes in assembling all solid-state Li–S batteries; solid-state Li–S batteries offer considerable potential for the next-generation energy storage systems due to their anticipated high specific capacity, low cost, and eco-friendly features obviating the use of flammable liquid organic electrolytes. Solid-state electrolytes also prevent the solubility of polysulfides, a major deterrent to liquid electrolyte-based Li–S cells as explained above. Chapter 3, coauthored by Si-Cheng Wan and Prof. Fu-Sheng Ke from the College of Chemistry and Molecular Sciences of Wuhan University, focuses on applications of metal–organic frameworks (MOFs) for Li–S batteries. Metal-organic frameworks are a novel class of porous crystalline materials; they have been the focus of significant recent study due to their large surface area, high porosity, tunable pore size, and potential for use in numerous chemical environments. These unique properties render them quite suitable for use in Li–S batteries. The chapter appraises recent development of MOF-based materials for Li–S batteries, including sulfur hosts, lithium interface, electrolyte, separators, or interlayers, as well as in situ characterization techniques. Furthermore, the benefits, challenges, and future prospects of the applications of MOF-based materials in Li–S batteries are assessed and discussed.

Part II titled *Modeling and characterization* comprises three chapters. Prof. Partha P. Mukherjee from Purdue University (West Lafayette, IN, United States) leads a team of coauthors presenting recent results on multiscale modeling of several physicochemical interactions in Li–S battery electrodes in Chapter 4. This complex system involves intricate spatial and temporal phenomena: stepwise electrochemical reactions, producing various intermediate products; deposition of short-chain polysulfides (PSs) at the electrolyte/cathode interface(s), migration of soluble PSs between the electrodes, and precipitation of insoluble phases (e.g., $Li_2S$). In Chapter 5, Prof. Deyang

Qu (from the University of Wisconsin-Milwaukee, United States) and coworkers summarize the use of reliable high-pressure liquid chromatography-mass spectrometry (HPLC-MS) methodology for the quantitative and qualitative analyses of the dissolved polysulfide ions during the operation of Li-S batteries. The advantages of HPLC-MS versus other spectroscopic techniques are discussed; also, assays of HPLC-MS for the polysulfide separation and determination are introduced. The use of HPLC-MS enables the distribution of polysulfide ions in electrolytes to be quantitatively and qualitatively determined. Coupled with an electrochemical method, redox mechanisms of sulfur cathode(s) are revealed through the determination of the distribution of PS ions at various stages of charge and discharge in the operation of a Li-S battery. In Chapter 6, Dr. Venkat Srinivasan and Dr. Aashutosh Mistry from Argonne National Laboratory Lemont, IL, United States, present an in-depth discussion of the modeling literature that examines various mechanisms such as morphology-dependent microstructure evolution, nucleation dynamics of the precipitate phase, electrolyte transport limitation, and PS shuttle effect that degrade the performance of Li-S batteries. Each of these mechanisms dominates the macroscopically measured electrochemical signatures for a certain subset of cell specifications and operation; thus, strategies for performance improvement differ. The chapter concludes with a detailed summary of opportunities for future investigations to elucidate the poorly understood interactions that would facilitate the rational design of such cells.

*Performance improvement* (of Li-S batteries) is the title and topic of Part III of this book; it contains three chapters. In Chapter 7, recent progress in fundamental understanding of selenium-doped sulfur cathodes ($Se_xS_y$) during charging and discharging with various electrolytes is summarized by a team of coauthors led by Dr. Khalil Amine, Chemical Science and Engineering Division, Argonne National Laboratory (also affiliated with Stanford University, CA, United States), that includes Prof. Tianshou Zhao from the Hong Kong University of Science and Technology. The chapter focuses on fundamental insights of the $Se_xS_y$ cathodes during charge/discharge in a variety of both liquid- and solid-state electrolytes. In addition, recent progress utilizing various cathode structures design for Li-$Se_xS_y$ batteries is also discussed; suggestions for future development are finally presented at the chapter's conclusion. In Chapter 8, the suppression of Li dendrite growth in Li-S batteries is addressed by Prof. Hao Wang and his coworker XiaoLong Xu from the College of Materials Science and Engineering, Beijing University of Technology, China. The chapter begins with a description of dendrite initiation and growth mechanisms. Some strategies for delaying and suppressing dendrite growth are described subsequently, typically by designing appropriate battery components in the separator, anode, and electrolyte. Finally, advantages and disadvantages of various dendrite-growth prevention strategies are discussed. In Chapter 9, the role of advanced host materials and binders for improving Li–S battery performance is discussed by a team of coauthors led by Prof. Shahid Hussain (School of Materials Science and Engineering, Jiangsu University, Zhenjiang, China) and Prof. Naseem Akhtar (College of Materials Science and Engineering, Beijing University of Chemical Technology, China). The chapter presents recent advances in polysulfide (PS) trapping–host materials and functional binders, which make the cathode more electronically conductive, thereby improving the cycle life of Li–S batteries. The chapter nicely summarizes how an ideal host material provides sufficient surface area for reaction(s), confinement of sulfur, as well as increasing the facility of movement of electrons and Li ions; also, how an ideal binder provides good adhesion, suitable swelling capacity, high Li-ion conductivity, and effective adsorption capability for PS.

Part IV titled *Future directions: Solid-state materials and novel battery architectures* comprises three chapters that discuss novel device designs and structures as well as anticipated future research directions, including issues to be addressed to enable the employment of advanced Li–S battery technology. Chapter 10 by Patrick Bonnick and John Muldoon of Toyota Research in Ann Arbor, Michigan, United States, delves into several key challenges with liquid electrolytes to be surmounted to enable the advent of robust and high-energy-density Li–S cells. The Li polysulfide shuttle and slow Li–PS deposition kinetics at low electrolyte/S ratios have proven to be practically insurmountable; this has therefore shifted research focus toward solid electrolytes. However, issues remain with solid electrolytes: repeated expansion and contraction of the active materials can lead to cracking, especially within the positive electrode as well as interface instabilities at both electrodes; also, Li metal electrodes must also overcome challenges, such as dendrite growth, when used with solid electrolytes at relevant current densities. The authors discuss strengths and weaknesses of many proposed solutions to meet these challenges with an eye toward the ever-present goal of competing with conventional Li-ion energy densities. Chapter 11, a contribution from a team from the University of Pittsburgh led by Prof. Kumta, addresses the problems of polysulfide formation, including loss in capacity and eventual cell failure in LSBs. The problem is further compounded by dendrite formation on the Li anode during electrochemical cycling, resulting in major safety hazards of flammability of the electrolyte and possible explosion due to short-circuiting of the battery. This chapter outlines various approaches to combat both cathode and anode issues. These involve synthesis and characterization of complex framework materials (CFMs) for confining the polysulfides and sulfur in the LSB cathodes. The CFMs include a confinement host and a coating applied to the CFM host, which includes one or more forms of an electrical conductor, a Li-ion conductor, and a functional electrocatalyst for electrocatalytically converting the soluble polysulfides to $Li_2S$. Furthermore, sulfur is infiltrated into the CFM host, creating a sulfur–carbon linkage serving as effective anchors for trapping the ensuing polysulfides. The systems have been tested in coin cells and pouch cells with metallic Li anodes under lean electrolyte conditions of electrolyte-to-sulfur ratios showing promise and feasibility. New emergent dendrite-free alloys have also been identified to test against pure Li and CFM cathodes in coin cell and pouch cell configurations under similar conditions. Chapter 12, coauthored by a team led by Prof. S. Hussain, School of Materials Science and Engineering, Jiangsu University, Zhenjiang, China, discusses all solid-state Li-S batteries (SSLiSBs); the use of traditional organic liquid electrolyte in lithium-sulfur batteries prevents them from being commercialized due to safety and several technical issues that are yet to be resolved. Critical challenges and problems in practical applications, such as chemical and electrochemical stability, compatibility of solid-state electrolytes with the sulfur-based active materials, and lithium metal, are discussed. Finally, this chapter addresses improved safety, extended working temperature window, and high energy/power densities that are several of the key advantages of SSLiSBs.

Part V, *Applications: System-level issues and challenging environments*, focuses on practical and systems considerations when utilizing advanced battery (especially Li-S) technologies for electric vehicle and aerospace transportation and exploration, energy storage during operation(s) which present significant challenges. Chapter 13 is a collaborative work of researchers from the School of Aerospace, Transport and Manufacturing, Cranfield University in Bedford, United Kingdom, led by Dr. Daniel Auger. Lithium-sulfur batteries offer particular promise for vehicle applications that need high gravimetric energy density. However, Li-S cells require careful management, as they behave differently than the well-known conventional lithium-ion systems and standard techniques do not work well. The chapter presents the state of the art in Li-S battery-state estimation, explaining the limitations of "standard" lithium-ion techniques and presenting two groups of techniques that have shown promise: recursive Bayesian and particle filters and adaptive neuro-fuzzy inference systems and classification techniques. The chapter also addresses these two advanced battery-state estimation methods in detail and concludes with brief remarks on the current and future research directions. Chapter 14 of the book is a collaboration between the four coeditors; a summary of energy storage options and issues for aeronautics and space exploration introduces this intriguing topic. Batteries have been successfully demonstrated for numerous exploration missions to several classes of solar system destinations over the past 50 years. Given the broad technology space of battery types and materials, the final sections of the chapter focus on a discussion of several instructive representative missions and practical aspects of batteries (with emphasis on rechargeable technologies) for space exploration. An important take-home lesson is the need to develop energy storage technologies and power systems that can withstand the radiation fluxes and temperature extremes encountered throughout the solar system; this will be critical for electronic devices, advanced instrumentation, and small off-world exploration vehicles. The chapter concludes with a consideration of the use of local resources for sustainable human subsistence that may one day enable off-world production of consumables, power components, and structural materials to facilitate the construction of settlements and the exploration of the farthest regions of the solar system.

The success of this edited book is the result of the full commitment of each contributing author. Without their availability and willingness in sharing their valuable knowledge and critical review of the respective topics, the book could not be published at such a high standard. Our heartfelt thanks are also extended to Rachel Pomery and Emily Thomson, our Editorial Project Managers. Christina Gifford, Acquisitions Editor, is particularly acknowledged for her constant support given throughout the entire publication process, especially during the pandemic. We sincerely hope that this edited compilation of exemplary work from established and world renowned researchers on an extremely high-profile, germane, and burgeoning system in the area of high-energy density, rechargeable Li-based batteries will be of significant value to the scientific community as well as practitioners, technologists and emerging researchers from materials science, electronics, space research and technology, battery technology as well as electrochemical industries. Finally, a further aim of this collection of cutting edge and practical topics is to enable the eventual implementation of this emerging and promising rechargeable energy storage technology.

# PART I

# Technology background and novel materials

# CHAPTER 1

# Introduction to the lithium-sulfur system: Technology and electric vehicle applications

**Tobias Glossmann[a,b], Abhi Raj[a], Tea Pajan[c], and Elizaveta Buch[c]**

[a]Mercedes-Benz Research & Development North America, Inc., Redford, MI, United States
[b]Oakland University, Rochester, MI, United States
[c]Mercedes-Benz AG, Stuttgart, Germany

## Contents

| | | |
|---|---|---|
| 1.1 | Introduction to lithium-sulfur battery | 3 |
| 1.2 | Electric vehicle batteries | 4 |
| 1.3 | Early lithium-sulfur batteries | 5 |
| 1.4 | Lithium-ion and lithium-sulfur batteries | 5 |
| 1.5 | Sulfur | 6 |
| 1.6 | Today's lithium-sulfur batteries | 7 |
| 1.7 | Cathodes | 8 |
| 1.8 | Anode and electrolyte | 9 |
| 1.9 | Fundamental challenge: Low cell voltage | 10 |
| 1.10 | Goal: Commercialized battery | 12 |
| References | | 12 |

## 1.1 Introduction to lithium-sulfur battery

Based on their promise of high specific energy ($\sim$2300 Wh/kg) as well as energy density ($\sim$2600 Wh/L) normalized by active materials alone, lithium-sulfur batteries (LSBs) have the potential to take a significant battery market share for various applications in the near future. In theory, LSBs can deliver much higher energy density at significantly lower cost than Li-ion batteries (LIBs), a key consideration in the emerging markets for grid storage and electric vehicles (EVs). Moreover, for these applications, LSBs are a highly anticipated alternative to Li-ion batteries as requirements are pushing the limits of the latter [1].

While significant research efforts have focused on LSB technology, notable commercialization has not been achieved. So far, only companies with

*Lithium-Sulfur Batteries*
https://doi.org/10.1016/B978-0-12-819676-2.00010-4

Copyright © 2022 Elsevier Inc.
All rights reserved.

smaller production volumes such as Oxis or Sion Power have manufactured LSBs and typically for rather specialized applications [2,3]. More generally, predictions of imminent market introduction have been made since the 1970s. Despite this, primarily due to a shortfall in cycle life, we have not seen major commercialization success [4]. It is not unusual for highly promising technologies to struggle with similar issues for many years until key discoveries enable a breakthrough. In this chapter, we will review the principles of LSBs and their history to understand why there is so much interest in LSBs and why they are not readily available.

## 1.2 Electric vehicle batteries

Today's battery packs for EVs need to store at least 40–120 kWh of energy, depending on the vehicle's size, energy efficiency, and target driving range. In some markets, EVs need to travel more than 320 km (200 miles) to find customer acceptance. While the packaging space for the battery system of small-sized vehicles may only be on the order of 200 L, longer wheelbase cars, especially sport utility vehicles (SUVs), may enable on the order of 500 L of volume for the battery pack. The battery cells, however, require a mechanical support structure, a battery management system, a cooling system, and pack-level safety features. Cables, connections, bus-bars, structural elements, cooling channels, and other thermal management materials require a significant amount of volume, depending on the cell characteristics. Cell energy density is the key characteristic in respect to packaging; however, cells with higher energy efficiency, higher specific energy (Wh/kg), and less violent abuse behavior than the state-of-the-art LIBs will allow more efficient packaging through savings in cooling system, mechanical structure, and measures to prevent thermal propagation between cells in case of a thermal runaway event. Final cell-level energy density development targets for LSBs will therefore depend at least on these three characteristics. Passenger cars, constituting the majority of the global vehicle market, are likely going to drive requirements for calendar or service life, cycle life, and fast charging as well. Vehicle manufacturers are expecting 10–15 years of service life for batteries [5]. Cycle life for any application will depend on the battery size and vehicle efficiency as energy throughput is the equivalent measure for mileage. A simplified example clarifies this relationship: A vehicle with a range of 200 miles fully utilized at every charge would have to support 1000 cycles to achieve 200,000 miles.

## 1.3 Early lithium-sulfur batteries

Significant research on lithium–metal/sulfide batteries was pioneered at Argonne National Laboratory in the 1960s and the 1970s, an effort that originated from the reactor program in the 1960s where electrochemical reactions of molten salts and liquid metal were investigated [6]. The assumption at the time was that sulfur or any other chalcogens had to be melted in order for the battery to operate [7]. These batteries were expected to achieve requirements for not only grid storage systems, but also electric vehicle applications [4]. The cells consisted of liquid lithium, sulfur, and electrolyte, a system that is not very practical for operation in a moving car due to complicated overhead associated with packaging. One of the three liquid components of the cell that had to be immobilized for use on the road and the electrolyte was chosen [7]. This cell design used rather high areal capacities to yield a compelling energy density, so large areal currents of about $1\,A/cm^2$ were required at practical discharge rates. Current densities in advanced LIBs for EVs may reach $20\,mA/cm^2$ for short intervals but are usually even lower.

## 1.4 Lithium-ion and lithium-sulfur batteries

The lithium-ion battery consists of stacks or rolls of two paired thin electrodes with a separator layer and liquid electrolyte between them. The negative electrode is usually referred to as anode and the positive as cathode due to their role in the discharge process. LIB technology, as it is common now, was a major development coming from high-temperature liquid cells aforementioned. The LIB manufacturing process produces thin coatings of dried slurry on current collector enabling better electron and ion transport due to the short distance between electrodes. The anode's active material is typically graphite, and the cathode's active material is typically a metal oxide or phosphate. These systems are called Li-ion batteries due to the fact that lithium nominally remains outside of the electrodes, meaning solvated in the electrolyte, in a $Li^+$ charged state. The Li-ions' locations, representing state-of-charge (SOC), are controlled by the electrical current through the cell. The energetically higher location is the anode where Li-ions are inserted into graphite. Volume expansion of graphite is about 13.1% upon anode lithiation (charge) but the state-of-the-art binders will accommodate this expansion [8,9]. Upon cell discharge, Li-ions travel to the energetically lower location, the transition metal oxide, and react there with one electron

Lithium-sulfur batteries

per lithium ion. Volume change of cathode active materials upon their lithiation (discharge) is manageable by binders as well and typically in the range of $-1.9\%$ for lithium cobalt oxide ($LiCoO_2$) to less than $+8\%$ depending on the type of cathode active material and voltage profile [9,10]. $LiCoO_2$ contracts with higher lithium content but most other materials would expand. The entire cell reaction is hence based on lithium intercalation reactions, a system often referred to as a rocking-chair battery. The cathode material is normally prelithiated, meaning the battery is assembled in a discharged state using active materials such as $LiCoO_2$ or $LiFePO_4$, to give a few examples.

In LSBs, by contrast, the cathode material is sulfur or sulfur-composite/S-containing materials which in its elemental form is typically $S_8$. During the discharge process, which will be discussed, sulfur reacts in some intermediate steps with lithium to lithium sulfide ($Li_2S$). $S_8$ doesn't contain lithium as compared to cathode materials in LIBs. The lithiated and de–lithiated active materials of the cathode are very different in structure; this is not merely an insertion of a lithium ion. Furthermore, lithium needs to be introduced to complete the cell.

Where could the lithium inventory in the cell come from if not from the cathode active material? Possibly from the anode? Fully prelithiated graphite or silicon anodes are highly reactive and difficult to produce. Mechanical expansion makes almost any lithiation outside a cell very difficult. Lithium inventory is therefore one important reason why metallic lithium was selected as preferred anode. Various efforts are underway to prelithiated cells, e.g., chemically or electrochemically, but Li-metal remains the most common anode for LSBs [11,12].

## 1.5 Sulfur

Besides the high theoretical capacity and thus the prospect of a battery with superior specific energy, sulfur offers several other interesting benefits. Sulfur is not only abundant in the earth's continental crust ($0.049\%$, $>10^{19}$ kg), it is also readily accessible in large amounts. Desulfurization of fuels, critical for the prevention of acid rain, generates massive amounts of unwanted sulfur. Currently, there are no concerns that sulfur could become rare and therefore expensive. In rubber, sulfur is literally the linking element between polymers. Gypsum ($CaSO_4 \cdot 2H_2O$), and hence drywall, is based on sulfate. Sulfates and sulfur are used as fertilizer, because sulfur is an essential ingredient of living organisms [13–16]. Elemental sulfur is an effective fungicide, sometimes produced by the plant itself [13].

Sulfur is a natural choice for batteries to be used on large scale. Should a substantial fraction of vehicles be electrified with one type of battery, millions of packs have to be produced every year and a sustainable material choice will be crucial. While cobalt and other transition metals are vital in organic molecules such as enzymes and vitamins as well, those metals are not the "drywall of the living world." As long as no lasting toxic sulfur compounds are formed, the element is not foreign to living things, in particular the human body that contains about 160 g sulfur but only 4.8 g iron and 1.6 mg cobalt [16].

Sulfur is in group 16 of the periodic table, the chalcogens, that also includes oxygen, selenium, tellurium, and the radioactive elements, polonium and livermorium. Compared to oxygen, due to sulfur's electron structure having an empty 4s and 3d orbital, it is more flexible in terms of the number of bonds with other atoms it can have. For example, in hexavalent oxidation state (+6), it can form six bonds, as in sulfur hexafluoride or the sulfate ion. In group 16, selenium and tellurium are the next two elements before the radioactive elements polonium and livermorium. The farther down in the periodic table from sulfur, the more toxic the elements are. The metallic character and thus conductive behavior also increase with higher atomic mass. Low electronic conductivity is surely one of the drawbacks of sulfur, one reason for early lithium–chalcogen batteries to be made from selenium and tellurium [7]. Conductivity was also the motivation for high-temperature batteries [17]. The opposite trend in the periodic table is true for the cell potential. Due to the higher electronegativity of sulfur compared with selenium and even more compared with tellurium, the voltage and thus the energy content are higher for sulfur.

## 1.6 Today's lithium-sulfur batteries

This introduction intentionally avoids detailed discussions of a specific technology in order to provide a big picture overview and motivate new innovative approaches. The main difference between LSBs in the 1970s and today's LSBs is that lithium, sulfur, and electrolyte are not in a molten state. Instead, sulfur is stored in some form in the cathode at room temperature and wetted with organic electrolyte. The electrode structure is designed to transport electrons to the active material and provide room for electrolyte, so

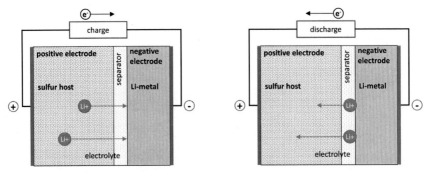

**Fig. 1.1** Generalized Li-S cell. *(Credit: Self-drawn, Tobias.)*

lithium ions can travel to the location of electron transfer reactions as illustrated in Fig. 1.1.

With disregard to sulfur's allotropes and all intermediates, the cathode reaction is simply:

$$S + 2\,Li^+ + 2\,e^- \rightarrow Li_2S \tag{1.1}$$

For the Li-metal anode, the electrode reaction is:

$$Li \rightarrow Li^+ + e^- \tag{1.2}$$

The cell reaction is:

$$S + 2\,Li \rightarrow Li_2S \tag{1.3}$$

In most LSBs, elemental sulfur ($S_8$) is the active cathode material. The electrochemical reaction will therefore happen in intermediate steps and the so-called polysulfides ($Li_2S_x$ with $3 < x < 6$) are formed before arriving at the final discharge product lithium sulfide ($Li_2S$). The cell discharge curve with plateaus in it indicates multiple reaction steps. Several polysulfide speciation studies have been done, but the chemistry is complex and depends on many factors [18,19]. Aurbach et al. pointed out that all reaction equations describing polysulfide intermediates are simplifications [18]. In reality, the electrolyte environment will impact the sequence of reactions.

## 1.7 Cathodes

As lithium sulfide requires significantly more volume than sulfur (~80% expansion), cathode host structures must be able to accommodate volume changes during lithiation and de-lithiation [20]. Its ability depends strongly on material design and binder; therefore, various binders and nanostructures have been studied [1]. Over the last decade, much effort has been invested in the optimization of cathode host structures with different pore sizes, shapes, dopants, coatings, encapsulations, shields, and more—literature is readily available [20–22]. We observe a trend from mere physical confinement of the active species to more sophisticated surface chemistry-based approaches [23,24]. Cathodes show increasingly improved performance by better conductivity and targeted adsorption of polysulfides with promising results [25,26].

But not all cathodes contain sulfur in the form of $S_8$. Some structures do not contain elemental sulfur but are covalently linked to sulfur or sulfur chains, often also cross-linked by sulfur through vulcanization as cp (S-TAR) demonstrated by Zeng et al. [27]. Poly(acrylonitrile) (PAN)-based conductive sulfur-containing polymer originally discovered by Wang et al. is one example that deserves attention [28]. The sulfur-containing PAN-based electrode material called SPAN was studied by several groups. For covalently bound sulfur, in theory, the reaction equation would have to be rewritten to be accurate, but for most systems, the cell equation is not fully understood [29]. Depending on the structure, an equation may be written as:

$$C_xN_yH_zS + 2\,Li \rightarrow Li_2C_xN_yH_zS \qquad (1.4)$$

Covalently sulfur-binding systems are attractive because they avoid the most challenging problems of LSBs, e.g., polysulfides in solution and non-conductive elemental sulfur, and with that eliminate a host of other inactive materials. Carbon and other elements can hardly be eliminated from LSBs, for the application it may not be significant how exactly these atoms interact with the sulfur. In conclusion, different strategies may lead to a good LSB, either a highly engineered micro/nanostructure that is able to confine the active species in their location or a chemical approach that avoids the use of elemental sulfur.

## 1.8 Anode and electrolyte

Due to their solubility in ether-based electrolytes, lithium-polysulfides are well accessible for their further electrochemical reaction with lithium ions as described in Eq. (1.1). The polysulfide-containing electrolyte will saturate throughout discharge of the cell, and its physical properties will change as well. Therefore, the electrolyte design and the load profile will impact the polysulfide species present in solution, reaction pathways, and precipitation. The system is complex if not impossible to gain complete control under dynamic operation and environmental conditions. The concept of sparingly solvating electrolytes was also referred to as nonsolvent, in an attempt to control polysulfide speciation with some success [30,31]. Dissolution of polysulfides in the electrolyte comes with a downside in that polysulfides will react with lithium metal. The mechanism leads to a loss of cell capacity as $Li_2S_2$ and $Li_2S$ do not dissolve well in the electrolyte and contaminate the anode surface. This leads to dramatic performance loss[32]. In an attempt to address this issue, electrolyte additives such as lithium nitrate ($LiNO_3$) and other passivating solvent molecules, e.g., TTE (1,1,2,2-tetrafluoroethyl-2,2,3,3-tetrafluoropropylether) and others, have been developed [18,33,34]. Indeed, Li-metal is very reactive, and with constant charging and discharging, the lithium surface will be changed. The passivating layer will open up and consume more additives, which is not a sustainable solution.

As a result, the idea to protect Li-metal from the liquid electrolyte and polysulfides has persisted for a long time. Several interlayers, membranes, shields, and solid electrolytes have been evaluated [20]. A solid electrolyte would truly separate the system, and it would also eliminate pitting corrosion for potentially more uniform lithium stripping. Most membranes, however, do not solve the polysulfide problem completely, impede Li–ion transport, and add inactive mass [20,35].

Carbonate-based electrolytes as used in lithium–ion batteries are not compatible with polysulfides, as chemical reactions between carbonates and polysulfides will occur and irreversibly damage the solvent [36,37]. Graphite is commonly used as anode material in LIBs; however, ethylene carbonate (EC) solvent is not compatible with polysulfides but is needed to form a stable SEI on graphite [38]. Anodes other than lithium itself also reduce the effective cell voltage that is already low with a sulfur cathode [18]. The need for electrochemical stability and higher cell voltages has driven researchers to mostly work on Li-metal anodes for LSBs. Nonmetallic anodes (such as SPANs) may enable LSBs without polysulfides in solution. Such systems work well with carbonate electrolytes, opening the possibility to use anodes common in LIBs. On the topic, Li et al. have elaborated on the "two schools of thought" for LSBs [39].

## 1.9 Fundamental challenge: Low cell voltage

The low cell voltage of LSBs would not be a problem if the cell impedance could be minimized as well. An example of a simplified electric vehicle

Introduction to the lithium-sulfur system 11

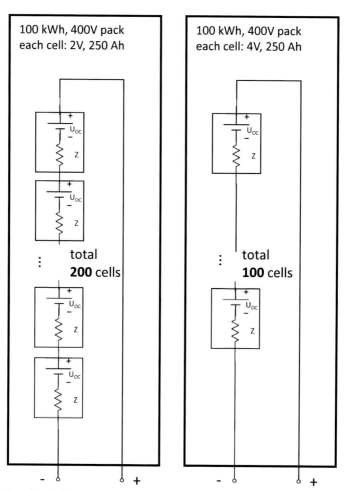

**Fig. 1.2** Simplified illustration of a battery pack with 2-V cells vs 4-V cells. *(Credit: Self-drawn, Tobias.)*

battery in Fig. 1.2 demonstrates the problem with a 2-V and a 4-V cell pack. For each cell, the open circuit potential $U_{OC}$ and the complex impedance Z are shown. With a larger number of cells in series, impedances add up and their ohmic contribution translates to increased heat loss upon an applied load. A battery pack with 2-V cells designed for the same charging or discharging current therefore requires cells with a resistance lower than needed for a comparable 4-V cell battery, not to mention losses due to additional cell connectors. In the cell, diffusion and reaction kinetics therefore play a critical role as has been discussed [35,40].

## 1.10 Goal: Commercialized battery

What are the key issues impeding the commercialization of LSBs? Energy density, cycle stability, and power are not where they need to be to compete with Li-ion batteries. The majority of research is done on a coin cell level so the performance is not necessarily representative of practical size cells as has been discussed by others [35,41,42]. Cycle life performance and coulombic efficiency do not demonstrate the actual performance when excess lithium is used [43]. Inactive materials such as host structures reduce the energy density, critical for large-scale applications to under LIB levels. Measures against polysulfide shuttle may add to the list of inactive materials and add resistance [20]. Reaction kinetics suffer in starved electrolytes, in most LSBs [44,45]. Polarization losses and kinetic limitations will become even bigger challenges with increased areal capacities in an effort to reduce inactive materials.

There is no shortness of challenges in LSB research and development but occasional discoveries and eye-opening fundamental studies underline the importance of continued work [40,46,47]. Significant progress has been made in the understanding of LSBs, hinting to possible solutions based on nanotechnology, solid-state, or advanced surface chemistry that will be able to not only solve the polysulfide problem but also produce a commercially viable battery for broad adoption.

## References

[1] M. Shaibani, M.S. Mirshekarloo, R. Singh, C.D. Easton, M.C.D. Cooray, N. Eshraghi, T. Abendroth, S. Dörfler, H. Althues, S. Kaskel, A.F. Hollenkamp, M.R. Hill, M. Majumder, Expansion-tolerant architectures for stable cycling of ultrahigh-loading sulfur cathodes in lithium-sulfur batteries, Sci. Adv. 6 (1) (2020) eaay2757.
[2] Sion Power's Lithium-Sulfur Batteries Power High Altitude Pseudo-Satellite Flight.n. d. https://sionpower.com/2014/sion-powers-lithium-sulfur-batteries-power-high-altitude-pseudo-satellite-flight/ (Accessed 9 October 2021).
[3] Press Release: n.d. OXIS Energy is Close to Achieving 500 Wh/kg and is targeting 600 Wh/kg with Solid State Lithium Sulfur technology https://45uevg34gwlltnbsf2plyua1-wpengine.netdna-ssl.com/wp-content/uploads/2020/01/500-and-600-whkg-pressor.pdf (Accessed 9 October 2021).
[4] J.R. Birk, R.K. Steunenberg, Chemical investigations of lithium-sulfur cells, in: New Uses of Sulfur, vol. 140, American Chemical Society, 1975, pp. 186–202.
[5] USCAR Energy Storage System Goals, http://uscar.org/usabc/#246-246-top (accessed 4 January 2022).
[6] J.M. Holl, A. Schriesheim, R.G. Hewlett, R.R. Harris, Argonne National Laboratory, 1946–96, University of Illinois Press, 1997.
[7] E.J. Cairns, H. Shimotake, High-temperature batteries, Science 164 (3886) (1969) 1347–1355.
[8] H. Chen, M. Ling, L. Hencz, H.Y. Ling, G. Li, Z. Lin, G. Liu, S. Zhang, Exploring chemical, mechanical, and electrical functionalities of binders for advanced energy-storage devices, Chem. Rev. 118 (18) (2018) 8936–8982.

Introduction to the lithium-sulfur system    **13**

[9] Y. Koyama, T.E. Chin, U. Rhyner, R.K. Holman, S.R. Hall, Y.M. Chiang, Harnessing the actuation potential of solid-state intercalation compounds, Adv. Funct. Mater. 16 (4) (2006) 492–498.

[10] C.D. Quilty, D.C. Bock, S. Yan, K.J. Takeuchi, E.S. Takeuchi, A.C. Marschilok, Probing sources of capacity fade in $LiNi_{0.6}Mn_{0.2}Co_{0.2}O_2$ (NMC622): an operando XRD study of Li/NMC622 batteries during extended cycling, J. Phys. Chem. C 124 (15) (2020) 8119–8128.

[11] Y. Shen, J. Zhang, Y. Pu, H. Wang, B. Wang, J. Qian, Y. Cao, F. Zhong, X. Ai, H. Yang, Effective chemical prelithiation strategy for building a silicon/sulfur li-ion battery, ACS Energy Lett. 4 (7) (2019) 1717–1724.

[12] F. Holtstiege, P. Bärmann, R. Nölle, M. Winter, T. Placke, Pre-lithiation strategies for rechargeable energy storage technologies: concepts, promises and challenges, Batteries 4 (1) (2018) 4, https://doi.org/10.3390/batteries4010004.

[13] J.S. Williams, R.M. Cooper, The oldest fungicide and newest phytoalexin—a reappraisal of the fungitoxicity of elemental sulphur, Plant Pathol. 53 (3) (2004) 263–279.

[14] E.J. Ko, Y. Ho Shin, H.N. Hyun, H.S. Song, J.K. Hong, Y.C. Jeun, Bio-sulfur pretreatment suppresses anthracnose on cucumber leaves inoculated with Colletotrichum orbiculare, Mycobiology 47 (3) (2019) 308–318.

[15] L.O. Fuentes-Lara, J. Medrano-Macias, F. Perez-Labrada, E.N. Rivas-Martinez, E.L. Garcia-Enciso, S. Gonzalez-Morales, A. Juarez-Maldonado, F. Rincon-Sanchez, A. - Benavides-Mendoza, From elemental sulfur to hydrogen sulfide in agricultural soils and plants, Molecules 24 (12) (2019) 2282, https://doi.org/10.3390/molecules24122282.

[16] P. Chellan, P.J. Sadler, The elements of life and medicines, Philos. Trans. A Math. Phys. Eng. Sci. 373 (2037) (2015) 20140182, https://doi.org/10.1098/rsta.2014.0182.

[17] D. Peramunage, S. Licht, A solid sulfur cathode for aqueous batteries, Science 261 (5124) (1993) 1029–1032.

[18] D. Aurbach, E. Pollak, R. Elazari, G. Salitra, C.S. Kelley, J. Affinito, On the surface chemical aspects of very high energy density, rechargeable Li–sulfur batteries, J. Electrochem. Soc. 156 (8) (2009).

[19] M. Cuisinier, P.-E. Cabelguen, S. Evers, G. He, M. Kolbeck, A. Garsuch, T. Bolin, M. Balasubramanian, L.F. Nazar, Sulfur speciation in Li–S batteries determined by operando X-ray absorption spectroscopy, J. Phys. Chem. Lett. 4 (19) (2013) 3227–3232.

[20] A. Manthiram, Y. Fu, S.-H. Chung, C. Zu, Y.-S. Su, Rechargeable lithium–sulfur batteries, Chem. Rev. 114 (23) (2014) 11751–11787.

[21] T.A. Pascal, I. Villaluenga, K.H. Wujcik, D. Devaux, X. Jiang, D.R. Wang, N. Balsara, D. Prendergast, Liquid sulfur impregnation of microporous carbon accelerated by nanoscale interfacial effects, Nano Lett. 17 (4) (2017) 2517–2523.

[22] T. Ould Ely, D. Kamzabek, D. Chakraborty, M.F. Doherty, Lithium–sulfur batteries: state of the art and future directions, ACS Appl. Energy Mater. 1 (5) (2018) 1783–1814.

[23] H.-J. Peng, Q. Zhang, Designing host materials for sulfur cathodes: from physical confinement to surface chemistry, Angew. Chem. Int. Ed. 54 (38) (2015) 11018–11020.

[24] R. Fang, J. Xu, D.-W. Wang, Covalent fixing of sulfur in metal–sulfur batteries, Energ. Environ. Sci. 13 (2) (2020) 432–471.

[25] Q. Pang, C.Y. Kwok, D. Kundu, X. Liang, L.F. Nazar, Lightweight metallic MgB2 mediates polysulfide redox and promises high-energy-density Lithium-sulfur batteries, Joule 3 (1) (2019) 136–148.

[26] B.-J. Lee, T.-H. Kang, H.-Y. Lee, J.S. Samdani, Y. Jung, C. Zhang, Z. Yu, G.-L. Xu, L. Cheng, S. Byun, Y.M. Lee, K. Amine, J.-S. Yu, Revisiting the role of conductivity and polarity of host materials for long-life Lithium–Sulfur battery, Adv. Energy Mater. 10 (22) (2020) 1903934.

[27] S. Zeng, L. Li, J. Yu, N. Wang, S. Chen, Highly crosslinked organosulfur copolymer nanosheets with abundant mesopores as cathode materials for efficient lithium-sulfur batteries, Electrochim. Acta 263 (2018) 53–59.

[28] J. Wang, J. Yang, J. Xie, N. Xu, A Novel conductive polymer–sulfur composite cathode material for rechargeable Lithium batteries, Adv. Mater. 14 (13–14) (2002) 963–965.

[29] W. Wang, Z. Cao, G.A. Elia, Y. Wu, W. Wahyudi, E. Abou-Hamad, A.-H. Emwas, L. Cavallo, L.-J. Li, J. Ming, Recognizing the mechanism of sulfurized polyacrylonitrile cathode materials for Li–S batteries and beyond in Al–S batteries, ACS Energy Lett. 3 (12) (2018) 2899–2907.

[30] L. Cheng, L.A. Curtiss, K.R. Zavadil, A.A. Gewirth, Y. Shao, K.G. Gallagher, Sparingly solvating electrolytes for high energy density lithium–sulfur batteries, ACS Energy Lett. 1 (3) (2016) 503–509.

[31] M. Cuisinier, P.E. Cabelguen, B.D. Adams, A. Garsuch, M. Balasubramanian, L.F. Nazar, Unique behaviour of nonsolvents for polysulphides in lithium–sulphur batteries, Energ. Environ. Sci. 7 (8) (2014) 2697.

[32] S.S. Zhang, J.A. Read, A new direction for the performance improvement of rechargeable lithium/sulfur batteries, J. Power Sources 200 (2012) 77–82.

[33] C. Zu, N. Azimi, Z. Zhang, A. Manthiram, Insight into lithium–metal anodes in lithium–sulfur batteries with a fluorinated ether electrolyte, J. Mater. Chem. A 3 (28) (2015) 14864–14870.

[34] Q.J. Meisner, T. Rojas, N.L. Dietz Rago, J. Cao, J. Bareño, T. Glossmann, A. Hintennach, P.C. Redfern, D. Pahls, L. Zhang, I.D. Bloom, A.T. Ngo, L.A. Curtiss, Z. Zhang, Lithium–sulfur battery with partially fluorinated ether electrolytes: interplay between capacity, coulombic efficiency and Li anode protection, J. Power Sources 438 (2019), 226939.

[35] D. Eroglu, K.R. Zavadil, K.G. Gallagher, Critical link between materials chemistry and cell-level design for high energy density and low cost Lithium-Sulfur transportation battery, J. Electrochem. Soc. 162 (6) (2015) A982–A990.

[36] J. Gao, M.A. Lowe, Y. Kiya, H.D. Abruña, Effects of liquid electrolytes on the charge–discharge performance of rechargeable Lithium/Sulfur batteries: electrochemical and in-situ X-ray absorption spectroscopic studies, J. Phys. Chem. C 115 (50) (2011) 25132–25137.

[37] R.S. Assary, L.A. Curtiss, J.S. Moore, Toward a molecular understanding of energetics in Li–S batteries using nonaqueous electrolytes: a high-level quantum chemical study, J. Phys. Chem. C 118 (22) (2014) 11545–11558.

[38] D. Aurbach, A. Zaban, Y. Ein-Eli, I. Weissman, O. Chusid, B. Markovsky, M. Levi, E. Levi, A. Schechter, E. Granot, Recent studies on the correlation between surface chemistry, morphology, three-dimensional structures and performance of Li and Li-C intercalation anodes in several important electrolyte systems, J. Power Sources 68 (1) (1997) 91–98.

[39] G. Li, S. Wang, Y. Zhang, M. Li, Z. Chen, J. Lu, Revisiting the role of polysulfides in lithium-sulfur batteries, Adv. Mater. 30 (22) (2018), e1705590.

[40] D.R. Wang, D.B. Shah, J.A. Maslyn, W.S. Loo, K.H. Wujcik, E.J. Nelson, M.J. Latimer, J. Feng, D. Prendergast, T.A. Pascal, N.P. Balsara, Rate constants of electrochemical reactions in a Lithium-sulfur cell determined by operando X-ray absorption spectroscopy, J. Electrochem. Soc. 165 (14) (2018) A3487–A3495.

[41] S. Chen, C. Niu, H. Lee, Q. Li, L. Yu, W. Xu, J.-G. Zhang, E.J. Dufek, M.S. Whittingham, S. Meng, J. Xiao, J. Liu, Critical parameters for evaluating coin cells and pouch cells of rechargeable Li-metal batteries, Joule 3 (4) (2019) 1094–1105.

[42] J. Xiao, Understanding the Lithium sulfur battery system at relevant scales, Adv. Energy Mater. 5 (16) (2015).

[43] J. Xiao, Q. Li, Y. Bi, M. Cai, B. Dunn, T. Glossmann, J. Liu, T. Osaka, R. Sugiura, B. Wu, J. Yang, J.-G. Zhang, M.S. Whittingham, Understanding and applying coulombic efficiency in lithium metal batteries, Nat. Energy 5 (2020) 561–568.

[44] S.S. Zhang, Liquid electrolyte lithium/sulfur battery: fundamental chemistry, problems, and solutions, J. Power Sources 231 (2013) 153–162.

[45] S.H. Chung, A. Manthiram, Rational design of statically and dynamically stable lithium-sulfur batteries with high sulfur loading and low electrolyte/sulfur ratio, Adv. Mater. 30 (6) (2018).

[46] Q. Pang, A. Shyamsunder, B. Narayanan, C.Y. Kwok, L.A. Curtiss, L.F. Nazar, Tuning the electrolyte network structure to invoke quasi-solid state sulfur conversion and suppress lithium dendrite formation in Li–S batteries, Nat. Energy 3 (9) (2018) 783–791.

[47] G. Zhou, A. Yang, G. Gao, X. Yu, J. Xu, C. Liu, Y. Ye, A. Pei, Y. Wu, Y. Peng, Y. Li, Z. Liang, K. Liu, L.-W. Wang, Y. Cui, Supercooled liquid sulfur maintained in three-dimensional current collector for high-performance Li-S batteries, Sci. Adv. 6 (21) (2020) eaay5098.

# CHAPTER 2

# Solid electrolytes for lithium-sulfur batteries

## Eleni Temeche and Richard M. Laine
Department of Materials Science and Engineering, University of Michigan, Ann Arbor, MI, United States

## Contents

| | |
|---|---|
| 2.1 Introduction to Li-S batteries | 18 |
| 2.2 Introduction to solid electrolytes | 19 |
| 2.3 Brief history of solid electrolytes | 19 |
| 2.4 Introduction to inorganic solid electrolytes | 20 |
|     2.4.1 Li-S batteries based on NASICON-type electrolytes | 22 |
|     2.4.2 Li-S battery based on garnet-type electrolytes | 29 |
|     2.4.3 Li-S batteries based on sulfide-type electrolytes | 32 |
| 2.5 Li-S batteries based on polymer electrolytes | 39 |
|     2.5.1 Solid polymer electrolytes | 39 |
|     2.5.2 Gel polymer electrolytes | 42 |
| 2.6 Summary | 43 |
| Acknowledgments | 44 |
| References | 45 |

## Abbreviations

| | |
|---|---|
| **ASSBs** | all solid-state batteries |
| **ASSLSBs** | all solid-state lithium–sulfur batteries |
| **GPE** | gel polymer electrolyte |
| **ISE** | inorganic solid electrolyte |
| **LIBs** | lithium-ion batteries |
| **LiSBs** | lithium-sulfur batteries |
| **LICGC** | lithium-conducting glass ceramic |
| **NASICON** | sodium superionic conductor |
| **PE** | polymer electrolyte |
| **SPE** | solid polymer electrolyte |
| **PAN** | polyacrylonitrile |

---

*Lithium-Sulfur Batteries*
https://doi.org/10.1016/B978-0-12-819676-2.00008-6

Copyright © 2022 Elsevier Inc.
All rights reserved.

## 2.1 Introduction to Li-S batteries

During the past decade, Li–S batteries (LiSBs) have drawn intense attention as candidates for next-generation energy storage technologies owing to their high theoretical capacity (1672 mAh/g) and specific energy (2600 Wh/kg) [1]. Part of the motivation comes from the fact that sulfur is low cost, abundant, and an environmentally friendly resource, making it a very promising cathode material [1]. Although LiSBs offer several advantages over conventional LIBs, some drawbacks hamper their implementation in large-scale applications [1,2]. The first drawback is associated with the electronic insulating nature of elemental sulfur, which results in inefficient utilization of sulfur-containing cathode materials and poor cyclability [1].

At present, most LiSBs have been assembled using organic liquid electrolytes. Thus, concerns about safety and cycling performance for any batteries employing liquid electrolytes limit their large-scale deployment and application. The most commonly discussed challenge is in controlling the formation of long–chain, linear lithium polysulfides ($Li_2S_x$, $4 \leq x \leq 8$) and/or blocking their diffusion toward the anode during cycling, which incurs multiple problems [1–3].

The continuous dissolution of polysulfides increases electrolyte viscosities, resulting in a decrease in electrolyte ionic conductivity. Cathode structures are also distorted as a result of polysulfide formation, again reducing full use of active material. Long–chain polysulfides are typically reduced to short-chain polysulfides at the lithium anode surface in a process that corrodes the lithium [1]. Polysulfides continuously migrate from the sulfur cathode to the Li anode and induce side reactions, resulting in low columbic efficiencies and high internal resistance. In addition, liquid electrolyte gradually gets consumed until it is exhausted, generating gas(es) that may cause the battery to expand and even explode [2]. This continuous migration from the cathode to the anode, and vice versa, is commonly referred as the polysulfide shuttle effect.

A separate issue centers on the formation of Li metal dendrites when cycling metallic Li as the anode [1], which reduces the available active material and columbic efficiency, and can cause internal short circuits.

A series of countermeasures to overcome these challenges are the focus of much of mainstream LiSB research [4–6], for example, by reducing the insulating nature of elemental sulfur and through the development of polysulfide confinement structures such as sulfur/carbon and sulfur/polymer composites [4]. A second strategy to improve capacity and cyclability is to protect the lithium anode to inhibit dendrite formation as well as solid electrolyte interphase formation using coatings [4]. A third approach targets the use of solid-state electrolytes, the subject of this chapter [7].

## 2.2 Introduction to solid electrolytes

Solid electrolytes enable several next-generation energy storage systems (designs), including solid oxide fuel cells, supercapacitors, and batteries [8]. State-of-the-art battery technologies depend highly on the discovery of electrically insulating solids with high ionic mobilities. Solid electrolytes provide exciting opportunities for developing new generations of Li-ion batteries (LIBs) such as Li-S and Li-air devices. Current LIBs, using traditional organic liquid electrolytes, suffer from poor electrochemical and thermal stabilities, leakage, and flammability. Hence, replacing liquids with solid electrolytes offers multiple possibilities for developing new battery chemistries and designs.

Section 2.3 briefly describes the discovery and history of solid electrolytes, while Section 2.4 categorizes the types of inorganic solid electrolytes as a function of crystal structures that enable fast ionic mobility. Recent progress in LiSBs based on the inorganic solid electrolytes is summarized.

Section 2.5 discusses issues concerned with the design and architecture of solid polymer electrolytes: crystalline, semicrystalline, or amorphous. The utility of polymer electrolytes with respect to ability to resist electrode volume variations, and to improve safety features, flexibility, and processability is also detailed. Recent progress in LiSBs based on the polymer electrolytes is also summarized.

## 2.3 Brief history of solid electrolytes

The transport of electrical charge by diffusion of ions in solids has been studied for almost two centuries. The history of solid-state ionic conductors dates back to the 1830s, with a report by Michael Faraday on ionic conduction in heated solids ($Ag_2S$ and $PbF_2$) [9]. A $ZrO_2/Y_2O_3$ composite was used as a glowing rod in a device known as the *Nernst glower* in the 1900s [10]. This composite conducts electrical charge via defect diffusion through the oxide lattice at high temperatures. The discovery of $Ag^+$ and $Na^+$ conductors with ionic conductivities comparable to liquid electrolytes led to a renewed interest in the development of solid electrolytes in the 1960s [11,12].

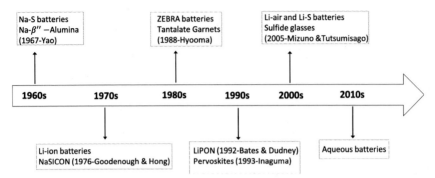

Fig. 2.1 Historical outline of progress in solid electrolytes.

Most common alkali halides exhibit low ionic conductivity and high activation energies, in which the electrical charge transport proceeds through temperature-dependent motion of vacancies in the crystal structure [13]. However, the discovery of silver and copper halides with high ionic conductivities and low activation enthalpies wherein ionic conductivity is via motion of interstitial species resulted in a breakthrough in the field of superionic conductors.

To date, a huge family of solid electrolytes has been developed that offers high ionic conductivities ($>10^{-4}$ S/cm) at ambient conditions, low activation energies ($<0.4$ eV), thermal stability ($>200°C$), and wide electrochemical stability windows (5 V). The discovery of ionic transport in poly(ethylene oxide) (PEO) in 1973 broadened the scope of solid electrolytes from inorganic materials to solid polymer materials [14]. Fig. 2.1 timeline illustrates key developments in solid electrolyte batteries. Solid electrolytes with different crystallographic or amorphous morphologies provide several specific advantages. Solid electrolytes can be divided into inorganic solid (ISEs) and polymer electrolytes (PEs).

## 2.4 Introduction to inorganic solid electrolytes

Key properties that determine the utility of solid electrolytes include high ionic conductivity ($>10^{-6}$ S/cm), high transference number ($\approx 1$), low electrical conductivity ($>10^{-8}$ S/cm), low ionic area-specific resistance ($<20$ $\Omega/cm^2$), wide electrochemical stability window (0–5 V vs Li/Li$^+$), good chemical and thermal stability, excellent mechanical properties, low cost, ease of fabrication, eco-friendliness, and simple device integration [15]. Much progress has been made to improve initially discovered ISEs to meet these properties.

ISEs demonstrate high lithium transference numbers ($\sim$1) compared to liquid electrolytes, unlike the coupled transport (cations and anions) in liquid electrolytes, (transference number: 0.25–0.5) [15]. ISEs enhanced thermal stabilities provide opportunities to design new architectures that simplify battery configurations and reduce the peripheral mass of traditional LIBs. For example, the battery pack can be redesigned to minimize thermal management systems and overpressure vents typically installed to overcome the challenges of using flammable liquid electrolytes [16]. Furthermore, ISEs facilitate the adoption of newer battery chemistries. The development of state-of-the-art Li-S and Li-air batteries will benefit greatly from the use of solid electrolytes [17].

Most ISE materials rely on solid-state diffusion through well-defined crystallographic structures wherein the ionic framework permits fast ion mobility via vacant and/or interstitial sites. Mostly, ISE crystal structures consist of a stationary, polyhedral metal oxide framework matrix with open channels that allow diffusion of mobile ions [15]. Mobile ions include $H^+$, $Li^+$, $Na^+$, $K^+$, $Ag^+$, $Cu^+$, $Mg^{2+}$, $F^-$, $Cl^-$, and $O^{2-}$. A number of framework metals and nonmetals can form polyhedral networks with the latter typically coming from elements in groups VA, VIA, and VIIA [9]. Ionic conductivity in ISEs can be tuned (enhanced) via selective doping. To do this effectively requires an extensive understanding of their crystal chemistries. Fundamental criteria that have been used to (greatly) improve fast ionic conduction in ISEs include:

**(1)** The number of sites that mobile ions can occupy should be larger than the number of mobile species.

**(2)** The mobile ion should fit through the smallest cross-sectional area of the conduction channel (bottleneck).

**(3)** The concentration of vacant and interstitial sites should be optimal to minimize the migration energy.

**(4)** The interaction forces between mobile ions and the main framework should be weak.

**(5)** The available sites for mobile ions to occupy should be interconnected to allow continuous diffusion pathways.

**(6)** The framework moieties should consist of stable ions with different coordination numbers to minimize the activation energy.

### 2.4.1 Li-S batteries based on NASICON-type electrolytes

Sodium (Na) Super Ionic CONductor (NASICON) electrolytes have been studied extensively because of their high ionic conductivities at room temperature, wide electrochemical potentials ($\sim$7 V), and high moisture stability. The main drawback to using $Li_{1+x}Al_xTi_{2-x}(PO_4)_3$ (LATP) materials is the irreversible reduction of $Ti^{4+}$ to $Ti^{3+}$ when in contact with the metallic Li at low potentials (2.5 V vs $Li/Li^+$), restricting their use [18]. The reduction of LATP by Li metal may induce electrical conductivity in the solid electrolyte causing short-circuits. Nevertheless, NASICON-type electrolytes continue to be studied intensely for use in LiSBs, as discussed below.

One issue with LiSBs involves the use of a metallic Li anode, which is not stable in organic electrolytes. Hence, fully lithiated $Li_2S$ cathodes can be used as a Li source to potentially couple with Li-free anodes [7]. However, $Li_2S$ still suffers from the polysulfide shuttle effect. Solid electrolytes offer the chance to eliminate or at least impede polysulfide shuttle effect. However, most solid electrolytes exhibit poor ionic conductivities compared to liquid electrolytes at ambient conditions. In addition, solid-solid ionic diffusion through electrode/ISE interfaces is sluggish, ascribed to poor mating of surfaces.

An alternate option is to use hybrid systems consisting of both solid and liquid electrolytes. In hybrid electrolyte systems, ISEs can play multiple roles. They can eliminate the need for a porous polymer separator, thereby eliminating polysulfide shuttling [7]. They can also provide an ionic pathway to sustain electrochemical reactions at the cathode and anode. On the other hand, liquid electrolytes in hybrid systems play an important role in serving as an ionic medium for sulfur-polysulfide-sulfide redox reactions within the cathode and coincidentally maintaining ionic paths at electrode/ISE interfaces [1].

Wang et al. [7] reported using hybrid electrolytes to assemble LiSBs composed of $Li_2S$ as a cathode and Li metal as an anode. Fig. 2.2 presents an example of this LiSB architecture composed of Cu foil/Li metal/organic electrolyte/LATP/organic electrolyte/$Li_2S$/Super P carbon/Ti foil.

The Li metal anode and $Li_2S$ cathode are separated by a protective LATP membrane (Ohara, Corp.). Wang et al. [7] reported that the unique features of this LiSB result in high specific capacities of 1100 mAh/g cycled at a rate of 1/30 C, high columbic efficiency, and no self-discharge. The thick but friable LATP membrane (150 μm) may decrease the Li-S cell performance; however, to be useful in commercial applications, thicknesses should be <25 μm to minimize mechanical failure [19].

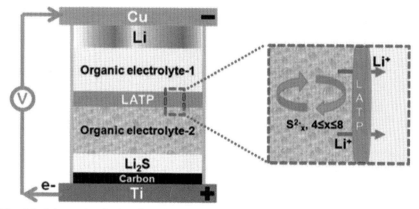

**Fig. 2.2** Schematic of a Li-S battery composed of Li/organic electrolyte/LATP/organic electrolyte/Li$_2$S. *(Reproduced with permission from L. Wang, Y. Wang, Y. Xia, A high performance lithium-ion sulfur battery based on a Li$_2$S cathode using a dual-phase electrolyte, Energy Environ. Sci. 8 (5) (March 11, 2015) 1551–1558, Copyright (2015) Royal Society of Chemistry.)*

The implementation of sulfide-impermeable solid electrolytes allows the development of new LiSB architectures. For example, a liquid-phase electrode (catholyte solution) containing active materials could be linked to a metallic Li anode using a hybrid electrolyte [20]. The main advantage of this architecture is that it can increase energy densities as a function of the volume of catholyte solution. Wang et al. [20] demonstrated the use of a sulfide-impermeable LATP electrolyte (Ohara) as depicted in Fig. 2.3 using a carbon film current collector.

In this work, the catholyte solution was prepared by dissolving elemental sulfur and LiClO$_4$ in anhydrous tetrahydrofuran. This LiSB system exhibits a specific capacity of 1200 mAh/g at 0.1 C for 70 cycles without significant capacity fading. At a current rate of 0.2 C, the specific capacity decreased to 1000 mAh/g around the 50th cycle. The decrease in specific capacity at high C-rate is attributed to the low ionic conductivity ($10^{-4}$ S/cm) of the LATP electrolyte at ambient conditions [20].

A moderate elevation in operating temperature could enhance the ionic conductivity of the ISE electrolyte and improve the diffusion of the active material increasing the specific capacity at high C-rates. However, operating hybrid electrolyte systems at elevated temperature is problematic as the liquid electrolyte poses safety issues due to the flammability of the organic solvent.

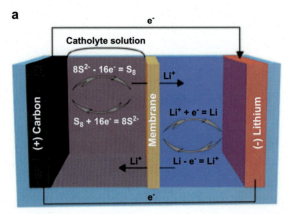

**Fig. 2.3** Schematic of a Li-S battery composed of Li/organic electrolyte/LATP/organic electrolyte/catholyte solution. *(Modified from L. Wang, Y. Zhao, M.L. Thomas, A. Dutta, H.R. Byon, Sulfur-based catholyte solution with a glass-ceramic membrane for Li-S batteries, ChemElectroChem. 3 (1) (October 01, 2015) 152–157, Copyright (2016) John Wiley & Sons)*

The energy densities of LiSBs using catholyte solutions may be improved by increasing the concentration of active materials while maintaining appropriate contact with the current collector. Hence, mixing elemental sulfur with $Li_2S$ is suggested to increase the concentration of the active material as a result of long-ordered ($Li_2S_x$, $4 \leq x \leq 8$) polysulfide formation [20].

Wang et al. [21] explored the electrochemical stability of LATP by assembling LiSBs per the schematic shown in Fig. 2.4A. Fig. 2.4B illustrates the cycling performance of a LiSB with high columbic efficiency (~100%)

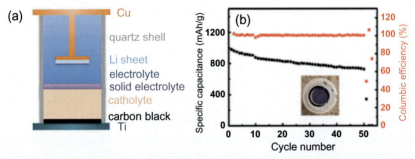

**Fig. 2.4** (A) Schematic of a Li-S battery composed of Li/liquid electrolyte/LATP/catholyte. (B) Cycling performance of the Li-S battery. *(Modified from S. Wang, Y. Ding, G. Zhou, G. Yu, A. Manthiram, Durability of the $Li_{1+x}Ti_{2-x}Al_x(PO_4)_3$ solid electrolyte in lithium-sulfur batteries, ACS Energy Lett. 1 (6) (October 31, 2016) 1080–1085, Copyright (2016) American Chemical Society.)*

and an initial discharge capacity of 978 mAh/g at 0.1 C rate. The high columbic efficiency is ascribed to LATP preventing the diffusion of polysulfide species to the anode side. However, the LiSB showed a gradual decrease in specific capacity leading to abrupt degradation after 50 cycles. Wang et al. [21] reported that the color of the LATP was blue after disassembling the cell as shown in the insert of Fig. 2.4B, indicating that the ISE is reduced by the polysulfides during cycling. The $Li_{1.3}Al_{0.2}Ti(PO_4)_3$ phase converts to Li-rich $Li_3Al_{0.3}Ti_{1.7}(PO_4)_3$ likely by reaction with dissolved polysulfides as $Ti^{4+}$ is reduced to $Ti^{3+}$. In addition, LATP grain boundaries seem to corrode in contact with polysulfides on long-term cycling forming pores that facilitate polysulfide diffusion. Thus, even with ISEs, the polysulfide shuttle effect can give rise to a sudden degradation of LiSBs [21].

LATP ISEs demonstrate relatively high ionic conductivities ($10^{-4}$ S/cm) at room temperature and are commercially available (Ohara, 50–150 μm). However, the relatively narrow electrochemical stability window and irreversible reduction of $Ti^{4+}$ in LATP by polysulfides raise both chemical and electrochemical incompatibility concerns for use in hybrid electrolyte systems for LiSBs [22]. Hence, the development of Ti-free NASICON electrolytes might overcome these challenges.

Yu et al. [22] explored the use of NASICON membranes with the formulae $Li_{1+x}Y_xZr(PO_4)_3$ (LYZP) ($x = 0$–0.15) as ISEs/separators to suppress polysulfide shuttling in LiSBs. The LYZP electrolyte (150 μm) provides ionic conductivities of ~$2.5 \times 10^{-5}$ S/cm at ambient conditions, one order of magnitude lower than LATP. Fig. 2.5 depicts the LiSB architecture composed of Li metal anode/liquid electrolyte-soaked polypropylene (PP) thin-film separator/LYZP ceramic electrolyte/liquid electrolyte/$Li_2S_6$ catholyte.

A PP separator containing a thin layer of liquid electrolyte is reported to play an important role as it helps build a facile ionic path at the Li metal/LYZP electrolyte interface. The LYZP membrane in the hybrid electrolyte system is reported to serve multiple functions, including as a $Li^+$ conductor and an electrical insulator to separate Li metal from the $Li_2S_6$ catholyte and to shield lithium from polysulfide shuttling.

Yu et al. [22] tested the stability of LYZP ISEs with Celgard membranes soaked in polysulfide solutions. Fig. 2.6 displays LYZP polysulfide retention on bonding to Celgard membranes. The LYZP ISE effectively inhibits polysulfide diffusion. Thus, there is no noticeable color change in the DME/DOL solution even after 30 days (Fig. 2.6C). In contrast, the Celgard membrane alone does not prevent polysulfide migration. Hence, the color changes (to light yellow) in the DME/DOL solution after 1 h (Fig. 2.6B).

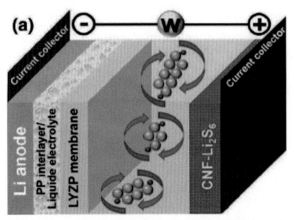

**Fig. 2.5** Schematic of a Li-S battery composed of Li/PP + liquid electrolyte/LYZP/liquid electrolyte/Li$_2$S$_6$ catholyte. *(Reproduced with permission from X. Yu, Z. Bi, F. Zhao, A. Manthiram, Polysulfide-shuttle control in lithium-sulfur batteries with a chemically/electrochemically compatible NaSICON-type solid electrolyte, Adv. Energy Mater. 6 (24) (August 30, 2016) 1601392, Copyright (2016) John Wiley & Sons.)*

Integration of the LYZP ISE with a liquid electrolyte in assembling a Li-Li$_2$S$_6$ battery resulted in an initial discharge capacity of 950 mAh/g with 89.5% capacity retention after 150 cycles at 0.2 C based on active sulfur. In comparison, a Li-Li$_2$S$_6$ battery assembled with a Celgard membrane alone demonstrates an initial capacity of 1180 mAh/g; however, the capacity retention was only 20% after 150 cycles. LYZP ISEs are reported to exhibit both chemical compatibility with the battery components and electrochemical stability after extensive cycling. The main drawbacks of this cell are the thickness of the LYZP electrolyte (150 μm) and its poor ionic conductivity (~2.5 × 10$^{-5}$ S/cm), limiting its potential utility in ASSLSBs [22].

Wang et al. [23] explored using Li$_{1.5}$Al$_{0.5}$Ge$_{1.5}$(PO$_4$)$_3$ (LAGP) ISEs separators in hybrid electrolyte LiSBs consisting of a Li metal anode and a Ketjen black-sulfur (KB-S) composite cathode as shown in Fig. 2.7. The 1 M LiN(O$_2$SCF$_3$)$_2$ in (1,1 v/v LiTFSI/DOL/DME) liquid electrolyte was used to wet the electrode and reduce the interfacial impedance at the LAGP/electrode interface. An LiSB with a hybrid electrolyte exhibited initial discharge capacities of ~1528, 1386, and 1341 mAh/g at C/20, C/5, and C/2 rates, respectively. The reversible capacity of the hybrid cell remained at 720 mAh/g after 40 cycles at the C rate of C/5. The gradual decrease in capacity is attributed to the progressive irreversible formation of insulating Li$_2$S discharge product [23].

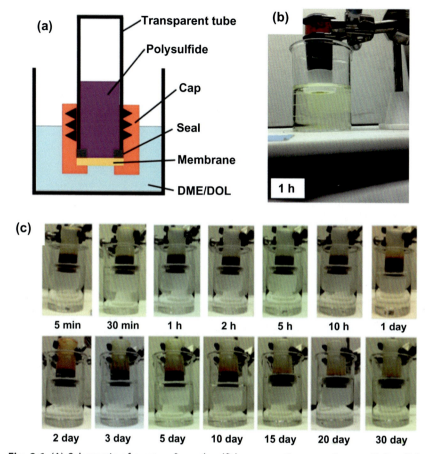

Fig. 2.6 (A) Schematic of a setup for polysulfide permeation experiments. Polysulfide diffusion tests of the (B). Celgard membrane after 1 h. (C) LYZP membrane after various resting times. *(Modified from X. Yu, Z. Bi, F. Zhao, A. Manthiram, Polysulfide-shuttle control in lithium-sulfur batteries with a chemically/electrochemically compatible NaSICON-type solid electrolyte, Adv. Energy Mater. 6 (24) (August 30, 2016) 1601392, Copyright (2016) John Wiley & Sons.)*

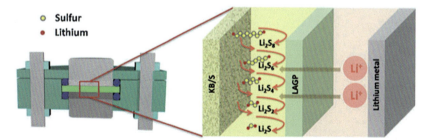

Fig. 2.7 Schematic of a Li-S battery composed of Li/liquid electrolyte/LAGP/liquid electrolyte/KB-S. *(Reproduced with permission from Q. Wang, J. Jin, X. Wu, G. Ma, J. Yang, Z. Wen, A shuttle effect free lithium sulfur battery based on a hybrid electrolyte, Phys. Chem. Chem. Phys. 16 (39) (September 1, 2014) 21225–21229, Copyright (2014) Royal Society of Chemistry.)*

During the first few cycles, the hybrid Li–S cell showed columbic efficiencies below 100%, but >96%. The decrease in columbic efficiency is ascribed to the formation of a surface passivating film, requiring excess lithium to compensate for lithium loss. Columbic efficiency gradually increases to 100% after the first few cycles, indicating that LAGP prevents polysulfide shuttling. In comparison, an LiSB assembled with liquid electrolyte-impregnated Celgard membrane alone demonstrated initial discharge capacities of ~719, 578, and 467 mAh/g at C/20, C/5, and C/2 rates, respectively. The decrease in discharge capacity is associated with polysulfide shuttling through the Celgard membrane [23].

Even though polysulfide shuttle effects are suppressed by ISEs, LiSB reversibility and rate capability seem to decrease gradually in hybrid electrolyte systems. The main reasons for this capacity loss are attributed to the formation of inactive discharge products and the loss of intimate contact between the sulfur and the carbon matrix as reported by Wang et al. [24]. Electronically isolated and electrochemically inactive $Li_2S$ does not participate in electrochemical oxidation during charging in the absence of a carbon matrix. Hence, irreversible formation of $Li_2S$ becomes severe with long-term cycling, leading to the loss of active material and decreases in discharge capacity. To fully use the dissolved sulfur-containing active materials, Wang et al. [24] used a carbon coating on the side of the solid electrolyte facing the sulfur cathode (Fig. 2.8). The carbon coating (5 μm thickness) improves the wettability of the LAGP electrolyte (0.7 mm thickness) by the liquid electrolyte.

The LiSB with carbon-coated LAGP demonstrated discharge capacities of 1509, 1453, 1381, and 1039 mAh/g at 0.05, 0.1, 0.2, and 0.5 C rates, respectively. In contrast, the LiSB with a pristine-uncoated LAGP electrolyte exhibited 1316, 1276, 1172, and 387 mAh/g, respectively, at the same rates. The carbon coating increases the conducting surface between the sulfur cathode and the LAGP electrolyte (0.7 mm thickness) where high reuse of dissolved polysulfide species is realized, leading to enhanced reversibility and rate capability [24]. This is an example where the electrochemical performance of a LiSB is improved by enhancing electron transport for the conversion of dissolved sulfur-containing active materials by careful ISE interface modification.

LiSBs benefit from ISEs such as LATP/LAGP that can protect the Li metal from polysulfide side reactions. In addition, these electrolytes widen the choice of solvents to be used in hybrid-electrolyte systems as the Li metal is no longer directly exposed. However, to date all the NASICON electrolytes described above are relatively thick (at least 150 μm in the case of

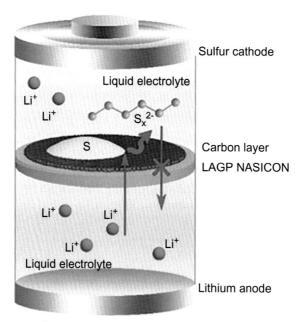

**Fig. 2.8** Schematic of a Li-S battery composed of Li/liquid electrolyte/LAGP with carbon coating/liquid electrolyte/S. *(Modified from Q. Wang, J. Guo, T. Wu, J. Jin, J. Yang, Z. Wen, Improved performance of Li-S battery with hybrid electrolyte by interface modification, Solid State Ionics 300 (February 2017) 67–72, Copyright (2017) Elsevier.)*

LATP), which adds unwanted mass and volume to the battery and thereby reducing the volumetric and galvanometric energy densities. It is thus important to design much thinner and mechanically robust ISEs to minimize the LiSB package size. Moreover, NASICON-type ceramics are unstable in contact with metallic Li. Typically, to solve this problem, a protective Li$^+$-conducting layer (polymer, gel, nonaqueous electrolyte) is placed between the Li anode and the ceramic [25]. This is discussed in more detail in Section 2.5.

### 2.4.2 Li-S battery based on garnet-type electrolytes

The thermal and mechanical stabilities of LLZO (Li$_7$La$_3$Zr$_2$O$_{12}$) electrolytes against Li metal make them well suited for use in Li metal-based batteries. In addition, features such as low electrical conductivity, good chemical stability, and the same order of magnitude inter- and intra-grain conductivities allow garnet electrolytes to satisfy the above-noted requirements for polycrystalline ISEs. The main drawback to garnet-type materials is that they

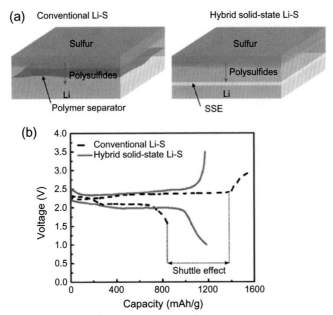

**Fig. 2.9** (A) Schematic of conventional Li-S and hybrid solid-state Li-S battery with garnet LLZO ISE. (B) Voltages profiles of conventional and hybrid Li-S cells. *(Modified from K. Fu, Y. Gong, S. Xu, Y. Zhu, Y. Li, J. Dai, Stabilizing the garnet solid-electrolyte/polysulfide interface in Li-S batteries, Chem. Mater. 29 (19) (September 20, 2017) 8037–8041, Copyright (2017) American Chemical Society.)*

absorb $H_2O$ and $CO_2$, forming surface $Li_2CO_3$ that degrades ionic conductivity [26]. Thus, efforts to reduce this susceptibility to reaction in ambient atmosphere are needed.

Fu et al. [27] compared conventional and hybrid LiSBs based on a porous polymer separator and a garnet LLZO membrane, respectively, as shown in Fig. 2.9. Soluble polysulfides diffuse through the porous polymer separator and migrate to the Li metal anode. In contrast, polysulfide shuttling is inhibited by dense LLZO. Thus, side reactions at the Li metal surface and corrosion are prevented. The potential vs. capacity plot shows that the LiSB with a hybrid electrolyte system offers enhanced reversible capacity compared to a conventional LiSB (~800 mAh/g) counterpart. The hybrid cell showed a specific capacity of ~1000 mAh/g at a current density of 200 mA/g and 100% columbic efficiency [27]. This is the first study demonstrating the electrochemical stability of garnet-type solid electrolytes in a LiSBs (Fig. 2.10).

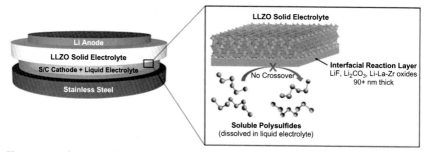

Fig. 2.10 Schematic for the Li/LLZO/S-C hybrid cell. (Modified from M. Naguib, A. Sharafi, E.C. Self, H.M. Meyer, J. Sakamoto, J. Nanda, Interfacial reactions and performance of Li$_7$La$_3$Zr$_2$O$_{12}$-stabilized li-sulfur hybrid cell, ACS Appl. Mater. Interfaces. 11 (45) (October 16, 2019) 42042–42048, Copyright (2019) American Chemical Society.)

Naguib et al. [28] reported the electrochemical performance of a LiSB assembled with LLZO and a liquid electrolyte. A sulfur/carbon composite cathode was prepared by infiltrating sulfur into the Ketjen black carbon at 150°C for 10 h. A 1 M of LiTFSI (1:1 v/v DOL/DME) liquid electrolyte was used to wet the LLZO electrolyte (1.2 mm thick) and the sulfur/carbon composite cathode. The initial discharge capacity is reported to be ~1013 mAh/g with a corresponding columbic efficiency of 87.8%. The columbic efficiency increases to >99% after 10 cycles, indicating that the LLZO ISE effectively prevents polysulfide shuttling during cycling. An improvement in the ISE electrolyte thickness is required to be considered for commercial applications.

Also, Naguib et al. [28] report that the specific capacity of the hybrid cell fades from the 1154 mAh/g in the first cycle to 604 mAh/g after 50 cycles at a current density of 100 mAh/g. This capacity fade is attributed to the sulfur/carbon cathode architecture wherein not all the polysulfides are trapped within the carbon matrix during cycling, resulting in a decrease in accessible active material. It appears that further improvements in reversible cyclability of the hybrid Li-S cell require optimization of the cathode architecture. In addition, this study shows that a spontaneous chemical reaction occurs between the liquid electrolyte and the LLZO ISE, forming an interfacial reaction layer (LiF, Li$_2$CO$_3$, Li-La-Zr oxides). This could also explain the lower capacity utilization and retention of the LiSB. Hence, optimizing the interfacial resistance between the ISE and the liquid electrolyte is crucial to enhancing the performance of hybrid electrolyte systems [28].

In summary, the implementation of sulfide-impermeable ISE electrolytes can eliminate or greatly reduce concerns with polysulfide shuttling

32  Lithium-sulfur batteries

and protect Li electrode surfaces. We again note that the use of solid electrolytes allows the development of new battery architectures for LiSBs.

## 2.4.3 Li-S batteries based on sulfide-type electrolytes

Sulfide-type electrolytes are structurally similar to LIhium Super Ionic CONductor (LISICON) solid electrolytes. Replacing $O^{2-}$ by $S^{2-}$ in $Li_3PO_4$ results in weaker interactions between $S^{2-}$ and $Li^+$, leading to higher $Li^+$ mobilities and conductivities compared to oxides [29]. This is one main reason why sulfides are preferable to oxide ISEs. $Li^+$ is more mobile in sulfide frameworks, because $S^{2-}$ is larger than $O^{2-}$; furthermore, the charge is more diffuse, reducing electrostatic interactions. Moreover, $S^{2-}$ introduction generates wider channels, allowing faster ion conduction. Finally, sulfide-based frameworks show higher polarizability due to weaker interactions between $Li^+$ and the anions, allowing more facile $Li^+$ diffusion [29].

Four types of sulfide-based electrolytes can be classified by chemical composition: thiophosphates, argyrodite, glasses, and glass ceramics. These sulfide electrolytes show superior ionic conductivities among ISEs, comparable to organic liquid electrolytes. In the following, we explore LiSBs based on sulfide-type electrolytes.

A major advantage of sulfide solid electrolytes is that they are relatively soft and deformable. They pack densely on cold pressing; thus, sintering is not as energy intensive as most oxide-based systems [30]. This feature is highly desirable for assembling ASSBs as good electrode-electrolyte interfaces can be achieved at low temperatures, limiting side reactions and impurities at the interface. ASSBs are more stable and safer than batteries with hybrid electrolyte systems; they have the potential to be used in high-energy density systems. Noninflammable sulfide-based electrolytes, in principle, permit very high capacities (>1000 mAh/g) to be realized.

### 2.4.3.1 Li-S batteries based on thiophosphates

The thiophosphates can be divided into two subgroups: (Li-P-S) and (Li-X-P-S) where X=Si, Ge, Sn, Al, and Y [30]. Both subgroups belong to the thio-LISICON family. The main advantage of Li-P-S systems is that they do not contain metals easily reduced or oxidized during electrochemical reactions. The stoichiometric compounds, such as $Li_3PS_4$, $Li_4P_2S_4$, $Li_7P_3S_{11}$, and $Li_7PS_6$, are part of the Li-P-S group. Among them, the most widely studied is $Li_3PS_4$ with an ionic conductivity of $3 \times 10^{-7}$ S/cm at room temperature [31].

Work by Yamada et al. [32] describes ASSLBs based on a $Li_3PS_4$ electrolyte (100 mg). The $Li/Li_3P_4/S$ battery exhibited a high initial capacity of $\sim$1600 mAh/g at 0.05 C rate with 99% columbic efficiency, ascribed to the inhibition of polysulfide shuttling by the sulfide-based solid electrolyte. As with oxide-based ISEs, sulfide-based electrolytes physically block polysulfide migration from the cathode to the anode. The cycle test shows that the $Li/Li_3P_4/S$ electrolyte is stable for 10 cycles delivering a specific capacity of 1500 mAh/g at 0.05 C rate. However, more extensive cycling studies of $Li/Li_3P_4/S$ battery are required to further understand the electrochemical performance of this system.

One other thiophosphate system with the Li-X-P-S formula also offers high ionic conductivity at room temperature. $Li_{10}GeP_2S_{12}$ (LGPS) is a primary example of the thio-LISICON family [33]. LGPS offers attractive features for the state-of-the-art ASSLSBs such as high ionic conductivity (12 mS/cm), safety, and electrochemical stability (0–5 V vs $Li/Li^+$).

Kobayashi et al. [34] used thio-LISCION ($Li_{3.25}Ge_{0.25}P_{0.75}S_4$) to assemble an ASSLSB. Two methods were used to prepare the cathode, composed of sulfur and acetylene black, by mechanical mixing and gas/solid mixing. The gas/solid method involves heating a mixture of sulfur and acetylene black (50:50 wt.%) in a quartz tube under vacuum at 300°C. The cathode mixture (5 mg) is then pressed (500 MPa) onto one side of the electrolyte pellet (70 mg).

The mechanically mixed electrode system gave specific discharge capacities of 120 mAh/g, whereas the vapor system gave 590 mAh/g. The deposition of sulfur from the gas phase is reported to lead to the formation of small particles, whereas the size of the sulfur particles after the mechanical mixing remained unchanged (1–10 μm). Thus, the gas/solid processed composite electrode improves charge-discharge characteristics. The resulting nano-sized particles (1–10 nm) play an important role in the reported high specific capacities. This is ascribed to the close contact between the sulfur and acetylene black, which acts as conducting matrix, resulting in a composite cathode with low resistivity [34]. This is one example in which cathode preparation plays an important role in the electrochemical performance of the ASSLSB by ensuring intimate contact between the active material and the conducting matrix.

The contact between the sulfur cathode and the ISE electrolyte can be further improved by optimizing the conducting matrix. The composite cathode based on an acetylene black framework offers a random network of macropores between constituents. In contrast, the mesoporous carbon

CMK-3 framework exhibits high electrical conductivity, providing a new electrode structure for LiSBs [33]. However, the value of the electrical conductivity was not reported in this work. A comparative conductivity study on the carbon/polymer additives is important to understand their role in enhancing the electrode stability. The CMK-3 framework is suggested to increase the electrode/ISE contact and improve the electrical conduction and hence charge/discharge characteristics [33].

Nagao et al. [33] described the electrochemical performance of an ASSLSB composed of a sulfur/CMK-3 composite cathode, Li-Al alloy anode, and $Li_{3.25}Ge_{0.25}P_{0.75}S_4$ electrolyte. This cell exhibited an initial discharge capacity of 1600 mAh/g and a reversible capacity $\sim$700 mAh/g after 10 cycles. The 2D carbon rod structure of CMK-3 provides a fixed pore size in the composite electrode framework, and the resulting cathode matrix has a high carbon/sulfur ratio, which reduces the total capacity for practical cells [33].

Three-dimensional, highly ordered, mesoporous carbon replicas with small pore sizes (8–100 nm) have been described as promising electrode matrices providing high electrical conductivity and high sulfur utilization (30 wt.%) [35]. ASSLSBs assembled with sulfur/carbon replicas with various pore sizes showed first discharge capacities of 1282, 1944, and 906 mAh/g-sulfur corresponding to carbon replicas with 10, 8, and 6 nm pore sizes, respectively.

Although the use of ISEs can overcome safety and the polysulfide shuttle challenges, they also bring new challenges by increasing interfacial resistance and excessive stress/strain during extensive cycling. Many studies indicate that good contact of the active sulfur to the ISE is necessary to ensure good electrochemical performance of the ASSLSB. Moreover, uniformly dispersing the active sulfur into the electronically conductive cathode matrix helps reduce stress/strain at the electrolyte/electrode interface. Yao et al. [36] demonstrated that the interface resistance and stress/strain processes were reduced by conformal coatings of sulfur on reduced graphene oxide (rGO). The rGO is reported to improve the structural stability of the sulfur by serving as electronic conduction framework and buffering the volume expansion during lithiation/delithiation. Fig. 2.11 shows a schematic of an ASSLSB constructed with a rGO@S/$Li_{10}GeP_2S_{12}$ composite cathode, $Li_{10}GeP_2S_{12}$ and 75%$Li_2$S-24@%$P_2S_5$-1%$P_2O_5$ bilayer electrolyte, and a Li metal anode.

Yao et al. [36] investigated the impact of operating temperatures (25°C, 60°C, or 100°C) on the cycling performance of the ASSLB as shown in Fig. 2.11. The Li-S cell at 25°C exhibited high ionic resistance ($\sim$100 $\Omega$)

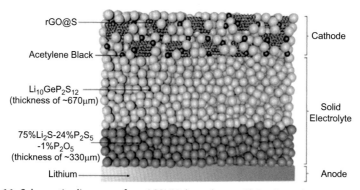

**Fig. 2.11** Schematic diagram of an ASSLSB based on sulfide electrolytes. *(Reproduced with permission from X. Yao, N. Huang, F. Han, Q. Zhang, H. Wan, J.P. Mwizerwa, High-performance all-solid-state lithium–sulfur batteries enabled by amorphous sulfur-coated reduced graphene oxide cathodes, Adv. Energy Mater. 7 (11) (May 11, 2017) 1602923, Copyright (2017) John Wiley & Sons.)*

attributed to poor ionic conductivity of the bilayer electrolyte. At 100°C, the 75%Li$_2$S-24@%P$_2$S$_5$-1%P$_2$O$_5$ electrolyte is suggested to decompose to sulfur and phosphorous, resulting in a decrease in electrolyte ionic conductivity.

This decrease in ionic conductivity and the side reactions between the electrodes and the electrolyte are reported to cause a rapid capacity decay at 100°C. At 60°C, the cell delivers high rate capacities of ~1525, 1384, 1336, 903, 502, and 205 mAh/g at 0.05, 0.1, 0.5, 1.0, 2.0, and 5.0 C, respectively [36]. Thus, the stress/strain induced by the sulfur volume changes seems to be effectively controlled. Two key factors for long ASSLSB cycle lives are minimization of stress/strain effects and reduction of interfacial resistance between the electrolyte and the electrode. The main drawbacks of this cell are the thickness of the overall solid electrolyte (~1 mm) and the 60°C operating temperature, which are not practical for large-scale applications.

### 2.4.3.2 Li-S batteries based on argyrodite

One of the most important criteria in assembling a LiSBs is the electrochemical stability window of the solid electrolyte. Most argyrodites exhibit wide electrochemical stabilities (~5 V vs Li/Li$^+$) [30]. The general formula for Li-argyrodite is Li$_6$PS$_5$X (X=Cl, Br, I) [37]. The fast Li-ion mobilities in Li$_6$PS$_5$Cl (1.9 × 10$^{-3}$ S/cm) and Li$_6$PS$_5$Br (6.8 × 10$^{-3}$ S/cm) are ascribed to disorder in S$^{2-}$/Cl$^-$ and S$^{2-}$/Br$^-$, respectively. However, I$^-$ does not

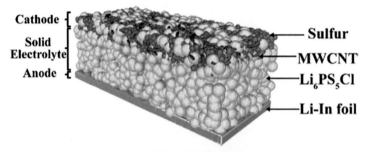

**Fig. 2.12** Schematic of a Li-In/Li$_6$PS$_5$Cl/S-MWCNT cell. (Reproduced with permission from S. Wang, Y. Zhang, X. Zhang, T. Liu, Y.H. Lin, Y. Shen, High-conductivity argyrodite Li$_6$PS$_5$Cl solid electrolytes prepared via optimized sintering processes for all-solid-state lithium-sulfur batteries, ACS Appl. Mater. Interfaces 10 (49) (November 19, 2018) 42279–42285, Copyright (2018) American Chemical Society.)

exchange with $S^{2-}$ due to its larger size, resulting in poor $Li^+$ mobility in Li$_6$PS$_5$I (4.6 × 10$^{-7}$ S/cm) [37]. Chen et al. [38] assembled an ASSLSB using a composite cathode, Li$_6$PS$_5$Br electrolyte, and an In-Li anode with a maximum capacity of 1460 mAh/g and a reversible capacity of 1080 mAh/g after 50 cycles. However, the sulfur content in the composite cathode is low (20–40 wt.%). Additionally, In-Li alloy is not economically practical for large-scale applications.

Wang et al. [39] reported high ionic conductivities for Li$_6$PS$_5$Cl (3.15 × 10$^{-3}$ S/cm) at room temperature, making it a good candidate for assembling ASSLSBs. Li-S cells based on Li$_6$PS$_5$Cl were assembled using nano-sulfur/multiwall carbon nanotube (MWCNT) composites combined with Li$_6$PS$_5$Cl as the cathode and Li-In alloy as the anode (see Fig. 2.12). This cell was reported to deliver a high initial discharge capacity of 1850 mAh/g at 0.1 C and retained a capacity of 1390 mAh/g after 50 cycles. The reversible capacity of the ASSLSB in the first 35 cycles is higher than the theoretical capacity of sulfur (1672 mAh/g). The extra capacity was attributed to the Li$_6$PS$_5$Cl electrolyte in the composite cathode, which may serve as an active cathode material. It has been reported that Li$_6$PS$_5$Cl electrolyte demonstrates reversible electrochemical activity, which makes it a good additive for the sulfur-based electrodes [39].

### 2.4.3.3 Li-S batteries based on glass and glass-ceramics

Glassy electrolytes offer multiple attractive features such as isotropic ion diffusion, negligible grain boundary resistance, ease of processing into films, and a wide compositional range. In general, glassy electrolytes can be divided into oxide- and sulfide-based ISEs. Ionic conductivities of glassy oxides are typically too low ($10^{-6}$–$10^{-8}$ S/cm) for practically high-energy batteries [30]. However, poor conductivity can be compensated by using thin, defect-free, and dense films.

Glassy sulfide electrolytes can be derived from binary $Li_2S$-$P_2S_5$ solid electrolytes doped with LiI. The composition $0.6(0.4SiS_2$-$0.6Li_2S)$-$0.4LiI$ demonstrates the highest ionic conductivity of $1.8 \times 10^{-3}$ S/cm and a low activation energy of $0.28$ eV. Most glass sulfide electrolytes are based on $xLi_2S$-$(1-x)P_2S_5$ and $xLi_2S$-$(1-x)Si_2S_2$ materials [30].

The first $Li_2S$-$SiS_2$ electrolyte was synthesized in 1986 by melt-quenching [40]. This electrolyte exhibited ionic conductivities of $10^{-3}$–$10^{-6}$ S/cm. Since the discovery of glassy sulfides, numerous efforts have been made to improve their ionic conductivity. One way is to optimize the synthesis route. High-energy ball milling of $Li_2S$-$SiS_2$ mixtures results in ionic conductivities of $1.5 \times 10^{-4}$ S/cm [41]. In addition, combining sulfide and oxide glasses causes the precipitation of superionic metastable crystals, improving ionic conductivity of glass-based electrolytes. For example, doping $Li_2S$-$SiS_2$ with $Li_3PO_4$ provides ionic conductivities of $\sim 7 \times 10^{-4}$ S/cm and high resistance to electrochemical reduction resulting from the introduction of an oxide glass component. The $Li^+$ concentration increases as a result of $Li_3PO_4$ doping, which is reported to improve the ionic conductivity. Moreover, $Li_3PO_4$ is stable in contact with lithium metal [41].

The conductivities of $Li_2S$-$SiS_2$ electrolytes are reported to improve through the formation of crystallites. Thus, glass–ceramic electrolytes based on $Li_2S_x(P_2S_5)_{1-x}$ ($x = 75$ mol%) have attracted growing interest for LiSBs [2]. Crystallization of precursor glasses can be expected to decrease ionic conductivities by decreasing $Li^+$ mobilities; however, precipitation can enhance ionic conductivities. The explanation is that grain boundary interfaces in glass–ceramic electrolytes are partially amorphous, lowering grain boundary resistances compared to polycrystalline systems [30]. Heat-treated $Li_2S$-$P_2S_5$ glass–ceramic electrolytes exhibit high conductivities of $\approx 1.7 \times 10^{-2}$ S/cm and low activation energies of $17$ kJ/mol at room temperature [42]. The optimum conditions for heat treating ($280°C$) are reported to reduce grain boundary resistance and improve densification. Glass–ceramic electrolytes offer advantages over crystalline systems in terms of low heat treatment temperatures to process thin films.

Many studies report high initial capacities for LiSBs using ISEs. However, these studies indicate that the sulfur content in the cathode is often insufficient or that the electrode is too thinly layered for practical use [38,43]. As a result, energy densities were relatively low. Due to the low electronic conductivity of sulfur, maintaining high sulfur utilization efficiency becomes more difficult as sulfur contents increase. Most efforts to improve the performance of ASSLSBs focus on improving the electronic conductive paths to sulfur by introducing carbon-based additives and increasing the ionic conductivity.

A few studies focus on improving the reactivity of sulfur by introducing ISEs into the cathode formulation. Nagata et al. [44] demonstrated a correlation between the P/S (0.2–0.8) ratio in a sulfide-based electrolyte and sulfur reactivity based on the obtained sulfur capacity. It is reported that when the P/S ratio is high, the activity in a positive composite electrode increases and the sulfur capacity decreases. Thus, understanding the influence of sulfur reactivity on battery performance is as important as improving the ionic conductivity of the ISEs. Recent advances in the use of glass and glass–ceramic sulfide-based electrolytes to assemble LiSBs with various composite cathode structures are summarized in Table 2.1.

Sulfide-based electrolytes are very promising for assembling ASSLSBs as they demonstrate high ionic conductivity at room temperature. However, sulfide-based electrolytes cannot be exposed to moisture as they form highly toxic $H_2S$, which restricts their potential for large-scale assembly and recycle. Advanced and effective battery construction technologies are required to resolve these stability problems.

**Table 2.1** The electrochemical performance of LiSBs based on glass and glass-ceramic electrolytes at various current densities.

| Electrolyte | $\sigma_t$ (S/cm) | Composition of cathode | Capacity (mAh/g) | Ref. |
|---|---|---|---|---|
| $60Li_2S\text{-}40P_2S_5$ | $2 \times 10^{-5}$ | S/C/SE = 50/10/40 | 1096 (1st) | [44] |
| $80Li_2S\text{-}20P_2S_5$ | $5 \times 10^{-4}$ | S/C/SE = 50/10/40 | 565 (1st) | [44] |
| $80Li_2S\text{-}20P_2S_5$ | $2.2 \times 10^{-4}$ | S/C/SE = 5/15/2 | $\sim$400 (20th) | [45] |
| $80Li_2S\text{-}20P_2S_5$ | $\sim 10^{-3}$ | $Li_2S/C/$ SE = 25/25/50 | $\sim$700 (1st) | [46] |
| $80Li_2S\text{-}20P_2S_5$ | $\sim 10^{-3}$ | S/C/SE = 25/25/50 | 850 (200th) | [47] |
| $80Li_2S\text{-}20P_2S_5$ | $\sim 10^{-3}$ | S/C/SE = 50/21/29 | 1050 (50th) | [48] |

C = additive, S = sulfur, SE = solid electrolyte.

## 2.5 Li-S batteries based on polymer electrolytes

A primary issue of LiSBs centers on the insulating nature of sulfur and the solubility of polysulfides in liquid organic electrolytes. One method of reducing sulfur loss by the dissolution of polysulfides in liquid electrolytes is to modify the electrolyte structure and composition [49]. Replacing traditional organic liquids with polymer electrolytes (PEs) is a promising alternative for an efficient operation of LiSBs [4].

PEs offer several advantages over ISEs, such as enhanced resistance to variations in electrode volumes during cycling, excellent flexibility, and low-cost processability. In addition, Li metal dendrite growth may be suppressed in solvent-free PEs under certain conditions [49]. A large number of PEs have been prepared and characterized since their discovery in 1973 [14]. Substantial efforts have targeted the elucidation of Li-ion transport mechanisms, and the physical and chemical properties of new polymer electrolytes. In this section, PEs are divided into solid polymer (SPEs) and gel polymer electrolytes (GPEs).

### 2.5.1 Solid polymer electrolytes

SPEs comprising polymer matrices and lithium salts are classified as dry solid polymer electrolytes (dry-SPEs). Lithium salts are dissolved in a polymer matrix to provide Li ionic conductivity. Polyethers dissolve Li-salts by complexation of $Li^+$ via binding interactions with ether oxygens. Thus, polyethylene oxide (PEO)-based solid electrolytes have received considerable attention because of their superior ability to solvate $Li^+$, combined with excellent segmental motion, allowing rapid $Li^+$ transport by hopping between segments but typically working best near their glass transition temperature ($T_g$) of 65°C. A further advantage is that PEOs are commercially available in relatively pure states with different molecular weights and at reasonable costs [49].

Dry-SPEs usually exhibit low ionic conductivities at ambient ($10^{-7}$–$10^{-6}$ S/cm), which represents a significant barrier to practical applications [4]. The ionic conduction mechanism is affected by two factors: one is the fraction of amorphous component in the polymer matrix, and the other is the glass transition temperature $T_g$. It is generally accepted that $Li^+$ transport (diffusion) occurs preferentially in amorphous regions of solid PEO, as segmental motion is not inhibited by crystallographic packing. Significant efforts to improve ionic conductivities of dry-SPEs have focused on reducing crystallinity. Such efforts include modifying the polymer matrix by copolymerization, crosslinking, and blending; increasing salt concentrations; adding inorganic fillers; and immobilizing anions as pendant groups [49].

SPEs have better processability and offer possibilities of achieving higher gravimetric energy densities than ISEs due to their low specific densities, e.g., $1.2\,g/cm^3$ for PEO-based dry-SPEs vs $5.0\,g/cm^3$ for garnet electrolytes. Judez et al. [49] developed polymer-rich PEO/lithium bis(fluorosulfonyl) imide composite electrolytes containing inorganic fillers [either $Al_2O_3$ or a lithium-conducting glass ceramic (LICGC)] as hybrid electrolytes for LiSBs. The LICGC is based on a NASICON crystal structure with a composition based on $Li_2O$-$Al_2O_3$-$SiO_2$-$P_2O_5$-$TiO_2$-$GeO_2$. A Li-S cell with an LICGC filler demonstrated a specific capacity of 1111 mAh/g. Although the cell assembled with this electrolyte increased the sulfur utilization and cell areal capacity, its reactivity with Li metal remains an issue. The composite polymer electrolyte with $Al_2O_3$ fillers was reported to improve the stability of the Li/electrolyte interface; however, the cell only delivered a specific capacity of 255 mAh/g. The cell assembled with bilayer electrolytes (a combination of LICGC and $Al_2O_3$), as shown in Fig. 2.13, delivered 993 mAh/g at 70°C [50].

As discussed in Section 2.4, even though ISEs have high conductivities $\sim10^{-4}\,S/cm$ at room temperature and good chemical stability, they are often too rigid and brittle, and show poor contact with electrodes. In contrast, polymer electrolytes are promising candidates for assembling LiSBs due to their low cost, mechanical stability, good compatibility with electrodes, and ease of fabrication using methods already used in processing LIBs. However, as noted just above, ASSLSBs based on PEO electrolytes usually require operating temperatures ranging from 60°C to 90°C owing to the poor ionic conductivity at room temperature. Tao et al. [50] fabricated composite electrolytes made of LLZO nanoparticles (15 wt.%) and PEO-$LiClO_4$ (EO:Li = 8:1) combining the advantages of both polymer and ceramic electrolytes. A cathode composed of sulfur and LLZO nanoparticle (30–200 nm)-decorated carbon was formulated to reduce interfacial resistance and improve conductivities. The composite cathode based on LLZO@C is shown in Fig. 2.14, which delivered $\sim900\,mAh/g$ at 37°C.

An alternate method to enhance LiSB performance is to improve the conductivity of the polymer electrolyte used. Temeche et al. [51] reported a maximum ionic conductivity of $2.8 \times 10^{-3}\,S/cm$ for 60PEO:$Li_3$SiPON films at ambient temperature. The increase in conductivity of this electrolyte is attributed to the suppression of PEO crystallinity and an increase in the N/P ratio. In addition to enhanced ionic conductivities in comparison with

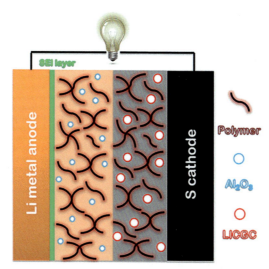

**Fig. 2.13** Schematic of the Li-S cell with bilayer electrolyte configuration. *(Modified from X. Judez, H. Zhang, C. Li, G.G. Eshetu, Y. Zhang, J.A. González-Marcos, Polymer-rich composite electrolytes for all-solid-state Li-S, Cells 8 (15) (July 11, 2017) 3473–3477, Copyright (2017) American Chemical Society.)*

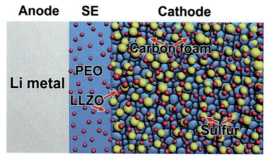

**Fig. 2.14** Schematic of the Li-S battery based on LLZO/PEO electrolytes. *(Modified from X. Tao, Y. Liu, W. Liu, G. Zhou, J. Zhao, D. Lin, Solid-state lithium-sulfur batteries operated at 37°C with composites of nanostructured Li$_7$La$_3$Zr$_2$O$_{12}$/carbon foam and polymer, Nano Lett. 17(5) (April 7, 2017) 2967–2972, Copyright (2017) American Chemical Society.)*

traditional PEO electrolytes, these active polymer precursor electrolytes (Li$_x$SiPON) offer an improved stability against lithium metal at higher current densities (3.75 mA/cm$^2$). Galvanostatic charge/discharge cycling of SPAN/60PEO:Li$_6$SiPON/Li cell offered discharge capacities of 1000 mAh/g$_{sulfur}$ at 0.25 C and 800 mAh/g$_{sulfur}$ at 1 C, as shown in Fig. 2.15.

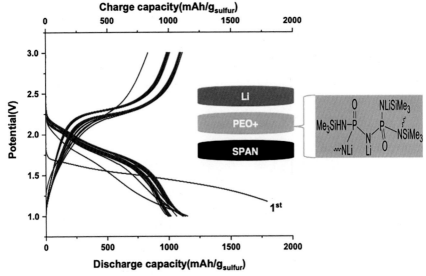

**Fig. 2.15** Schematic of the Li-SPAN battery based on PEO/Li$_x$SiPON electrolytes. (Reproduced with permission from E. Temeche, X. Zhang, R.M. Laine., Solid solutions of poly(ethylene oxide) with Li$_x$PON and Li$_x$SiPON based polymers, ACS Appl. Mater. Interfaces 12 (2020) 30353–30364, Copyright (2020) American Chemical Society.)

## 2.5.2 Gel polymer electrolytes

A main limitation of SPEs is poor ionic conductivity at room temperature (generally $<10^{-6}$ S/cm), which limits their use in practical applications. As a result, GPEs were developed and can be considered as electrolyte intermediates between liquid electrolytes and dry-SPEs [52]. GPEs are obtained by introducing liquid plasticizer and/or solvent into a polymer–salt system. The transport of Li$^+$ ions in GPEs is not dominated by polymer chain segmental motion as transport occurs in the gel-swollen liquid phase. The pore structure of the polymer membrane and the properties of the trapped liquid electrolyte are the key components for determining ionic conductivities of GPEs [52]. One can suggest target properties that would enable the use of GPEs for practical applications, including good mechanical strength, ability to isolate the liquid electrolyte minimizing leakage, high ionic conductivity, and wide electrochemical stability [4].

Solvent-free, solid electrolytes have been recognized as the ultimate approach to suppress polysulfide shuttle effects. However, it is critical and still a challenge to maintain good interfacial contact between the ISE or dry-SPE and the solid electrode in LiSBs. A feasible approach is to use GPEs

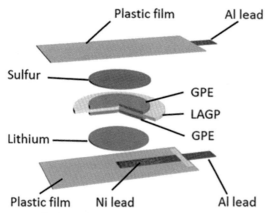

**Fig. 2.16** Schematic of the Li-S battery based on GPE and LAGP electrolytes. (Reproduced with permission from Q. Wang, Z. Wen, J. Jin, J. Guo, X. Huang, J. Yang, C. Chen, A gel-ceramic multi-layer electrolyte for long-life lithium sulfur batteries, Chem. Commun. 52 (8) (January 28, 2016) 1637–1640, Copyright (2016) Royal Society of Chemistry.)

as buffer layers between solid electrodes to reduce interfacial resistance and protect the Li metal. GPEs can stick to solid electrolytes and trap liquid electrolyte between the electrode and the solid electrolyte. Wang et al. [53] fabricated gel-ceramic multilayer electrolytes composed of LAGP and PEO-1 M LiTFSI TEGDME, as shown in Fig. 2.16. The cell delivered an initial capacity of 725 mAh/g and remained at 700 mAh/g after 300 cycles. This cell configuration ultimately paves the way for the construction of ASSLSBs with long cycle stability.

## 2.6 Summary

Some of the most important concepts outlined in the present chapter can be briefly described as follows:

(1) The implementation of sulfide-impermeable membranes such as polymer and ISE electrolytes can eliminate or greatly reduce concerns with the negative effects of the polysulfide shuttle and protect Li electrode surfaces. In addition, the use of solid electrolytes allows the development of new battery architectures for LiSBs.

(2) The main drawback to LATP and LAGP materials is irreversible reduction of $Ti^{4+}$ and $Ge^{4+}$ in contact with metallic Li at low potentials (2.5 V vs Li/Li$^+$), restricting their use with Li metal and low potential anodes. Hence, the development of Ti- and Ge-free

NASICON electrolytes might overcome these challenges. To date, multiple NASICON electrolytes, which are polysulfide-impermeable membranes, are used in hybrid-electrolyte systems to assemble LiSBs.

(3) Efforts to inhibit the $Ti^{4+}$ and $Ge^{4+}$ irreversible reduction resulted in the development of LYZP ISE. This electrolyte is reported to exhibit both chemical compatibility with the battery components and electrochemical stability under extensive cycling. The decrease in the thickness of ISEs has to be examined to be considered for practical applications.

(4) The thermal and mechanical stabilities of LLZO electrolytes against Li metal make them well suited for use in Li metal-based batteries. The main drawback to garnet-type materials is that they absorb $H_2O$ and $CO_2$, forming surface $Li_2CO_3$ that degrades ionic conductivity. Thus, efforts to reduce this susceptibility to reaction in ambient atmosphere are needed.

(5) A major advantage of sulfide solid electrolytes is that they are relatively soft and deformable. They pack densely on cold pressing; thus, sintering is not as energy intensive as most oxide-based systems.

(6) Sulfide-based electrolytes have drawbacks such as poor thermal stabilities, brittleness, and costly raw materials like $Li_2S$, side reaction with electrodes (Li metal), and moisture sensitivity.

(7) SPEs have better processability and offer possibilities of achieving higher gravimetric energy density of LiSBs due to their low specific densities. The main limitation of dry-SPE is the low conductivity at room temperature, which limits their operating temperature.

(8) One feasible approach is to use GPEs as buffer layers between solid electrodes to reduce the interfacial resistance and protect the Li metal.

## Acknowledgments

This work was supported by DOE through Batt500 Seedling Project DE-EE0008235. We are grateful for the support and the gift from Mercedes Benz Research & Development North America (MBRDNA). A portion of this work was also supported by a DMR NSF Grant No. DMR 099217. We also like to thank the University of Michigan Rackham Merit Fellowship (RMF) program. In memory of Prof. Dr. Andreas Hintennach, who passed away May 10, 2020; he was always the nicest and smartest person in any room that he was in.

# References

[1] H.C. Wang, X. Cao, W. Liu, X. Sun, Research progress of the solid state lithium-sulfur batteries, Front. Energy Res. 7 (2019) 112.

[2] X. Yu, A. Manthiram, Electrode-electrolyte interfaces in lithium-sulfur batteries with liquid or inorganic solid electrolytes, Acc. Chem. Res. 50 (11) (2017) 2653–2660.

[3] Y.Z. Sun, J.Q. Huang, C.Z. Zhao, Q. Zhang, A review of solid electrolytes for safe lithium-sulfur batteries, Sci. China Chem. 60 (12) (2017) 1508–1526.

[4] Y. Zhao, Y. Zhang, D. Gosselink, T.N.L. Doan, M. Sadhu, H.J. Cheang, et al., Polymer electrolytes for lithium/sulfur batteries, Membranes 2 (3) (2012) 553–564.

[5] S. Warneke, R.K. Zenn, T. Lebherz, K. Müller, A. Hintennach, U. Starke, et al., Hybrid Li/S battery based on dimethyl trisulfide and sulfurized poly(acrylonitrile), Adv. Sustain. Syst. 2 (2) (2018) 1700144.

[6] S. Wei, L. Ma, K.E. Hendrickson, Z. Tu, L.A. Archer, Metal-sulfur battery cathodes based on PAN-sulfur composites, J. Am. Chem. Soc. 137 (37) (2015) 12143–12152.

[7] L. Wang, Y. Wang, Y. Xia, A high performance lithium-ion sulfur battery based on a $Li_2S$ cathode using a dual-phase electrolyte, Energy Environ. Sci. 8 (5) (2015) 1551–1558.

[8] J.M. Tarascon, M. Armand, Issues and challenges facing rechargeable lithium batteries, in: Materials for Sustainable Energy: A Collection of Peer-Reviewed Research and Review Articles from Nature Publishing Group, 2010, pp. 171–179.

[9] A. Manthiram, X. Yu, S. Wang, Lithium battery chemistries enabled by solid-state electrolytes, Nat. Rev. Mater. 2 (4) (2017) 1–16.

[10] J.A. Kilner, Feel the strain, Nat. Mater. 7 (11) (2008) 838–839.

[11] J.N. Bradley, P.D. Greene, Solids with high ionic conductivity in group 1 halide systems, Trans. Faraday Soc. 63 (1967) 424–430.

[12] J.B. Goodenough, H.Y.P. Hong, J.A. Kafalas, Fast Na +-ion transport in skeleton structures, Mater. Res. Bull. 11 (2) (1976) 203–220.

[13] J. PWM, Ionic conductivity, in: NATO ASI Series, Series B: Physics, 1983.

[14] D.E. Fenton, J.M. Parker, P.V. Wright, Complexes of alkali metal ions with poly(ethylene oxide), Polymer 14 (11) (1973) 589.

[15] J.C. Bachman, S. Muy, A. Grimaud, H.H. Chang, N. Pour, S.F. Lux, et al., Inorganic solid-state electrolytes for lithium batteries: mechanisms and properties governing ion conduction, Chem. Rev. 116 (1) (2016) 140–162.

[16] P.H.L. Notten, F. Roozeboom, R.A.H. Niessen, L. Baggetto, 3-D integrated all-solid-state rechargeable batteries, Adv. Mater. 19 (24) (2007) 4564–4567.

[17] Y. Liu, P. He, H. Zhou, Rechargeable solid-state Li–air and Li–S batteries: materials, construction, and challenges, Adv. Energy Mater. 8 (4) (2018) 1701602.

[18] E. Zhao, F. Ma, Y. Guo, Y. Jin, Stable LATP/LAGP double-layer solid electrolyte prepared: via a simple dry-pressing method for solid state lithium ion batteries, RSC Adv. 6 (95) (2016) 92579–92585.

[19] E. Yi, W. Wang, S. Mohanty, J. Kieffer, R. Tamaki, R.M. Laine, Materials that can replace liquid electrolytes in Li batteries: superionic conductivities in $Li_{1.7}Al_{0.3}Ti_{1.7-}Si_{0.4}P_{2.6}O_{12}$. Processing combustion synthesized nanopowders to free standing thin films, J. Power Sources 269 (2014) 577–588.

[20] L. Wang, Y. Zhao, M.L. Thomas, A. Dutta, H.R. Byon, Sulfur-based catholyte solution with a glass-ceramic membrane for Li-S batteries, ChemElectroChem 3 (1) (2016) 152–157.

[21] S. Wang, Y. Ding, G. Zhou, G. Yu, A. Manthiram, Durability of the $Li_{1+x}Ti_{2-x}Al_x(PO_4)_3$ solid electrolyte in lithium-sulfur batteries, ACS Energy Lett. 1 (6) (2016) 1080–1085.

[22] X. Yu, Z. Bi, F. Zhao, A. Manthiram, Polysulfide-shuttle control in lithium-sulfur batteries with a chemically/electrochemically compatible NaSICON-type solid electrolyte, Adv. Energy Mater. 6 (24) (2016) 1601392.

[23] Q. Wang, J. Jin, X. Wu, G. Ma, J. Yang, Z. Wen, A shuttle effect free lithium sulfur battery based on a hybrid electrolyte, Phys. Chem. Chem. Phys. 16 (39) (2014) 21225–21229.

[24] Q. Wang, J. Guo, T. Wu, J. Jin, J. Yang, Z. Wen, Improved performance of Li-S battery with hybrid electrolyte by interface modification, Solid State Ionics 300 (2017) 67–72.

[25] P.G. Bruce, S.A. Freunberger, L.J. Hardwick, J.M. Tarascon, Li-$O_2$ and Li-S batteries with high energy storage, Nat. Mater. 11 (1) (2012) 19–29.

[26] R. Chen, W. Qu, X. Guo, L. Li, F. Wu, The pursuit of solid-state electrolytes for lithium batteries: from comprehensive insight to emerging horizons, Mater. Horiz. 3 (6) (2016) 487–516.

[27] K. Fu, Y. Gong, S. Xu, Y. Zhu, Y. Li, J. Dai, et al., Stabilizing the garnet solid-electrolyte/polysulfide interface in Li-S batteries, Chem. Mater. 29 (19) (2017) 8037–8041.

[28] M. Naguib, A. Sharafi, E.C. Self, H.M. Meyer, J. Sakamoto, J. Nanda, Interfacial reactions and performance of $Li_7La_3Zr_2O_{12}$-stabilized Li-sulfur hybrid cell, ACS Appl. Mater. Interfaces 11 (45) (2019) 42042–42048.

[29] F. Zheng, M. Kotobuki, S. Song, M.O. Lai, L. Lu, Review on solid electrolytes for all-solid-state lithium-ion batteries, J. Power Sources 389 (2018) 198–213.

[30] Z. Ma, H.G. Xue, S.P. Guo, Recent achievements on sulfide-type solid electrolytes: crystal structures and electrochemical performance, J. Mater. Sci. 53 (6) (2018) 3927–3938.

[31] M. Tachez, J.P. Malugani, R. Mercier, G. Robert, Ionic conductivity of and phase transition in lithium thiophosphate $Li_3PS_4$, Solid State Ionics 14 (3) (1984) 181–185.

[32] T. Yamada, S. Ito, R. Omoda, T. Watanabe, Y. Aihara, M. Agostini, et al., All solid-state lithium–sulfur battery using a glass-type $P_2S_5$ –$Li_2S$ electrolyte: benefits on anode kinetics, J. Electrochem. Soc. 162 (4) (2015) A646.

[33] M. Nagao, Y. Imade, H. Narisawa, T. Kobayashi, R. Watanabe, T. Yokoi, et al., All-solid-state Li-sulfur batteries with mesoporous electrode and thio-LISICON solid electrolyte, J. Power Sources 222 (2013) 237–242.

[34] T. Kobayashi, Y. Imade, D. Shishihara, K. Homma, M. Nagao, R. Watanabe, et al., All solid-state battery with sulfur electrode and thio-LISICON electrolyte, J. Power Sources 182 (2) (2008) 621–625.

[35] M. Nagao, K. Suzuki, Y. Imade, M. Tateishi, R. Watanabe, T. Yokoi, et al., All-solid-state lithium–sulfur batteries with three-dimensional mesoporous electrode structures, J. Power Sources 330 (2016) 120–126.

[36] X. Yao, N. Huang, F. Han, Q. Zhang, H. Wan, J.P. Mwizerwa, et al., High-performance all-solid-state lithium–sulfur batteries enabled by amorphous sulfur-coated reduced graphene oxide cathodes, Adv. Energy Mater. 7 (17) (2017) 1602923.

[37] P.R. Rayavarapu, N. Sharma, V.K. Peterson, S. Adams, Variation in structure and $Li^+$-ion migration in argyrodite-type $Li_6PS_5X$ (X = Cl, Br, I) solid electrolytes, J. Solid State Electrochem. 16 (5) (2012) 1807–1813.

[38] M. Chen, S. Adams, High performance all-solid-state lithium/sulfur batteries using lithium argyrodite electrolyte, J. Solid State Electrochem. 19 (3) (2015) 697–702.

[39] S. Wang, Y. Zhang, X. Zhang, T. Liu, Y.H. Lin, Y. Shen, et al., High-conductivity argyrodite $Li_6PS_5Cl$ solid electrolytes prepared via optimized sintering processes for all-solid-state lithium-sulfur batteries, ACS Appl. Mater. Interfaces 10 (49) (2018) 42279–42285.

[40] J. Kim, Y. Yoon, M. Eom, D. Shin, Characterization of amorphous and crystalline $Li_2S-P_2S_5-P_2Se_5$ solid electrolytes for all-solid-state lithium ion batteries, Solid State Ionics 225 (2012) 626–630.

[41] H. Morimoto, H. Yamashita, M. Tatsumisago, T. Minami, Mechanochemical synthesis of new amorphous materials of $60Li_2S \cdot 40SiS_2$ with high lithium ion conductivity, J. Am. Ceram. Soc. 82 (5) (1999) 1352–1354.

[42] Y. Seino, T. Ota, K. Takada, A. Hayashi, M. Tatsumisago, A sulphide lithium super ion conductor is superior to liquid ion conductors for use in rechargeable batteries, Energy Environ. Sci. 7 (2) (2014) 627–631.

[43] S. Boulineau, M. Courty, J.M. Tarascon, V. Viallet, Mechanochemical synthesis of Li-argyrodite $Li_6PS_5X$ (X = Cl, Br, I) as sulfur-based solid electrolytes for all solid state batteries application, Solid State Ionics 221 (2012) 1–5.

[44] H. Nagata, Y. Chikusa, Activation of sulfur active material in an all-solid-state lithium-sulfur battery, J. Power Sources 263 (2014) 141–144.

[45] M. Agostini, Y. Aihara, T. Yamada, B. Scrosati, J. Hassoun, A lithium-sulfur battery using a solid, glass-type $P_2S_5-Li_2S$ electrolyte, Solid State Ionics 244 (2013) 48–51.

[46] M. Nagao, A. Hayashi, M. Tatsumisago, High-capacity $Li_2S$-nanocarbon composite electrode for all-solid-state rechargeable lithium batteries, J. Mater. Chem. 22 (19) (2012) 10015–10020.

[47] M. Nagao, A. Hayashi, M. Tatsumisago, Sulfur-carbon composite electrode for all-solid-state Li/S battery with $Li_2S-P_2S_5$ solid electrolyte, Electrochim. Acta 56 (17) (2011) 6055–6059.

[48] M. Nagao, A. Hayashi, M. Tatsumisago, Electrochemical performance of all-solid-state Li/S batteries with sulfur-based composite electrodes prepared by mechanical milling at high temperature, Energy Technol. 1 (2-3) (2013) 186–192.

[49] X. Judez, H. Zhang, C. Li, G.G. Eshetu, Y. Zhang, J.A. González-Marcos, et al., Polymer-rich composite electrolytes for all-solid-state Li-S cells, J. Phys. Chem. Lett. 8 (15) (2017) 3473–3477.

[50] X. Tao, Y. Liu, W. Liu, G. Zhou, J. Zhao, D. Lin, et al., Solid-state lithium-sulfur batteries operated at 37°C with composites of nanostructured $Li_7La_3Zr_2O_{12}$/carbon foam and polymer, Nano Lett. 17 (5) (2017) 2967–2972.

[51] E. Temeche, X. Zhang, M.L. Richard, Solid electrolytes for Li-S batteries. Solid solutions of polyethylene oxide with $Li_xPON$ and $Li_xSiPON$ based polymers, ACS Appl. Mater. Interfaces 12 (2020) 30353–30364.

[52] L. Long, S. Wang, M. Xiao, Y. Meng, Polymer electrolytes for lithium polymer batteries, J. Mater. Chem. A 4 (26) (2016) 10038–10069.

[53] Q. Wang, Z. Wen, J. Jin, J. Guo, X. Huang, J. Yang, et al., A gel-ceramic multi-layer electrolyte for long-life lithium sulfur batteries, Chem. Commun. 52 (8) (2016) 1637–1640.

# CHAPTER 3

# Applications of metal-organic frameworks for lithium-sulfur batteries

**Fu-Sheng Ke and Si-Cheng Wan**

Sauvage Center for Molecular Sciences, College of Chemistry and Molecular Sciences, Wuhan University, Wuhan, China

## Contents

| | |
|---|---|
| 3.1 Introduction | 49 |
| 3.2 MOFs for lithium-sulfur batteries | 54 |
|    3.2.1 MOFs as sulfur hosts for lithium-sulfur batteries | 54 |
|    3.2.2 MOF-based separator/interlayer for Li-S batteries | 82 |
|    3.2.3 MOF-based electrolytes for Li-S batteries | 85 |
|    3.2.4 MOF-based anode for Li-S batteries | 88 |
| 3.3 Characterization techniques | 92 |
|    3.3.1 In situ X-ray techniques | 93 |
|    3.3.2 In situ optical spectroscopic techniques | 100 |
| 3.4 Summary and outlook | 104 |
|    3.4.1 Cathode | 105 |
|    3.4.2 Interlayers/separators | 106 |
|    3.4.3 Electrolyte | 106 |
|    3.4.4 Anode | 107 |
|    3.4.5 Characterization | 107 |
| Acknowledgments | 107 |
| References | 108 |

## 3.1 Introduction

To satisfy the explosively growing demands of portable electronic devices and pure/hybrid electric vehicles, exploration of next-generation lithium-ion batteries (LIBs) with a high energy density and safety is urgent [1]. However, electrode materials of present commercialized LIBs hardly meet the requirements of the next-generation LIBs due to their limited energy density, which relies on lithium-ion intercalation into cathode and anode

---

*Lithium-Sulfur Batteries*
https://doi.org/10.1016/B978-0-12-819676-2.00007-4

Copyright © 2022 Elsevier Inc.
All rights reserved.

**49**

materials [2]. Compared to conventional petroleum-based fuels, the commercial LIBs have been hindered by low energy density, resulting in a limited mileage range, as shown in Fig. 3.1A [3]. It is normally named "range anxiety," which corresponds to the fundamental parameters of LIBs, i.e., energy density and power density. Therefore, it is critical to develop electrode materials and new energy storage systems with high energy and power densities. According to Fig. 3.1A, lithium-sulfur (Li-S) batteries, as an alternative choice, have extremely high theoretical capacity of $1675\,mAh\,g^{-1}$ and energy density of $2600\,Wh\,kg^{-1}$ based on the sulfur. Meanwhile, large natural abundance and low cost of sulfur also benefit the development of Li-S batteries. The energy density of Li-S batteries is larger than $500\,Wh\,kg^{-1}$ (Fig. 3.1B) [4], using sulfur/C and Li metal as cathode and anode, respectively. Corresponding mileage can be achieved to 400 km, which is close to that of conventional petroleum-based fuels. Therefore, Li-S batteries have been considered as one of the most promising next-generation lithium batteries.

In comparison with $Li^+$ insertion-desertion mechanism of the commercial LIBs, a series of complicated redox reactions between lithium and sulfur takes place in Li-S batteries, as shown in Fig. 3.2 [5]. Typically, during the discharge process in liquid-state electrolyte-based Li-S batteries, the crown sulfur species ($S_8$) firstly open the loop and form lithium polysulfides (LiPSs), i.e., $Li_nS_8^{2-}$ and $Li_2S_6^{2-}$. With dissolution process of the LiPSs, the potential generally decreases and generates $Li_2S_4^{2-}$ during 2.3–2.1 V. Then, the long-chain LiPSs are transformed to $Li_2S_2$ at 2.1 V, which are insoluble in the

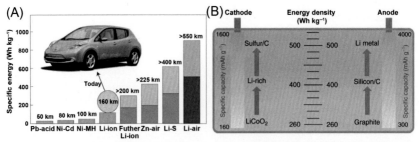

**Fig. 3.1** (A) Practical specific energies for some rechargeable batteries. (B) Specific energy densities of lithium batteries based on different cathode and anode materials. *(Reproduced with permission from: panel (A) P.G. Bruce, S.A. Freunberger, L.J. Hardwick, J.-M. Tarascon, Li–O₂ and Li–S batteries with high energy storage, Nat. Mater. 11 (2012) 19–29, Copyright (2012) Nature Publishing Group, and (B) Z. Lin, T. Liu, X. Ai, C. Liang, Aligning academia and industry for unified battery performance metrics, Nat. Commun. 9 (2018) 5262, Copyright (2018) Nature Publishing Group.)*

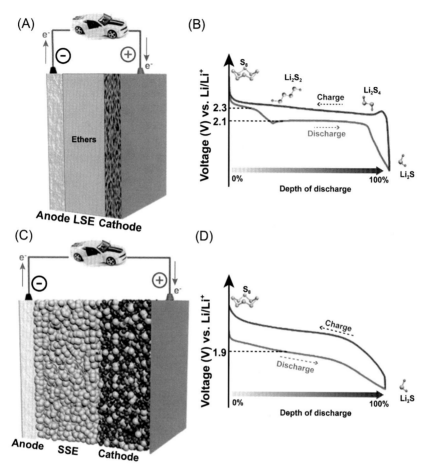

**Fig. 3.2** Schematic configurations of Li-S batteries using (A) liquid-state electrolyte and (C) solid-state electrolyte. Typical charge-discharge voltage profiles of (B) solid-liquid dual-phase Li-S reactions and (D) solid-phase Li-S reactions. *(Reproduced with permission from X. Yang, J. Luo, X. Sun, Towards high-performance solid-state Li-S batteries: from fundamental understanding to engineering design, Chem. Soc. Rev. 49 (2020) 2140–2195, Copyright (2020) Royal Society of Chemistry.)*

ether electrolyte, and finally form the insulated $Li_2S$ at 1.6–2.1 V. During the charge process, a reversible reaction occurs, in which $Li_2S$ is transformed to $S_8$ via insoluble short-chain LiPSs and soluble long-chain LiPSs [6]. There are two typical potential plateaus during the discharge process in the ether(s) electrolyte, i.e., solid-liquid dual-phase Li-S reactions (Fig. 3.2B). Recently, a solid electrolyte was used to replace the liquid electrolyte in Li-S batteries [7,8], to overcome the electrochemical and thermal instabilities of the liquid electrolyte system. Therefore, the reaction mechanism is changed to solid–solid-phase Li–S reactions (Fig. 3.2D), which exhibits a single-discharge plateau. It means that direct conversion of sulfur to $Li_2S$ occurs without the formation of intermediate LiPSs [5,9,10]. Therefore, reaction mechanism is much more complicated in the Li–S batteries compared to the commercial LIBs.

Although Li–S batteries have some unique properties, such as high energy density and low cost, some critical challenges still impede their practical applications. These challenges include LiPSs' shuttling effects, poor electric conductivity of sulfur and its discharge products, large volumetric variation, and Li dendrite growth. These challenges are associated with the physical properties and the reaction mechanisms of sulfur. The specific discussion of those challenges is as follows:

(1) Shuttling effect. Soluble long-chain LiPSs produced during the discharge-charge process can be dissolved into ether electrolyte, penetrated through the separator to the anode, and then reacted with lithium anode. It resulted in causing the loss of sulfur species, passivation of metallic Li surface, and self-discharging. This phenomenon is known as the "shuttling effect," which is one of the major challenges affecting the electrochemical performance of Li–S batteries.

(2) Sluggish reaction kinetics. The poor electric conductivity of both reactant (sulfur) and product ($Li_2S$) as well as the complex chemical reactions during reduction of $S_8$ and formation of $Li_2S$ results in slow reaction kinetics and high electrochemical polarization. Therefore, conductive additives or hosts are also helpful to overcome this challenge.

(3) Large volume variation. Volume expansion ($\sim$80%) accompanies the conversion from $S_8$ to $Li_2S$, leading to instability and irreversibility of the cathode. Therefore, porous hosts are beneficial to relieve the stress of volume change and prevent the pulverization of the cathode.

(4) Dendrite of metallic Li anode. Li–S batteries belong to lithium metal battery system. Metal Li as anode suffered from Li dendrite issues during the repeat plating/stripping. The Li dendrite induced a short circuit very easily, which led to safety problems, especially in organic liquid electrolyte systems. The nonuniform deposition of metallic Li was caused by the nonuniform exchange current density, resulted in metallic Li dendrites, which could pierce through the separator and cause a short circuit. Thus, it is in need to develop strategies to control the exchange current density and avoid Li dendrites during cycling.

Various strategies have been explored to address the aforementioned challenges. For example, there are three main strategies that have been developed to control the shuttling effect of LiPSs. The first strategy is to develop hosts or separator interlayers with strong physical and chemical adsorption to suppress the diffusion of long-chain LiPSs to anode, including a variety of porous carbon materials [11–16], metal-organic frameworks (MOFs) [17–20], covalent organic frameworks (COFs) [21–25], polymers [26–29], metal oxides [30–32], metal sulfides [33,34], metal nitrides [35,36], and metallic compounds [37–39]. Some of them also can utilize heteroatom and metal ions to accelerate reaction kinetics. The second strategy is to block the dissolution of long-chain LiPSs into electrolyte [40]. For example, the use of highly concentrated electrolytes can significantly reduce the number of free-state solvent molecules and suppress the dissolution of long-chain LiPSs into electrolyte. When the concentration is larger than a threshold (usually $> \sim 3$–$5\,M$ depending on the salt-solvent combinations), free solvent molecules will disappear and a particular three-dimensional (3D) solution structure will form [41]. Thus, long-chain LiPSs cannot dissolve into this electrolyte. However, there are still some disadvantages of the concentrated electrolyte, including precipitation of the Li salts at low temperature, difficulty of wetting cell separators and thick electrodes, and higher cost [42,43]. Another alternative strategy is exploring solid electrolyte interface (SEI) [44] or using small sulfur molecules [45–48] to suppress the production of long-chain LiPSs. It means that no long-chain LiPSs would appear in these systems. For example, using small sulfur molecules of $S_{2-4}$ can totally prevent the formation of long-chain LiPSs to enhance the electrochemical performance of Li-S batteries [49].

Different types of materials, especially porous structured materials, have been utilized as sulfur hosts, separators, and electrolytes. Among them, MOFs with large surface area, ultrahigh porosity, tunable pore size, and chemical environment have received huge attention in the past decades. A wide range of applications including gas separation/storage, catalysis, drug delivery, and electrochemical energy storage have been explored. Until now, MOFs and MOF-based materials have been used as electrodes, electrolytes, and separators for batteries [50,51], fuel cells [52,53], supercapacitors [54,55], and other electrochemical storage systems. Herein, the current chapter appraises recent and significant development of MOFs and MOF-derived materials for Li-S batteries, including sulfur hosts, lithium interface, electrolyte, separators, or interlayers. Furthermore, the benefits, challenges, and future prospects of the applications of MOF-based materials in Li-S batteries are assessed and discussed. We hope that this review could provide some beneficial opinions for the direction of MOFs in high-performance Li-S batteries.

## 3.2 MOFs for lithium-sulfur batteries

As a new type of organic–inorganic hybrid porous crystalline materials, MOFs have experienced explosive development during the past two decades. These materials are fabricated by linking metal-containing units secondary building units (SBUs), which include metal components, with organic linkers, using strong bonds to create porous crystalline materials. SBUs and organic linkers are organized with various patterns (such as post-synthetic modification and multivariate MOFs), which have rendered thousands of MOFs being prepared and studied each year [56]. MOFs possess unique properties, i.e., large surface area ($>7000\,\mathrm{m^2\,g^{-1}}$), ultrahigh porosity, tunable pore size, and geometry, controllable chemical environment. All of those unique properties are suitable for applications in Li-S batteries. Up to now, different MOFs have been employed as four primary components of Li-S batteries, including cathode, anode, electrolyte, separator/interlayer (Fig. 3.3). For example, MOFs with (1) large surface area and ultrahigh porosity and (2) strong adsorption of LiPSs by SBUs and organic linkers are suitable as sulfur hosts to increase sulfur loading and alleviate the "shuttling effect." In addition, small pore size MOFs suit to separate and block the LiPS diffusion to the anode side. The specific review and discussion of MOFs for sulfur host, lithium metal, separator/interlayer, and electrolyte of Li-S batteries are in the following.

### 3.2.1 MOFs as sulfur hosts for lithium-sulfur batteries

Due to the nonconductivity of sulfur, the hosts are anticipated to load sulfur with molecular or nanosized level, which can accelerate the reaction kinetics. As far as we know, MOFs with high surface area and large pore volume are capable of loading sulfur as high as 80 wt%. Meanwhile, the organic linker and SBU with polarity strongly adsorb long-chain LiPSs and can thus alleviate the "shuttling effect." Furthermore, the conductivity can be highly enhanced by calcination or combination with conductive composites. We will review and discuss the MOF/MOF-based composites as sulfur hosts here (Fig. 3.4).

Applications of metal-organic frameworks for lithium-sulfur batteries 55

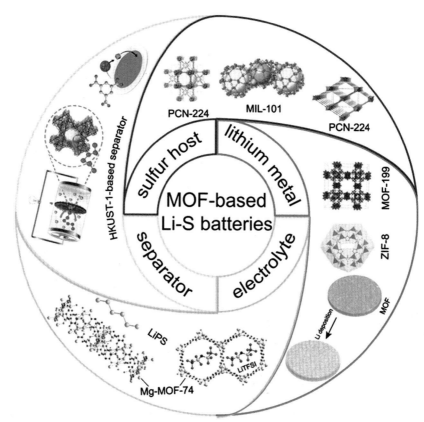

**Fig. 3.3** Typical examples of MOFs applied as sulfur hosts, separators, solid electrolytes, and lithium metal. *(Reproduced with permission from: H. Jiang, X.C. Liu, Y. Wu, Y. Shu, X. Gong, F.S. Ke, H. Deng, Metal-organic frameworks for high charge-discharge rates in lithium-sulfur batteries, Angew. Chem. Int. Ed. 57 (2018) 3916–3921, Copyright (2018) Wiley-VCH; S. Bai, X. Liu, K. Zhu, S. Wu, H. Zhou, Metal-organic framework-based separator for lithium-sulfur batteries, Nat. Energy 1 (2016) 16094, Copyright (2016) Nature Publishing Group; D.D. Han, Z.Y. Wang, G.L. Pan, X.P. Gao, Metal-organic-framework-based gel polymer electrolyte with immobilized anions to stabilize a lithium anode for a quasi-solid-state lithium-sulfur battery, ACS Appl. Mater. Interfaces 11 (2019) 18427–18435; Copyright (2019) American Chemical Society; J. Qian, Y. Li, M. Zhang, R. Luo, F. Wang, Y. Ye, Y. Xing, W. Li, W. Qu, L. Wang, L. Li, Y. Li, F. Wu, R. Chen, Protecting lithium/sodium metal anode with metal-organic framework based compact and robust shield, Nano Energy 60 (2019) 866–874, Copyright (2019) Elsevier; M. Zhu, B. Li, S. Li, Z. Du, Y. Gong, S. Yang, Dendrite-free metallic lithium in lithiophilic carbonized metal-organic frameworks, Adv. Energy Mater. 8 (2018) 1703505, Copyright (2018) Wiley-VCH.)*

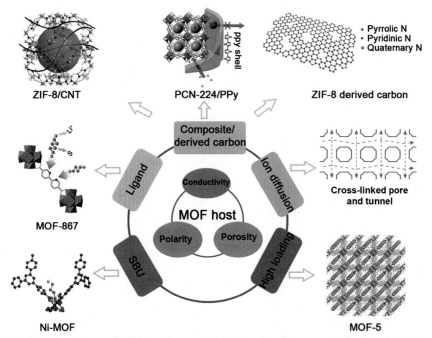

Fig. 3.4 Advantages of MOF structure and the strategies of the nonconductivity in MOF-based sulfur hosts in Li-S batteries. *(Reproduced with permission from: H. Jiang, X.C. Liu, Y. Wu, Y. Shu, X. Gong, F.S. Ke, H. Deng, Metal-organic frameworks for high charge-discharge rates in lithium-sulfur batteries, Angew. Chem. Int. Ed. 57 (2018) 3916–3921, Copyright (2018) Wiley-VCH; J. Zheng, J. Tian, D. Wu, M. Gu, W. Xu, C. Wang, F. Gao, M.H. Engelhard, J.-G. Zhang, J. Liu, J. Xiao, Lewis acid-base interactions between polysulfides and metal organic framework in lithium sulfur batteries, Nano Lett. 14 (2014) 2345–2352, Copyright (2014) American Chemical Society; Y. Mao, G. Li, Y. Guo, Z. Li, C. Liang, X. Peng, Z. Lin, Foldable interpenetrated metal-organic frameworks/carbon nanotubes thin film for lithium-sulfur batteries, Nat. Commun. 8 (2017) 14628, Copyright (2017) Nature Publishing Group; F. Zheng, Y. Yang, Q. Chen, High lithium anodic performance of highly nitrogen-doped porous carbon prepared from a metal-organic framework, Nat. Commun. 5 (2014) 5261, Copyright (2014) Nature Publishing Group; P.M. Shanthi, P.J. Hanumantha, B. Gattu, M. Sweeney, M.K. Datta, P.N. Kumta, Understanding the origin of irreversible capacity loss in non-carbonized carbonate-based metal organic framework (MOF) sulfur hosts for lithium-sulfur battery, Electrochim. Acta 229 (2017) 208–218, Copyright (2017) Elsevier; J. Park, K. Choi, D. Lee, B. Moon, S. Shin, M. Song, J. Kang, Encapsulation of redox polysulphides via chemical interaction with nitrogen atoms in the organic linkers of metal-organic framework nanocrystals, Sci. Rep. 6 (2016) 25555, Creative Commons Attribution 4.0) (http://creativecommons.org/licenses/by/4.0/.)*

### 3.2.1.1 Pristine MOFs as sulfur hosts
Pore structure

The pore structure, including pore size, pore aperture, and pore geometry, is an important factor to affect the ion diffusion in the MOFs. It is worthwhile to note that pore size is not the same as pore aperture. For example, the pore geometry of MIL-101 is a three-dimensional hierarchical nanocages, including large and medium pore sizes of 34.0 and 29.0 Å. In addition, there are pore apertures (7 Å) in between the pores, as shown in Fig. 3.5A [57]. The pore apertures will affect the ion transfer process, especially at a high charge-discharge rate. Meanwhile, they also can block the LiPSs from leaking out of MOFs to some extent. All MOFs can be divided into two types according to the pore size, i.e., microporous ($\sim$2 nm) and mesoporous (2–50 nm) MOFs. Generally, the microporous MOFs have smaller pore size and aperture, which can restrict the migration of LiPSs when the size is lower than that of LiPSs. However, the microporous MOFs have less sulfur loading due to their smaller pore size [58]. On the contrary, mesoporous MOFs can load more sulfur than microporous MOFs, but LiPSs can leak out of the MOFs easily [59]. Therefore, cage-typed MOFs with relatively large inner space but small apertures possess high loading capacity and high LiPS confinement efficiency, which exhibit excellent cycling performance at low current density. MOFs with large pore apertures possess high sulfur loading and facilitate fast ion diffusion, exhibiting excellent electrochemical performance at high current density.

To study the electrochemical performance of MOFs with different pore apertures, four different MOFs, namely ZIF-8, HKUST-1, $NH_2$-MIL-53 (Al), and MIL-53(Al), corresponding to pore apertures of 3.4, 6.9, 7.5, and 8.5 Å, respectively, were used as sulfur hosts [58]. The electrochemical results indicated that the average decay rates of ZIF-8, HKUST-1, $NH_2$-MIL-53(Al), and MIL-53(Al) were 0.08%, 0.11%, 0.14%, and 0.19% per cycle at 0.5 C (1 C = 1675 mA g$^{-1}$), respectively (Fig. 3.5C and D). This result shows that the capacity fading is associated with the pore aperture, i.e., with the increase of pore aperture of MOF, the capacity decay rate becomes severer. It is illustrated that small pore aperture can alleviate the LiPS leaking out of the sulfur host. However, this result is limited to the relatively low current density measurements. For high rate performance, MOFs can also serve as hosts to construct three dimensionally ordered macro- and microporous metal-organic frameworks (3DOM ZIF-8) via a self-templated coordination-replication method [60], as shown in Fig. 3.5E. The 3D hierarchical architecture can facilitate the electrolyte

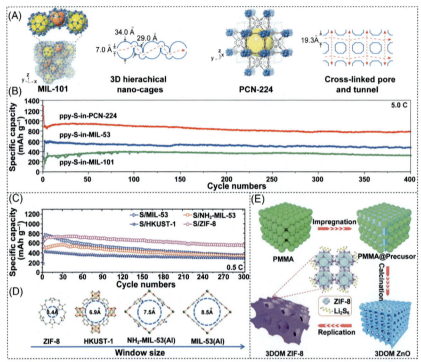

Fig. 3.5 (A) Crystal structures and illustrations of the corresponding ion diffusion pathway in pores of PCN-224 and MIL-101, (B) Cycling performance of different ppy-S-in-MOFs electrodes at 5.0 C, (C) Discharge capacities of four MOF/S electrodes at 0.5 C. (D) Schematic of the largest apertures of the four MOFs, (E) Schematic illustration of 3DOM ZIF-8 synthesis process. (*Reproduced with permission from: panel (B) H. Jiang, X.C. Liu, Y. Wu, Y. Shu, X. Gong, F.S. Ke, H. Deng, Metal-organic frameworks for high charge-discharge rates in lithium-sulfur batteries, Angew. Chem. Int. Ed. 57 (2018) 3916–3921, Copyright (2018) Wiley-VCH; panel (D) J.W. Zhou, R. Li, X.X. Fan, Y.F. Chen, R.D. Han, W. Li, J. Zheng, B. Wang, X.G. Li, Rational design of a metal-organic framework host for sulfur storage in fast, long-cycle Li-S batteries, Energ. Environ. Sci. 7 (2014) 2715–2724, Copyright (2014) Royal Society of Chemistry; panel (E) G. Cui, G. Li, D. Luo, Y. Zhang, Y. Zhao, D. Wang, J. Wang, Z. Zhang, X. Wang, Z. Chen, Three-dimensionally ordered macro-microporous metal organic frameworks with strong sulfur immobilization and catalyzation for high-performance lithium-sulfur batteries, Nano Energy 72 (2020) 104685, Copyright (2020) Elsevier.*)

infiltration and accelerate ion/mass diffusion in electrodes to improve the rate performance. Meanwhile, this structure can increase the specific area and expose more active interfaces. The Li-S batteries based on the 3DOM ZIF-8 exhibited a low–capacity decay of 0.028% per cycle over 500 cycles under a high rate of 2.0 C. These results illustrate that the pore aperture affects the electrochemical performance.

Next, we discuss why pore geometry affects the electrochemical properties. Deng and coauthors employed three MOFs with different pore geometries, i.e., one-dimensional channels of MIL-53, three-dimensional hierarchical nanocages of MIL-101, and cross-linked pores and tunnels of PCN-224 (Fig. 3.5A) [57]. To minimize the impact of other factors, such as conductivity and surface area, the three MOFs were fully filled by sulfur into their all pores and coated with polypyrrole (ppy). The cycling performance of the PCN-224, MIL-53, and MIL-101-based electrodes are 780, 480, and 330 $mAh g^{-1}$ after 400 cycles under the high rate of 5.0 C, as shown in Fig. 3.5B. The clear difference in the electrochemical performance of the three MOFs should be mainly attributed to their distinct difference in pore geometry. Based on the earlier discussion of pore aperture and pore geometry, it revealed that pore structure can influence the electrochemical performance of MOF-based Li-S batteries. Therefore, rational design of pore structure of MOFs is one of the key factors to obtain high-performance Li-S batteries.

### Metal-containing units

According to previous research, the nonpolar nature of carbon, which was used as sulfur host, is unfavorable for trapping the polar LiPSs. Polar inorganics, such as metal oxide and metal sulfide, were chosen as hosts to trap LiPSs during the charge-discharge process. Metal-containing units, i.e., SBUs, are a key constituent of MOFs. Although those SBUs were linked to organic ligands, MOFs always contain open metal sites or coordinatively unsaturated sites, which are one of the main parts for the confinement of polysulfides as Lewis acid centers in MOFs. The metal centers with free d-orbitals are able to coordinate to nucleophilic $S_x^{2-}$ anion clusters. Theoretical calculation indicates a higher binding energy between LiPSs and open metal sites in MOFs than other hosts like nonpolar carbon (0.1–0.7 eV) [61].

To clearly observe the attraction between SBUs and sulfur species, Avery et al. employed the X-ray absorption spectroscopy (XAS) to check the local Cu structure changes with different sulfur loadings [62]. The results indicated a notable decrease in the white line intensity with increasing the S: MOF ratios. It illustrated that a comparative lack of electron occupancy in the 3d states and increased S-Cu interactions. They also proved a high density of Cu-rich surface defects drastically improved polysulfide retention by adjusting MOF particle size, resulted in significantly enhancing the cycling performance. Recently, Zheng and coauthors used Ni-MOF ($Ni_6(BTB)_4(BP)_3$, BTB = benzene-1,3,5-tribenzoate and BP = 4,4'-bipyridyl) as a sulfur host and found the strong interactions between Lewis acidic

Ni(II) center and the LiPSs [63], which was confirmed by the results of first-principles calculations (Fig. 3.6A). The electrochemical test results exhibited high-capacity retention of 89% at 0.1 C after 100 cycles, as demonstrated in Fig. 3.6 Bespecially at high charge. It illustrated that the SBUs of MOFs can trap the soluble LiPS species within the MOF scaffold, as reflected by the minimum changes of the first-discharge potential plateau, i.e., $Li_2S_8$/$Li_2S_6$ at ca. 2.3 V, after 200 cycles. Similarly, Liu. et al. obtained a remaining capacity of about 990 mAh g$^{-1}$ after 200 cycles at 0.2 C by a novel manganese cluster-based MOF [64].

The interaction between sulfur species and metal sites is often beneficial to anchor LiPSs. However, too strong interaction may cause adverse effects. When the metal sites react with sulfur strongly and form sulfides, the structure of MOFs will be destroyed, and then, the MOFs cannot work as host materials to prevent the shuttling effect. For instance, Cu-based HKUST-1 MOFs often show the worst performance in the previous reports [58,65]. The sulfur cathode reacted with Cu ions from the HKUST-1 cluster to form CuS, which can further react with Li$^+$ and transform into $Cu_2S$ and $Li_2S$ at 2.1 V. To solve this problem, one strategy is the synthesis of metal Lewis acid sites in the ligands of MOFs to enhance adsorption effect. Meanwhile, inert metal-containing units, such as Zr-based oxides, were selected as SBUs to prevent the reaction between clusters and sulfur, to maintain the stability of MOFs during the charge-discharge process [66].

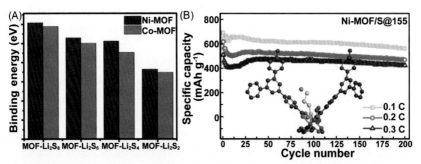

**Fig. 3.6** (A) Binding energies of LiPSs to Ni/Co-MOFs and (B) cycling performance of Ni-MOF/S@155 electrodes. (Permission from J. Zheng, J. Tian, D. Wu, M. Gu, W. Xu, C. Wang, F. Gao, M.H. Engelhard, J.-G. Zhang, J. Liu, J. Xiao, Lewis acid-base interactions between polysulfides and metal organic framework in lithium sulfur batteries, Nano Lett. 14 (2014) 2345–2352, Copyright (2014) American Chemical Society.)

## Organic ligands

Besides the metal-containing units, organic ligands also can provide strong adsorption with LiPSs to enhance the electrochemical performance of MOF-based Li-S batteries. These organic ligands normally include diversified functional Lewis basic groups, such as N, S, and P [67,68]. For example, ligands with N groups, such as pyridinic and pyrrolic, can strongly interact with the $Li^+$ of soluble polysulfides to form N-Li bond. Recently, Yin et al. investigated the interaction of LiPSs with different N configurations by density functional theory (DFT) calculations [69]. The results indicated that pyridinic-N was the best anchoring group for LiPSs among the amino-, pyridinic-, graphitic-, and pyrrolic-N species. The interaction between nitrogen and $Li^+$ of LiPSs has been also utilized in carbon [70,71], Mxene [72], and polymer [73] host materials extensively. As a result, these organic ligands with N configurations can enhance absorption and slow down the migration of soluble polysulfides.

To further block the diffusion of LiPSs, the combination of organic ligands and SBUs might be a much better choice. Recently, Hong et al. synthesized a bifunctional MOF Cu-TDPAT, which has Lewis basic sites from the nitrogen atoms of the ligand $H_6TDPAT$ and Lewis acid sites from Cu(II) open metal sites simultaneously (Fig. 3.7A) [74]. The multiple synergistic effects from this bi-functional MOF contributed to a high reversible capacity of $745\,mAh\,g^{-1}$ at 1.0 C after 500 cycles (Fig. 3.7B). Furthermore, metal ions also can be coordinated with some heterocycles, such as porphyrin, as Lewis acid centers to tune the interaction of LiPSs. The artificial Lewis acid centers increase active reaction sites with sulfur species. More recently, Wang and coauthors investigated the impacts of different local environments at the centers of the porphyrin moieties via comparing a series of MMOF-MOF-525 (2H), MOF-525 (FeCl), and MOF-525 (Cu) [75]. Among them, MOF-525 (Cu) was considered as a very powerful MOF host owing to that it can offer two Lewis acidic sites per Cu sites. The MOF-525 (Cu)-based sulfur cathode showed a reversible capacity of $704\,mAh\,g^{-1}$ after 200 cycles along with trivial fade rate of 0.07% per cycle after the 10th cycle. However, this strategy has not caught much attention. The metal atoms doped in carbon materials have been proved the property of accelerating redox reaction of LiPSs, and the metal sites in ligands can act as a catalyst to decompose sulfur species [76]. Consequently, it was postulated that artificial metal Lewis acid centers can improve reaction kinetics even in the nonconductivity hosts, especially at low charge-discharge rates ($\leq 1.0\,C$).

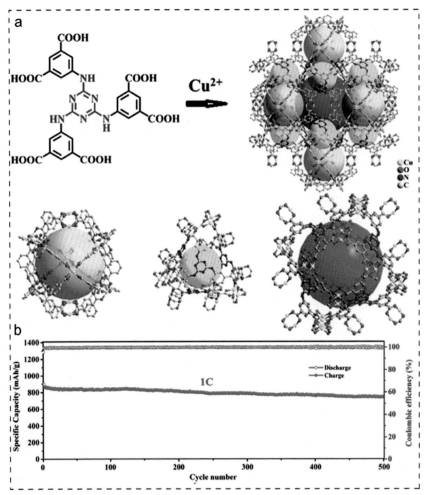

**Fig. 3.7** (A) Crystal structure of Cu-TDPAT, (B) cycling stability of a S@Cu-TDPAT. *(Reproduced with permission from X.J. Hong, T.X. Tan, Y.K. Guo, X.Y. Tang, J.Y. Wang, W. Qin, Y.P. Cai, Confinement of polysulfides within bi-functional metal-organic frameworks for high performance lithium-sulfur batteries, Nanoscale 10 (2018) 2774–2780, Copyright (2018) Royal Society of Chemistry.)*

### 3.2.1.2 MOF composites as sulfur hosts

According to the previous discussions, pristine MOFs with ideally sufficient active sites can function as chemical anchors of LiPSs and alleviate the shuttling effect. However, the innate nonconductivity of MOFs will aggravate the low electronic conductivity of sulfur species and cause severe polarization, which even leads to block electrode reaction, especially at high charge–discharge rates (>1.0 C). Therefore, it is crucial to enhance the conductivity of MOF materials. At present, two main strategies have been developed to improve the conductivity of MOFs for sulfur hosts. The first strategy is to combine MOF with other conductive materials, such as graphenes, carbon nanotubes (CNTs), and conductive polymers. The other strategy is to calcinate MOFs at high temperature to produce the conductive MOF-derived materials. We will review and discuss both of them in the following.

### MOF with carbon-based composites

**MOF/graphene composites** Graphene, a single layer of carbon atoms in $sp^2$ bonding configuration, has unique properties, i.e., excellent conductivity, high carrier mobility, high chemical stability, high mechanical strength, and flexibility [77,78]. These unique properties endow graphene and reduced graphene oxide (rGO) as one of the most commonly used coating materials to form core-shell or 3D hierarchical structured composites [79], aiming to enhance the conductivity and stability of those composites. This strategy is particularly suitable for MOFs to get rid of their disadvantages, i.e., low conductivity, low chemical and thermal stability, in electrochemical energy storage systems. Graphene can accelerate electron transit from current collector to the MOF materials. Meanwhile, it can also accelerate the heat diffusion during the charge-discharge process in battery systems. We will take several examples here to illustrate the function of graphene in MOF/graphene composites for Li-S batteries.

For the core-shell structure, Bao et al. synthesized the MIL-101(Cr)@rGO/S composite successfully by a facile and flexible two-step liquid-phase method [80]. The electrochemical test results showed the discharge capacity and capacity retention rate of MIL-101(Cr)@rGO/S composite cathode are, respectively, $650\,mAh\,g^{-1}$ and 66.6% at the 50th cycle at 0.2 C. Correspondingly, the MIL-101(Cr)/S mixed sulfur cathode only delivered the capacity of $458\,mAh\,g^{-1}$ and the capacity retention rate of 37.3%, which is much lower than MIL-101(Cr)@rGO/S composite cathode obviously. Similarly, Zhao and coauthors also combined rGO and MIL-101(Cr) to form a graphene/chromium-MOF (MIL-101) composite, named GNS-MIL-101(Cr)/S (Fig. 3.8A) [81]. The capacity of this composite cathode decayed slowly from the reversible $1190\,mAh\,g^{-1}$ at 0.1 C to 1045, 870, 650, and $500\,mAh\,g^{-1}$ at 0.2, 0.4, 0.8, 2.4, and 3.0 C, respectively. When the rate decreased from 3.0 C to 0.1 C, the capacity still retained $1123\,mAh\,g^{-1}$, which showed great rate capability. For the cycling stability test, the GNS-MIL-101(Cr)/S cathode exhibited the capacity of

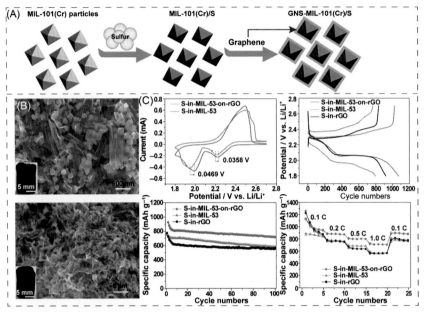

**Fig. 3.8** (A) Schematic illustration of the graphene-MIL-101(Cr)/S composite preparation, (B) Optical images and (C) electrochemical characterization of MOF-on-rGO composites. *(Reproduced with permission from: panel (A) Z. Zhao, S. Wang, R. Liang, Z. Li, Z. Shi, G. Chen, Graphene-wrapped chromium-MOF(MIL-101)/sulfur composite for performance improvement of high-rate rechargeable Li-S batteries, J. Mater. Chem. A 2 (2014) 13509–13512, Copyright (2014) Royal Society of Chemistry; panel (C) Y. S. Wu, H.Q. Jiang, F.S. Ke, H.X. Deng, Three-dimensional hierarchical constructs of MOF-on-reduced graphene oxide for lithium-sulfur batteries, Chem. Asian J. 14 (2019) 3577–3582, Copyright (2019) Wiley-VCH.)*

847 mAh g$^{-1}$ at 0.8 C after 100 cycles and 395.9 mAh g$^{-1}$ at a high rate of 2.4 C over 300 cycles. As for the control experiment, the capacity of MIL-101(Cr)/S cathode is only 512 mAh g$^{-1}$ at 0.8 C. It illustrated that the graphene-wrapped MIL-101(Cr)/S composite showed a much faster Li$^+$ diffusion and lower charge transfer resistance compared to the MIL-101(Cr)/S composite, confirmed by the electrochemical impedance spectroscopy (EIS). However, it is worth noting that the wrapped 2D layer graphene and rGO will affect the Li-ion diffusion to some extent, especially at the high current densities.

As for the 3D hierarchical structure, Deng et al. developed an in situ reduced method to combine graphene oxide and MOFs [82], to form a 3D hierarchical MOF-on-rGO compartment (Fig. 3.8B), which combines the polarity and porous features of MOFs and the high conductivity of rGO. This method is suitable for different hydrostable MOFs, such as MIL-101 and MIL-53. The electrochemical performance test delivered a high specific capacity of 601 mAh g$^{-1}$ at 1.0 C after 400 cycles, which could be attributed to the synergistic effect between MOF and rGO. It illustrated that both the hierarchical structures of rGO and the polar pore environment of MOF restrict the leak and migration of LiPSs. In addition, the spongy-layered rGO can buffer the volume expansion and contraction, effectively suppressing the destruction of the structure. Nevertheless, graphene and rGO are 2D layered and rigid materials, which cannot contact with MOFs compactly; thus, the interfacial resistance between MOFs and graphene needs to be taken into consideration.

**MOF/carbon nanotubes composites** Similar to graphene, CNTs also display significant conductivity and carrier mobility [83]. More importantly, the one-dimensional (1D) CNTs can contact with MOFs more flexibly and compactly compared to 2D layered graphene. Meanwhile, CNTs can easily form segregated networks with the Li storage materials, as both binder and conductive additive, to boost the electrical and mechanical properties. These networks allow the fabrication of thick electrodes and maintain high electronic conductivity [84]. Therefore, the CNTs are another choice to enhance the electronic conductivity and mechanism of MOFs compared to graphene.

In recent decades, there has been an increasing research effort geared at tackling the nonconductivity of MOFs by using CNTs [85,86]. Compared to physical mixture of CNTs and MOFs as sulfur hosts [87], there are two impressive examples, which interpenetrated CNTs in MOFs for Li-S batteries. For the first one, Mao et al. reported a strategy of using foldable interpenetrated MOFs/CNTs thin films as sulfur hosts (Fig. 3.9A) [88]. The CNTs interpenetrate through the MOFs crystal and interweave into a 3D hierarchical structure, providing both conductivity and structural integrity, as shown in SEM images (Fig. 3.9B). It is noted that the electrode consisted of 40 wt% sulfur loading, 36 wt% HKUST-1, and 24 wt% CNTs, without any binders and other conductive agents. The resultant Li-S batteries demonstrated a high initial capacity of 1263 mAhg$^{-1}$ and excellent cyclability over 500 cycles with a fading rate of 0.08% per cycle at 0.2 C. It also exhibited excellent rate capability, i.e., a highly reversible capacity of 880 mAh g$^{-1}$ at 2.0 C and 449 mAh g$^{-1}$ at a rate of 10.0 C. More interestingly, the fabricated Li-S batteries achieved a capacity of $\sim$7.45 mAh cm$^{-2}$ and favorably worked over 50 cycles, with sulfur loading reached up to 11.33 mg cm$^{-2}$. These superior performances are hard to achieve completely for pristine MOFs.

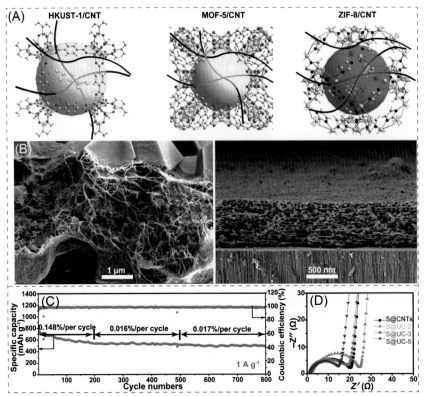

**Fig. 3.9** (A) Schematic illustration of MOFs/CNT composite, (B) SEM images before sulfur loading, (C) Long cycling performance and (D) Nyquist plots of the UiO-66 with CNT. *(Reproduced with permission from: panel (B) Y. Mao, G. Li, Y. Guo, Z. Li, C. Liang, X. Peng, Z. Lin, Foldable interpenetrated metal-organic frameworks/carbon nanotubes thin film for lithium-sulfur batteries, Nat. Commun. 8 (2017) 14628, Copyright (2017) Nature Publishing Group, and panel (D) Y. Pu, W. Wu, J. Liu, T. Liu, F. Ding, J. Zhang, Z. Tang, A defective MOF architecture threaded by interlaced carbon nanotubes for high-cycling lithium-sulfur batteries, RSC Adv. 8 (2018) 18604–18612, Copyright (2018) Royal Society of Chemistry.)*

The other example, a cross-link conductive network by UiO-66 and CNTs, was constructed by using a competitive coordination principle [89]. In this method, CNTs threaded through the pores of UiO-66 and constructed a reliable 3D-conductive network. Meanwhile, abundant liker-missing defects were introduced to increase the number of active sites in the MOF. The conductivities of the UiO-66/CNTs composite (13.68 S cm$^{-1}$) increased by 11 order of magnitudes compared to pristine UiO-66 ($< 10^{-10}$ S cm$^{-1}$). The UiO-66/CNTs composite electrodes demonstrated a capacity retention of $765\,\mathrm{mAh\,g^{-1}}$ after 300 cycles at a current density of $0.5\,\mathrm{A\,g^{-1}}$. When the current density was high up to $1\,\mathrm{A\,g^{-1}}$, the battery demonstrated a very low fading rate of 0.07% per cycle after 800 cycles (Fig. 3.9C). It was worth noting that the CNTs entered into the pores of MOFs and contacted with sulfur species directly, which was confirmed by the EIS results (Fig. 3.9D). Compared to HKUST-1 and UiO-66, other MOFs with high specific surface area and high pore volume can probably further enhance the MOF-based Li-S batteries in the future.

CNTs with 1D features easily thread through the pores of MOFs and construct uniform conductive networks, which decrease the length of transit path, resulting in fast transit of both electrons and $Li^+$ according to the relationship $\tau_d \sim L_2/D_{Li}$. Compared to MOFs/graphene composites, the 3D MOF/CNTs compartments displayed much better electrochemical performance in Li-S batteries, especially at high current densities and with high area capacity. Although graphenes and CNTs can improve the conductivity of MOF-based composites, they occupy part of total mass as inert matter leading to decrease in the energy densities of Li-S batteries. According to most of the reports [86,88], the ratio of conductive graphene or CNTs is larger than 20 wt%, but the calculation of specific capacity often did not consider their weight to evaluate the performance of electrodes. Therefore, the evaluation of electrochemical performance of electrodes needs to consider the total mass of them.

### MOF/conductive polymer composites

Except for the aforementioned graphenes and CNTs, some organic polymers with high conductivity $(10^2–10^3\,\mathrm{S\,cm^{-1}})$ are another choice to improve the conductivity of MOFs. Conductive polymers are known for their variable electrical conductivity, high electrochemical stability, and good flexibility when compositing with other materials and show optimistic application potential in electrochemistry energy storage devices. For example, conductive polymer wrapped hollow sulfur nanospheres were synthesized and demonstrated excellent electrochemical performance [90,91]. Compared to the nonpolar graphene and CNTs, the polar conductive polymers had a strong interaction with the polar LiPSs. Meanwhile, these conductive polymers, including polypyrrole (ppy), polyaniline (PAN), poly(3,4-ethylenedioxythiophene) (PEDOT), and polythiophene (PTh), have a large number of positively charged groups, which can adsorb negatively charged polysulfide anions by electrostatic attraction [92]. Therefore, conductive polymers can be even directly used as sulfur hosts [93–95]. As the coating materials, polymers have several advantages, i.e., flexibility, controllable thickness, easy coating on MOFs, and better compactness compared to graphene [96,97]. More importantly, the polymer layer can serve as an interlayer to block the sulfur species from leaking out of pores (Fig. 3.10A).

Conductive polymers have been explored to coat S-in-MOFs (sulfur loaded in MOFs) for sulfur species entrapment and conductivity enhancement. Notable examples include the work by Deng et al., who reported the use of ppy layer for coating different S-in-MOFs to prepare ppy-S-in-MOFs composites [57]. These composites are the fulfillment of three criteria, i.e., polarity, porosity, and conductivity, in Li-S batteries. Among those ppy-S-in-MOFs composites, the ppy-S-in-PCN-224 electrode with cross-linked pores and tunnels stood out, with a high capacity of 640 and 440 mAh g$^{-1}$ at 10.0 C after 400 and 1000 cycles (Fig. 3.10B), respectively. This excellent result was attributed to the union of the polarity and porosity advantages of MOFs with the conductive feature of conductive polymers. A similar structure was fabricated by using PPy and ZIF-67, forming ZIF-67-S-PPy composite [98], which also demonstrated excellent performance in Li-S batteries. Another example, Jin et al. synthesized doped-PEDOT:PSS-coated MIL-101/S multicore-shell-structured composite [99]. EIS measurement revealed the improvement of conductivity of this composite, but its performance in capacity and stability was not very good. A possible reason is the 3D hierarchical structure of MIL-101, which is composed of a small pore aperture of 7.0 Å in between the alternating large and small pores (Fig. 3.5A) restricted the ions transfer process. On the contrary, the PCN-224 with large pores and cross-linked tunnels (Fig. 3.5A), exhibited the fast ion transfer capability. The results mentioned earlier demonstrated that the electrochemical performance of MOFs/conductive polymer composite electrodes was obviously improved by designing MOFs and conductive polymers.

### 3.2.1.3 MOF-derived materials

Porous carbon is one of the important materials for electrochemical energy storage systems, such as LIBs [100,101], supercapacitors [102], and fuel cells [103], due to its porosity and high conductivity. Various materials have been selected as precursors for the synthesis of porous carbon. Among them, MOFs originally own ordered pores and highly consistent array. Therefore, MOFs as precursors to prepare porous carbon, metal compounds, and their composites are another strategy to use them in electrochemical energy storage systems (Fig. 3.11). The MOF-derived materials possess high surface area, high pore volume, and various metal compounds, because of the enriched organic ligands and metal-containing units in MOFs structure.

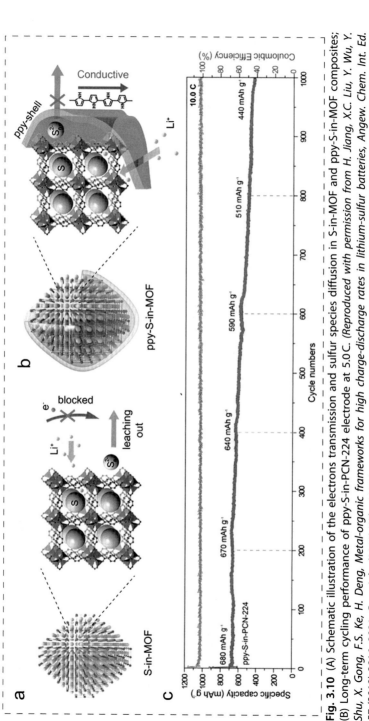

**Fig. 3.10** (A) Schematic illustration of the electrons transmission and sulfur species diffusion in S-in-MOF and ppy-S-in-MOF composites; (B) Long-term cycling performance of ppy-S-in-PCN-224 electrode at 5.0C. (*Reproduced with permission from H. Jiang, X.C. Liu, Y. Wu, Y. Shu, X. Gong, F.S. Ke, H. Deng, Metal-organic frameworks for high charge-discharge rates in lithium-sulfur batteries, Angew. Chem. Int. Ed. 57 (2018) 3916–3921, Copyright (2018) Wiley-VCH.*)

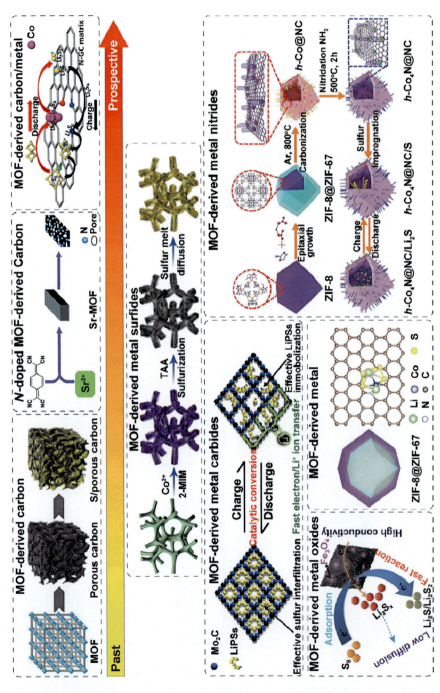

Fig. 3.11 See figure legend on opposite page

(Continued)

The organic ligands and metal–containing units can be converted into carbon and metal compounds, respectively, by thermochemical methods. Due to the vast choice of ligands, metal-containing units, and morphologies in MOFs, there have been a larger number of works focusing on using MOFs as precursors to prepare materials for electrochemical energy storage systems [104–109]. Specifically, these include (1) ligands with different functional groups and multiple organic linkers in a single framework to obtain MOF–derived carbon or heteroatom doped carbon; (2) various SBUs and multiple metal-containing units to get MOF-derived metal compounds; (3) loading MOFs to acquire more abundant MOF–derived materials. In addition, the thermochemical methods also include calcination and laser. Compared to pristine MOFs, those MOF–derived materials have higher conductivity and more enriched metal compounds, which are suitable for energy storage systems, especially Li–S batteries. Herein, we will appraise recent and significant development of MOF–derived materials as sulfur hosts at this part.

---

**Fig. 3.11, cont'd** Schematic illustration of the development of MOF-derived carbon and the composite with different metal compounds. *(Reproduced with permission from: K. Xi, S. Cao, X. Peng, C. Ducati, R.V. Kumar, A.K. Cheetham, Carbon with hierarchical pores from carbonized metal-organic frameworks for lithium Sulphur batteries, Chem. Commun. 49 (2013) 2192–2194, Copyright (2013) Royal Society of Chemistry; G.L. Chen, Y.J. Li, W.T. Zhong, F.H. Zheng, J. Hu, X.H. Ji, W.Z. Liu, C.H. Yang, Z. Lin, M.L. Liu, MOFs-derived porous Mo C-C nano-octahedrons enable high-performance lithium-sulfur batteries, Energy Storage Mater. 25 (2020) 547–554, Copyright (2020) Elsevier; F. Zhou, Z. Qiao, Y. Zhang, W. Xu, H. Zheng, Q. Xie, Q. Luo, L. Wang, B. Qu, D.L. Peng, Bimetallic MOF-derived CNTs-grafted carbon nanocages as sulfur host for high-performance lithium-sulfur batteries, Electrochim. Acta 349 (2020) 136378, Copyright (2020) Elsevier; G. Liu, K. Feng, H. Cui, J. Li, Y. Liu, M. Wang, MOF derived in-situ carbon-encapsulated Fe O @C to mediate polysulfides redox for ultrastable lithium-sulfur batteries, Chem. Eng. J. 381 (2020) 122652, Copyright (2020) Elsevier; J. He, Y. Chen, A. Manthiram, MOF-derived cobalt sulfide grown on 3D graphene foam as an efficient sulfur host for long-life lithium-sulfur batteries, iScience 4 (2018) 36–43, Copyright (2018) Elsevier; Z.X. Sun, S. Vijay, H.H. Heenen, A.Y.S. Eng, W.G. Tu, Y. X. Zhao, S.W. Koh, P.Q. Gao, Z.W. Seh, K.R. Chan, H. Li, Catalytic polysulfide conversion and physiochemical confinement for lithium-sulfur batteries, Adv. Energy Mater. 10 (2020) 1904010, Copyright (2020) Wiley-VCH; Y. Tong, D. Ji, P. Wang, H. Zhou, K. Akhtar, X. Shen, J. Zhang, A. Yuan, Nitrogen-doped carbon composites derived from 7,7,8,8-tetracyanoquinodimethane-based metal-organic frameworks for supercapacitors and lithium-ion batteries, RSC Adv. 7 (2017) 25182, Copyright (2017) Royal Society of Chemistry; Y.J. Li, J.M. Fan, M.S. Zheng, Q.F. Dong, A novel synergistic composite with multi-functional effects for high-performance Li-S batteries, Energ. Environ. Sci. 9 (2016) 1998–2004, Copyright (2016) Royal Society of Chemistry.)*

## MOF-derived carbon

At the initial stage, the researchers focused on employing the pore structure of MOFs to synthesize porous carbon materials, which was used as sulfur hosts and compared with traditional porous carbon. As we know, when the MOF is calcined at high temperature, the product is carbon and metal oxide. The metal oxides were removed by acid or high temperature. Among them, the Zn–MOFs were frequently selected as the precursors of derived porous carbon because the Zn atoms can be vaporized over 900°C.

One typical example, hierarchically porous carbon was prepared through a simple method, i.e., direct carbonization of MOF-5. Xi et al. reported tunable hierarchical MOF-derived carbon as sulfur host materials via carbonized Zn–MOFs [110], included room temperature synthesized MOF-5 (RT-MOF-5), solvothermally synthesized MOF-5 (solvo-MOF-5), and $Zn_3(fumarate)_3(dmf)_2$ (ZnFumarate). The BET surface area of MOF-derived carbon by RT-MOF-5, solvo-MOF-5, and ZnFumarate was 1945, 2372, and 4793 $m^2\,g^{-1}$, corresponding to micropore volume of 0.31, 0.78, and 1.27 $cm^3\,g^{-1}$, and mesopore volume of 1.41, 0.30, and 2.24 $cm^3\,g^{-1}$, respectively. The initial capacity of the sulfur electrodes of carbonized RT-MOF-5, solvo-MOF-5, and ZnFumarate was 1294.1, ~980, and 1471.8 $mAh\,g^{-1}$, respectively. It illustrated that mesopores were responsible for delivering high initial capacities. They also found that carbon hosts with higher micropore volumes were suitable to cathode materials with better cycling stability. This is because the micropores restrict electrolyte and ion mobilities, but mesopores go against the confinement of LiPSs. These results illustrated preliminarily the relation between the pore size distribution and volume and the performance of MOF-derived carbon. Similarly, Xu and the coworkers also synthesized the hierarchically porous carbon nanoplates by one-step pyrolysis of MOF-5 [111]. The battery test demonstrated that a more stable cycling performance was observed at the rate of 0.5 C than at 0.1 C. It may be attributed to shorter diffusion time of LiPSs at a high current rate. This sulfur cathode delivered a discharge capacity of 730 $mAh\,g^{-1}$ after 50 cycles at 0.5 C. In fact, the pure MOF-derived carbon hosts suffered the same issue as common porous carbon materials, i.e., nonpolarity with weak interaction with LiPSs, resulted in poor electrochemical performance in Li–S batteries.

To overcome the aforementioned problems, heteroatom-doped carbon materials were developed to improve the interaction between LiPSs and sulfur hosts. These heteroatoms, which include nitrogen, sulfur, boron, and so on, can improve the chemical and physical properties of carbon. Among them, N-doped carbon materials were widely used in electrochemical energy storage and conversion devices, such as fuel cells and LIBs. Doped nitrogen atoms as Lewis bases can interact with the $Li^+$ in LiPSs to form N-Li bond and thus anchor the LiPSs. For example, Song et al. revealed the detailed mechanism of doped N in carbon via X-ray absorption near-edge structure (SANES) spectroscopy and DFT calculation [112]. An obvious change in O-coordination structure could be observed after sulfur loading. In addition, there were more ether-type oxygen functional groups and less carboxyl-type and/or carbonyl-type functional groups in N-doped carbon after loading sulfur via linear combination fitting of XANES spectra. These both results proved oxygen-sulfur bonding between carbonyl and carboxyl groups. Meanwhile, DFT calculations indicated that more electronegative nitrogen could polarize the nearby oxygen-containing groups. As a result, it demonstrated that the electron-withdrawing N atoms cannot adsorb the S atoms but can enhance the O affinity for sulfur. In another example, Wang et al. characterized the interaction between Li and N using XPS [113]. There are two peaks at 54.6 and 55.4 eV in the Li 1 s spectrum after full discharge of cells, which correspond to Li-N and Li-S bonds, respectively. Meanwhile, the binding energy of N 1 s peak changed from 399.7 eV to 399.2 eV, attributed to the interaction with the less electronegative atoms, i.e., Li. The results provided previously fully proved the interaction between LiPSs and N during the cycling.

Various MOFs with N-ligands, such as porphyrin and imidazole, are suitable as a precursor to fabricate N-doped porous carbon to deal with the issue of weak interaction with LiPSs. For example, the ZIF-derived N-doped porous carbons/rGO (NPC/G) hybrids were synthesized by Chen et al. (Fig. 3.12A) A C-N binding energy of 285.8 eV could be observed in XPS [114]. The strong polysulfide adsorption capability of N atoms in carbon can improve the cycling performance, i.e., a specific capacity of $608 \, mAh \, g^{-1}$ after 300 cycles with a capacity decay of 0.043% per cycle. Similarly, Walle and coworkers prepared N-doped cubic carbon (NC) with CNT composite derived from ZIFs [115]. The sulfur loading was as high as 89 wt%. It delivered $674.4 \, mAh \, g^{-1}$ with a capacity retention of 61.4% after 120 cycles at 0.5 C. This illustrated that the high-active sites and strong adsorption sites for the soluble polysulfides from N atoms are in favor of the confinement of LiPSs efficiently.

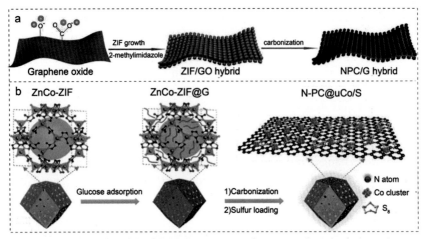

**Fig. 3.12** (A) Schematic of the fabrication process of nitrogen-doped porous carbon on graphene (NPC/G), (B) Schematic diagram of the synthesis route for N-PC@Co/S composite. *(Reproduced with permission from: panel (A) K. Chen, Z. Sun, R. Fang, Y. Shi, H.M. Cheng, F. Li, Metal-organic frameworks (MOFs)-derived nitrogen-doped porous carbon anchored on graphene with multifunctional effects for lithium-sulfur batteries, Adv. Funct. Mater. 28 (2018) 1707592, Copyright (2018) Wiley-VCH; panel (B) R. Wang, J. Yang, X. Chen, Y. Zhao, W. Zhao, G. Qian, S. Li, Y. Xiao, H. Chen, Y. Ye, G. Zhou, F. Pan, Highly dispersed cobalt clusters in nitrogen-doped porous carbon enable multiple effects for high-performance Li-S battery, Adv. Energy Mater. 10 (2020) 1903550, Copyright (2020) Wiley-VCH.)*

## MOF-derived metal compounds/carbon

Recently, metal-based compounds, including metal/C, metal oxide/C, metal sulfide/C, metal carbide/C, and metal nitride/C, have been employed as sulfur hosts [116–120]. Those compounds not only carried strong adsorption sites for long-chain LiPSs, but also are deemed as catalysts to accelerate the redox process of $Li_2S_2/Li_2S$ exhibiting slow reaction dynamics. MOFs, composite of metal and organic ligands, were used as precursors for those metal-based compounds under different calcined conditions. As follows, we will appraise the development of metal-based compounds, which were derived from MOFs, as sulfur hosts at this part.

**MOF-derived metal/C composites** During the calcination process, the product of carbon can continue reducing the SBUs of MOFs to metal atoms or clusters, obtained MOF-derived metal/C composites. Wang et al. used ZnCo-ZIF as precursors and synthesized nitrogen-doped porous carbon (N-PC) with highly dispersed cobalt catalysts (Fig. 3.12B) [121]. The

XPS exhibited two characteristic peaks of metallic cobalt in Co 2p spectrum. When using this compound as a sulfur host, they discovered that highly dispersed cobalt atoms showed better electrochemical performance than aggregated cobalt. The cathode of N-PC-uniform Co exhibited $1150 \, \text{mAh} \, \text{g}^{-1}$, but the N-PC and N-PC-aggregate Co only delivered 832 and $567 \, \text{mAh} \, \text{g}^{-1}$ after 100 cycles, respectively. It was illustrated that highly dispersed Co clusters can accelerate $Li^+$ diffusion dynamics and enhance the conversion capability of LiPSs.

Furthermore, to build conductive network, Zhou et al. obtained CNT-grafted nitrogen-doped carbon@graphitic carbon nanocages (CNT-NC@GC) by using bimetallic core-shell MOFs, i.e., ZIF-8@ZIF-67, as precursor [122]. During the annealing process, the ZIF-67 shells were transformed into CoO/C and ZIF-8 cores contracted outward to form hollow structure. When the temperature increased to 800°C, the CoO was further reduced to metallic Co, which could serve as a catalyst to grow CNTs, similar to Ni-MOF for CNTs [123]. DFT calculation shows higher binding energy between $Li_2S_6$ and N-Co-Carbon ($-3.72 \, \text{eV}$) than between $Li_2S_6$ and carbon ($-1.24 \, \text{eV}$). This result demonstrated that metallic Co enhanced adsorption interaction with LiPSs and weakened the shuttling effect. The CNT-NC@GC/S electrode delivered a reversible capacity of $743 \, \text{mAh} \, \text{g}^{-1}$ after 500 cycles at 1.0 C. According to the cyclic voltammetry (CV) results, the metallic Co in the composite could be played the chemisorption and electrocatalytic sites for sulfur species during cycling.

**MOF-derived metal oxide/C composites** Metal oxides, such as $MnO_2$ and $Fe_3O_4$, were employed as sulfur hosts to immobilize the sulfur and LiPSs [124–128]. MOFs with porous structure and enriched SBUs are suitable as precursors to fabricate metal oxide/C for sulfur hosts. Yang et al. constructed a V-MOF (MIL-47) and pyrolyzed to $V_2O_3$@C hollow microcuboid with a hierarchical lasagna-like structure (Fig. 3.13A) [129], which provided plentiful ion-transport channels. Meanwhile, the lasagna-like structure allows high sulfur loading and alleviates the volume change during cycling. After 1000 cycles, the capacity of $V_2O_3$@C electrode is $598 \, \text{mAh} \, \text{g}^{-1}$ at 1.0 C, which is 62.3% of the initial capacity. According to the CVs in $Li_2S_6$ electrolyte, the $V_2O_3$@C electrode revealed much larger current than that of the amorphous carbon electrode, which proved the $V_2O_3$ catalytic effect to accelerate the conversion of LiPSs during cycling.

In another study, a $Fe_3O_4$@C composite host was fabricated by sintering a Fe-based MIL-53 [130]. The morphology of $Fe_3O_4$@C composite is like

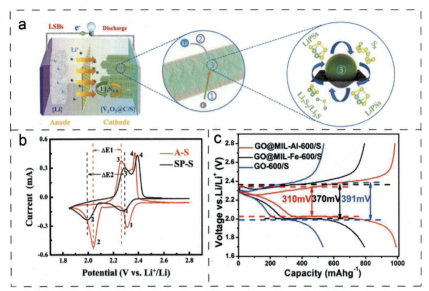

Fig. 3.13 (A) Illustration of the main superior properties of V$_2$O$_3$@C/S cathode, (B) CV curves of Fe$_3$O$_4$@Super C and Super C@S at a scan rate of 0.1 mV s$^{-1}$, (C) Galvanostatic charge-discharge curves of GO@MIL-Al-600/S, GO@MIL-Fe-600/S, and GO-600/S cathodes for the first cycle at 1.0 C. *(Reproduced with permission from: panel (A) J. Yang, B. Wang, F. Jin, Y. Ning, H. Luo, J. Zhang, F. Wang, D. Wang, Y. Zhou, A MIL-47(V) derived hierarchical lasagna-structured V$_2$O$_3$@C hollow microcuboid as an efficient sulfur host for high-performance lithium-sulfur batteries, Nanoscale 12 (2020) 4552–4561, Copyright (2020) Royal Society of Chemistry; panel (B) G. Liu, K. Feng, H. Cui, J. Li, Y. Liu, M. Wang, MOF derived in-situ carbon-encapsulated Fe$_3$O$_4$@C to mediate polysulfides redox for ultrastable lithium-sulfur batteries, Chem. Eng. J. 381 (2020) 122652, Copyright (2020) Elsevier; panel (C) Y. Yan, L. Wei, X. Su, S. Deng, J. Feng, J. Yang, M. Chi, H. Lei, Z. Li, M. Wu, The crystallinity of metal oxide in carbonized metal organic frameworks and the effect on restricting polysulfides, ChemNanoMat 6 (2020) 274–279, Copyright (2020) Wiley-VCH.)*

MIL-53, i.e., a 1-D honeycomb structure that is favorable for Li$^+$ transportation. The polar Fe$_3$O$_4$ can trap the LiPSs and promote the redox of LiPSs. A clear voltage difference between Super P and Fe$_3$O$_4$@C hosts was observed in the CV test (Fig. 3.13B). A Fe$_3$O$_4$@C-S electrode demonstrated lower overpotential; this indicated lower polarization and faster redox reaction kinetics. The cell for Fe$_3$O$_4$@C-S composite delivered 755 mAh g$^{-1}$ after 300 cycles at 1.0 C, corresponding to the capacity retention of 99.3%. To further check the function of different metal oxides with the same morphology, GO@MIL-Al and GO@MIL-Fe were chosen as precursors, which were transformed into rGO@MIL-Al-600 and rGO@-MIL-Fe-600, after 600°C thermal treatment [131]. Both of those hosts have the same morphology, with different crystallinity of metal oxides (amorphous $Al_2O_3$ and crystalline $Fe_3O_4$). The galvanostatic discharge-charge curves of the initial cycle at a rate of 1.0 C show different voltage hysteresis (Fig. 3.13C), i.e., at 310, 370, and 391 mV for rGO@MIL-Al-600/S, rGO@MIL-Fe-600/S, and rGO-600/S cathodes, respectively. This result demonstrated kinetically efficient reaction and high compatibility with electrolyte for rGO@MIL-Al-600/S electrodes. This illustrated that the amorphous structure can accelerate ion diffusion and charge transfer compared to crystalline structure.

**MOF-derived metal sulfide/C composites** Compared to metal oxides, metal sulfides with higher conductivity are another promising sulfur hosts for Li-S batteries [132]. The MOF-derived metal sulfide/C composites were obtained by sulfurization of MOF-derived metal/C compounds. Luo et al. reported a facile method of synthesis of a polar $CoS_2$ embedded in porous nitrogen-doped carbon frameworks ($CoS_2$-N-C) through carbonization and sulfurization of ZIF-67 (Fig. 3.14A) [133]. XPS test showed that the Co $2p_{3/2}$ peaks shifted from 778.9, 781.1, and 783.1 eV to 777.4, 779.3, and 781.2 eV, respectively, which illustrated the electron transfer from LiPSs to Co atoms. This result confirmed the strong interaction between Co and LiPSs. Meanwhile, CV and galvanostatic discharge-charge (GDC) curves revealed the catalytic effect of $CoS_2$-N-C during cycling. The electrochemical test shows the reversible capacity of $572 \, mAh \, g^{-1}$ after 500 cycles at 1.0 C, corresponding to a capacity decay rate as low as 0.064% per cycle.

Another strategy is to fabricate MOF-derived metal sulfide/C without a carbonizing process. He and coworkers used thioacetamide (TAA) as a sulfur source, directly transforming Co-MOF to a $Co_9S_8$ array onto a three-dimensional graphene foam (3DGR), i.e., a $Co_9S_8$-3DGR composite (Fig. 3.14B) [134]. The $Co_9S_8$-3DGR can serve as an efficient sulfur host, and sulfur loading is as high as 86.9 wt%. It is worth noting that the $Co_9S_8$-3DGR/S electrode was produced without adding any polymeric binders, conductive additives, and current collectors. Therefore, the $Co_9S_8$-3DGR/S electrode achieved a high areal sulfur loading ($10.4 \, mg \, cm^{-2}$) and a high areal capacity ($10.9 \, mAh \, cm^{-2}$). It also exhibited excellent cycling stability, i.e., $736 \, mAh \, g^{-1}$ after 500 cycles at the rate of 1.0 C. The excellent results were attributed to the unique structure, which

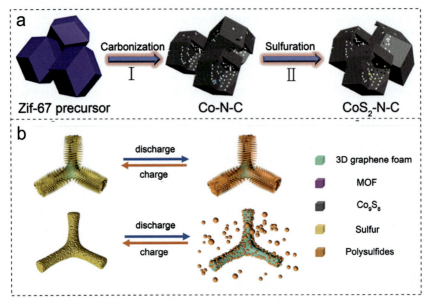

Fig. 3.14 (A) Schematic illustration of the synthesis process of CoS$_2$-N-C frameworks, (B) Advantages of the Co$_9$S$_8$-3DGF/S composite over 3D graphene foam/S. (Reproduced with permission from: panel (A) S. Luo, C. Zheng, W. Sun, Y. Wang, J. Ke, Q. Guo, S. Liu, X. Hong, Y. Li, W. Xie, Multi-functional CoS$_2$-N-C porous carbon composite derived from metal-organic frameworks for high performance lithium-sulfur batteries, Electrochim. Acta 289 (2018) 94–103, Copyright (2018) Elsevier; panel (B) J. He, Y. Chen, A. Manthiram, MOF-derived cobalt sulfide grown on 3D graphene foam as an efficient sulfur host for long-life lithium-sulfur batteries, iScience 4 (2018) 36–43, Copyright (2018) Elsevier.)

favors fast diffusion of LiPSs on the nanowall arrays and is beneficial for the conversion and prevention of LiPSs' accumulation. This illustrated that the rational design of electrode can further promote the electrochemical performance.

**MOF-derived metal carbide/C composites** Transition metal carbides (TMCs) own unique physical and chemical properties, such as high conductivity and high catalytic activity, which have been applied in the field of catalysis and energy storage [135,136]. The TMC composites have been successfully fabricated using MOFs as precursors [118,137,138]. Among them, one example reported that MOF-derived Mo$_2$C was applied as a sulfur host [118]. The porous nanooctahedron Mo$_2$C crystallites encapsulated in porous carbon (Mo$_2$C-C-NOs) were prepared by the pyrolysis of a

Mo-MOF (NENU-5) and posttreatment in $FeCl_3$ aqueous solution (Fig. 3.15). When using the $Mo_2C$-C-NOs as a sulfur host, the CV test showed that the $Mo_2C$-C-NOs cathode exhibited lower polarization of voltage than CNTs@S and commercial $Mo_2C$@S. Meanwhile, the $Mo_2C$-C-NOs electrode demonstrated high reactivity with $Li_2S_6$ compared to CNTs. This illustrated that the $Mo_2C$ is an efficient electrochemical catalyst for rapid LiPS conversion. The EIS test also showed the lowest charge transfer resistance of $Mo_2C$-C-NOs cathode than commercial $Mo_2C$@S. As expected, the $Mo_2C$-C-NOs@S cathode maintained the capacity of $762\,mAh\,g^{-1}$ after 600 cycles at 1.0 C with the average specific capacity decay rate of 0.0457% per cycle, which was much lower than that of CNTs@S (0.143% per cycles) and commercial $Mo_2C$@S (0.101% per cycle). This result demonstrated that the metal carbides have electrocatalytically accelerated reaction kinetics toward LiPSs' conversion.

**MOF-derived metal nitride/C composites** Metal nitride/C composites are promising nonprecious metal catalysts in fuel cells. It also has a catalytic function in Li-S batteries, such as decomposition of $Li_2S_2/Li_2S$. Thus, various metal nitride/C composites have been used as sulfur hosts [139–142]. MOFs can also be used as precursors to prepare metal nitrides through two steps, i.e., carbonization and nitridation. Recently, there are several reports that using MOF-derived $Co_4N$ as sulfur hosts improves the performance of Li-S batteries. $Co_4N$ is considered to have better catalytic activity than metallic Co that has better performance to suppress the shuttling effect. Xiao et al. utilized $NH_3$ as N sources for nitriding the MOF-derived Co/C composite and obtained MOF-derived $Co_4N$ composite on the carbon cloth [143]. The vertical nanowalls and open porous structures could be observed in SEM (Fig. 3.16A). Meanwhile, the nanosized $Co_4N$ particles were uniformly dispersed on the nanowalls. This structure offered fast ions and electron transfer passages, and effective catalytic sites to the redox reaction of LiPSs. Notably, as shown in Fig. 3.16B on the GDC curves, there are no obvious voltage differences between MOF-$Co_4N$/S and MOF-Co/S in the discharge plateau, but lower charge potential plateau for MOF-$Co_4N$/S electrode, which directly proved the lower polarization and better catalytic activity to decompose $Li_2S_2/Li_2S$. Furthermore, according to the fitting curves of the Warburg resistance from the EIS test, MOF-$Co_4N$/S displayed lowest $Li^+$ diffusion resistance and largest diffusion coefficient. The MOF-$Co_4N$/S cathode delivered a specific capacity of

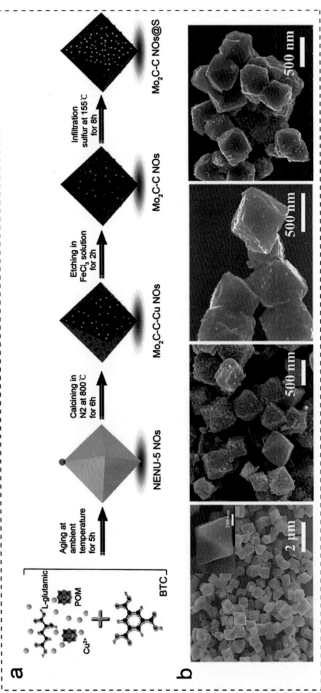

**Fig. 3.15** (A) Schematic illustration of the sequential preparation strategy for the MoC$_2$-C nanooctahedrons(NOs)@S composite; (B) SEM images of the NENU-5 NOs, Mo$_2$C-C-Cu-C NOs, Mo$_2$C-C NOs, and Mo$_2$C-C NOs@S. *(Reproduced with permission from G.L. Chen, Y.J. Li, W.T. Zhong, F.H. Zheng, J. Hu, X.H. Ji, W.Z. Liu, C.H. Yang, Z. Lin, M.L. Liu, MOFs-derived porous Mo$_2$C-C nano-octahedrons enable high-performance lithium-sulfur batteries, Energy Storage Mater. 25 (2020) 547–554, Copyright (2020) Elsevier.)*

Applications of metal-organic frameworks for lithium-sulfur batteries 81

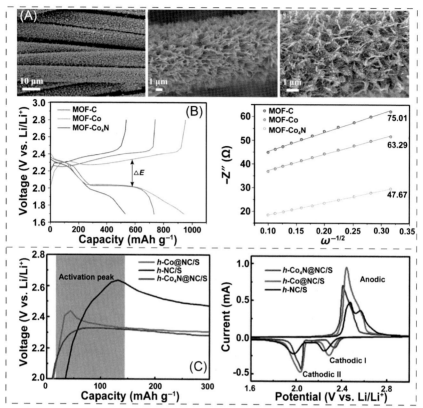

Fig. 3.16 (A) SEM images of MOF-Co$_4$N, (B) Nyquist plot of MOF-Co$_4$N, MOF-Co, and MOF-C at the states of discharge for the second cycle and fitting curves of the Warburg resistance, (C) First cycle charge voltage profiles of the h-NC/S, h-Co@NC/S, and h-Co$_4$N@NC/S electrodes. *(Reproduced with permission from: panels (A and B) K.K. Xiao, J. Wang, Z. Chen, Y.H. Qian, Z. Liu, L.L. Zhang, X.H. Chen, J.L. Liu, X.F. Fan, Z.X. Shen, Improving polysulfides adsorption and redox kinetics by the Co4N nanoparticle/N-doped carbon composites for lithium-sulfur batteries, Small 15 (2019) 1901454, Copyright (2019) Wiley-VCH; and panel (C) Z.X. Sun, S. Vijay, H.H. Heenen, A.Y.S. Eng, W.G. Tu, Y.X. Zhao, S.W. Koh, P.Q. Gao, Z. W. Seh, K.R. Chan, H. Li, Catalytic polysulfide conversion and physiochemical confinement for lithium-sulfur batteries, Adv. Energy Mater. 10 (2020) 1904010, Copyright (2020) Wiley-VCH.)*

745 mAh g$^{-1}$ after 400 cycles at 1.0 C, corresponding to a capacity retention of 82.5%.

Similarly, Co$_4$N nanoparticles embedded in nitrogen-doped carbon in the outer shell (h-Co$_4$N@NC) have been reported by Sun et al. [144]. The h-Co$_4$N@NC composite was also prepared in two steps: carbonization of the core-shelled ZIF-8@ZIF-67 and further nitridation in NH$_3$. From the initial cycle charge voltage profiles of h–NC/S, h–Co@NC/S, and h–Co$_4$N@NC/S electrodes, both h–NC/S and h–Co@NC/S electrodes showed clearly overpotential peaks (Fig. 3.16C), which are attributed to active energy barrier for Li$_2$S$_2$/Li$_2$S. More interestingly, this energy barrier disappeared for h–Co$_4$N@NC/S electrode, which demonstrated decreased polarization and higher catalytic ability of Co$_4$N. This was also supported by the CV test. These results illustrated that the MOF-derived metal nitride/C composites have higher catalytic activity toward the decomposition of Li$_2$S$_2$/Li$_2$S. When tested in Li-S batteries, the h–Co$_4$N@NC/S cathode exhibited great cycling stability. After 400 cycles, the capacities of the h–Co$_4$N@NC/S electrode at high current rates of 5.0 and 8.0 C retained 658 and 481 mAh g$^{-1}$, respectively.

## 3.2.2 MOF-based separator/interlayer for Li-S batteries

Except for the polar sulfur hosts, separators/interlayers provide the strategy to confine LiPSs into the cathode region. According to the resistance mechanism, the interlayers are classified into adsorbing and repelling mechanisms. The adsorbing mechanism is that LiPSs are physically or chemically captured in polar adsorbents to suppress the migration toward anodes. These adsorbents include doped carbon–based materials and polar inorganics. However, all of the adsorbents have limited adsorbability. It means that the LiPS-blocking effect will degrade and even disappear when the adsorbents are fully adsorbed. For the repelling mechanisms, the separators/interlayers have the effect of chemically/electrostatically repelling of the anions of LiPSs and block the LiPSs from leaking out of the cathode region. The repelling separators/interlayers include polymers and some inorganics, such as Nafion and VOPO$_4$ [145–148]. Compared to the adsorbing separators, the repelling separators are more suitable for high-sulfur-loading Li-S batteries. It is worth noting that the separators should be electron insulator, which is fulfilled for most of the MOF materials.

MOFs have tunable pore size and chemical environment and have been considered as separators/interlayers to block LiPSs diffusion to anode [149]. According to DFT calculation of the LiPSs molecular structure, the diameters of LiPSs (Li$_2$S$_n$, $4 < n \leq 8$) are larger than 0.9 nm [150]. There are various MOFs met this demand of the pore size. It is worth noting that the pore size is not the smaller the better principle because Li-ions need to pass through the MOFs to anode [151]. Zhou's group firstly used HKUST-1

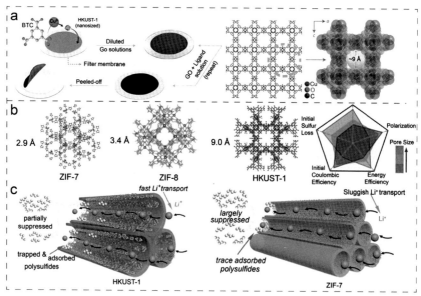

**Fig. 3.17** (A) Schematic for the preparation steps of HKUST-1-based separator, (B, C) The influence of pore size on the diffusion of both polysulfides and lithium ions. *(Reproduced with permission from: Panel (A) S. Bai, X. Liu, K. Zhu, S. Wu, H. Zhou, Metal-organic framework-based separator for lithium-sulfur batteries, Nat. Energy 1 (2016) 16094, Copyright (2016) Nature Publishing Group, and panels (B and C) Z. Chang, Y. Qiao, J. Wang, H. Deng, P. He, H. Zhou, Fabricating better metal-organic frameworks separators for Li-S batteries: pore sizes effects inspired channel modification strategy, Energy Storage Mater. 25 (2020) 164–171, Copyright (2020) Elsevier.)*

and GO to fabricate a separator (Fig. 3.17A) [152]. The pore size of HKUST-1 was ~0.9 nm, which is smaller than the diameter of LiPSs (Li$_2$S$_n$, 4 < n ≤ 8). The Li-S battery with the HKUST-1@GO separator displayed 855 mAh g$^{-1}$ after 1500 cycles at the rate of 1.0 C, with a degrading rate of 0.019% per cycle. Subsequently, the same group studied three MOF-based separators with different pore sizes for Li-S batteries [153]. The three MOFs are ZIF-7, ZIF-8, and HKUST-1, corresponding to pore sizes of 0.29, 0.34, and 0.90 nm (Fig. 3.17B), respectively. For the large pore size, too much adsorption of LiPSs can lead to higher initial "sulfur loss" and lower initial coulombic efficiency. On the contrary, small pores can hinder the migration of Li$^+$ and cause severe polarization. Therefore, introducing the negatively charged sulfonic polymer (NSP) in the pores can block the diffusion of S$_x^{2-}$ and promote the transportation of Li-ions (Fig. 3.17C).

The battery with HKUST-1-NSP separator achieved a high initial specific capacity of $589\,mAh\,g^{-1}$ after 1000 cycles at 2.0 C. This result illustrated that the rational design is an important strategy to use MOFs as separators, and even as solid electrolytes.

Tuning function between LiPSs and MOF is another strategy to improve the performance of MOF-based separators. Recently, Han et al. improved the wettability of the separator by utilizing the polar functional groups of –COOH and –OH on the UiO-66, which was a mechanically and chemically stable Zr-MOF [154]. The high wettability of interface promoted the $Li^+$ diffusion and decreased the charge transfer resistance ($R_{ct}$), which was supported by the EIS results. Therefore, the battery with UiO-66-based separator delivered a high capacity of $964.1\,mAh\,g^{-1}$ after 200 cycles at a rate of 0.5 C, corresponding to an average capacity fading of 0.08% per cycle. This result demonstrated that the chemical environment of MOF will influence the migration of LiPSs.

Except for pore size and chemical environment of MOFs, other effects need to be considered, such as particle size and packing density. Li et al. used four different MOFs [155], including Y-FTZB, HKUST-1, ZIF-8, and ZIF-7, and CNTs to modify the glass fiber as separators for Li-S batteries. When using them in the cells under the same conditions, the battery with the Y-FTZB separator displayed the best cycling performance among the five separators, i.e., $557\,mAh\,g^{-1}$ after 300 cycles at 0.25 C. Compared to the CNT separator, most MOFs demonstrated the function to suppress the shuttling effect of LiPSs and enhance cycling stability. However, only $197\,mAh\,g^{-1}$ for the battery using HKUST-1 separator was obtained at the same rate and cycling number. This result was attributed to the stacking morphology of MOF particles and the interfacial structures, which are dependent on the sizes and shapes of MOF particles. Among the four MOFs, in Y-FTZB, large particles occupied most of the surface area; meanwhile, the small particles filled the interval space; therefore, this separator looked denser. The densely packed structure is beneficial for better electrochemical performance because the soluble short-chain LiPSs were able to pass through the gap between the particles. Another reason for the poor performance of the HKUST-1 separator is the side reactions between LiPSs and SBUs of HKUST-1. This result illustrated that chemical stability and the aggregation morphology of MOF particles are two key points for separators/interlayers. It is worth noting that the introduction of interlayers may decrease the diffusive rate of $Li^+$, leading to poor rate performance.

### 3.2.3 MOF-based electrolytes for Li-S batteries

The electrolyte is an essential component of Li-S batteries. However, traditional liquid organic electrolytes act as a medium to transport LiPSs to Li-metal region, leading to losing active materials and poisoning the Li anodes. In addition, the high flammability, easy volatilization, and poor electrochemical stability of organic solvents are harmful to the safety of the systems. Quasisolid-state electrolytes (QSSEs)/solid-state electrolytes (SSEs) have been regarded as promising electrolytes to solve the problems of organic liquid electrolytes and improve the safety and electrochemical performance of the Li-S batteries. Similar to the separators and sulfur hosts, MOFs with open channels can provide the ion transportation passages to enhance the diffusion of $Li^+$. Recently, MOF-based QSSEs/SSEs were developed in different types of energy storage devices. At here, we will appraise recent and significant developments of MOF-based electrolytes for Li-S batteries at this part. Pores are the key part for MOFs to play a role in QSSEs.

The huge pore volume of MOFs can provide space to capture liquid electrolyte or ionic liquid, especially for the larger anion groups. Therefore, MOF-based QSSEs have much higher cation transfer numbers and confine LiPSs better than the traditional liquid electrolytes. To maximize the ion mobility in MOFs, the ligands can be grafted with various ion-conductive groups like lithium sulfonate (-SO$_3$Li). The first MOF-based solid electrolyte was reported by Wiers et al. [156]. The lithium isopropoxide was found to preferentially integrate into the $Mg_2$(dobdc) framework. This interaction between alkoxide anions and open cation sites might deliver higher ionic conductivity, which hypothetically facilitate $Li^+$ to go along the pore channels. As a result, the ionic conductivity of $3.1 \times 10^{-4}\,S\,cm^{-1}$ has been obtained at room temperature. Meanwhile, with temperature increase, the ionic conductivity increases according to the Arrhenius plots of ionic conductivity data. The activation energy of $Mg_2$(dobdc) was only $0.15\,eV$, which opened the possibility of MOF-based solid electrolytes as a superionic conductor. Subsequently, various MOF-based QSSEs were developed. For example, Han et al. utilized the MOF-74 and PVDF to synthesize a gel polymer electrolyte [157]. The MOFs can be employed as the cages to capture electrolyte anions and promoted the uniform $Li^+$ flux and deposition. The pore size of $10.2\,\text{Å}$ cannot only suppress the shuttling effect, but also capture the $TFSI^-$ anions $(7.9\,\text{Å})$ to increase the transport number. Meanwhile, the $Mg^{2+}$ as the Lewis acid can adsorb the $TFSI^-$, and the transport number is increased from 0.37 to 0.66. Then, the $Li^+$ is easier to move from the electrolyte to the lithium anode through the 1D channel of MOF-74. The Li-S battery achieved $778.4\,mAh\,g^{-1}$ with a low capacity fading of 0.09% per cycle. Notably, the Li anode displays a fairly smooth morphology after 100 cycles. More importantly, as low as 3.50 wt% S and as high as 12.45 wt% F were determined on the surface of Li anode, demonstrating the uniform SEI on Li surface and confinement of LiPSs.

Compared to the normal liquid electrolyte with flammable solvent, ionic liquid electrolytes have unique properties, i.e., higher safety and more stability. Recently, Chiochan et al. synthesized a UiO-based electrolyte by grafting the lithium sulfonate group on the UiO-66 [158]. The lithium sulfonate group was able to enhance the ionic conductivity and the cation transference number. To enhance the ionic conductivity, a lithium-ionic liquid (Li-IL) was utilized to get rid of proton from the $SO_3H$ group. The uniform distribution of Li-IL in the pores can be observed by the EDS mapping, which indicated the localization of Li-IL in the micropores of UiO-66. Then, the effect of the ratio between MOFs and Li-IL was investigated. Results revealed that the best ratio was 1 to 1 because low content of Li-IL will decrease the ion conductivity, but too high content of Li-IL will deteriorate the mechanical stability. Therefore, this electrolyte shows a high ionic conductivity of $3.3 \times 10^{-4}\,S\,cm^{-1}$ at room temperature. When using it in a Li-S battery, the capacity retains as high as 84% at 0.2 C after 250 cycles, corresponding to a capacity-fading rate of 0.06% per cycle. Similarly, Wang et al. used MOF-525(Cu) with the Li-IL in the QSSE [159]. The aperture size of MOF-525(Cu) is $\sim7\,\text{Å}$, and the anions of Li-IL were confined in the pores. This electrolyte displayed a high ionic conductivity of $3.0 \times 10^{-4}\,S\,cm^{-1}$ at room temperature (Fig. 3.18A).

In addition, MOFs can use as additives in polymer electrolyte to promote the confinement of LiPSs and the ions transport number. Polymers have excellent ion conductivity at a high temperature, such as PEO. However, the crystallinity at the low temperature will affect ion conduction. MOFs as the additive in polymers can decrease the crystallinity at low temperature and enhance ion conductivity at the room temperature. Huo et al. synthesized a novel cationic UiO-66-$NH_2$ (CMOF) [160]. The -$NH_2$ group can form a hydrogen bond with the ether oxygen of PEO and permit the electrochemical windows to 4.97 V. And, the cationic center can enhance anion

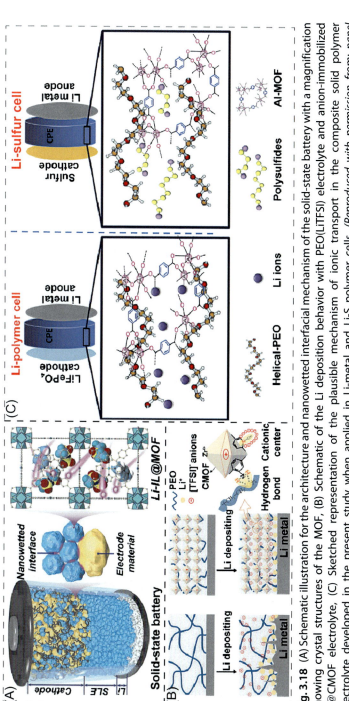

**Fig. 3.18** (A) Schematic illustration for the architecture and nanowetted interfacial mechanism of the solid-state battery with a magnification showing crystal structures of the MOF, (B) Schematic of the Li deposition behavior with PEO(LiTFSI) electrolyte and anion-immobilized P@CMOF electrolyte, (C) Sketched representation of the plausible mechanism of ionic transport in the composite solid polymer electrolyte developed in the present study when applied in Li-metal and Li-S polymer cells. *(Reproduced with permission from: panel (A) Z. Wang, R. Tan, H. Wang, L. Yang, J. Hu, H. Chen, F. Pan, A metal-organic-framework-based electrolyte with nanowetted interfaces for high-energy-density solid-state lithium battery, Adv. Mater. 30 (2018) 1704436, Copyright (2018) Wiley-VCH; panel (B) H. Huo, B. Wu, T. Zhang, X. Zheng, L. Ge, T. Xu, X. Guo, X. Sun, Anion-immobilized polymer electrolyte achieved by cationic metal-organic framework filler for dendrite-free solid-state batteries, Energy Storage Mater. 18 (2019) 59–67, Copyright (2019) Elsevier; panel (C) S. Suriyakumar, S. Gopi, M. Kathiresan, S. Bose, E.B. Gowd, J.R. Nair, N. Angulakshmi, G. Meligrana, F. Bella, C. Gerbaldi, A.M. Stephan, Metal organic framework laden poly(ethylene oxide) based composite electrolytes for all-solid-state Li-S and Li-metal polymer batteries, Electrochim. Acta 285 (2018) 355–364, Copyright (2018) Elsevier.)*

absorption and increase the $Li^+$ transference number to 0.72 (Fig. 3.18B). Then, the Li symmetrical cells successfully operate after 400 h at $0.1\,mA\,cm^{-2}$ and 200 h at $0.5\,mA\,cm^{-2}$ without obvious lithium dendrites. In addition, CMOF can decrease the PEO crystallinity to improve the ionic conductivity at room temperature. Only 10 vol% CMOF decreases the melting temperature from 62.1°C to 56.3°C. However, when the content of CMOFs is higher than 12.5 vol%, the agglomeration of CMOF particles will lead to the obstruction of interfacial pathway and decrease the ionic conductivity. This result illustrated the addition of MOFs that can enhance the electrochemical performance of the QSSE. Similarly, a composite polymer electrolyte (CPE) was also obtained by combining Al-TPA-MOF and PEO (Fig. 3.18C) [161]. The Al-TPA-MOF was well dispersed into the PEO-LiFTSI and hot-pressed into films. At low concentration, the interaction between MOFs and $Li^+$ can restrict the diffusion effect and enhance the ionic conductivity. When the concentration was increased, the MOFs diluted the fraction of PEO and led to ionic conductivity decrease. Solid-state electrolyte with 10% Al-TPA-MOF is the best composition to achieve the highest ionic conductivity and compatibility, providing the ionic conductivity of $10^{-1}\,mS\,cm^{-1}$. The laboratory-scale all-solid-state Li–S batteries delivered a capacity as high as $800\,mAh\,g^{-1}$.

### 3.2.4 MOF-based anode for Li-S batteries

For the Li-S batteries, the Li metal and its alloy were chosen as anodes to offer Li sources. Therefore, except for the challenge of the shuttling effect of LiPSs in the cathodes, the issue of lithium dendrites of anodes is another major obstacle to develop Li-S batteries. There have been various methods to enhance the electrochemical performance of lithium metal anode during the past decade [162–166]. According to previous studies, MOFs have advantages of fast ion diffusion passage, tunable chemical environment, and abundant pores. These properties are favorable for lithium metal anodes. Similar to SSEs and separators, MOFs have also been used as coating materials, to uniformize $Li^+$ distribution. Meanwhile, the rigid structure of MOFs with a high Young's modulus (>32 GPa) can prevent large volume expansion and Li dendrite growth [167]. There are several researches using MOFs for lithium metal anodes.

The MOFs with rigid structure can buffer the volume expansion of the Li deposition and restrict the Li dendrite growth. Recently, Zheng et al.

constructed a 3D conducting scaffold with MOF–808 and carbon fibers by electrospinning technique [167]. As Fig. 3.19A and B shows, the anode with MOF–808 shows smaller volume changes than that of bare Cu electrode. The thicknesses of Li metal on MOF–808 electrodes are 12.8, 24.9, and 51.3 μm after 2, 4, and 8 mAh cm$^{-2}$ amount of Li deposition at a current density of 0.5 mA cm$^{-2}$, which are close to the theoretical value and much thinner than that of bare Cu electrodes. This result illustrated that the MOFs can change the Li deposition behavior. The Li plating/stripping shows a stable voltage profile with only 20 mV overpotential and longer cycle life of over 1050 h (Fig. 3.19C). The pores can also serve as ion sieves to adjust the distribution and diffusion of Li$^+$ in the anode.

To further study the effect of particle size of MOFs, Qian et al. fabricated artificial protective layers by binding PVDF and MOFs, i.e., MOF–199 and ZIF-8 with different sizes [168]. In situ XRD results displayed that no obvious MOF structure change occurred after the initial Li deposition-stripping cycle, which confirmed the stability of MOFs during cycling. According to the cross-sectional SEM images, a sandwich structure was formed, i.e., Li growth between MOF layer and Cu current collector. It was attributed that the MOFs were only lithium-ion conductive due to their electronic insulation. Compared to bare Cu electrode, the MOFs with polarity feature have affinity to Li-ions, in which Li-ions can uniformly distribute and diffuse. The coulombic efficiency of MOF–199 coated Cu electrode remains 96% after 350 cycles for 1 mAh cm$^{-2}$ amount of Li deposition at the current density of 3 mA cm$^{-2}$, but only 8% for bare Cu electrode. Another MOF (nano-ZIF-8) with a similar particle size of MOF–199 displayed the similar electrochemical performance of MOF–199. However, when the particle size of ZIF-8 increased from nanosize to microsize, the coulombic efficiency of micro-ZIF-8 exhibited an obvious decline after 250 cycles at the same conditions. More importantly, it is difficult to observe sandwich structure for the micro-ZIF-8 electrodes, which illustrated the Li dendrites penetrated through this MOF layer, confirmed by EDS elemental maps. It demonstrated that lithium metal penetrated the micro-ZIF-8 layer through the space between microparticles. These results illustrate the following: when spaces between MOF particles are smaller than the critical nucleation sizes of Li dendrite, the penetration of Li dendrites will be blocked. Therefore, the particle size of MOF is very important, which determines the spaces between the particles.

Except for the particle size of MOFs, the binder among MOFs is another important factor. Especially, the polar functional groups of binders have a

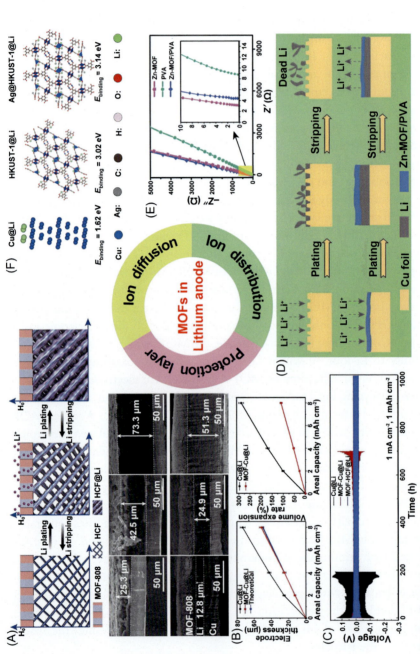

Fig. 3.19 See figure legend on opposite page

(Continued)

strong affinity with MOFs, thus making MOF stacks closer together. Meanwhile, the polar groups, such as $-OH$ and $-COO^-$, can adsorb and uniformly distribute the $Li^+$ ions. Recently, Fan and coworkers constructed an artificial SEI film by PVA-bonded Zn-MOF (Fig. 3.19D) [169]. The $-OH$ group of PVA and $-CN$ group of Zn-MOF can efficiently accelerate $Li^+$ diffusion into SEI layer (Fig. 3.19E). Moreover, the excellent wettability of Zn-MOF can eliminate the surface concentration gradient and enhance the uniform distribution of $Li^+$. When using it in a battery, the coulombic efficiency of Zn-MOF/PVA maintained 97.7% at a high current density of $3\,mA\,cm^{-2}$ after 250 cycles. In addition, carboxymethyl cellulose (CMC) binder with rich $-OH$ group was chosen to bind 2D Cu-MOF and coated on Cu sheet [170]. The contact angle of Cu-MOF electrode was $14°$, which was much smaller than that of the bare Cu electrode ($48°$). The excellent affinity with electrolytes played an important role to promote uniform lithium deposition and inhibit dendrite growth. Meanwhile, the electrochemical window was as wide as $3.6\,V$, which was suitable for Li-S batteries. The Cu-MOFs electrode achieved a steady-voltage hysteresis even after 180 cycles at $2\,mA\,cm^{-2}$. It was believed that the MOF-based electrode can be used in high-performance Li-metal batteries.

In order to further enhance the performance of MOF-based Li anode, the MOF-loaded lithiophilic materials provided an effective strategy. The lithiophilic materials can decrease the nucleation energy of Li and prevent dendrite growth. Recently, Yuan et al. utilized the HKUST-1 loaded

**Fig. 3.19, cont'd** (A) Side-SEM image of planar Cu after plating 2, 4, and $8\,mAh\,cm_{-2}$ of Li metal on Cu foil at $0.5\,mA\,cm_{-2}$, (B) The electrode thickness comparison of Cu, MOF-Cu and the volume expansion rate of Cu and MOF-Cu at different deposition capacities, (C) Voltage profiles of metallic Li plating/stripping in Cu@Li, MOF-Cu@Li, and MOF-HCF@Li symmetric cell, (D) Schematic illustration of the Li plating/stripping behavior on bare Cu foil, and the artificial SEI film protected Cu foil, (E) impedance plots estimating Li conductivity of Zn-MOF, PVA, and Zn-MOF/PVA, (F) PBE-D3-BJ computed averaged binding energy between Li atom and Cu substrate, HKUST-1 substrate, and Ag@HKUST-1 substrate. *(Reproduced with permission from: panel (C) Z.J. Zheng, Q. Su, Q. Zhang, X.C. Hu, Y.X. Yin, R. Wen, H. Ye, Z.B. Wang, Y.G. Guo, Low volume change composite lithium metal anodes, Nano Energy 64 (2019) 103910, Copyright (2019) Elsevier; panel (E) from L. Fan, Z. Guo, Y. Zhang, X. Wu, C. Zhao, X. Sun, G. Yang, Y. Feng, N. Zhang, Stable artificial solid electrolyte interphase films for lithium metal anode via metal-organic frameworks cemented by polyvinyl alcohol, J. Mater. Chem. A 8 (2020) 251–258, Copyright (2020) Royal Society of Chemistry; and panel (F) S. Yuan, J.L. Bao, C. Li, Y. Xia, D.G. Truhlar, Y. Wang, Dual Lithiophilic structure for uniform Li deposition, ACS Appl. Mater. Interfaces 11 (2019) 10616–10623, Copyright (2019) American Chemical Society.)*

lithiophilic Ag (Ag@HKUST-1) as anode [171]. The Ag not only decreased the nucleation overpotential of Li but also improved the conductivity of HKUST-1. On the contrary, the HKUST-1 provided enough space and surface area for the initial Li nucleation and decreasing local current density. According to the DFT calculations (Fig. 3.19F), the binding energy between $Li^+$ of HKUST-1(3.02 eV) and Ag @ HKUST-1 (3.14 eV) is much higher than that for bare Cu substrate, which is only 1.62 eV. This Ag@HKUST-1 structure can decrease the overpotential of Li nucleation to nearly zero and significantly promote the uniform deposition of Li. When using the Ag@HKUST-1 in batteries, a high CE of above 98.5% was achieved over 50 cycles for 5 mAh cm$^{-2}$. This value readily dropped for HKUST-1 electrode after 40 cycles at the same conditions. This result demonstrated that loading lithiophilic materials in MOFs can further improve the Li plating and stripping performances. Growth of MOFs on current collectors is another strategy to further enhance MOF-based Li-metal anodes. Recently, MOFs grown on carbon fiber cloth (CFC) were used as Li-metal anode [172]. For this structure, the CFC facilitated the electron passage and the MOFs served as the nucleation seeds to inhibit the local fast growth of lithium. Meanwhile, the lithium was deposited on the surface of CFC and trapped by the MOFs, which can guide the uniform nucleation. In the electrochemical tests, the CFC-MOF electrode displayed the coulombic efficiency of 99% after 500 cycles at 1.0 mAh cm$^{-2}$.

## 3.3 Characterization techniques

Until now, there are four major scientific issues in the study of Li–S batteries: (i) shuttling effect, (ii) sluggish reaction kinetics, (iii) clarifying the relationship between host and sulfur species, and (iv) dendrite of metallic Li. These issues can be understood by various characterization techniques and then guided rationally designing advanced materials for Li–S batteries. Various characterization techniques, including ex situ and in situ/operando, have been successfully developed and utilized to analyze the structure and chemical evolution of materials and interfaces during operation, such as PXRD, XAS, Raman spectroscopy, and FTIR spectroscopy. These characterization techniques, including principles, experimental equipment, and applications, have been reviewed in references [173,174]. Compared to the ex situ characterizations, in situ/operando characterization techniques play more important roles in comprehensive understanding of the physical and chemical evolution of electrode and electrolyte materials in real-time measurements. For example, the morphology and compositions for sensitive products of SEI layer and LiPSs could change when they transferred from batteries to characterization equipment. As follows, we will focus on the present development and future expectations regarding the applications of the in situ X-ray and optical spectroscopic techniques in Li–S batteries in this part (Fig. 3.20).

## 3.3.1 In situ X-ray techniques

There are several in situ/operando characterization techniques based on X-ray beam, including powder X-ray diffraction (PXRD), X-ray microscopy (XRM), X-ray absorption spectroscopy (XAS), X-ray photoelectron spectroscopy (XPS), which have been developed for Li–S batteries. Specifically, PXRD can provide the structural information of sulfur species and host materials; XAS and XPS can offer chemical bonding information of the electrodes and electrode/electrolyte interfaces and XRM can detect the morphology change of electrode materials and products during operation.

### 3.3.1.1 In situ powder X-ray diffraction

Powder X-ray diffraction (PXRD) is a powerful characterization technique to determine the crystal structure and space groups. Moreover, in situ/operando PXRD, which can be carried out in widespread laboratory sources, has been widely used to research the structural change of electrode materials during the charge-discharge processes. Recently, we first used in situ PXRD to track the phase change of sulfur species and MOF hosts by a reflection mode on PXRD equipment with the power of 9.0 kW (Fig. 3.21). Three minutes were required to record each in situ PXRD pattern. More interestingly, peaks of PCN-224 were detected during the cycles from the home-made experimental setup. During the initial discharge process, the intensity of sulfur diffraction peaks decreased and gradually disappeared during the first 1.0 h, corresponding to the 2.3 V discharge plateau. Meanwhile, a broad peak arising from $Li_2S$ appeared at 26.9 degree. It meant that the sulfur converted to crystalline $Li_2S$. Upon charging, this $Li_2S$ peak disappeared. Moreover, no obvious signal of peak assigned to sulfur was detected, implying the sulfur was still confined within the framework of PCN-224 in molecular form. Similar behavior was also observed in subsequent cycles. Interestingly, the diffraction peaks of PCN-224 were retained during the cycles. Furthermore, all of the PNC-224 peak positions stayed the same after the initial discharge process. This in situ PXRD result illustrated that the MOF is stable

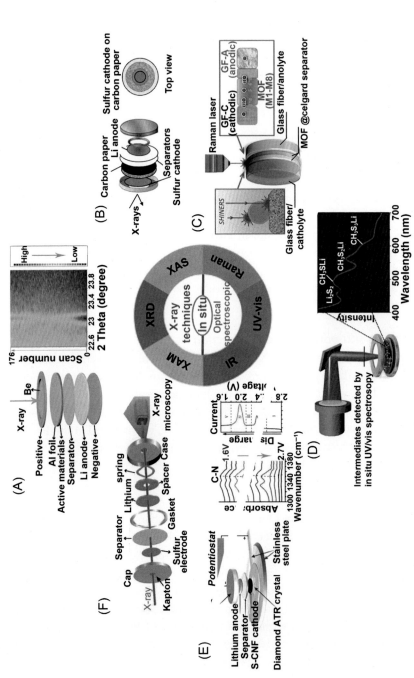

**Fig. 3.20** See figure legend on opposite page

*(Continued)*

Applications of metal-organic frameworks for lithium-sulfur batteries **95**

and suitable to confine LiPSs in the Li–S batteries. The formation and dissolution of crystal Li$_2$S associated with the discharge and charge processes were also observed in S/C and TiS$_2$/S cathodes by in situ PXRD [175,176].

Compared to the Li$_2$S, the intermediate products of LiPSs were very difficult to detect by the normal PXRD technique because of the poor crystallinity and solubility in the ether-based electrolyte of LiPSs. Very recently, Conder and coworkers reported the direct observation of LiPSs by using in operando PXRD in a reflection mode [177]. The fumed SiO$_2$ was chosen as an electrolyte additive. When adding the SiO$_2$ into the electrolyte, both cycling performance and coulombic efficiency were significantly improved in Li–S batteries. To investigate the role of the SiO$_2$, the positive electrode was monitored through in operando PXRD during the cycles. The intermediate products of LiPSs were adsorbed by SiO$_2$, which buffered the shuttle effect of LiPSs. The diffraction peaks of LiPSs appeared at the end of first plateau, corresponding to the formation of long-chain LiPSs. The intensity of the LiPSs peaks continuously decreased during further lithiation. At the end of the discharge process, crystal Li$_2$S formed. Interestingly, the LiPSs were detected in 8th and 33rd cycles

---

**Fig. 3.20, cont'd** Schematic illustration of the devices of (A) XRD, (B) XAS, (C) Raman, (D) UV-Vis, (E) IR, and (F) XRM. *(Reproduced with permission from: panel (A) X. Shang, T. Qin, P. Guo, K. Sun, H. Su, K. Tao, D. He, A novel strategy for the selection of polysulfide adsorbents toward high-performance lithium-sulfur batteries, Adv. Mater. Interfaces 6 (2019) 1900393, Copyright (2019) Wiley-VCH; panel (B) M. Cuisinier, P.-E. Cabelguen, S. Evers, G. He, M. Kolbeck, A. Garsuch, T. Bolin, M. Balasubramanian, L.F. Nazar, Sulfur speciation in Li-S batteries determined by operando X-ray absorption spectroscopy, J. Phys. Chem. Lett. 4 (2013) 3227–3232, Copyright (2013) American Chemical Society; panel (C) Z. Chang, Y. Qiao, J. Wang, H. Deng, P. He, H. Zhou, Fabricating better metal-organic frameworks separators for Li-S batteries: pore sizes effects inspired channel modification strategy, Energy Storage Mater. 25 (2020) 164–171, Copyright (2020) Elsevier; panel (D) J. Liu, T. Qian, N. Xu, M. Wang, J. Zhou, X. Shen, C. Yan, Dendrite-free and ultra-high energy lithium sulfur battery enabled by dimethyl polysulfide intermediates, Energy Storage Mater. 24 (2020) 265–271, Copyright (2020) Elsevier; panel (E) C. Dillard, A. Singh, V. Kalra, Polysulfide speciation and electrolyte interactions in lithium-sulfur batteries with in situ infrared spectroelectrochemistry, J. Phys. Chem. C 122 (2018) 18195–18203, Copyright (2018) American Chemical Society and X. Ding, S. Yang, S. Zhou, Y. Zhan, Y. Lai, X. Zhou, X. Xu, H. Nie, S. Huang, Z. Yang, Biomimetic molecule catalysts to promote the conversion of polysulfides for advanced lithium-sulfur batteries, Adv. Funct. Mater. 30 (2020) 2003354, Copyright (2020) Wiley-VCH; and panel (F) S.-H. Yu, X. Huang, K. Schwarz, R. Huang, T.A. Arias, J.D. Brock, H.D. Abruna, Direct visualization of sulfur cathodes: new insights into Li-S batteries via operando X-ray based methods, Energ. Environ. Sci. 11 (2018) 202–210, Copyright (2018) Royal Society of Chemistry.)*

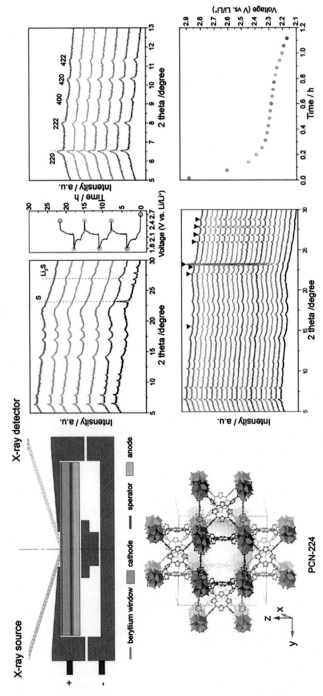

**Fig. 3.21** Schematic illustration of the devices of in situ XRD and the detection result of ppy-S-in-PCN-224 cathode. *(Source: Authors)*.

by the in operando PXRD. This result demonstrated that the polysulfide adsorbents not only enhance the detective sensitivity of LiPSs, but also improve the electrochemical performances of Li-S batteries. Similarly, Fe-MOF-derived $Fe_3O_4/C$ composites have a strong interaction with LiPSs by their $Fe_3O_4$ constitute, resulting in situ PXRD detecting the LiPSs signal during the cycling [178]. Synchrotron-based XAS is another strategy to detect LiPSs during the discharge-charge processes [179]. We will discuss it in the in situ/operando XAS part.

### 3.3.1.2 In situ X-ray microscopy

In situ/operando PXRD can provide information of the structural change and intermediate formation during the cycling. This information was usually averaged/accumulated from relatively large area electrode materials. However, the local/tiny differences happening in the electrodes, such as morphology and compositions, were difficult to detect. Therefore, a direct visual characterization technique was necessary to track the morphology changes including the deformation, crack, and degradation of electrode materials during the operation. However, the traditional high-resolution morphology characterization techniques, such as SEM and TEM, use electronic beam to detect the materials. Moreover, these electronic beams with high energy led to component and morphology change for the sensitive intermediates. Meanwhile, the electron microscope techniques have strict requirements for samples, such as conductive samples for SEM and thin samples for TEM. Compared to electron microscope techniques, X-ray microscopy (XRM) is more suitable for tracking the morphology of electrode materials in real-time during the discharge-charge process. The experimental setups for in operando XRM are presented in Fig. 3.20F. The X-ray beam was passed through the battery and collected on a detector.

Toney's group firstly used the in operando XRM to study the morphology of sulfur species in Li-S batteries during cycling [180]. The morphology of a $\sim$10-$\mu$m S/C particle was recorded via the in operando XRM during the initial discharge-charge process. The particle size has changed a little; meanwhile, it lost some sulfur species through the formation of LiPSs and created small particles with porous structure during the first discharge plateau. The S/C particle was not found to dissolve significantly in this whole process. Upon charging, the trapped LiPSs may promote the formation of crystal sulfur at the end of the charging. This result illustrated that the complete encapsulation of sulfur can significantly enhance the long-time cycling performance of Li-S batteries. Albeit the morphology of sulfur observed by the XRM technique, the image resolution was not enough to offer more information in this study.

To improve the resolution of the in operando XRM, Abruna and coworkers employed two types of contrast information, i.e., absorption and phase, which can clearly observe the sulfur clusters and the current collector of carbon fibers [181]. The dissolution and reformation of sulfur clusters were clearly observed in discharge and charge processes, respectively. Interestingly, it can characterize a large area at the same time, i.e., several hundreds of particles in an image. Combined with the in operando PXRD, different sulfur phases and $Li_2S$ were identified. According to the in operando XRM results, the size of $Li_2S$ and reformed sulfur were dependent on the current densities and temperature. Small size and uniform distribution were obtained at high current densities. It was attributed to a higher nucleation density. Larger particle size of sulfur and $Li_2S$ formed at low current density. In addition, $Li_2S$ was formed in smaller clusters at lower temperature. Based on this study, it provided new insights for the development of Li-S batteries. This in operando XRM technique will be employed to confirm the sulfur particles inside/outside of the MOFs during operation.

### 3.3.1.3 In situ X-ray absorption spectroscopy

The interaction between hosts and sulfur species during cycling is another key point to develop excellent-performance Li-S batteries. This information cannot be obtained from the in situ XRD and XRM, which focus only on the structure and morphology of active materials. X-ray absorption spectroscopy (XAS) is based on electron excitation to unoccupied states through absorbing the X-ray photon energy and provides chemical states and local structural analysis, such as chemical interaction between sulfur species and hosts in Li-S batteries. X-ray absorption spectroscopy can be classified into X-ray absorption near-edge structure (XANES) and extended X-ray absorption fine structure (EXAFS) categories. The former is often used for measuring the valence states of transition metal and the latter is for detecting the local chemical bond changes and coordination numbers [174]. Therefore, XAS is sensitive enough to detect crystallized or amorphous species and obtain a variety of valuable information including chemical environment, type of bonding, and site symmetry.

For the Li-S batteries, XAS can detect intermediates of sulfur species and chemical environment, including hosts, binder, and Li anode. Concerning sulfur species, Nazar group first directly monitored sulfur speciation within the battery during operation by in situ XAS on the S K-edge [179]. Sulfur species $S_6^{2-}$, $S_4^{2-}$, and $Li_2S$ in the discharge process and $S_4^{2-}$, $S_6^{2-}$, and $S_8^0$ in the charge process were detected. This result revealed the first detailed evidence of the redox chemistry in Li-S batteries. For binders, Ling et al. utilized the XAS to investigate the nucleophilic substitution reaction between LiPSs and a sulfate-based binder [182]. The typical peaks of C-S and S-S bond could be clearly observed, which demonstrated the LiPSs reaction with the polymer binder. This result was also supported by in situ XAS experiment, which detected fast LiPSs nucleophilic substitution reaction with the chemical leaving groups of polymer binders. Similarly, the polar PEI polymer was chosen as the functional binder for sulfur cathode [183]. The in situ XAS demonstrated that the amino groups of PEI polymer interacted with LiPSs by a strong electrostatic force, which effectively restrict the shuttle effect of LiPSs.

The in situ XAS technique has been employed to study the Li-metal side and sulfur hosts in Li–S batteries. $LiNO_3$ is an effective electrolyte additive to suppress the LiPSs shuttle effect and improve the stability of Li-metal anode. To study the influence of $LiNO_3$, Guo and coworkers used the in situ XAS to investigate SEI layer on Li anode. Without the addition of $LiNO_3$ into the electrolyte, the intensity of $Li_2S$ signal on the anode increased with the proceeding of discharge [184]. However, a stable SEI layer composed of $Li_2SO_3$ and $Li_2SO_4$ coated the Li anode when $LiNO_3$ was added. With the discharge process, the peak intensity of $SO_4^{2-}$ increased at first and then remained constant, whereas the peak intensity of $SO_3^{2-}$ increased initially and then decreased noticeably. In addition, a typical strong N–O peak from the insoluble $LiNO_2$ occurred in the EXAFS result. Therefore, the $LiNO_3$ additive can restrict the shuttle effect of LiPSs by forming a stable SEI layer, which promoted the cycling stability of Li–S batteries. Sulfur hosts are key parts for the cathode electrode, but the interaction mechanism is still unclear. The in situ XAS technique is a powerful tool to study the mechanisms. Recently, Sun et al. utilized the in situ XAS to investigate the reaction mechanism of CuS hosts [185], which own obvious structure and valence state. The in situ XAS showed an edge jump shift to lower energies, which corresponded to the consumption of sulfur species. It meant that the transformation of CuS to $Cu_{1+x}S$ due to interaction with the LiPSs had a place. This result illustrated that the interaction between sulfur hosts and LiPSs was directly observed by in situ XAS. For the MOF-based Li–S batteries, the interaction between LiPSs and SBUs/function groups in ligands could also be detected by in situ XAS technique.

## 3.3.2 In situ optical spectroscopic techniques

The reaction mechanisms of Li-S batteries are extremely complicated based on the "solid-liquid-solid" reaction routes. There are various intermediates, including polysulfides ($S_8^{2-}$, $S_7^{2-}$, $S_6^{2-}$, $S_5^{2-}$, $S_4^{2-}$, $S_3^{2-}$, $S_8^{2-}$, $S_2^{2-}$, etc.) and radicals ($S_3^-$, $S_4^-$, etc.). Among characteristic techniques, the optical spectroscopic techniques, including infrared (IR), Raman, and ultraviolet-visible (UV-Vis), are considered as powerful characterization tools for chemical structure analysis, chemical, and structural fingerprint recognition to identify the chemical intermediates. Meanwhile, these optical spectroscopic techniques have unique advantages, such as nondestructive nature, high time, and spatial resolutions, facile operation under electrochemical test conditions. Therefore, the in situ optical spectroscopic techniques have been employed for qualitative and quantitative analysis of LiPSs in a Li-S battery during operation (Fig. 3.22).

### 3.3.2.1 In situ UV-Vis spectroscopy

UV-Vis spectroscopy is a typical absorption spectroscopy, which can detect molecules and ions that contain $\pi$-electron and nonbonding electrons. For Li-S batteries, LiPSs and sulfur-based radicals could be detected by in situ UV-Vis spectroscopy. Meanwhile, this technique can provide a quantitative analysis of LiPSs in Li-S batteries. Patel et al. demonstrated that the qualitative and quantitative analyses of soluble polysulfides in Li-S batteries were obtained by in situ UV-Vis spectroscopy [186]. During the discharge process, various LiPSs can be detected from $Li_2S_8$ to $Li_2S$ in which the concentration of long-chain LiPSs was 5–10 mM and 1–2 mM for short-chain LiPSs. The low concentration of short-chain LiPSs was attributed to the poor dissolution of $Li_2S_2$ and $Li_2S$. Upon charging, the concentration of LiPSs was similar to that during the discharge process except for a slight shift into lower wavelength. Recently, Liu et al. revealed a new reaction mechanism with the absence of LiPSs in the dimethyl trisulfide (DMTS)-modified electrolyte through the in situ UV-Vis spectroscopy [187]. Six typical peaks, corresponding to $Li_2S_2$ and $CH_3S_nLi$ ($n = 1$–5), were detected between 480 and 570 nm. Meanwhile, the peaks of long-chain LiPSs were not observed. This result demonstrated that $CH_3S_nLi$ replaced the long-chain LiPSs and restrained the shuttling effect of LiPSs. Another additive, biphenyl-4,4′-dithiol (BPD), was added to the electrolyte [188]. When using the in situ UV-Vis spectroscopy to detect the sulfur species, the intensity of peaks associated with short-chain polysulfide ($S_n^{2-}$, $4 \geq n \geq 1$) and $S_3^-$ was clearly weaker than the corresponding peaks in electrolyte without BPD additive. This

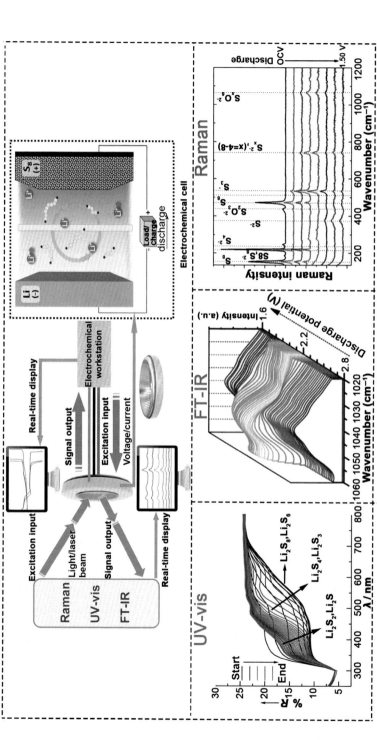

**Fig. 3.22** Schematic illustration of common devices of the in situ optical spectroscopic techniques and the relevant spectra. (Reproduced (top row, then bottom row, left to right) with permission from: L. Zhang, T. Qian, T. Zhu, Z. Hu, M. Wang, L. Zhang, T. Jiang, J.-H. Tian, C. Yan, In situ optical spectroscopy characterization for optimal design of lithium-sulfur batteries, Chem. Soc. Rev. 48 (2019) 5432–5453, Copyright (2019) Royal Society of Chemistry; M.U.M. Patel, R. Demir-Cakan, M. Morcrette, J.-M. Tarascon, M. Gaberscek, R. Dominko, Li-S battery analyzed by UV/Vis in operando mode, ChemSusChem 6 (2013) 1177–1181, Copyright (2013) Wiley-VCH; C. Dillard, A. Singh, V. Kalra, Polysulfide speciation and electrolyte interactions in lithium-sulfur batteries with in situ infrared spectroelectrochemistry, J. Phys. Chem. C 122 (2018) 18195–18203, Copyright (2018) American Chemical Society; B.P. Vinayan, T. Diemant, X.M. Lin, M.A. Cambaz, U. Golla-Schindler, U. Kaiser, R.J. Behm, M. Fichtner, Nitrogen rich hierarchically organized porous carbon/sulfur composite cathode electrode for high performance Li/S battery: a mechanistic investigation by operando spectroscopic studies, Adv. Mater. Interfaces 3 (2016) 1600372, Copyright (2020) Wiley-VCH.)

result demonstrated that the dissolution of LiPSs was significantly decreased. Combined with Raman and mass spectrometry, various BPD–polysulfide complexes were determined to form and prevent the LiPS dissolution. These studies illustrated that the in situ UV–Vis spectroscopic technique was a powerful tool for analysis of soluble LiPSs, but it was insensitive to the less soluble of $S_8$ and $Li_2S$.

### 3.3.2.2 In situ infrared spectroscopy

Similar to UV–Vis spectroscopy, IR spectroscopy is another typical absorption spectroscopy, which can detect the vibrations and rotations of molecular bonds. In particular, Fourier transform infrared (FTIR) spectroscopy and attenuated total reflection FTIR (ATR–FTIR) spectroscopy have been successfully applied to investigate the chemical bond information of electrode/electrolyte interfaces and the liquid electrolyte phase. Both of them are sensitive to sulfur species containing the S–S bonds in Li–S batteries. Recently, Saqib et al. utilized in situ ATR–FTIR spectroscopy to detect and analyze the quantity of LiPSs in a Li–S battery under operation [189]. The reduction of S to $Li_2S_8$, then to small polysulfides, and finally to $Li_2S_2$ and $Li_2S$, was observed during the discharge process. Compared to in situ UV–Vis spectroscopy, the in situ ATR–FTIR was more sensitive to solid products on the electrode/electrolyte interface. In addition, sulfur was observed at the end of the initial charging, but this sulfur signal disappeared after multiple cycles. Meanwhile, the concentration of LiPSs decreased with cycling number, corresponding to the capacity fading of the Li–S batteries. This result directly illustrated that the capacity fading was resulted from the shuttling effect of LiPSs.

To understand the dissolution process of LiPSs, Dillard et al. employed in situ ATR–FTIR to investigate the LiPS evolution and coordination state of electrolyte anions [190]. The S–S vibrational mode ($\sim 500\,cm^{-1}$) and the symmetric $SO_3^{2-}$ stretch of $LiCF_3SO_3$ vibrational mode ($\sim 1050$–$1030\,cm^{-1}$) have provided insightful redox information. The observed redshift of the S–S wave numbers during the discharge process corresponded to the reduction of sulfur species. In addition, there were three peaks of $SO_3^{2-}$, which were assigned to coordinated ion pairs ($1040\,cm^{-1}$), free triflate ion ($1030\,cm^{-1}$), and ion aggregates ($1050\,cm^{-1}$). When the concentration of LiPSs increased, the free triflate ions decreased. After the insoluble LiPSs had deposited, the intensity of the peak associated with free triflate ions returned to the normal value. It illustrated the interaction between LiPSs and electrolyte salt. This result could be instrumental in the design of new electrolytes for Li-S batteries.

Furthermore, the in situ IR spectroscopy can detect the chemical bond information on the interface between sulfur hosts and electrolytes, especially for the molecular catalysis-grafted hosts. Ding and coworkers studied three hemin molecular-grafted functionalized CNTs as sulfur hosts [191]. Among them, the CNT-COOH@hemin cathode exhibited the best electrochemical performance, i.e., $1637.8\,\mathrm{mAh\,g^{-1}}$ for initial cycle and longest cycling durability. According to the in situ FTIR results, the positions and intensity of the C-N band ($\sim1330$ and $1350\,\mathrm{cm^{-1}}$) changed during the discharge-charge process. This phenomenon could be explained as a strong electronic interaction between Fe and LiPSs, resulting in a stronger Fe-S bond and a weaker C-N bond. As a result, both Fe and N atoms provided electrons to LiPSs and promoted the reduction of LiPSs.

### 3.3.2.3 In situ Raman spectroscopy

Raman spectroscopy is utilized to differentiate molecules through their structural fingerprints, mainly vibrational but also some rotational, and other low-frequency modes [192]. Simultaneous detection of sulfur species in electrolyte and in the solid state can be accomplished by in situ Raman spectroscopy. More importantly, the phase change on the interface of solid/electrolyte could be observed by in situ Raman spectroscopy. To investigate the function of MOF-derived sulfur hosts, Li et al. utilized in situ Raman spectroscopy to track the conversion of sulfur species in Li-S batteries [193]. During the discharge process, $S_8$, located at $150, 219, 474\,\mathrm{cm^{-1}}$, transformed to $L_2S_6$ ($398\,\mathrm{cm^{-1}}$), $Li_2S_4$ ($202\,\mathrm{cm^{-1}}$), and $Li_2S_2$ ($452\,\mathrm{cm^{-1}}$) gradually. Upon charging, $Li_2S_2$ was oxidized to $Li_2S_4$ and $L_2S_6$ and finally converted to $S_8$ at the potential of $2.80\,\mathrm{V}$. The intensity of $S_8$ peaks at the end of charging was as high as the initial state for the ZIF-67-derived sulfur hosts. This result demonstrated a high conversion rate of LiPSs to sulfur for this host. This illustrated the confinement effect and the sufficient exposure of catalytically active sites of the MOF-derived sulfur hosts. Similarly, in situ Raman spectroscopy was used to check the interaction between LiPSs and N-doped porous carbon hosts [153]. Except for LiPSs, the thiosulfate ($S_2O_3^{2-}$) species, located at $442\,\mathrm{cm^{-1}}$, was detected on the electrode surface. The thiosulfate reacted with long-chain LiPSs producing polythionate complexes (peak at $1065\,\mathrm{cm^{-1}}$), which restrained the shuttle effect of LiPSs.

For the lithium metal anode, in situ Raman spectroscopy was carried out on Cu foil in a Li|Cu battery [194]. In the normal Li-S batteries electrolyte, the intensity of peaks at $741 \, cm^{-1}$ and $942 \, cm^{-1}$, corresponding to the S-N stretching in TFSI$^-$ and Li-coordinated solvent band, decreased dramatically during the Li deposition process. This illustrated a sharp consumption of lithium ions at the electrode/electrolyte interface, resulted in a nonuniform ion flux and unstable dendritic growth. When adding graphene quantum dots (GQDs) in the electrolyte, two new peaks of $\sim 1350 \, cm^{-1}$ and $1535 \, cm^{-1}$, assigned to the D and G bands of the GQDs, appeared during the Li plating. This demonstrated that the GQD accumulated on the surface of Cu electrode. Interestingly, the intensity peaks of S-N and Li-coordinated solvent band significantly increased. This result showed that the in situ formed GQDs layer could serve as a regulator to continually smooth the local electrical field, which enabled a stable ion flux near the electrochemical interface and uniform lithium deposition.

For the separators, in situ Raman spectroscopy with shell-isolated nanoparticle-enhanced technique was employed to check the interaction of MOF-based separator with sulfur species in Li-S batteries under operation [153]. Except for S-S stretches in polysulfides, two new peaks, located at 323 and $349 \, cm^{-1}$ corresponding to $CuS_{7/8}^{2-}$ and $CuS_{4/6}^{2-}$, could be detected on the surface of MOFs separators. This result directly proved the interaction between LiPSs and SBU of MOFs in the separators. In addition, the peaks of LiPSs can be observed until $2 \, \mu m$ deep, which suggested that MOFs with large pore size can trap LiPS molecules. It further proved the restraining of the shuttling effect of LiPSs by MOF-based separators. As these results showed, MOFs own strong capabilities to confine the LiPSs and thus restrain the shuttling effect.

## 3.4 Summary and outlook

To meet the requirements for electric vehicles and grid storage, the exploration of advanced batteries with high energy density, low cost, long cycle life, and high safety is an urgent task. Among energy storage systems, the Li-S battery has become a promising candidate to achieve the energy density beyond the current state-of-the-art LIBs. With over 15 years of development, there are huge improvements in electrochemical performance and understanding of the reaction mechanisms of Li-S batteries. Especially, various new materials for sulfur hosts, electrolytes, separators, and lithium anodes have been explored. Among them, MOFs belong to a new kind of crystal porous materials with unique properties, including large surface area, high porosity, low density, controllable structure, tunable pore size, and tunable chemical environment. Until now, MOFs, MOF-based composites, and MOF-derived composites as sulfur hosts, electrolytes, separators, and lithium anodes for Li-S batteries have been widely reported. Meanwhile, various in situ characterization techniques, especially X-ray and optical spectroscopy, have been employed to deeply understand the electrochemical reaction mechanisms and interactions in Li-S batteries. These results from in situ characterization techniques could promote a further optimal design of materials for the advanced Li-S batteries. Nevertheless, there is still a large room for imperative improvement to use MOFs in the future commercial Li-S batteries. Several key challenges and directions are highlighted in the following sections.

### 3.4.1 Cathode

MOFs and MOF-derived composites have been widely used as sulfur hosts due to MOFs with high surface area and large pore volume to be capable of high sulfur loading. In addition, the organic ligands and SBUs with polarity strongly adsorb LiPSs to alleviate the "shuttling effect." However, the non-conductivity of MOFs resulted in large potential polarization during cycling, especially at high current densities. Although the conductivity could be improved by pyrolysis to carbon, both polarity and pore geometries were sacrificed. The other strategy is to combine conductive composites, including graphene, conductive polymers, and CNTs, to form conductive MOF-based composites. Nevertheless, these conductive MOF-based composites will decrease the sulfur loading. For example, the maximum sulfur loading of PCN-224 is as high as 70 wt%, but that of PCN-224-ppy is less 50 wt% [57]. Therefore, the design and fabrication of MOFs with certain electrical conductivity will be important in the development of excellent MOF-based sulfur hosts. These novel conductive MOF hosts have enough electronic conductivity and high $Li^+$ conducting channels. The second strategy is introducing metal ions on the ligands of MOFs. These metal ions could provide the electron hopping and redox hopping for electron transfer in MOFs [195]. Meanwhile, metal ions can act as a catalyst to decompose sulfur species during the cycling. Furthermore, the third strategy is introducing a small amount of electron conductive elements in sulfur to enhance the electric transport [196]. These electron conductive elements include Se and Te, which are also active to lithium during Li-S batteries operation.

### 3.4.2 Interlayers/separators

To restrain the shuttling effect of LiPSs, MOFs with unique properties, including nonconductivity, tunable pore size, and chemical environment, are considered as interlayers/separators for Li-S batteries. Various MOFs were reported as separators to suppress the LiPS diffusion [149,151–155]. Among these MOFs, the HKUST-1 with pore size matched to the diameters of LiPSs ($Li_2S_n$, $4 < n \leq 8$), enabled excellent electrochemical performance of Li-S batteries incorporating it as a separator. However, when discharged to 1.5 V, the HKUST-1 was unstable and formed $CuS/Cu_2S$ species [155]. Therefore, the design and fabrication of stable MOFs with suitable pore size will be key to develop MOF-based separators. Meanwhile, multivariate-MOFs and grafting function groups on the stable MOFs provide other strategies to restrain the shuttling effect of LiPSs. Another challenge is the space between MOF particles in the separator still allowing LiPS diffusion to Li-metal anodes. Thus, it is expected to develop techniques that can fabricate larger size stable MOF films to be directly used as separators, eliminating the spaces among MOF particles.

### 3.4.3 Electrolyte

Replacing the liquid electrolytes with solid-state electrolytes was considered as the effective strategy to solve the issues of Li-S batteries, which include the shuttling effect of LiPSs and Li dendrite formation. An ideal solid-state electrolyte should fulfill the requirements of high ionic conductivity, high chemical stability, enough mechanical strength, and resistance to LiPSs' shuttling [197]. MOFs almost met all those requirements. Until now, there have been some MOFs used as carriers for ion liquid/liquid electrolyte and additives for polymer in solid-state electrolyte, which demonstrated the advantages of MOFs to some degree. In the future, the design of MOFs with Li-ion conductivity is an important route to develop the real MOF-based solid-state electrolytes for Li-S batteries, even for other batteries. For example, functional groups/polymers with high Li-ion conductivity, such as - $SO_3Li$ and PEO, are grafting on the ligands of MOFs. When the PEO is grated on the MOFs, it can dramatically decrease the electrode/electrolyte interfacial resistance. Furthermore, similarly to separators, the growth of larger size MOFs films that can be directly utilized as solid-state electrolytes for Li-S batteries is also called upon.

### 3.4.4 Anode

The challenges for lithium metal anodes are the Li dendrite formation and highly active reactions between Li metal and electrolyte. MOFs with high Young's modulus (>32 GPa), large porous volume, and tunable chemical environment with lithiophilic interface are suitable to suppress the Li dendrites and facilitate uniform ion flux. Compared to the cathodes of Li-S batteries, there are relatively fewer reports in MOFs for Li-metal anodes. There is a significant room for the improvement of the stability and safety of Li-metal anode by MOFs, including: (1) coating stable MOFs with high ion conductivity on Li metal to tune the ions flux and uniform distribution; (2) coating MOFs with enriched Li salts on Li metal to increase the electrolyte concentration at the electrode/electrolyte interface; and (3) loading lithium metal/Li-metal alloy in MOF-based composites to make homogeneous $Li^+/e^-$ conductive interfaces.

### 3.4.5 Characterization

Advanced characterization techniques, especially in situ ones, are necessary to understand the real reaction routes and mechanisms because of their real-time features. Various in situ characterization techniques have been utilized in Li-ion and Li-S batteries. For MOF-based Li-S batteries, MOFs display different features compared to other electrode materials, mainly diffraction at a low angle and enhanced sensitivity to laser stimulation. Therefore, it is more difficult to design in situ characterization techniques for MOF-based Li-S batteries. The practical chemical reactions and interactions in Li-S batteries are very complicated, which include the conversion of sulfur to different sulfur species and interaction between LiPSs and MOFs. It should combine two or even more characterization techniques to study those reactions and interactions. In addition, quantitative analysis of different polysulfides is still difficult to achieve during operation because of disproportionation reactions of sulfur species. Thus, it is necessary to develop new or couple of existing techniques in the near future.

### Acknowledgments

The authors would like to thank Dr. Lian-Jie Xue, Cai-Hong Zhang, Dr. Xuan Gong, Prof. Hexiang Deng for his invaluable help. This work was sponsored by the National Nature Science Foundation of China (22172116, 21773176, 21403157).

# References

[1] X. Liu, D. Ren, H. Hsu, X. Feng, G.-L. Xu, M. Zhuang, H. Gao, L. Lu, X. Han, Z. Chu, J. Li, X. He, K. Amine, M. Ouyang, Thermal runaway of lithium-ion batteries without internal short circuit, Joule 2 (2018) 2047–2064.

[2] Z.W. Seh, Y. Sun, Q. Zhang, Y. Cui, Designing high-energy lithium-sulfur batteries, Chem. Soc. Rev. 45 (2016) 5605–5634.

[3] P.G. Bruce, S.A. Freunberger, L.J. Hardwick, J.-M. Tarascon, Li–$O_2$ and Li–S batteries with high energy storage, Nat. Mater. 11 (2012) 19–29.

[4] Z. Lin, T. Liu, X. Ai, C. Liang, Aligning academia and industry for unified battery performance metrics, Nat. Commun. 9 (2018) 5262.

[5] X. Yang, J. Luo, X. Sun, Towards high-performance solid-state Li-S batteries: from fundamental understanding to engineering design, Chem. Soc. Rev. 49 (2020) 2140–2195.

[6] J. Liang, Z.H. Sun, F. Li, H.M. Cheng, Carbon materials for Li-S batteries: functional evolution and performance improvement, Energy Storage Mater. 2 (2016) 76–106.

[7] S. Xu, C. Kwok, L. Zhou, Z. Zhang, I. Kochetkov, L. Nazar, A high capacity all solid-state Li-sulfur battery enabled by conversion-intercalation hybrid cathode architecture, Adv. Funct. Mater. 31 (2020) 2004239.

[8] D. Wang, Y. Wu, X. Zheng, S. Tang, Z. Gong, Y. Yang, $Li_2S$@NC composite enable high active material loading and high $Li_2S$ utilization for all-solid-state lithium sulfur batteries, J. Power Sources 479 (2020) 228792.

[9] A. Manthiram, S.H. Chung, C.X. Zu, Lithium-sulfur batteries: progress and prospects, Adv. Mater. 27 (2015) 1980–2006.

[10] S.L. Li, W.F. Zhang, J.F. Zheng, M.Y. Lv, H.Y. Song, L. Du, Inhibition of polysulfide shuttles in Li-S batteries: modified separators and solid-state electrolytes, Adv. Energy Mater. 11 (2020) 2000779.

[11] N. Jayaprakash, J. Shen, S.S. Moganty, A. Corona, L.A. Archer, Porous hollow carbon@sulfur composites for high-power lithium-sulfur batteries, Angew. Chem. Int. Ed. 50 (2011) 5904–5908.

[12] Y.-S. Su, A. Manthiram, Lithium-Sulphur batteries with a microporous carbon paper as a bifunctional interlayer, Nat. Commun. 3 (2012) 1166.

[13] X. Hong, Y. Liu, J. Fu, X. Wang, T. Zhang, S. Wang, F. Hou, J. Liang, A wheat flour derived hierarchical porous carbon/graphitic carbon nitride composite for high-performance lithium-sulfur batteries, Carbon 170 (2020) 119–126.

[14] K. Zhang, F.R. Qin, Y.Q. Lai, J. Li, X.K. Lei, M.R. Wang, H. Lu, J. Fang, Efficient fabrication of hierarchically porous graphene–derived aerogel and its application in lithium sulfur battery, ACS Appl. Mater. Interfaces 8 (2016) 6072–6081.

[15] Y. Wei, B. Wang, Y. Zhang, M. Zhang, Q. Wang, Y. Zhang, H. Wu, Rational design of multifunctional integrated host configuration with lithiophilicity-sulfiphilicity toward high-performance Li-S full batteries, Adv. Funct. Mater. 31 (2020) 2006033.

[16] S. Zeng, G.M. Arumugam, X. Liu, Y. Yang, X. Li, H. Zhong, F. Guo, Y. Mai, Encapsulation of sulfur into N–doped porous carbon cages by a facile, template-free method for stable lithium-sulfur cathode, Small 16 (2020) 2001027.

[17] F. Zhu, Y. Tao, H. Bao, X. Wu, C. Qin, X. Wang, Z. Su, Ferroelectric metal–organic framework as a host material for sulfur to alleviate the shuttle effect of lithium-sulfur battery, Chem. A Eur. J. 26 (2020) 13779–13782.

[18] G.K. Gao, Y.R. Wang, H.J. Zhu, Y. Chen, R.X. Yang, C. Jiang, H. Ma, Y.Q. Lan, Rapid production of metal-organic frameworks based separators in industrial-level efficiency, Adv. Sci. 7 (2020) 2002190.

[19] D. Capkova, M. Almasi, T. Kazda, O. Cech, N. Kiraly, P. Cudek, A.S. Fedorkova, V. Hornebecq, Metal-organic framework MIL-101(Fe)-$NH_2$ as an efficient host for Sulphur storage in long-cycle Li-S batteries, Electrochim. Acta 354 (2020) 136640.

[20] A.E. Baumann, J.R. Downing, D.A. Burns, M.C. Hersam, V.S. Thoi, Graphene-metal-organic framework composite sulfur electrodes for Li-S batteries with high volumetric capacity, ACS Appl. Mater. Interfaces 12 (2020) 37173–37181.

[21] X.D. Song, F.Y. Zhou, M. Yao, C. Hao, J.S. Qiu, Insights into the anchoring of polysulfides and catalytic performance by metal phthalocyanine covalent organic frameworks as the cathode in lithium-sulfur batteries, ACS Sustain. Chem. Eng. 8 (2020) 10185–10192.

[22] X.H. Hu, J.H. Jian, Z.S. Fang, L.F. Zhong, Z.K. Yuan, M.J. Yang, S.J. Ren, Q. Zhang, X.D. Chen, D.S. Yu, Hierarchical assemblies of conjugated ultrathin COF nanosheets for high-sulfur-loading and long-lifespan lithium-sulfur batteries: fully-exposed porphyrin matters, Energy Storage Mater. 22 (2019) 40–47.

[23] J.Y. Wang, L.P. Si, Q. Wei, X.J. Hong, L.G. Lin, X. Li, J.Y. Chen, P.B. Wen, Y.P. Cai, An imine-linked covalent organic framework as the host material for sulfur loading in lithium-sulfur batteries, J. Energy Chem. 28 (2019) 54–60.

[24] X.D. Song, M.R. Zhang, M. Yao, C. Hao, J.S. Qiu, New insights into the anchoring mechanism of polysulfides inside nanoporous covalent organic frameworks for lithium-sulfur batteries, ACS Appl. Mater. Interfaces 10 (2018) 43896–43903.

[25] D.G. Wang, N. Li, Y.M. Hu, S. Wan, M. Song, G.P. Yu, Y.H. Jin, W.F. Wei, K. Han, G.C. Kuang, W. Zhang, Highly fluoro-substituted covalent organic framework and its application in lithium-sulfur batteries, ACS Appl. Mater. Interfaces 10 (2018) 42233–42240.

[26] S. Lim, R.L. Thankamony, T. Yim, H. Chu, Y.-J. Kim, J. Mun, T.-H. Kim, Surface modification of sulfur electrodes by chemically anchored cross-linked polymer coating for lithium-sulfur batteries, ACS Appl. Mater. Interfaces 7 (2015) 1401–1405.

[27] Z.B. Cheng, H. Pan, H. Zhong, Z.B. Xiao, X.J. Li, R.H. Wang, Porous organic polymers for polysulfide trapping in lithium-sulfur batteries, Adv. Funct. Mater. 28 (2018) 1707597.

[28] T. Lei, W. Chen, Y. Hu, W. Lv, X. Lv, Y. Yan, J. Huang, Y. Jiao, J. Chu, C. Yan, C. Wu, Q. Li, W. He, J. Xiong, A nonflammable and thermotolerant separator suppresses polysulfide dissolution for safe and long-cycle lithium-sulfur batteries, Adv. Energy Mater. 8 (2018) 1802441.

[29] X. Yu, H. Wu, J.H. Koo, A. Manthiram, Tailoring the pore size of a polypropylene separator with a polymer having intrinsic nanoporosity for suppressing the polysulfide shuttle in lithium-sulfur batteries, Adv. Energy Mater. 10 (2020) 1902872.

[30] J. Wang, G. Li, D. Luo, Y. Zhang, Y. Zhao, G. Zhou, L. Shui, X. Wang, Z. Chen, Engineering the conductive network of metal oxide-based sulfur cathode toward efficient and longevous lithium-sulfur batteries, Adv. Energy Mater. 10 (2020) 2002076.

[31] X.Y. Tao, J.G. Wang, C. Liu, H.T. Wang, H.B. Yao, G.Y. Zheng, Z.W. Seh, Q.X. Cai, W.Y. Li, G.M. Zhou, C.X. Zu, Y. Cui, Balancing surface adsorption and diffusion of lithium-polysulfides on nonconductive oxides for lithium-sulfur battery design, Nat. Commun. 7 (2016) 11203.

[32] Y.T. Liu, S. Liu, G.R. Li, T.Y. Yan, X.P. Gao, High volumetric energy density sulfur cathode with heavy and catalytic metal oxide host for lithium-sulfur battery, Adv. Sci. 7 (2020) 1903693.

[33] X. Zhang, C. Shang, E.M. Akinoglu, X. Wang, G. Zhou, Constructing $Co_3S_4$ nanosheets coating N-doped carbon nanofibers as freestanding sulfur host for high-performance lithium-sulfur batteries, Adv. Sci. 7 (2020) 2002037.

[34] Q. Zhang, X.F. Zhang, M. Li, J.Q. Liu, Y.C. Wu, Sulfur-deficient $MoS_{2-x}$ promoted lithium polysulfides conversion in lithium-sulfur battery: a first-principles study, Appl. Surf. Sci. 487 (2019) 452–463.

[35] B. Qi, X. Zhao, S. Wang, K. Chen, Y. Wei, G. Chen, Y. Gao, D. Zhang, Z. Sun, F. Li, Mesoporous TiN microspheres as an efficient polysulfide barrier for lithium-sulfur batteries, J. Mater. Chem. A 6 (2018) 14359–14366.

[36] J.R. He, A. Manthiram, 3D CoSe@C aerogel as a host for dendrite-free lithium-metal snode and efficient sulfur cathode in Li-S full cells, Adv. Energy Mater. 10 (2020) 1903241.

[37] J. Li, L. Zhang, F.R. Qin, B. Hong, Q. Xiang, K. Zhang, J. Fang, Y.Q. Lai, $ZrO(NO_3)_2$ as a functional additive to suppress the diffusion of polysulfides in lithium - Sulfur batteries, J. Power Sources 442 (2019) 227232.

[38] M. Li, W. Feng, X. Wang, The dual-play of carbon nanotube embedded with CoNi N codoped porous polyhedra toward superior Lithium–Sulfur batteries, J. Alloys Compd. 853 (2021) 157194.

[39] Q. Cheng, Z.H. Yin, S.Y. Pan, G.Z. Zhang, Z.X. Pan, X.Y. Yu, Y.P. Fang, H.S. Rao, X.H. Zhong, Enhancing adsorption and reaction kinetics of polysulfides using CoP-coated N-doped mesoporous carbon for high-energy-density lithium–sulfur batteries, ACS Appl. Mater. Interfaces 12 (2020) 43844–43853.

[40] R. Amine, J.Z. Liu, I. Acznik, T. Sheng, K. Lota, H. Sun, C.J. Sun, K. Fic, X.B. Zuo, Y. Ren, D. Abd El-Hady, W. Alshitari, A.S. Al-Bogami, Z.H. Chen, K. Amine, G. L., Xu, regulating the hidden solvation-ion-exchange in concentrated electrolytes for stable and safe lithium metal batteries, Adv. Energy Mater. 10 (2020) 2000901.

[41] Y. Yamada, J. Wang, S. Ko, E. Watanabe, A. Yamada, Advances and issues in developing salt-concentrated battery electrolytes, Nat. Energy 4 (2019) 269–280.

[42] H. Lu, Y. Yuan, Q.H. Yang, Z.Z. Hou, Y.Q. Lai, Concentrated lithium salt-monoglyme complex for restraining polysulfide dissolution and improving electrochemical performance of lithium-sulfur batteries, Ionics 22 (2016) 997–1002.

[43] Y.Z. Zhang, S. Liu, G.C. Li, G.R. Li, X.P. Gao, Sulfur/polyacrylonitrile/carbon multi-composites as cathode materials for lithium/sulfur battery in the concentrated electrolyte, J. Mater. Chem. A 2 (2014) 4652–4659.

[44] F. He, X. Wu, J. Qian, Y. Cao, H. Yang, X. Ai, D. Xia, Building a cycle- stable sulphur cathode by tailoring its redox reaction into a solid-phase conversion mechanism, J. Mater. Chem. A 6 (2018) 23396–23407.

[45] Y. Yuan, G. Tan, J. Wen, J. Lu, L. Ma, C. Liu, X. Zuo, R. Shahbazian-Yassar, T. Wu, K. Amine, Encapsulating various sulfur allotropes within graphene nanocages for long-lasting lithium storage, Adv. Funct. Mater. 28 (2018) 1706443.

[46] Q. Zhao, Q. Zhu, J. Miao, Z. Guan, H. Liu, R. Chen, Y. An, F. Wu, B. Xu, Three-dimensional carbon current collector promises small sulfur molecule cathode with high areal loading for lithium-sulfur batteries, ACS Appl. Mater. Interfaces 10 (2018) 10882–10889.

[47] Q. Zhu, Q. Zhao, Y. An, B. Anasori, H. Wang, B. Xu, Ultra-microporous carbons encapsulate small sulfur molecules for high performance lithium-sulfur battery, Nano Energy 33 (2017) 402–409.

[48] S.Z. Niu, G.M. Zhou, W. Lv, H.F. Shi, C. Luo, Y.B. He, B.H. Li, Q.H. Yang, F.Y. Kang, Sulfur confined in nitrogen-doped microporous carbon used in a carbonate-based electrolyte for long-life, safe lithium-sulfur batteries, Carbon 109 (2016) 1–6.

[49] S. Xin, L. Gu, N.H. Zhao, Y.X. Yin, L.J. Zhou, Y.G. Guo, L.J. Wan, Smaller sulfur molecules promise better lithium-sulfur batteries, J. Am. Chem. Soc. 134 (2012) 18510–18513.

[50] B.-J. Chae, Y.E. Jung, C.Y. Lee, T. Yim, Metal-organic framework as a multifunctional additive for selectively trapping transition-metal components in lithium-ion batteries, ACS Sustain. Chem. Eng. 6 (2018) 8547–8553.

[51] X. Kang, I. Di Bernardo, H. Yang, J.F. Torres, L. Zhang, Metal–organic framework microdomains in 3D conductive host as polysulfide inhibitor for fast, long-cycle Li–S batteries, Appl. Surf. Sci. 535 (2021) 147680.

[52] L.F. Yang, S. Kinoshita, T. Yamada, S. Kanda, H. Kitagawa, M. Tokunaga, T. Ishimoto, T. Ogura, R. Nagumo, A. Miyamoto, M. Koyama, A metal–organic framework as an electrocatalyst for ethanol oxidation, Angew. Chem. Int. Ed. 49 (2010) 5348–5351.

[53] M. Yoon, K. Suh, S. Natarajan, K. Kim, Proton conduction in metal-organic frameworks and related modularly built porous solids, Angew. Chem. Int. Ed. 52 (2013) 2688–2700.

[54] S.H. Kazemi, B. Hosseinzadeh, H. Kazemi, M.A. Kiani, S. Hajati, Facile synthesis of mixed metal-organic frameworks: electrode materials for supercapacitors with excellent areal capacitance and operational stability, ACS Appl. Mater. Interfaces 10 (2018) 23063–23073.

[55] T. Deng, W. Zhang, O. Arcelus, D. Wang, X. Shi, X. Zhang, J. Carrasco, T. Rojo, W. Zheng, Vertically co-oriented two dimensional metal-organic frameworks for packaging enhanced supercapacitive performance, Commun. Chem. 1 (2018) 6.

[56] H. Furukawa, K.E. Cordova, M. O'Keeffe, O.M. Yaghi, The chemistry and applications of metal-organic frameworks, Science 341 (2013) 974.

[57] H. Jiang, X.C. Liu, X. Wu, Y. Shu, X. Gong, F.S. Ke, H. Deng, Metal-organic frameworks for high charge-discharge rates in lithium-sulfur batteries, Angew. Chem. Int. Ed. 57 (2018) 3916–3921.

[58] J.W. Zhou, R. Li, X.X. Fan, Y.F. Chen, R.D. Han, W. Li, J. Zheng, B. Wang, X.G. Li, Rational design of a metal-organic framework host for sulfur storage in fast, long-cycle Li-S batteries, Energ. Environ. Sci. 7 (2014) 2715–2724.

[59] Z. Liang, C. Qu, C. Wang, W. Guo, R. Zou, Q. Xu, Pristine metal-organic frameworks and their composites for energy storage and conversion, Adv. Mater. 30 (2018) 1702891.

[60] G. Cui, G. Li, D. Luo, Y. Zhang, Y. Zhao, D. Wang, J. Wang, Z. Zhang, X. Wang, Z. Chen, Three-dimensionally ordered macro-microporous metal organic frameworks with strong sulfur immobilization and catalyzation for high-performance lithium-sulfur batteries, Nano Energy 72 (2020) 104685.

[61] Q. Pang, X. Liang, C.Y. Kwok, L.F. Nazar, Advances in lithium-sulfur batteries based on multifunctional cathodes and electrolytes, Nat. Energy 1 (2016) 16132.

[62] A.E. Baumann, G.E. Aversa, A. Roy, M.L. Falk, N.M. Bedford, V.S. Thoi, Promoting sulfur adsorption using surface cu sites in metal-organic frameworks for lithium sulfur batteries, J. Mater. Chem. A 6 (2018) 4811–4821.

[63] J. Zheng, J. Tian, D. Wu, M. Gu, W. Xu, C. Wang, F. Gao, M.H. Engelhard, J.-G. Zhang, J. Liu, J. Xiao, Lewis acid-base interactions between polysulfides and metal organic framework in lithium sulfur batteries, Nano Lett. 14 (2014) 2345–2352.

[64] X.F. Liu, X.Q. Guo, R. Wang, Q.C. Liu, Z.J. Li, S.Q. Zang, T.C.W. Mak, Manganese cluster-based MOF as efficient polysulfide-trapping platform for high-performance lithium-sulfur batteries, J. Mater. Chem. A 7 (2019) 2838–2844.

[65] B. Liu, M. Taheri, J.F. Torres, Z. Fusco, T. Lu, Y. Liu, T. Tsuzuki, G. Yu, A. Tricoli, Janus conductive/insulating microporous ion-sieving membranes for stable Li-S batteries, ACS Nano 14 (2020) 13852–13864.

[66] A.E. Baumann, X. Han, M.M. Butala, V.S. Thoi, Lithium thiophosphate functionalized zirconium MOFs for Li-S batteries with enhanced rate capabilities, J. Am. Chem. Soc. 141 (2019) 17891–17899.

[67] B.-Q. Li, S.-Y. Zhang, L. Kong, H.-J. Peng, Q. Zhang, Porphyrin organic framework hollow spheres and their applications in lithium-sulfur batteries, Adv. Mater. 30 (2018) 1707483.

[68] L. Chen, L. Han, X. Liu, Y. Li, M. Wei, General synthesis of sulfonate-based metal-organic framework derived composite of MxSy@N/S-doped carbon for high-performance lithium/sodium ion batteries, Chem.Eur. J. 27 (2021) 2104–2111.

[69] L.C. Yin, J. Liang, G.M. Zhou, F. Li, R. Saito, H.M. Cheng, Understanding the interactions between lithium polysulfides and N-doped graphene using density functional theory calculations, Nano Energy 25 (2016) 203–210.

[70] T.-Z. Hou, H.-J. Peng, J.-Q. Huang, Q. Zhang, B. Li, The formation of strong-couple interactions between nitrogen-doped graphene and sulfur/lithium (poly)sulfides in lithium-sulfur batteries, 2D Mater. 2 (2015) 014011.

[71] J. Song, M.L. Gordin, T. Xu, S. Chen, Z. Yu, H. Sohn, J. Lu, Y. Ren, Y. Duan, D. Wang, Strong lithium polysulfide chemisorption on electroactive stes of nitrogen-doped carbon composites for high-performance lithium–sulfur battery cathodes, Angew. Chem. Int. Ed. 54 (2015) 4325–4329.

[72] W. Bao, L. Liu, C. Wang, S. Choi, D. Wang, G. Wang, Facile synthesis of crumpled nitrogen-doped MXene nanosheets as a new sulfur host for lithium-sulfur batteries, Adv. Energy Mater. 8 (2018) 1702485.

[73] X. Liu, P. Chen, J. Chen, Q. Zeng, Z. Wang, Z. Li, L. Zhang, A nitrogen-rich hyper-branched polymer as cathode encapsulated material for superior long-cycling lithium-sulfur batteries, Electrochim. Acta 330 (2020) 135337.

[74] X.J. Hong, T.X. Tan, Y.K. Guo, X.Y. Tang, J.Y. Wang, W. Qin, Y.P. Cai, Confinement of polysulfides within bi-functional metal-organic frameworks for high performance lithium-sulfur batteries, Nanoscale 10 (2018) 2774–2780.

[75] Z.Q. Wang, B.X. Wang, Y. Yang, Y.J. Cui, Z.Y. Wang, B.L. Chen, G.D. Qian, Mixed-metal–organic framework with effective Lewis acidic sites for sulfur confinement in high-performance lithium–sulfur batteries, ACS Appl. Mater. Interfaces 7 (2015) 20999–21004.

[76] T. Wang, G. Cui, Y. Zhao, A. Nurpeissova, Z. Bakenov, Porous carbon nanotubes microspheres decorated with strong catalyst cobalt nanoparticles as an effective sulfur host for lithium-sulfur battery, J. Alloys Compd. 853 (2021) 157268.

[77] K.S. Novoselov, V.I. Fal'ko, L. Colombo, P.R. Gellert, M.G. Schwab, K. Kim, A roadmap for graphene, Nature 490 (2012) 192–200.

[78] A.K. Geim, Graphene: status and prospects, Science 324 (2009) 1530–1534.

[79] J. Rong, M. Ge, X. Fang, C. Zhou, Solution ionic strength engineering as a generic strategy to coat graphene oxide (GO) on various functional particles and its application in high-performance lithium–sulfur (Li–S) batteries, Nano Lett. 14 (2014) 473–479.

[80] W. Bao, Z. Zhang, Y. Qu, C. Zhou, X. Wang, J. Li, Confine sulfur in mesoporous metal–organic framework @ reduced graphene oxide for lithium sulfur battery, J. Alloys Compd. 582 (2014) 334–340.

[81] Z. Zhao, S. Wang, R. Liang, Z. Li, Z. Shi, G. Chen, Graphene-wrapped chromium-MOF(MIL-101)/sulfur composite for performance improvement of high-rate rechargeable Li-S batteries, J. Mater. Chem. A 2 (2014) 13509–13512.

[82] Y.S. Wu, H.Q. Jiang, F.S. Ke, H.X. Deng, Three-dimensional hierarchical constructs of MOF-on-reduced graphene oxide for lithium-sulfur batteries, Chem. Asian J. 14 (2019) 3577–3582.

[83] J. Ren, Z. Song, X. Zhou, Y. Chai, X. Lu, Q. Zheng, C. Xu, D. Lin, A porous carbon polyhedron/carbon nanotube based hybrid material as multifunctional sulfur host for high-performance lithium-sulfur batteries, ChemElectroChem 6 (2019) 3410–3419.

[84] S.H. Park, P.J. King, R. Tian, C.S. Boland, J. Coelho, C. Zhang, P. McBean, N. McEvoy, M.P. Kremer, D. Daly, J.N. Coleman, V. Nicolosi, High areal capacity

battery electrodes enabled by segregated nanotube networks, Nat. Energy 4 (2019) 560–567.

[85] H. Zhang, W. Zhao, Y. Wu, Y. Wang, M. Zou, A. Cao, Dense monolithic MOF and carbon nanotube hybrid with enhanced volumetric and areal capacities for lithium-sulfur battery, J. Mater. Chem. A 7 (2019) 9195–9201.

[86] G. Xu, Y. Zuo, B. Huang, Metal-organic framework-74-Ni/carbon nanotube composite as sulfur host for high performance lithium-sulfur batteries, J. Electroanal. Chem. 830 (2018) 43–49.

[87] G. Gao, W. Feng, W. Su, S. Wang, L. Chen, M. Li, C. Song, Preparation and modification of MIL-101(Cr) metal organic framework and its application in lithium-sulfur batteries, Int. J. Electrochem. Sci. 15 (2020) 1426–1436.

[88] Y. Mao, G. Li, Y. Guo, Z. Li, C. Liang, X. Peng, Z. Lin, Foldable interpenetrated metal-organic frameworks/carbon nanotubes thin film for lithium-sulfur batteries, Nat. Commun. 8 (2017) 14628.

[89] Y. Pu, W. Wu, J. Liu, T. Liu, F. Ding, J. Zhang, Z. Tang, A defective MOF architecture threaded by interlaced carbon nanotubes for high-cycling lithium-sulfur batteries, RSC Adv. 8 (2018) 18604–18612.

[90] W. Li, G. Zheng, Y. Yang, Z.W. Seh, N. Liu, Y. Cui, High-performance hollow sulfur nanostructured battery cathode through a scalable, room temperature, one-step, bottom-up approach, Proc. Natl Acad. Sci. USA 110 (2013) 7148.

[91] W. Li, Q. Zhang, G. Zheng, Z.W. Seh, H. Yao, Y. Cui, Understanding the role of different conductive polymers in improving the nanostructured sulfur cathode performance, Nano Lett. 13 (2013) 5534.

[92] X. Hong, R. Wang, Y. Liu, J. Fu, J. Liang, S. Dou, Recent advances in chemical adsorption and catalytic conversion materials for Li-S batteries, J. Energy Chem. 42 (2020) 144–168.

[93] J. Fanous, M. Wegner, M.B.M. Spera, M.R. Buchmeiser, High energy density poly (acrylonitrile)-sulfur composite-based lithium-sulfur batteries, J. Electrochem. Soc. 160 (2013) A1169–A1170.

[94] T. Zhu, J.E. Mueller, M. Hanauer, U. Sauter, T. Jacob, Structural motifs for modeling sulfur-poly(acrylonitrile) composite materials in sulfur-lithium batteries, ChemElectroChem 4 (2017) 2494–2499.

[95] H. Hu, H. Cheng, Z. Liu, G. Li, Q. Zhu, Y. Yu, In situ polymerized PAN-assisted S/C nanosphere with enhanced high-power performance as cathode for lithium/sulfur batteries, Nano Lett. 15 (2015) 5116–5123.

[96] J. Wu, S. Li, P. Yang, H. Zhang, C. Du, J. Xu, K. Song, S@$TiO_2$ nanospheres loaded on PPy matrix for enhanced lithium-sulfur batteries, J. Alloys Compd. 783 (2019) 279–285.

[97] F. Yin, J. Ren, Y. Zhang, T. Tan, Z. Chen, A PPy/ZnO functional interlayer to enhance electrochemical performance of lithium/sulfur batteries, Nanoscale Res. Lett. 13 (2018) 307.

[98] P. Geng, S. Cao, X. Guo, J. Ding, S. Zhang, M. Zheng, H. Pang, Polypyrrole coated hollow metal–organic framework composites for lithium–sulfur batteries, J. Mater. Chem. A 7 (2019) 19465–19470.

[99] W.W. Jin, H.J. Li, J.Z. Zou, S.Z. Zeng, Q.D. Li, G.Z. Xu, H.C. Sheng, B.B. Wang, Y.H. Si, L. Yu, X.R. Zeng, Conducting polymer-coated MIL-101/S composite with scale-like shell structure for improving Li-S batteries, RSC Adv. 8 (2018) 4786–4793.

[100] Z. Zhao, S. Das, G. Xing, P. Fayon, P. Heasman, M. Jay, S. Bailey, C. Lambert, H. Yamada, T. Wakihara, A. Trewin, T. Ben, S. Qiu, V. Valtchev, A 3D organically synthesized porous carbon material for lithium-ion batteries, Angew. Chem. Int. Ed. 57 (2018) 11952–11956.

[101] Y. Xu, Y. Zhu, Y. Liu, C. Wang, Electrochemical performance of porous carbon/tin composite anodes for sodium-ion and lithium-ion batteries, Adv. Energy Mater. 3 (2013) 128–133.

[102] J. Zhao, Y. Jiang, H. Fan, M. Liu, O. Zhuo, X. Wang, Q. Wu, L. Yang, Y. Ma, Z. Hu, Porous 3D few-layer graphene-like carbon for ultrahigh-power supercapacitors with well-defined structure-performance relationship, Adv. Mater. 29 (2017) 1604569.

[103] L. Liu, G. Zeng, J. Chen, L. Bi, L. Dai, Z. Wen, N-doped porous carbon nanosheets as pH-universal ORR electrocatalyst in various fuel cell devices, Nano Energy 49 (2018) 393–402.

[104] B.Y. Xia, Y. Yan, N. Li, H.B. Wu, X.W.D. Lou, X. Wang, A metal-organic framework-derived bifunctional oxygen electrocatalyst, Nat. Energy 1 (2016) 15006.

[105] Z. Liang, C. Qu, D. Xia, R. Zou, Q. Xu, Atomically dispersed metal sites in MOF-based baterials for electrocatalytic and photocatalytic energy conversion, Angew. Chem. Int. Ed. 57 (2018) 9604–9633.

[106] X. Xu, J. Liu, J. Liu, L. Ouyang, R. Hu, H. Wang, L. Yang, M. Zhu, A general metal-organic framework (MOF)-derived selenidation strategy for in situ carbon-encapsulated metal selenides as high-rate anodes for Na-ion batteries, Adv. Funct. Mater. 28 (2018) 1707573.

[107] F. Zou, X. Hu, Z. Li, L. Qie, C. Hu, R. Zeng, Y. Jiang, Y. Huang, MOF-derived porous $ZnO/ZnFe_2O_4/C$ octahedra with hollow interiors for high-rate lithium-ion batteries, Adv. Mater. 26 (2014) 6622–6628.

[108] F. Zou, K. Liu, C.F. Cheng, Y. Ji, Y. Zhu, Metal-organic frameworks (MOFs) derived carbon-coated NiS nanoparticles anchored on graphene layers for high-performance Li-S cathode material, Nanotechnology 31 (2020) 485404.

[109] X. Wei, Y. Li, H. Peng, D. Gao, Y. Ou, Y. Yang, J. Hu, Y. Zhang, P. Xiao, A novel functional material of $Co_3O_4/Fe_2O_3$ nanocubes derived from a MOF precursor for high-performance electrochemical energy storage and conversion application, Chem. Eng. J. 355 (2019) 336–340.

[110] K. Xi, S. Cao, X. Peng, C. Ducati, R.V. Kumar, A.K. Cheetham, Carbon with hierarchical pores from carbonized metal-organic frameworks for lithium Sulphur batteries, Chem. Commun. 49 (2013) 2192–2194.

[111] G.Y. Xu, B. Ding, L.F. Shen, P. Nie, J.P. Han, X.G. Zhang, Sulfur embedded in metal organic framework-derived hierarchically porous carbon nanoplates for high performance lithium-sulfur battery, J. Mater. Chem. A 1 (2013) 4490.

[112] J. Song, T. Xu, M.L. Gordin, P. Zhu, D. Lv, Y.-B. Jiang, Y. Chen, Y. Duan, D. Wang, Nitrogen-doped mesoporous carbon promoted chemical adsorption of sulfur and fabrication of high- areal- capacity sulfur cathode with exceptional cycling stability for lithium- sulfur batteries, Adv. Funct. Mater. 24 (2014) 1243–1250.

[113] Z. Wang, Y. Dong, H. Li, Z. Zhao, H.B. Wu, C. Hao, S. Liu, J. Qiu, X.W. Lou, Enhancing lithium-Sulphur battery performance by strongly binding the discharge products on amino-functionalized reduced graphene oxide, Nat. Commun. 5 (2014) 5002.

[114] K. Chen, Z. Sun, R. Fang, Y. Shi, H.M. Cheng, F. Li, Metal-organic frameworks (MOFs)-derived nitrogen-doped porous carbon anchored on graphene with multifunctional effects for lithium-sulfur batteries, Adv. Funct. Mater. 28 (2018) 1707592.

[115] M.D. Walle, M. Zhang, K. Zeng, Y. Li, Y.-N. Liu, MOFs-derived nitrogen-doped carbon interwoven with carbon nanotubes for high sulfur content lithium-sulfur batteries, Appl. Surf. Sci. 497 (2019) 143773.

[116] R. Liu, Z. Liu, W. Liu, Y. Liu, X. Lin, Y. Li, P. Li, Z. Huang, X. Feng, L. Yu, D. Wang, Y. Ma, W. Huang, $TiO_2$ and co nanoparticle-decorated carbon polyhedra as efficient sulfur host for high-performance lithium-sulfur batteries, Small 15 (2019) 1804533.

[117] H. Zhang, J. Ma, M. Huang, H. Shang, J. Jiang, Y. Qiao, W. Liu, H. Zhou, T. Li, X. Zhou, G. Peng, M. Ye, M. Qu, MOF-derived $Co_9S_8$/C hollow polyhedra grown on 3D graphene aerogel as efficient polysulfide mediator for long-life Li-S batteries, Mater. Lett. 277 (2020) 128331.

[118] G.L. Chen, Y.J. Li, W.T. Zhong, F.H. Zheng, J. Hu, X.H. Ji, W.Z. Liu, C.H. Yang, Z. Lin, M.L. Liu, MOFs-derived porous $Mo_2C$-C nano-octahedrons enable high-performance lithium-sulfur batteries, Energy Storage Mater. 25 (2020) 547–554.

[119] H. Li, Y. Ma, H. Zhang, T. Diemant, R.J. Behm, A. Varzi, S. Passerini, Metal-organic framework derived $Fe_7S_8$ nanoparticles embedded in heteroatom-doped carbon with lithium and sodium storage capability, Small Methods 4 (2020) 2000637.

[120] W. Lei, X. Wang, Y. Zhang, Z. Luo, P. Xia, Y. Zou, Z. Ma, Y. Pan, S. Lin, Facile synthesis of $Fe_3C$ nano-particles/porous biochar cathode materials for lithium sulfur battery, J. Alloys Compd. 853 (2021) 157024.

[121] R. Wang, J. Yang, X. Chen, Y. Zhao, W. Zhao, G. Qian, S. Li, Y. Xiao, H. Chen, Y. Ye, G. Zhou, F. Pan, Highly dispersed cobalt clusters in nitrogen-doped porous carbon enable multiple effects for high-performance Li-S battery, Adv. Energy Mater. 10 (2020) 1903550.

[122] F. Zhou, Z. Qiao, Y. Zhang, W. Xu, H. Zheng, Q. Xie, Q. Luo, L. Wang, B. Qu, D.L. Peng, Bimetallic MOF-derived CNTs-grafted carbon nanocages as sulfur host for high-performance lithium-sulfur batteries, Electrochim. Acta 349 (2020) 136378.

[123] P. Zuo, H. Zhang, M. He, Q. Li, Y. Ma, C. Du, X. Cheng, H. Huo, G. Gao, G. Yin, Clew-like N-doped multiwalled carbon nanotube aggregates derived from metal-organic complexes for lithium-sulfur batteries, Carbon 122 (2017) 635–642.

[124] W. Qiu, J. Li, Y. Zhang, G. Kalimuldina, Z. Bakenov, Carbon nanotubes assembled on porous TiO2 matrix doped with $Co_3O_4$ as sulfur host for lithium-sulfur batteries, Nanotechnology 32 (2021) 075403.

[125] L.L. Gu, J. Gao, C. Wang, S.Y. Qiu, K.X. Wang, X.T. Gao, K.N. Sun, P.J. Zuo, X.D. Zhu, Thin-carbon-layer-enveloped cobalt-iron oxide nanocages as a high-efficiency sulfur container for Li-S batteries, J. Mater. Chem. A 8 (2020) 20604–20611.

[126] Z. Li, MnO2-graphene nanosheets wrapped mesoporous carbon/sulfur composite for lithium-sulfur batteries, R. Soc. Open Sci. 5 (2018) 171824.

[127] K. Chen, X. Kong, X. Xie, J. Chen, X. Cao, S. Liang, A. Pan, Sulfur-doped carbon-wrapped heterogeneous $Fe_3O_4$/$Fe_7S_8$/C nanoplates as stable anode for lithium-ion batteries, Batteries Supercaps 3 (2020) 344–353.

[128] X. Liang, C. Hart, Q. Pang, A. Garsuch, T. Weiss, L.F. Nazar, A highly efficient polysulfide mediator for lithium-sulfur batteries, Nat. Commun. 6 (2015) 5682.

[129] J. Yang, B. Wang, F. Jin, Y. Ning, H. Luo, J. Zhang, F. Wang, D. Wang, Y. Zhou, A MIL-47(V) derived hierarchical lasagna-structured $V_2O_3$@C hollow microcuboid as an efficient sulfur host for high-performance lithium-sulfur batteries, Nanoscale 12 (2020) 4552–4561.

[130] G. Liu, K. Feng, H. Cui, J. Li, Y. Liu, M. Wang, MOF derived in-situ carbon-encapsulated $Fe_3O_4$@C to mediate polysulfides redox for ultrastable lithium-sulfur batteries, Chem. Eng. J. 381 (2020) 122652.

[131] Y. Yan, L. Wei, X. Su, S. Deng, J. Feng, J. Yang, M. Chi, H. Lei, Z. Li, M. Wu, The crystallinity of metal oxide in carbonized metal organic frameworks and the effect on restricting polysulfides, ChemNanoMat 6 (2020) 274–279.

[132] G. Zhou, H. Tian, Y. Jin, X. Tao, B. Liu, R. Zhang, Z.W. Seh, D. Zhuo, Y. Liu, J. Sun, J. Zhao, C. Zu, D.S. Wu, Q. Zhang, Y. Cui, Catalytic oxidation of $Li_2S$ on the surface of metal sulfides for Li-S batteries, Proc. Natl Acad. Sci. USA 114 (2017) 840–845.

[133] S. Luo, C. Zheng, W. Sun, Y. Wang, J. Ke, Q. Guo, S. Liu, X. Hong, Y. Li, W. Xie, Multi-functional $CoS_2$-N-C porous carbon composite derived from metal-organic frameworks for high performance lithium-sulfur batteries, Electrochim. Acta 289 (2018) 94–103.

[134] J. He, Y. Chen, A. Manthiram, MOF-derived cobalt sulfide grown on 3D graphene foam as an efficient sulfur host for long-life lithium-sulfur batteries, iScience 4 (2018) 36–43.

[135] H.B. Wu, B.Y. Xia, L. Yu, X.Y. Yu, X.W. Lou, Porous molybdenum carbide nano-octahedrons synthesized via confined carburization in metal-organic frameworks for efficient hydrogen production, Nat. Commun. 6 (2015) 6512.

[136] M.R. Lukatskaya, O. Mashtalir, C.E. Ren, Y. Dall'Agnese, P. Rozier, P.L. Taberna, M. Naguib, P. Simon, M.W. Barsoum, Y. Gogotsi, Cation intercalation and high volumetric capacitance of two-dimensional titanium carbide, Science 341 (2013) 1502–1505.

[137] Z. Yu, Y. Bai, S. Zhang, Y. Liu, N. Zhang, G. Wang, J. Wei, Q. Wu, K. Sun, Metal-organic framework-derived $Co_3ZnC/Co$ embedded in nitrogen-doped carbon nanotube-grafted carbon polyhedra as a high-performance electrocatalyst for water splitting, ACS Appl. Mater. Interfaces 10 (2018) 6245–6252.

[138] X. Li, L. Yang, T. Su, X. Wang, C. Sun, Z. Su, Graphene-coated hybrid electrocatalysts derived from bimetallic metal-organic frameworks for efficient hydrogen generation, J. Mater. Chem. A 5 (2017) 5000–5006.

[139] X. Xia, Carbon nitride nanosheets as efficient bifunctional host materials for high-performance lithium-sulfur batteries, Mater. Res. Express 6 (2019) 105519.

[140] Z. Sun, J. Zhang, L. Yin, G. Hu, R. Fang, H.-M. Cheng, F. Li, Conductive porous vanadium nitride/graphene composite as chemical anchor of polysulfides for lithium-sulfur batteries, Nat. Commun. 8 (2017) 14627.

[141] N. Mosavati, S.O. Salley, K.Y.S. Ng, Characterization and electrochemical activities of nanostructured transition metal nitrides as cathode materials for lithium sulfur batteries, J. Power Sources 340 (2017) 210–216.

[142] Y. Lu, Y. Wang, W. Wang, Y. Guo, Y. Zhang, R. Luo, X. Liu, T. Peng, Uniform honeycomb-like microspheres constructed with titanium nitride to confine polysulfides for extremely stable lithium-sulfur batteries, J. Phys. D Appl. Phys. 52 (2019) 025502.

[143] K.K. Xiao, J. Wang, Z. Chen, Y.H. Qian, Z. Liu, L.L. Zhang, X.H. Chen, J.L. Liu, X.F. Fan, Z.X. Shen, Improving polysulfides adsorption and redox kinetics by the $Co_4N$ nanoparticle/N-doped carbon composites for lithium-sulfur batteries, Small 15 (2019) 1901454.

[144] Z.X. Sun, S. Vijay, H.H. Heenen, A.Y.S. Eng, W.G. Tu, Y.X. Zhao, S.W. Koh, P.Q. Gao, Z.W. Seh, K.R. Chan, H. Li, Catalytic polysulfide conversion and physiochemical confinement for lithium-sulfur batteries, Adv. Energy Mater. 10 (2020) 1904010.

[145] M. Rana, M. Li, Q. He, B. Luo, L. Wang, I. Gentle, R. Knibbe, Separator coatings as efficient physical and chemical hosts of polysulfides for high-sulfur-loaded rechargeable lithium-sulfur batteries, J. Energy Chem. 44 (2020) 51–60.

[146] S.H. Kim, J.S. Yeon, R. Kim, K.M. Choi, H.S. Park, A functional separator coated with sulfonated metal- organic framework/Nafion hybrids for Li- S batteries, J. Mater. Chem. A 6 (2018) 24971–24978.

[147] Y. He, S. Wu, Q. Li, H. Zhou, Designing a multifunctional separator for high-performance Li-S batteries at elevated temperature, Small 15 (2019) 1904332.

[148] Y. He, Y. Qiao, Z. Chang, X. Cao, M. Jia, P. He, H. Zhou, Developing a "polysulfide-phobic" strategy to restrain shuttle effect in lithium-sulfur batteries, Angew. Chem. Int. Ed. 58 (2019) 11774–11778.

[149] X.J. Hong, C.L. Song, Y. Yang, H.C. Tan, G.H. Li, Y.P. Cai, H. Wang, Cerium based metal-organic frameworks as an efficient separator coating catalyzing the conversion of polysulfides for high performance lithium-sulfur batteries, ACS Nano 13 (2019) 1923–1931.

[150] M. Vijayakumar, N. Govind, E. Walter, S.D. Burton, A. Shukla, A. Devaraj, J. Xiao, J. Liu, C. Wang, A. Karim, S. Thevuthasan, Molecular structure and stability of dissolved lithium polysulfide species, Phys. Chem. Chem. Phys. 16 (2014) 10923–10932.

[151] Z. Chang, Y. Qiao, H. Yang, H. Deng, X. Zhu, P. He, H. Zhou, A stable high-voltage lithium-ion battery realized by an in-built water scavenger, Energ. Environ. Sci. 13 (2020) 4122–4131.

[152] S. Bai, X. Liu, K. Zhu, S. Wu, H. Zhou, Metal-organic framework-based separator for lithium-sulfur batteries, Nat. Energy 1 (2016) 16094.

[153] Z. Chang, Y. Qiao, J. Wang, H. Deng, P. He, H. Zhou, Fabricating better metal-organic frameworks separators for Li-S batteries: pore sizes effects inspired channel modification strategy, Energy Storage Mater. 25 (2020) 164–171.

[154] J. Han, S. Gao, R. Wang, K. Wang, M. Jiang, J. Yan, Q. Jin, K. Jiang, Investigation of the mechanism of metal-organic frameworks preventing polysulfide shuttling from the perspective of composition and structure, J. Mater. Chem. A 8 (2020) 6661–6669.

[155] M. Li, Y. Wan, J.-K. Huang, A.H. Assen, C.-E. Hsiung, H. Jiang, Y. Han, M. Eddaoudi, Z. Lai, J. Ming, L.-J. Li, Metal-organic framework-based separators for enhancing Li-S battery stability: mechanism of mitigating polysulfide diffusion, ACS Energy Lett. 2 (2017) 2362–2367.

[156] B.M. Wiers, M.L. Foo, A solid lithium electrolyte via addition of lithium isopropoxide to a metal-organic framework with open metal sites, J. Am. Chem. Soc. 133 (2011) 14522–14525.

[157] D.D. Han, Z.Y. Wang, G.L. Pan, X.P. Gao, Metal-organic-framework-based gel polymer electrolyte with immobilized anions to stabilize a lithium anode for a quasi-solid-state lithium-sulfur battery, ACS Appl. Mater. Interfaces 11 (2019) 18427–18435.

[158] P. Chiochan, X. Yu, M. Sawangphruk, A. Manthiram, A metal organic framework derived solid electrolyte for lithium-sulfur batteries, Adv. Energy Mater. 10 (2020) 2001285.

[159] Z. Wang, R. Tan, H. Wang, L. Yang, J. Hu, H. Chen, F. Pan, A metal-organic-framework-based electrolyte with nanowetted interfaces for high-energy-density solid-state lithium battery, Adv. Mater. 30 (2018) 1704436.

[160] H. Huo, B. Wu, T. Zhang, X. Zheng, L. Ge, T. Xu, X. Guo, X. Sun, Anion-immobilized polymer electrolyte achieved by cationic metal-organic framework filler for dendrite-free solid-state batteries, Energy Storage Mater. 18 (2019) 59–67.

[161] S. Suriyakumar, S. Gopi, M. Kathiresan, S. Bose, E.B. Gowd, J.R. Nair, N. Angulakshmi, G. Meligrana, F. Bella, C. Gerbaldi, A.M. Stephan, Metal organic framework laden poly(ethylene oxide) based composite electrolytes for all-solid-state Li-S and Li-metal polymer batteries, Electrochim. Acta 285 (2018) 355–364.

[162] F. Liu, Q. Xiao, H.B. Wu, L. Shen, D. Xu, M. Cai, Y. Lu, Fabrication of hybrid silicate coatings by a simple vapor deposition method for lithium metal anodes, Adv. Energy Mater. 8 (2018) 1701744.

[163] F. Zhou, Z. Li, Y.-Y. Lu, B. Shen, Y. Guan, X.-X. Wang, Y.-C. Yin, B.-S. Zhu, L.-L. Lu, Y. Ni, Y. Cui, H.-B. Yao, S.-H. Yu, Diatomite derived hierarchical hybrid anode for high performance all-solid-state lithium metal batteries, Nat. Commun. 10 (2019) 2482.

[164] S.J. Zhang, Z.G. Gao, W.W. Wang, Y.Q. Lu, Y.P. Deng, J.H. You, J.T. Li, Y. Zhou, L. Huang, X.D. Zhou, S.G. Sun, A natural biopolymer film as a robust protective layer to effectively stabilize lithium-metal anodes, Small 14 (2018) 1801054.

[165] R. Zhang, N.W. Li, X.B. Cheng, Y.X. Yin, Q. Zhang, Y.G. Guo, Advanced micro/nanostructures for lithium metal anodes, Adv. Sci. 4 (2017) 1600445.

[166] R. Du, Y. Jie, Y. Chen, F. Huang, W. Cai, Y. Liu, X. Li, S. Wang, Z. Lei, R. Cao, G. Zhang, S. Jiao, Modulating lithium nucleation behavior through ultrathin interfacial layer for superior lithium metal batteries, ACS Appl. Energy Mater. 3 (2020) 6692–6699.

[167] Z.J. Zheng, Q. Su, Q. Zhang, X.C. Hu, Y.X. Yin, R. Wen, H. Ye, Z.B. Wang, Y.G. Guo, Low volume change composite lithium metal anodes, Nano Energy 64 (2019), 103910.

[168] J. Qian, Y. Li, M. Zhang, R. Luo, F. Wang, Y. Ye, Y. Xing, W. Li, W. Qu, L. Wang, L. Li, Y. Li, F. Wu, R. Chen, Protecting lithium/sodium metal anode with metal-organic framework based compact and robust shield, Nano Energy 60 (2019) 866–874.

[169] L. Fan, Z. Guo, Y. Zhang, X. Wu, C. Zhao, X. Sun, G. Yang, Y. Feng, N. Zhang, Stable artificial solid electrolyte interphase films for lithium metal anode via metal-organic frameworks cemented by polyvinyl alcohol, J. Mater. Chem. A 8 (2020) 251–258.

[170] Z. Jiang, T. Liu, L. Yan, J. Liu, F. Dong, M. Ling, C. Liang, Z. Lin, Metal-organic framework nanosheets-guided uniform lithium deposition for metallic lithium batteries, Energy Storage Mater. 11 (2018) 267–273.

[171] S. Yuan, J.L. Bao, C. Li, Y. Xia, D.G. Truhlar, Y. Wang, Dual Lithiophilic structure for uniform Li deposition, ACS Appl. Mater. Interfaces 11 (2019) 10616–10623.

[172] M. Zhu, J. Zhang, Y. Ma, Y. Nan, S. Li, Guiding confined deposition of lithium through the conductivity changing interface within a hierarchical heterostructure toward dendrite-free lithium anodes, Carbon 168 (2020) 633–639.

[173] L. Zhang, T. Qian, X. Zhu, Z. Hu, M. Wang, L. Zhang, T. Jiang, J.-H. Tian, C. Yan, In situ optical spectroscopy characterization for optimal design of lithium-sulfur batteries, Chem. Soc. Rev. 48 (2019) 5432–5453.

[174] W. Zuo, M. Luo, X. Liu, J. Wu, H. Liu, J. Li, M. Winter, R. Fu, W. Yang, Y. Yang, Li-rich cathodes for rechargeable Li-based batteries: reaction mechanisms and advanced characterization techniques, Energ. Environ. Sci. 13 (2020) 4450–4497.

[175] N.A. Canas, S. Wolf, N. Wagner, K.A. Friedrich, In-situ X-ray diffraction studies of lithium-sulfur batteries, J. Power Sources 226 (2013) 313–319.

[176] X.-C. Liu, Y. Yang, J. Wu, M. Liu, S.P. Zhou, B.D.A. Levi, X.-D. Zhou, H. Cong, D.A. Muller, P.M. Ajayan, H.D. Abruna, F.-S. Ke, Dynamic hosts for high-performance Li-S batteries studied by cryogenic transmission electron microscopy and in situ X-ray diffraction, ACS Energy Lett. 3 (2018) 1325–1330.

[177] J. Conder, R. Bouchet, S. Trabesinger, C. Marino, L. Gubler, C. Villevieille, Direct observation of lithium polysulfides in lithium-sulfur batteries using operando X-ray diffraction, Nat. Energy 2 (2017) 69.

[178] M. Ding, S. Huang, Y. Wang, J. Hu, M.E. Pam, S. Fan, Y. Shi, Q. Ge, H.Y. Yang, Promoting polysulfide conversion by catalytic ternary $Fe_3O_4$/carbon/graphene composites with ordered microchannels for ultrahigh-rate lithium-sulfur batteries, J. Mater. Chem. A 7 (2019) 25078–25087.

[179] M. Cuisinier, P.-E. Cabelguen, S. Evers, G. He, M. Kolbeck, A. Garsuch, T. Bolin, M. Balasubramanian, L.F. Nazar, Sulfur speciation in Li-S batteries determined by operando X-ray absorption spectroscopy, J. Phys. Chem. Lett. 4 (2013) 3227–3232.

[180] J. Nelson, S. Misra, Y. Yang, A. Jackson, Y. Liu, H. Wang, H. Dai, J.C. Andrews, Y. Cui, M.F. Toney, In operando X-ray diffraction and transmission X-ray microscopy of lithium sulfur batteries, J. Am. Chem. Soc. 134 (2012) 6337–6343.

[181] S.-H. Yu, X. Huang, K. Schwarz, R. Huang, T.A. Arias, J.D. Brock, H.D. Abruna, Direct visualization of sulfur cathodes: new insights into Li-S batteries via operando X-ray based methods, Energ. Environ. Sci. 11 (2018) 202–210.

[182] M. Ling, L. Zhang, T. Zheng, J. Feng, J. Guo, L. Mai, G. Liu, Nucleophilic substitution between polysulfides and binders unexpectedly stabilizing lithium sulfur battery, Nano Energy 38 (2017) 82–90.

[183] L. Zhang, M. Ling, J. Feng, G. Liu, J. Guo, Effective electrostatic confinement of polysulfides in lithium/sulfur batteries by a functional binder, Nano Energy 40 (2017) 559–565.

[184] L. Zhang, M. Ling, J. Feng, L. Mai, G. Liu, J. Guo, The synergetic interaction between $LiNO_3$ and lithium polysulfides for suppressing shuttle effect of lithium-sulfur batteries, Energy Storage Mater. 11 (2018) 24–29.

[185] K. Sun, C. Zhao, C.-H. Lin, E. Stavitski, G.J. Williams, J. Bai, E. Dooryhee, K. Attenkofer, J. Thieme, Y.-C.K. Chen-Wiegart, H. Gan, Operando multi-modal synchrotron investigation for structural and chemical evolution of cupric sulfide (CuS) additive in Li-S battery, Sci. Rep. 7 12976 (2017).

[186] M.U.M. Patel, R. Demir-Cakan, M. Morcrette, J.-M. Tarascon, M. Gaberscek, R. Dominko, Li-S battery analyzed by UV/Vis in operando mode, ChemSusChem 6 (2013) 1177–1181.

[187] J. Liu, T. Qian, N. Xu, M. Wang, J. Zhou, X. Shen, C. Yan, Dendrite-free and ultrahigh energy lithium sulfur battery enabled by dimethyl polysulfide intermediates, Energy Storage Mater. 24 (2020) 265–271.

[188] H.-L. Wu, M. Shin, Y.-M. Liu, K.A. See, A.A. Gewirth, Thiol-based electrolyte additives for high-performance lithium-sulfur batteries, Nano Energy 32 (2017) 50–58.

[189] N. Saqib, G.M. Ohlhausen, J.M. Porter, In operando infrared spectroscopy of lithium polysulfides using a novel spectro-electrochemical cell, J. Power Sources 364 (2017) 266–271.

[190] C. Dillard, A. Singh, V. Kalra, Polysulfide speciation and electrolyte interactions in lithium-sulfur batteries with in situ infrared spectroelectrochemistry, J. Phys. Chem. C 122 (2018) 18195–18203.

[191] X. Ding, S. Yang, S. Zhou, Y. Zhan, Y. Lai, X. Zhou, X. Xu, H. Nie, S. Huang, Z. Yang, Biomimetic molecule catalysts to promote the conversion of polysulfides for advanced lithium-sulfur batteries, Adv. Funct. Mater. 30 (2020) 2003354.

[192] Z.Q. Tian, B. Ren, Adsorption and reaction at electrochemical interfaces as probed by surface-enhanced Raman spectroscopy, Annu. Rev. Phys. Chem. 55 (2004) 197–229.

[193] J. Li, C. Chen, Y. Chen, Z. Li, W. Xie, X. Zhang, M. Shao, M. Wei, Polysulfide confinement and highly efficient conversion on hierarchical mesoporous carbon nanosheets for Li-S batteries, Adv. Energy Mater. 9 (2019) 1901935.

[194] Y. Hu, W. Chen, T. Lei, Y. Jiao, H. Wang, X. Wang, G. Rao, X. Wang, B. Chen, J. Xiong, Graphene quantum dots as the nucleation sites and interfacial regulator to suppress lithium dendrites for high-loading lithium-sulfur battery, Nano Energy 68 (2020) 104373.

[195] C.-W. Kung, S. Goswami, I. Hod, T.C. Wang, J. Duan, O.K. Farha, J.T. Hupp, Charge transport in zirconium-based metal-organic frameworks, Acc. Chem. Res. 53 (2020) 1187–1195.

[196] C. Zhao, G.-L. Xu, T. Zhao, K. Amine, Beyond the polysulfide shuttle and lithium dendrite formation: addressing the sluggish sulfur redox kinetics for practical high-energy Li-S batteries, Angew. Chem. Int. Ed. 59 (2020) 17634–17640.

[197] H. Wang, X. Cao, W. Liu, X. Sun, Research progress of the solid state lithium-sulfur batteries, Front. Energy Res. 7 (2019) 112.

**PART II**

# Modeling and characterization

# CHAPTER 4

# Multiscale modeling of physicochemical interactions in lithium-sulfur battery electrodes

**Partha P. Mukherjee[a], Zhixiao Liu[b], Feng Hao[c], and Bairav S. Vishnugopi[a]**

[a]School of Mechanical Engineering, Purdue University, West Lafayette, IN, United States
[b]College of Materials Science and Engineering, Hunan University, Changsa, Hunan, China
[c]Department of Engineering Mechanics, Shandong University, Jinan, Shandong, China

## Contents

| | |
|---|---|
| 4.1 Introduction | 123 |
| 4.2 The growth of crystalline Li$_2$S film in cathode | 125 |
|    4.2.1 Exposed surface of solid Li$_2$S film | 125 |
|    4.2.2 Atomistic insights into the growth process | 127 |
|    4.2.3 Formation of Li$_2$S/graphite interface | 129 |
|    4.2.4 Interfacial model for Li$_2$S film growth | 134 |
| 4.3 Parasitic reactions in anode | 144 |
|    4.3.1 Passivation of metallic Li anode | 144 |
|    4.3.2 Mesoscale model for analyzing self-discharge | 147 |
| 4.4 Summary and outlook | 152 |
| Acknowledgment | 154 |
| References | 154 |

## 4.1 Introduction

Lithium–ion batteries (LIBs) are widely used in portable devices and vehicular transportation [1–4]. The development of electric vehicles (EVs) requires energy storage systems with higher specific energy density and lower cost. Going beyond LIBs based on intercalation mechanism, the lithium-sulfur (Li—S) battery based on conversion chemistry is a promising energy storage device due to a high-energy density of $2567\,\mathrm{Wh\,kg^{-1}}$. In addition, the abundance of sulfur resources on the earth can decrease the

---

*Lithium-Sulfur Batteries*
https://doi.org/10.1016/B978-0-12-819676-2.00001-3

Copyright © 2022 Elsevier Inc.
All rights reserved.

material cost, which can potentially make the price of Li—S battery competitive in the energy-storage market [5,6].

However, the commercialization of Li—S batteries still has a far way to go. A key challenge for Li—S battery is the internal shuttle effect [7]. During the discharge, solid sulfur is dissolved into the electrolyte in the form of $S_8$ molecule, and then $S_8$ is gradually reduced to insoluble $Li_2S$ with soluble polysulfides (PSs) as intermediate discharge products. When operating the cell, PSs can diffuse to the anode side due to the potential and concentration gradients, which is named as "shuttle effect." The metallic Li anode is very reactive to PSs, leading to form an insulating $Li_2S$ on the anode surface [8]. Therefore, the shuttle effect reduces the utilization of active materials and leads to an irreversible capacity loss, poor cycling stability, and high self-discharge rate [9]. Parasitic reactions between PSs and Li anode also occur during cell resting. As the resting time increases, it can be found to decrease the open–circuit voltage (OCV) and the discharge capacity resulted from the self-discharge [10]. The surface decoration is a promising method to trap PSs. It has been found that the heteroatoms and functional groups on the carbon surface can significantly enhance the affinity to PSs [7,11–29]. Researchers are also interested in finding new materials to replace the carbon-based cathode. It is found that transition metal dichalcogenides [30–32] and MXene [33] as cathode host materials can reduce the capacity loss. Understanding the interaction mechanisms between PSs and substrates is essential for designing novel materials, and the density functional theory (DFT) calculation would be the probe to decrypt the gene code of cathode materials [34,35].

The performance of the Li—S battery is not only affected by the shuttle effect. Another issue is the insulating nature of the solid discharge product, $Li_2S$ [36]. The theoretical indirect bandgap of $Li_2S$ is 3.297 eV [37], and its electronic resistivity is larger than $10^{14}$ cm$\Omega$. In addition, the growth of the insulating $Li_2S$ film can cause a "sudden death" during the discharge process before achieving the theoretical capacity [38]. The growth of $Li_2S$ on the substrate affects the porosity of the cathode microstructure, which increases the tortuosity and decreases the effective ionic conductivity and diffusivity. Understanding the deposition mechanism of $Li_2S$ is of significant importance for the rational design of novel cathode architectures that can improve the performance of Li—S batteries.

As discussed above, the Li—S battery is a complex system involving multistep electrochemical reactions, insoluble PS deposition in the cathode, mass transport between two electrodes, and self-discharge with the formation of $Li_2S$ film in the anode. These phenomena happen at different temporal and spatial scales. The multiscale modeling approach could be a probe to gauge physico-electrochemical phenomena in the Li—S batteries.

## 4.2 The growth of crystalline $Li_2S$ film in cathode

Surface passivation is attributed to the deposition of insoluble $Li_2S$ during the discharge process. It is known that crystalline $Li_2S$ is an electronic insulator [36,39], hence the electrochemical reactions for PSs reduction are difficult to happen at the electrolyte/$Li_2S$ interface. The lateral growth of $Li_2S$ precipitation can reduce the fresh cathode surface area which supplies electrons for electrochemical reactions. In this regard, it is necessary to control the surface passivation of the carbon-based substrate caused by $Li_2S$ precipitation during the discharge process.

### 4.2.1 Exposed surface of solid $Li_2S$ film

To answer how the $Li_2S$ film grows at the atomistic scale, it is necessary to know the Miller index of the exposed surface. In this regard, the surface energies of the low Miller index plane are calculated by a first-principles approach. At the atomistic scale, the slab model is usually used to represent the surface structure as schematically shown in Fig. 4.1: for the (001) surface, either a Li layer or an S layer can be the center layer of the slab, and in each case, the termination can also be a Li layer or an S layer; if the (001) surface is terminated by a Li layer, the ratio of the Li atoms to the S atoms is larger than 2:1, making it a Li-rich structure; if the (001) surface is terminated by an S layer, it is an S-rich structure; for the (110) surface, the ratio of Li to S is always 2:1, so there is only a stoichiometric structure; for the (111) surface, the Li:S ratio is determined by the sequence of the atomic layers, and three structures (stoichiometric, Li-rich, and S-rich) are considered.

To find the most stable surface, the surface energy of each slab is calculated. The surface Gibbs energy can be estimated by Eq. (4.1):

$$\gamma = \frac{1}{2A} \left[ E^{slab} - N_s E^{bulk}_{Li_2S} - (N_{Li} - 2N_S) \times \mu_{Li} \right] \qquad (4.1)$$

Here $E^{slab}$ represents the total energy of a surface structure, $E^{bulk}_{Li2S}$ is the energy per $Li_2S$ formula unit in the bulk phase, and $\mu_{Li}$ is the chemical potential energy per Li atom in the *fcc* Li crystal. All of these energies are calculated by the first-principles DFT approach at 0 K. In Eq. (4.1), $A$ represents the surface area of the model, $N_S$ is the number of S atoms

126  Lithium-sulfur batteries

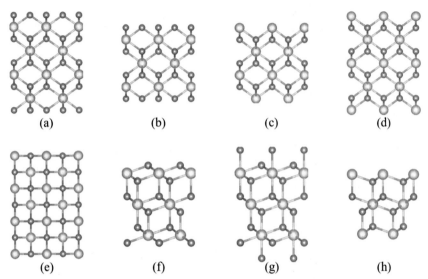

**Fig. 4.1** Side view of different slab models for (A) Li-rich (001) surface with Li-center layer, (B) Li-rich (001) surface with S-center layer, (C) S-rich (001) surface with Li-center layer, (D) S-rich (001) surface with S-center layer, (E) stoichiometric (110) surface, (F) stoichiometric (111) surface, (G) Li-rich (111) surface, (H) S-rich (111) surface. Here, larger size balls (yellow; light gray in print versions) in the atomic structure represent S atoms and the smaller size balls (violet; dark gray in print versions) represent Li atoms. *(Source: Authors)*.

in the model, and $N_{Li}$ is the number of the Li atoms in the model, respectively. To take into consideration realistic battery operation conditions, the applied potential should be taken into account and Eq. (4.2) would be rewritten as [40]

$$\gamma = \frac{1}{2A}\left[E^{slab} - N_s E^{bulk}_{Li_2S} - (N_{Li} - 2N_S) \times (\mu_{Li} - eU)\right] \quad (4.2)$$

where $U$ is the applied voltage.

According to Eq. (4.2), the theoretical equilibrium voltage is 2.2 eV [41]. Around this potential, the most stable surface structure is the stoichiometric (111) surface. This is in agreement with experimental observations. For example, X-ray diffraction measurements also showed that Li$_2$S (111) surface had higher intensities than surfaces with other Miller indices [42]. Zhang et al. synthesized Li$_2$S nanoparticles by chemical lithiation, in which the (111) plane with an interspacing of 3.2 Å was observed using high-resolution transmission electron microscopy (HRTEM) [43].

## 4.2.2 Atomistic insights into the growth process

The reduction of sulfur is a multistep process during the discharge, and the final product is $Li_2S$. Generally, solid sulfur ($S_8$) is reduced to soluble long-chain polysulfides, and then soluble polysulfides are reduced to the insoluble short-chain polysulfides ($Li_2S_2$ and $Li_2S$) which finally deposit on the substrate [44]. Solid-phase $Li_2S_2$, as an intermediate product, is converted to $Li_2S$ to lower the total energy of the system [45].

To model the growth process, an extra $Li_2S$ layer is added onto the stoichiometric (111) - (2 × 2) surface unit cell stepwise. From the initial state to the final state with a new layer, the reaction can be expressed as $8Li + 0.5S_8 + surface = 4Li_2S/surface$. Our present calculations predict that the theoretical discharge voltage according to the reaction is 2.02 eV for stoichiometric (111) surface. This voltage agrees very well with the lower plateau in the discharge profile of Li/S batteries, which corresponds to the formation of solid $Li_2S$ [46].

The free energy difference ($\Delta G$) induced by Li atoms and $Li_2S_x$ ($x = 1, 2$) adsorption on $Li_2S$ (111) surface is investigated. The energy difference is approximated by

$$\Delta G = E(mLi, nLi_2S_x) - m\mu_{Li} - nE_{Li_2S_x} - E^{slab} \tag{4.3}$$

Here $E(mLi, nLi_2S_x)$ is the total energy of the substrate with the adsorbate, and $m$ and $n$ represent the number of Li atoms and $Li_2S_x$ molecules which are deposited on the substrate, respectively. $ELi_2S_x$ is the energy of the isolated molecule.

Four reaction paths, shown in Fig. 4.2, are designed to understand the growth mechanism of the thermodynamically stable surface. In paths (I) and (II), $Li_2S_2$ molecules and Li atoms are alternatively deposited on

(I)    $Li_2S_2 + surface = Li_2S_2/surface$,
     $2Li + Li_2S_2/surface = Li_4S_2/surface$,
     $Li_2S_2 + Li_4S_2/surface = Li_6S_4/surface$,
     $2Li + Li_6S_4/surface = Li_8S_4/surface$;

(II)    $2Li + surface = Li_2/surface$,
     $Li_2S_2 + Li_2/surface = Li_4S_2/surface$,
     $2Li + Li_4S_2/surface = Li_6S_2/surface$,
     $Li_2S_2 + Li_6S_2/surface = Li_8S_4/surface$;

(III)    $Li_2S_2 + surface = Li_2S_2/surface$,
     $Li_2S_2 + Li_2S_2/surface = Li_4S_4/surface$,
     $2Li + Li_4S_4/surface = Li_6S_4/surface$,
     $2Li + Li_6S_4/surface = Li_8S_4/surface$;

(IV)    $Li_2S + surface = Li_2S/surface$,
     $Li_2S + Li_2S/surface = Li_4S_2/surface$,
     $Li_2S + Li_4S_2/surface = Li_6S_3/surface$,
     $Li_2S + Li_6S_3/surface = Li_8S_4/surface$;

**Fig. 4.2** Reaction paths for $Li_2S$ surface growth. *(Reproduced with permission from Z. Liu, D. Hubble, P.B. Balbuena, P.P. Mukherjee, Adsorption of insoluble polysulfides $Li_2S_x$ (x = 1, 2) on $Li_2S$ surfaces, Phys. Chem. Chem. Phys. 17 (14) (2015) 9032–9039, Copyright (2015) Royal Society of Chemistry.)*

the surface. The difference is that $Li_2S_2$ is first deposited on the surface in path (I), while Li is first deposited on the surface in path (II). In path (III), two $Li_2S_2$ molecules are deposited on the surface at the first two steps, and then $Li_2S_2$ deposition is reduced to $Li_2S$ by Li atoms. In path (IV), $Li_2S$ molecules are deposited on the surface step by step. The energy difference ($\Delta G$) of each intermediate state referencing the initial state is calculated according to Eq. (4.3) and shown in Fig. 4.3.

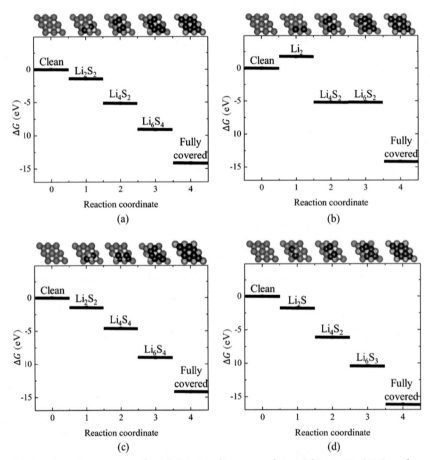

**Fig. 4.3** Atomic structure of each intermediate state for stoichiometric (111) surface growth and the corresponding energy difference. Violet spheres (dark gray in print versions) and yellow spheres (light gray in print versions) represent Li atoms and S atoms in the substrate, respectively. Blue spheres (dark gray in print versions) and green spheres (light gray in print versions) represent Li atoms and S atoms in the deposition. (Reproduced with permission from Z. Liu, D. Hubble, P.B. Balbuena, P.P. Mukherjee, Adsorption of insoluble polysulfides $Li_2S_x$ (x=1, 2) on $Li_2S$ surfaces, Phys. Chem. Chem. Phys. 17 (14) (2015) 9032–9039, Copyright (2015) Royal Society of Chemistry).

On Figs. 4.3A–C one can see that $Li_2S_2$ deposition on the (111) surface at the first step produces a negative Gibbs energy difference, while Li atom deposition at the first step produces a positive Gibbs energy difference. It is obvious that path (II) is blocked because Li deposition on the stoichiometric (111) surface is energetically unfavored. Oppositely, $Li_2S_2$ can spontaneously adsorb on the stoichiometric surface because of the negative Gibbs energy difference. The Gibbs energy difference between step 2 and step 1 in path (I) is −3.74 eV, which is lower than the difference between step 2 and step 1 in path (III) by 0.57 eV. According to this energy difference, it can be inferred that reducing $Li_2S_2$ deposition to $Li_2S$ [path (I)] is energetically more favored than the $Li_2S_2$ deposition growth [path (III)]. Fig. 4.3D shows the Gibbs energy profile of $Li_2S$ molecule deposition on the (111) surface in a stepwise manner [path (IV)]. For $Li_2S$ deposition, the Gibbs free energy difference at the first step is −1.78 eV, which is 0.34 eV lower than that of $Li_2S_2$ deposition at the first step. Additionally, the Gibbs free energy difference between the final state and initial state in path (IV) is also 2 eV lower than that in path (I).

### 4.2.3 Formation of $Li_2S$/graphite interface

Carbon-based materials are widely used for constructing cathode frameworks in Li—S batteries. In this section, the graphene monolayer is used to represent graphite-based substrate for understanding insoluble PS/substrate interactions. In order to fundamentally understand the $Li_2S$ growth on a carbon substrate, the adsorption of polysulfide $Li_2S_x$ ($x = 1, 2$) molecules on graphene is investigated. DFT calculations demonstrate that $Li_2S$ molecule adsorption is energetically favored over $Li_2S_2$ adsorption. The $Li_2S$ adsorption energy is −0.55 eV, while the $Li_2S_2$ adsorption energy is only −0.39 eV. As discussed in Section 4.2.2, the adsorption energy of a $Li_2S$ molecule on a crystalline (111) surface is −1.78 eV, which indicates that the crystalline $Li_2S$ surface is more favorable for $Li_2S$ deposition than graphene as a substrate.

The electronic structure of $Li_2S_x$ adsorption on graphene is investigated to deeply understand the $Li_2S_x$-graphene interaction. The distribution of charge density difference induced by $Li_2S_x$ adsorption is demonstrated in Fig. 4.4, which is calculated by the following equation

$$\Delta\rho(r) = \rho_{Li_2S_x@G}(r) - \left(\rho_{Li_2S_x}(r) + \rho_G(r)\right) \tag{4.4}$$

where $\rho_{Li2Sx@G}(r)$ is the charge density in $Li_2S_x$@graphene system, $\rho_{Li2Sx}(r)$ is the charge density of isolated $Li_2S_x$ and $\rho_G(r)$ is the charged density of clean graphene with atoms at the same positions as in the $Li_2S_x$@graphene system.

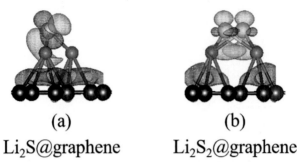

(a) Li$_2$S@graphene  (b) Li$_2$S$_2$@graphene

**Fig. 4.4** Difference charge density induced by (A) Li$_2$S molecule and (B) Li$_2$S$_2$ molecule adsorption. Green (light gray in print version) isosurface indicates electron depletion region and red (medium gray in print version) isosurface indicates electron accumulation region. *(Reproduced with permission from A.D. Dysart, J.C. Burgos, A. Mistry, C.-F. Chen, Z. Liu, C.N. Hong, P.B. Balbuena, P.P. Mukherjee, V.G. Pol, Towards next generation lithium-sulfur batteries: non-conventional carbon compartments/sulfur electrodes and multi-scale analysis, J. Electrochem. Soc. 163 (5) (2016) A730–A741, Copyright (2016) The Electrochemical Society.)*

Electron accumulation regions appear between Li atoms in the adsorbate and C atoms in the substrate as shown in Fig. 4.4. In addition, although interaction with graphene varies charge distribution around S atoms, the electron accumulation region is not observed between S atoms and C atoms. Thereby, it can be inferred that the adsorbate interacts with graphene via Li—C bonds.

The process of Li$_2$S layer formation on the carbon cathode is simulated by introducing more Li$_2$S molecules onto the graphene (3 × 3) supercell. The Li$_2$S coverage ($\Theta$) on graphene is defined as the ratio of the total number of Li and S atoms to the total number of hollow sites on the graphene. Hence, $\Theta = \frac{1}{3}$ ML represents a single Li$_2$S molecule adsorption, $\Theta = \frac{2}{3}$ ML means that two Li$_2$S molecules (or (Li$_2$S)$_2$) adsorb on graphene, and $\Theta = 1$ ML means that three Li$_2$S molecules (or (Li$_2$S)$_3$) adsorb on graphene.

For Li$_2$S adsorption with $\Theta = \frac{2}{3}$ ML, two different configurations (structure I and structure II as shown in Fig. 4.5) are predicted by the DFT simulations. Multiple (3 × 3) supercells are shown clearly demonstrating the atomic arrangement and periodicity of the resulting structure. It can be seen that S atoms form periodically repeated rectangles, and each S atom is located at the center of a small rectangle formed of four Li atoms. From the side view of structure I, it can be seen that all Li and S atoms are in the same plane, parallel to the graphene monolayer. It is obvious that the atomic structure of Li$_2$S layer formed by adsorbed Li$_2$S molecules with $\Theta = \frac{2}{3}$ ML is similar to the typical Li$_2$S (110) layer in Li$_2$S crystal with antifluorite structure

Multiscale modeling of lithium-sulfur battery electrodes 131

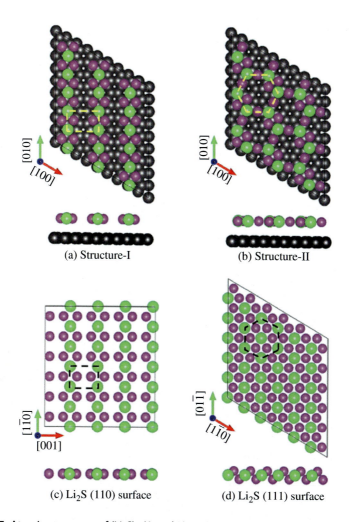

**Fig. 4.5** Atomic structures of (Li$_2$S)$_2$ (A and B) on (3 × 3) graphene supercell and typical Li$_2$S (110) surface (C) as well as Li$_2$S (111) surface (D). Two structures are observed from computational results. In Structure-I, the arrangement of S atoms is rectangle-like, which is similar to the arrangement of a typical Li$_2$S (110) surface. In Structure-II, the arrangement of S atoms is hexagonal-like, which is similar to the arrangement of a typical Li$_2$S (111) surface. *(Source: Authors).*

[Fig. 4.5C]. In structure I, the Li—S bond length is 2.21 Å, which is 0.27 Å shorter than the Li—S bond in Li$_2$S crystal [39]. The length of short S—S bridge ($D_{S-S}$) in structure I is 3.70 Å, which is 0.90 Å longer than the S—S distance in a typical Li$_2$S (110) layer. The variation of these geometric parameters is attributed to the lattice parameter mismatch between graphene and Li$_2$S (110) layer.

For structure II, it is interesting that S atom arrangement is hexagonal [Fig. 4.5B], which is similar to S atom arrangement in typical Li$_2$S (111) layer [Fig. 4.5D]. In structure II, each S atom coordinates with four Li atoms, and the S atom at the center of the S hexagon disappears compared to the typical Li$_2$S (111) surface. Hence, structure II is a defective Li$_2$S (111) surface with Li$_2$S vacancies. The side view of structure II also demonstrates that all Li and S atoms are in the same plane parallel to the graphene monolayer. In structure II, Li—S bond length is 2.24 Å, which is longer than the Li—S bond length in structure I, but still shorter than the Li—S bond length in Li$_2$S crystal. The distance between two neighboring S atoms is about 4.28 Å, which is close to the S—S distance 4.05 Å in the typical Li$_2$S (111) layer.

The first-principles atomistic thermodynamics [47] developed by Scheffler is employed to estimate the temperature-dependent stability of Structure-I and Structure-II. The Gibbs free energy of Structure-II is lower than that of Structure-I around room temperature (300 K). The Structure-I is only stable at an extremely low temperature (below 100 K).

When the Li$_2$S coverage Θ increases to 1 ML, the complete Li$_2$S (111) layer named Structure-III appears on graphene as shown in Fig. 4.6. It can be seen that the atomic arrangement is exactly the same as

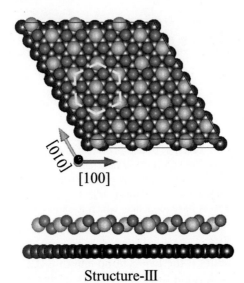

Structure-III

**Fig. 4.6** Atomic structure of (Li$_2$S)$_3$ adsorption on (3×3) graphene supercell. *(Source: Authors).*

the typical Li$_2$S (111) layer in crystal, and Li atoms are outside S plane. Li—S bond length in Structure-III is about 2.54 Å, which agrees well with Li—S bond length in Li$_2$S crystal 2.48 Å. The S—S distance in Structure-III is 4.27 Å, which is 0.22 Å longer than that in Li$_2$S crystal due to the lattice parameter mismatch between graphene and Li$_2$S (111) layer.

The process of Li$_2$S (111) formation on graphene is summarized in Fig. 4.7 and the corresponding energy profile is plotted. It can be seen that a single Li$_2$S molecule adsorbs on graphene first with a relatively small adsorption energy. The adsorbed Li$_2$S molecule interacts with graphene via strong covalent Li—C bonds. When introducing one more Li$_2$S molecule to the pre-adsorbed graphene, two configurations, Structure-I and Structure-II, are observed. Structure-I is similar to the Li$_2$S (110) layer and Structure-II is an incomplete Li$_2$S (111) layer. The thermal stability of Structures-I and II are examined, and it is found that Structure-II is more stable at room temperature. According to Fig. 4.7, it is found that the adsorption energy of introducing one more Li$_2$S to graphene with the pre-adsorbed single Li$_2$S molecule is more negative than −4 eV. Hence,

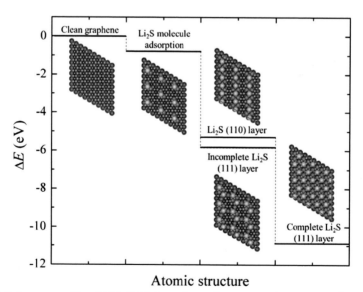

**Fig. 4.7** Energy profile of Li$_2$S (111) layer formation on graphene. *(Reproduced with permission from A.D. Dysart, J.C. Burgos, A. Mistry, C.-F. Chen, Z. Liu, C.N. Hong, P.B. Balbuena, P.P. Mukherjee, V.G. Pol, Towards next generation lithium-sulfur batteries: non-conventional carbon compartments/sulfur electrodes and multi-scale analysis, J. Electrochem. Soc. 163 (5) (2016) A730–A741, Copyright (2016) The Electrochemical Society.)*

it can be inferred that the pre-adsorbed single $Li_2S$ molecule is the seed for the formation of the $Li_2S$ layer on graphene.

## 4.2.4 Interfacial model for $Li_2S$ film growth

The formation of $Li_2S$ during discharge undergoes multistep reactions including: [48,49]

  **(i)** $S_8(s) = S_8(l)$,

  **(ii)** $S_8 + 2e^- = S_8^{2-}$,

  **(iii)** $S_8^{2-} + 2e^- = 2S_4^{2-}$,

  **(iv)** $S_4^{2-} + 2e^- = 2S_2^{2-}$,

  **(v)** $S_2^{2-} + 2e^- = 2S^{2-}$.

Reaction (i) represents the dissolution of $\alpha$-S into the electrolyte. Reactions (ii) $\sim$ (v) represent electrochemical reactions, in which long-chain polysulfides (PSs) are gradually reduced into short-chain PSs. Short-chain PSs tend to precipitate on the substrate due to the low solubility in the electrolyte, as shown in Reactions (vi) and (vii).

  **(vi)** $2Li^+ + S_2^{2-} = Li_2S_2(\downarrow)$,

  **(vii)** $2Li^+ + S^{2-} = Li_2S(\downarrow)$.

Besides the direct precipitation of $Li_2S$ [Reaction (vii)], solid $Li_2S_2$ can also be converted to $Li_2S$.

  **(viii)** $Li_2S_2 + 2Li^+ + 2e^- = 2Li_2S$.

There is a controversy about the composition of the discharge products in Li—S batteries. Barghamadi et al. reported that the direct formation of solid $Li_2S$ is the predominant reaction and the Reaction (viii) is kinetically slow [44]. Xiao et al. detected $Li_2S_2$ by using an in situ nuclear magnetic resonance (NMR) technique [50]. However, $Li_2S_2$ is not a thermodynamically stable phase according to experimental observations [51] and the first-principles calculations [52], and the XRD pattern of the final product matches the crystal structure of $Li_2S$ rather than the structure of $Li_2S_2$ predicted by the first-principles calculations [52]. Cuisinier et al. and Dominko et al. independently analyzed products during the discharge/charge cycling by operando X-ray absorption spectroscopy, and they found that $Li_2S$ is the only detectable crystalline phase among discharge products [48,53]. Cuisinier et al. also tracked the PS evolution during discharge with NMR, but they did not detect solid $Li_2S_2$ as reported by Xiao et al. [50] Cañas et al. analyzed discharge products with in situ XRD technique and they did not find solid $Li_2S_2$ [54]. Cañas et al. also found that

(111) surface dominates the facets of crystalline $Li_2S$, which is also confirmed by first-principles calculations [39,55,56].

Based on experimental and computational findings discussed above, a coarse-grained lattice-based mesoscale interfacial model is developed to represent the $Li_2S$ formation and growth with the following assumptions:

(1) $Li_2S$ is the only discharge product.

(2) The film grows along the normal direction of the $Li_2S$ (111) surface.

(3) The film growth is only attributed to the direct deposition of $Li_2S$ molecules rather than $Li_2S_2$ deposition and reduction.

(4) The structure of the $Li_2S$ is represented by a coarse-grained model. Each triatomic $Li_2S$ unit is simplified to a lattice site, and the position of a $Li_2S$ unit in the solid phase is represented by the position of the S atom. Therefore, the antifluorite structure of crystalline $Li_2S$ is converted to a face-centered cubic (fcc) structure. The coarse-grained model neglects the geometric parameters (i.e., bond length, bond angle, and molecule orientation) at the atomistic scale.

(5) The adsorption and diffusion of a $Li_2S$ unit on the solid substrate are restricted by a solid-on-solid model [57], in which an empty cell cannot accept a $Li_2S$ site unless this site coordinates with three occupied sites in the sublayer.

A kinetic Monte Carlo (KMC) algorithm is employed to implement transition events taking place at the electrolyte/solid substrate interface. Three transition events are considered in the present model, which are $Li_2S$ adsorption, desorption, and diffusion on the surface (schematically illustrated in Fig. 4.8). $Li_2S$ adsorption can only happen at an empty site cooperating with three occupied sites in the sublayer, the adsorbed $Li_2S$ can desorb from the substrate if it does not coordinate with other $Li_2S$ molecules, and the adsorbed $Li_2S$ molecule can also diffuse to the adjacent empty site on the carbon substrate. The adsorption rate at an available site is calculated by

$$R_{ads} = k_0 N_a V \frac{S_a}{S} \left( C_{Li^+}^2 C_{S^{2-}} - \Theta \right) \tag{4.5}$$

In Eq. (4.5), $k_0$ is the reaction rate constant, and $N_a$ is the Avogadro constant. $V$ and $S$ are the pore volume and cathode surface area in the porous cathode framework, respectively. $S_a$ is the area of a lattice site projected to the $Li_2S$ (111) surface. $C_i$ is the reactant concentration and $\Theta$ is the $Li_2S$ solubility term. The values of reaction rate constant $k_0$ and solubility $\Theta$ are adopted from previous simulation work [49]. The diffusion rate is calculated by

$$R_{dif} = \nu \exp \left( -\frac{E_b}{\kappa T} \right) \tag{4.6}$$

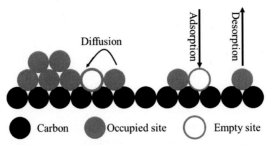

**Fig. 4.8** Schematic illustration of adsorption, desorption, and diffusion kinetics in the coarse-grained mesoscale model. *(Adapted with permission from Z. Liu, P.P. Mukherjee, Mesoscale elucidation of surface passivation in the Li–sulfur battery cathode, ACS Appl. Mater. Interfaces 9 (6) (2017) 5263–5271, Copyright (2017) American Chemical Society.)*

In Eq. (4.6), $R_{dif}$ is the number of diffusion attempts per second. The term "$\nu$" is the jumping frequency. $T$ is the temperature, and $\kappa$ is the Boltzmann constant. Previous first-principle calculation demonstrated that the chemical adsorption energy of a single $Li_2S$ molecule on graphene is only $-0.55\,eV$ [58], hence the desorption of a $Li_2S$ from the cathode surface should be considered and the desorption rate is calculated by

$$R_{des} = \frac{2\kappa T}{h} \exp\left(\frac{E_{ads}}{\kappa T}\right) \quad (4.7)$$

where $h$ represents the Planck constant. Table 1 lists the values of input parameters in Eq. (4.5)–(4.7). As discussed in Sections 4.2.2 and 4.2.3, first-principles calculation showed that there is a strong attractive interaction between a $Li_2S$ molecule and pre-adsorbed $Li_2S$ [39,58]. Thereby, the adsorbed $Li_2S$ molecule will not desorb or diffuse once it coordinates with other $Li_2S$ sites.

In KMC simulation, the procedure of $Li_2S$ film growth undergoes the following steps:

(a) *Calculate the total transition rate.* The total adsorption rate ($\Omega_{ads}$), diffusion rate ($\Omega_{dif}$), and desorption rate ($\Omega_{des}$) are calculated based on Eqs. (4.8)–(4.12).

$$\Omega_{ads} = \sum_{i=1}^{N} R_{ads}^{i} \quad (4.8)$$

$$\Omega_{dif} = \sum_{i=1}^{N} R_{dif}^{i} \quad (4.9)$$

$$\Omega_{des} = \sum_{i=1}^{N} R_{des}^{i} \quad (4.10)$$

Multiscale modeling of lithium-sulfur battery electrodes 137

**Table 1** The values of input parameters in Eqs. (4.4)–(4.6).

| Symbol | | Value |
|---|---|---|
| $k_0$ | Li$_2$S deposition rate constant | $6.875 \times 10^{-5}$ m$^6$ mol$^2$ s$^{-1}$ |
| $N_a$ | Avogadro constant | $6.02 \times 10^{23}$ mol$^{-1}$ |
| $V$ | Total pore volume of cathode microstructure | $1.57 \times 10^{-7}$ m$^3$ |
| $S$ | Total cathode/electrolyte interfacial area | $5.51 \times 10^{-2}$ m$^2$ |
| $S_a$ | Area per lattice site | $1.41 \times 10^{-19}$ m$^2$ |
| $E_b$ | Diffusion barrier of Li$_2$S molecule on cathode surface | 0.01 eV |
| $E_{ads}$ | Chemical adsorption energy of Li$_2$S on graphene | $-0.55$ eV |
| $C_{Li^+}$ | Concentration of Li$^+$ | $10^3$ mol m$^{-3}$ |
| $C_{S^{2-}}$ | Concentration of S$^{2-}$ | $10^{-5} \sim 10^{-3}$ mol m$^{-3}$ |
| $T$ | Operation temperature | $-40\,^\circ$C$\sim 80\,^\circ$C |
| $\kappa$ | Boltzmann constant | $8.617 \times 10^{-5}$ eV K$^{-1}$ |
| $h$ | Planck constant | $4.136 \times 10^{-15}$ eV s$^{-1}$ |

$$\Omega_{tot} = \Omega_{ads} + \Omega_{dif} + \Omega_{des} \tag{4.11}$$

The total transition event rate is the summation of $\Omega_{ads}$, $\Omega_{dif}$, and $\Omega_{dep}$. Here, $N$ is the total number of lattice site in the simulation domain, and $i$ is the $i$th lattice site.

**(b)** *Select a transition event.* A random number $\gamma_1$ uniformly distributed in $(0,1)$ is generated. In the condition of $\gamma_1 \Omega_{tot} < \Omega_{ads}$, adsorption event will happen; in the case of $\Omega_{ads} \leq \gamma_1 \Omega_{tot} < \Omega_{ads} + \Omega_{dif}$, the diffusion event will happen; and in the case of $\Omega_{ads} + \Omega_{dif} \leq \gamma_1 \Omega_{tot} < \Omega_{tot}$, the desorption event will happen. After determining the transition event, the position where the event will happen is determined by another random number $\gamma_2$. For adsorption event, the position is selected by

$$\sum_{i=1}^{k-1} R_{ads}^i < \gamma_2 \Omega_{ads} \leq \sum_{i=1}^{k} R_{ads}^i \tag{4.12}$$

For diffusion event, the position is selected by

$$\sum_{i=1}^{k-1} R_{dif}^i < \gamma_2 \Omega_{dif} \leq \sum_{i=1}^{k} R_{dif}^i \tag{4.13}$$

For desorption event, the position is selected by

$$\sum_{i=1}^{k-1} R_{des}^i < \gamma_2 \Omega_{des} \leq \sum_{i=1}^{k} R_{des}^i \tag{4.14}$$

Here, $k$ indicates the $k$th lattice site where the transition event happens.

**(c)** *Update structure and time.* The film structure is updated according to the transition event selected in the previous step. The time step of the selected event is estimated by a random number $\gamma_3$

$$\delta t = -\frac{1}{\Omega_{tot}} \ln \gamma_3 \qquad (4.15)$$

The data flow of the whole simulation process is shown in Fig. 4.9.

The interfacial model demonstrates that the formation of $Li_2S$ film undergoes three stages before the carbon surface is fully covered: cluster formation (nucleation), isolated island growth, and island coalescence. The formation of $Li_2S$ film cannot be observed in the first duration (the orange region, light gray in print versions, in Fig. 4.10). In this stage, $Li_2S$ desorption inhibits other transition events (adsorption and diffusion). The desorption rate dominates the total transition rates. The high desorption rate is attributed to the weak chemical interaction between the single $Li_2S$ molecule and the carbon substrate [58]. If a single adsorbed $Li_2S$ molecule collides with other adsorbed $Li_2S$ molecules, a stable $(Li_2S)_n$ forms on the surface of the carbon substrate. In this case, desorption of the $Li_2S$ is difficult to happen due to the strong chemical interaction. These clusters act as seeds for $Li_2S$ film growth. As shown in Fig. 4.10, the coverage keeps increasing in the second stage (pink region; dark gray in print versions) and the third stage (yellow region; light gray in print versions). The slope of the coverage curve represents the coverage growth rate, and the larger slope indicates a higher coverage growth rate (faster lateral growth). In the second stage, $Li_2S$ islands experience the isolated growth which associates an increase in coverage growth rate. In the third stage, the coverage growth rate gradually decreases to zero. The decrease in the coverage growth rate is attributed to the coalescence of $Li_2S$ islands. In Fig. 4.10, the solid black line represents the average thickness of the $Li_2S$ precipitation; and the slope represents the thickness growth rate. It is found that the thickness growth in the second stage (the stage isolated island growth) is slower than the thickness growth in the third stage (the stage of island coalescence). The adsorption only makes contributions to the lateral growth and the thickness growth of $Li_2S$ islands. In the second stage, the lateral growth is dominant, and the thickness growth is suppressed. In the third stage, the coalescence of islands eliminates the lateral growth, which leads to an increase in the thickness growth rate. The film thickness grows linearly after the cathode surface is fully covered by the discharge product.

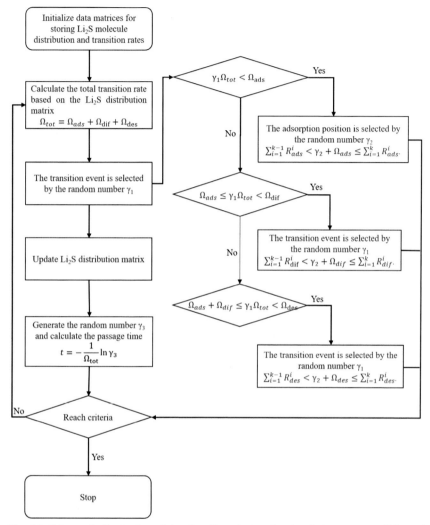

**Fig. 4.9** Schematic illustration of the data flow during the simulation process. *(Adapted with permission from Z. Liu, P.P. Mukherjee, Mesoscale elucidation of surface passivation in the Li–sulfur battery cathode, ACS Appl. Mater. Interfaces 9 (6) (2017) 5263–5271, Copyright (2017) American Chemical Society.)*

It is known that Li$_2$S is an electrical insulator and the resistivity is larger than $10^{14}$ $\Omega$ cm [36]. In the cathode, the three-phase boundary (the boundary between electrolyte, Li$_2$S, and carbon) is electrochemically active for PS reduction, and the surface of Li$_2$S film cannot allow the electrochemical reduction [59]. As discussed above, the island coalescence reduces the length

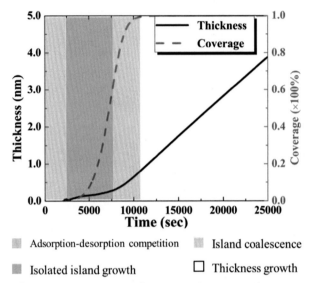

**Fig. 4.10** Surface coverage as a function of time with constant reaction $C_{Li^+} = 10^3$ mol m$^{-3}$, $C_{S^{2-}} = 10^{-4}$ mol m$^{-3}$, and $T = 20°C$. (Adapted with permission from Z. Liu, P.P. Mukherjee, Mesoscale elucidation of surface passivation in the Li–sulfur battery cathode, ACS Appl. Mater. Interfaces 9 (6) (2017) 5263–5271, Copyright (2017) American Chemical Society.)

of the three-phase boundary, which leads to the decrease of active sites for electrochemical reactions. Hence, it is necessary to control the passivation of the cathode surface to improve the battery performance.

Fig. 4.11 shows the effect of temperature on coverage variation. It is found that the adsorption–desorption competition duration is very short at $T = -20°C$. The reason is that the desorption kinetic rate is significantly reduced by the low temperature according to Arrhenius equation. For a given $S^{2-}$ concentration, the adsorption rate is a constant in the present model, and the desorption rate is proportional to $e^{-\frac{1}{T}}$. More desorption events can happen at a relatively higher temperature condition, which slows down Li$_2$S cluster formation. Hence, the adsorption–desorption competition duration with $T = 40°C$ is longer than that with $T = -20°C$. However, the high temperature ($T > 40°C$) also decreases the adsorption–desorption competition duration as shown in Fig. 4.11. The increase in temperature facilitates the diffusion of Li$_2$S molecules. Therefore, Li$_2$S molecules have more chances to collide to form clusters at $T = 80°C$; thus, the zero-coverage duration is reduced.

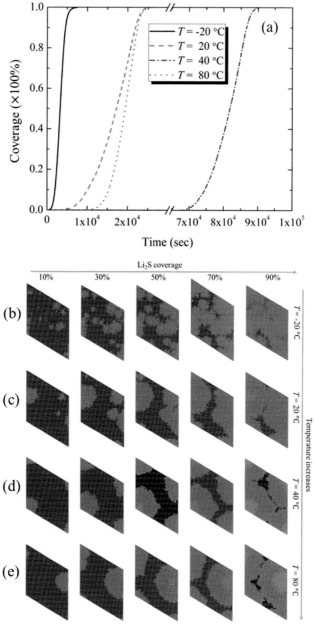

**Fig. 4.11** (A) Coverage variation vs. time with $C_{S^{2-}} = 1 \times 10^{-4}$ mol m$^{-3}$ at different temperatures. Snapshots demonstrate the Li$_2$S growth processes at different temperatures: (B) $T = -20°C$, (C) $T = 20°C$, (D) $T = 40°C$, and (E) $T = 80°C$. *(Adapted with permission from Z. Liu, P.P. Mukherjee, Mesoscale elucidation of surface passivation in the Li–sulfur battery cathode, ACS Appl. Mater. Interfaces 9 (6) (2017) 5263–5271, Copyright (2017) American Chemical Society.)*

Snapshots in Fig. 4.11 depict morphology evolution during deposition at different temperatures. Snapshots in the first column show the temperature effect on the Li$_2$S island distribution at 10% ML coverage. It is found that some small Li$_2$S nanoislands appear on the cathode surface at $T=-20°C$, while fewer nanoislands are observed at $T=20°C$. Only one nanoisland is found in the computational domain when the temperature is larger than 40°C. At low-temperature conditions, adsorbed Li$_2$S molecules can be easily stabilized on the cathode surface due to the low desorption rate. As temperature increases, more pre-adsorbed Li$_2$S molecules will desorb from the cathode surface to the ambient electrolyte environment, hence the number of islands decreases. As shown in Fig. 4.11, the temperature affects the morphology variation of the precipitation film. At $T=-20°C$ and $20°C$, the island coalescence happens at 50% of coverage. When the temperature is above 40°C, the island coalescence happens after 70% of coverage.

The density of the lateral sites of Li$_2$S islands is calculated to quantitatively show the morphology evolution of the precipitation as shown in Fig. 4.12. Lateral sites are empty sites adjacent to Li$_2$S islands. In the present interface model,

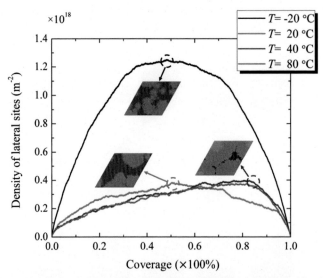

**Fig. 4.12** Density of lateral sites as a function of coverage. The lateral site of a Li$_2$S island is defined as the empty lattice site adjacent to the island. If a Li$_2$S molecule is placed at the lateral site, the molecule cannot desorb from the carbon substrate due to the strong chemical interaction between Li$_2$S molecules. The formation and isolated growth of islands will increase the density of lateral sites, while the coalescence of islands will decrease the density of lateral sites. *(Adapted with permission from Z. Liu, P.P. Mukherjee, Mesoscale elucidation of surface passivation in the Li–sulfur battery cathode, ACS Appl. Mater. Interfaces 9 (6) (2017) 5263–5271, Copyright (2017) American Chemical Society.)*

the Li$_2$S molecule placed at the lateral site cannot desorb from the carbon substrate due to the strong attraction between Li$_2$S molecules. The increase of density corresponds to the isolated island growth and the decrease of density corresponds to the island coalescence. The saddle point of the density curve is shifted to a high coverage by increasing the temperature. As discussed above, the low temperature produces smaller-sized Li$_2$S islands, which lead to a larger density of lateral sites. Once a Li$_2$S molecule is located at a lateral site, it is stabilized on the carbon surface due to the strong attraction between this adsorbed Li$_2$S molecule and the pre-deposited Li$_2$S island. Hence, the cathode surface with smaller islands (larger density of lateral sites) always has a higher coverage growth rate, which leads to a faster surface passivation.

For a carbon-based cathode, the cathode surface is usually decorated by functional groups to trap polysulfides [34,35,60]. The interfacial model can also evaluate the effect of the population of functional groups on the surface passivation. In the interfacial model, functional groups are randomly distributed on the cathode surface. It is also hypnotized that the Li$_2$S molecule cannot desorb or diffuse once it is located at a functionalized site. Fig. 4.13A

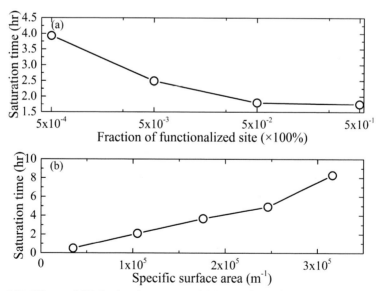

**Fig. 4.13** Effects of (A) factional groups and (B) specific surface area on the carbon cathode surface passivation. All simulations are performed with $T=20°C$, $C_{Li}=10^3$ mol m$^{-3}$, and $C_{S^{2-}}=10^{-4}$ mol m$^{-3}$. *(Adapted with permission from Z. Liu, P.P. Mukherjee, Mesoscale elucidation of surface passivation in the Li–sulfur battery cathode, ACS Appl. Mater. Interfaces 9 (6) (2017) 5263–5271, Copyright (2017) American Chemical Society.)*

clearly shows that saturation time is dependent on the population of the functionalized site. The cathode surface with more functional groups is passivated much faster. Functional groups can serve as nucleation agents, which can produce many small $Li_2S$ islands on the carbon surface and facilitate the lateral growth of the $Li_2S$ precipitation. Hence, the carbon cathode decorated by functional groups will also suffer from more severe passivation. In summary, a strong affinity of the cathode surface to polysulfides is required to mitigate the shuttle effect. However, the stronger affinity also associates with the faster surface passivation. One open question to the Li—S batteries is which effect (shuttle effect *vs.* surface passivation) dominates the cell performance and how to trade off these two effects.

The geometric properties (pore volume, surface area, and architecture) of the cathode microstructure also influence the $Li_2S$ film growth. Fig. 4.13B shows the effect of the specific surface area on the surface passivation. The specific surface area is defined as the ratio of cathode surface area to the pore volume. It is found that the increase in specific surface area is helpful for deferring the surface passivation. According to Eq. (4.5), the adsorption rate is inversely proportional to the specific surface area. Therefore, the high specific surface area can inhibit the $Li_2S$ precipitation.

## 4.3 Parasitic reactions in anode

### 4.3.1 Passivation of metallic Li anode

Parasitic reactions of polysulfides with Li anode could result in problematic issues, such as low cycling efficiency, poor safety, and high self-discharge rate [61]. During cycling, the reactions continuously consume active sulfur species and corrode Li anode, leading to a blocking surface of insoluble $Li_2S/Li_2S_2$ and increased charge transfer resistance [62,63].

The atomic structure evolution of $Li_2S$ film formation on the Li (110)-$(2 \times 2)$ surface unit cell (SUC) and on the Li (111)-$(2 \times 2)$ SUC is studied by DFT simulations. The stable atomic structure of two $Li_2S$ molecules co-adsorption on the $(2 \times 2)$ SUC is shown in Fig. 4.14A. It can be seen that $Li_2S$ columns appear along the [001] orientation. By periodically extending the atomic structure along the [001] and [1$\bar{1}$0] orientations, it can be seen that the arrangement of Li and S atoms is similar to that in a typical $Li_2S$ (110) plane. The Li—S bond length in the $Li_2S$ film is 2.37 Å, and the S—S distance is 3.44 Å, both of which are close to values in the crystalline $Li_2S$ (110) plane.

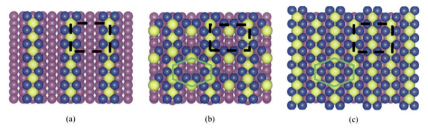

**Fig. 4.14** Top view of (A) two, (B) three, and (C) four Li$_2$S molecules' adsorption on the Li$_2$S (110)-(2 × 2) surface unit cell which is marked by a black dash square. *(Adapted with permission from Z. Liu, S. Bertolini, P.B. Balbuena, P.P. Mukherjee, Li$_2$S film formation on lithium anode surface of Li–S batteries, ACS Appl. Mater. Interfaces 8 (7) (2016) 4700–4708, Copyright (2016) American Chemical Society.)*

Based on the atomic structure shown in Fig. 4.14A, one more Li$_2$S molecule is placed on the surface, which means that three Li$_2$S molecules co-adsorb on the Li (110)-(2 × 2) SUC. The atom positions after structure optimization are depicted in Fig. 4.14B. It is interesting that the hexagon consisting of 6 S atoms linked by green lines shown in Fig. 4.14B appears in the deposited Li$_2$S film. An S hexagon with an S atom at the center is the feature of the typical crystalline Li$_2$S (111) plane. The atomic structure shown in Fig. 4.14B can be an intermediate state during the formation of the Li$_2$S (111) film. In this intermediate state, the distance between two adjacent S atoms of the hexagon varies from 3.91 Å to 5.20 Å, and the S—S distance in a perfect crystalline Li$_2$S (111) plane is 4.05 Å [64].

Fig. 4.14C depicts the top view of the stable atomic structure in which four Li$_2$S molecules are placed on the Li (110)-(2 × 2) SUC. It is obvious that the S positions projected onto the substrate follow the pattern of S arrangement in the crystalline Li$_2$S (111) plane as discussed above. In the atomic structure shown in Fig. 4.14C, each S atom is surrounded by six Li atoms and the Li—S distance varies from 2.53 Å to 4.25 Å. The side view of this fully covered Li (110) surface is shown in Fig. 4.16. The arrangement of atoms along the normal direction in the deposited Li$_2$S film is different from the crystalline Li$_2$S (111) plane. In the perfect Li$_2$S (111) plane, all S atoms are in one layer. However, in the Li$_2$S film on the Li (110) surface, S atoms are distributed into two layers. This Li$_2$S film can be treated as a Li$_2$S (111) plane distorted along the normal direction. Previous theoretical and experimental studies demonstrated that the facets of solid Li$_2$S are dominated by the (111) surface which has the lowest Gibbs free energy [39,42,43,56,65]. Hence,

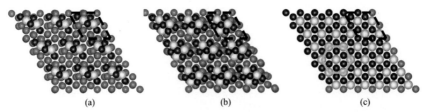

**Fig. 4.15** Top view of (A) two Li$_2$S molecules, (B) three molecules, and (C) four Li$_2$S molecules' adsorption on the Li$_2$S (111)-(2 × 2) surface unit cell which is marked by a black dash parallelogram. *(Adapted with permission from Z. Liu, S. Bertolini, P.B. Balbuena, P.P. Mukherjee, Li$_2$S film formation on lithium anode surface of Li–S batteries, ACS Appl. Mater. Interfaces 8 (7) (2016) 4700–4708, Copyright (2016) American Chemical Society.)*

the distorted Li$_2$S (111) film formed on the Li (110) surface can be the base for the growth of crystalline Li$_2$S film.

Snapshots in Fig. 4.15 demonstrate the mechanism of Li$_2$S film formation on the Li (111)-(2 × 2) SUC. Fig. 4.15A depicts the stable atomic structure of the two Li$_2$S molecules' co-adsorption on the surface. It is found that a (Li$_2$S)$_2$ cluster forms on the Li (111) surface. In the cluster, each Li$_2$S unit shares one Li atom with its partner, hence each S atom coordinates with 3 Li atoms. The Li—S bond length in the adsorbed (Li$_2$S)$_2$ varies from 2.30 Å to 2.51 Å, which is longer than the Li—S bond length of the free Li$_2$S$_x$ ($x$ = 1, 2) molecule. Fig. 4.15B depicts the optimized configuration of three Li$_2$S molecules' co-adsorption on Li (111)-(2 × 2) SUC. In this case, a (Li$_2$S)$_3$ cluster forms on the anode surface. Fig. 4.15C depicts the atomic structure of a fully covered Li (111) surface, which is represented by four Li$_2$S molecules' co-adsorption on the Li (111)-(2 × 2) SUC. It is clearly shown that the atom positions projected to the surface exactly follow the atomic arrangement in the crystalline Li$_2$S (111) plane. The hexagon consisting of six S atoms can be identified, and the center of the hexagon is occupied by another S atom. Each S atom is surrounded by six Li atoms and the averaged Li—S bond length is around 2.93 Å, which is 0.45 Å longer than the Li—S bond in the Li$_2$S crystal. The averaged distance between the two adjacent S atoms is around 4.86 Å, which is also longer than the S—S distance in the Li$_2$S crystal by 0.8 Å. These slight differences are attributed to the lattice mismatch between the Li (111) surface and the Li$_2$S (111) surface. The atomic positions in the Li$_2$S film along the normal direction are shown in Fig. 4.16B, which is the side view of the Li$_2$S film/Li (111) interface. It is obvious that the S atoms are in the same layer and the coordinating Li atoms are above and below the S layer alternatively.

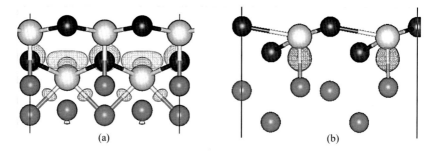

**Fig. 4.16** Difference charge density of Li$_2$S film adsorption on (A) Li (110) surface and (B) Li (111) surface. The red (cross-hatched gray in the print version) isosurface (3.5 × 10$^{-3}$ e/Å$^3$) represents electron accumulation and the green (light gray in the print version) isosurface (3.5 × 10$^{-3}$ e/Å$^3$) represents electron depletion. *(Adapted with permission from Z. Liu, S. Bertolini, P.B. Balbuena, P.P. Mukherjee, Li$_2$S film formation on lithium anode surface of Li–S batteries, ACS Appl. Mater. Interfaces 8 (7) (2016) 4700–4708, Copyright (2016) American Chemical Society.)*

The difference charge density of the Li$_2$S film/Li anode interface is shown in Fig. 4.16. Apparently, electron accumulation regions (red isosurface; cross-hatched gray in the print versions) appear between the Li surface and Li$_2$S film, and the bonds formed by S atoms and Li atoms in the substrate (violet sphere; dark gray in print versions) penetrate electron accumulation regions. The electronic structures demonstrate that the Li$_2$S film interacts with the Li anode surface via strong chemical bonds.

## 4.3.2 Mesoscale model for analyzing self-discharge

Polysulfide shuttle phenomenon deteriorates the electrochemical performance of Li—S batteries, leading to self-discharge and capacity fade. In the following, a mesoscale analysis is given to illustrate the mechanisms of self-discharge behavior during rest. The kinetic Monte Carlo method is employed to capture the self-discharge characteristics of the Li—S battery [66–71].

Sulfur molecules could be desorbed from the carbon host. Because of the concentration gradient across the system, the dissolved S$_8$ molecules diffuse out of the cathode through the electrolyte. In the vicinity of the Li anode, S$_8$ is reduced on the anode surface. To simplify the model, certain assumptions are made below. In addition to the dissolution of elemental sulfur into the electrolyte, two chemical reactions are considered on the anode surface.

$$S_{8(s)} \rightarrow S_{8(l)} \quad (4.16)$$

$$S_{8(l)} + 4Li \rightarrow 2Li_2S_4 \quad (4.17)$$

$$Li_2S_4 + 6Li \rightarrow 4Li_2S \quad (4.18)$$

Eq. (4.17) represents one cycloS$_8$ molecule is reduced to two lithium tetrasulfide Li$_2$S$_4$, which is responsible for the disappearance of the upper discharge plateau. In Eq. (4.18), Li$_2$S$_4$ is reduced to insoluble lithium sulfide Li$_2$S, which is directly deposited on the anode surface. In fact, there are more types of polysulfides involved. In this simplified model, soluble polysulfides are represented by Li$_2$S$_4$, and insoluble polysulfides are represented by Li$_2$S. The self-discharge originates from the chemical reactions on the anode surface (Fig. 4.17), and the capacity fade caused by the reactions between polysulfides and current collectors (or cathode) is not taken into account.

Fig. 4.18 illustrates a representative simulation at a capacity loss of 10%. Initially, all the sulfur molecules (blue) are adsorbed on the carbon backbone (brown). During the resting time, the desorption of sulfur continues until the electrolyte is saturated with S$_8$ molecules, approximately 21 S$_8$ molecules in the electrolyte corresponding to a solubility of 19 mol m$^{-3}$. The concentration gradient drives S$_8$ molecules to transport from the cathode to the Li anode. After the elemental sulfur is reduced at the Li anode–electrolyte interface, other S$_8$ molecules are desorbed from the cathode to replenish the dissolved S$_8$ in the electrolyte. Meanwhile, the reaction product of Li$_2$S$_4$ (gray) diffuses from the Li anode surface to the cathode under the concentration gradient. Alternatively, Li$_2$S$_4$ could be reduced to Li$_2$S (red), which is directly deposited over the anode surface.

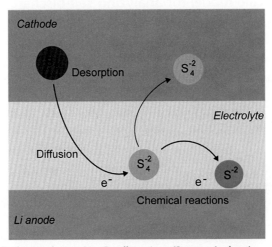

**Fig. 4.17** Self-discharge during Li—S cell resting. *(Source: Authors)*

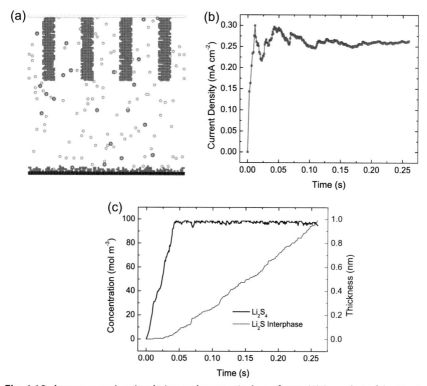

**Fig. 4.18** A representative simulation at the capacity loss of 10%. (A) Snapshot of the Li—S system. The system includes solid sulfur (blue), dissolved S$_8$ molecules (purple), Li$_2$S$_4$ (gray), Li$_2$S (red), carbon backbone (brown), and Li anode (black). (B) "Imaginary current density" with time. (C) Profiles of the Li$_2$S$_4$ concentration and average thickness of the Li$_2$S interphase. *(Adapted with permission from F. Hao, Z. Liu, P.B. Balbuena, P.P. Mukherjee, Mesoscale elucidation of self-discharge-induced performance decay in lithium–sulfur batteries, ACS Appl. Mater. Interfaces 11 (14) (2019) 13326–13333, Copyright (2019) American Chemical Society.)*

Though no current runs through the system, charge transfer takes place between the active materials and the anode. Herein, the involved electrons in the chemical reactions are converted into the "imaginary current density," which is calculated as

$$i = \frac{Nq}{tS} \quad (4.19)$$

where $N$ is the total number of electrons transferred in the reactions during a resting time $t$, $S$ is the area of the electrolyte-anode interface, and $q$ is the elemental charge. Fig. 4.18B shows the "imaginary current density" as the function of time. In the early stage, the current density remarkably increases. During this period, $S_8$ molecules rapidly diffuse to the anode. After the concentrations of $S_8$ molecules and $Li_2S_4$ reach the equilibrium states, the current density is stable at $0.25\,mA\,cm^{-2}$. As indicated in Fig. 4.18C, the concentration of $Li_2S_4$ gradually increases with time. Meanwhile, the precipitation of insoluble $Li_2S$ results in a passivation layer on the anode surface. Fig. 4.18C shows that the average thickness of the passivation layer progressively increases with the resting time. In addition, the interphase layer features a porous $Li_2S$ structure, as illustrated in Fig. 4.18A, which is consistent with the recently reported results. In the experiments, the failure of lithium-sulfur battery was attributed to the continued growth of a porous passivation layer on a Li metal anode, with a thickness up to $440\,\mu m$, which was identified as $Li_2S$ by X-ray diffraction. The formation of $Li_2S$ interphase degrades the performance by means of consuming the electrolyte and Li anode, eventually leading to a dead cell.

Because of the nanoscale system used in the model, the resting time is only $0.26\,s$ when the capacity loss is 10% in Fig. 4.18. To scale up to the system comparable to a practical lithium-sulfur system, the dimensionless time and "imaginary current density" are employed

$$\bar{t} = \frac{Dt}{L^2}, \bar{i} = \frac{iL}{Dc_0F} \tag{4.20}$$

where $L$ is the system height, $c_0$ is the concentration, and $D$ is the diffusivity. We have $D = d^2 k_d$, where $k_d$ is the diffusion rate, and $d$ is the distance between two adjacent sites, i.e., $6.0\,\text{Å}$. The calculated reference diffusivities are $D_{10} = 2 \times 10^{-8}\,cm^2/s$ and $D_{20} = 1 \times 10^{-9}\,cm^2/s$ for $S_8$ and $Li_2S_4$, respectively. Using the dimensionless time and current density, the diffusion equation can be transformed into nondimensional forms. For the capacity loss of 10%, the resting time of $0.26\,s$ and "imaginary current density" of $0.25\,mA\,cm^{-2}$ in our model are equivalent to 3 days and $0.0025\,mA\,cm^{-2}$ for a practical system with a thickness of $600\,\mu m$.

Finally, the effect of interlayer on self-discharge is explored. In our model, the interlayer is assumed to have a higher diffusion barrier for sulfur and polysulfides than the electrolyte. Hence, we have $D_i/D_0 < 1$, where $D_i$ is the diffusivity in the interlayer, $D_0$ is the diffusivity in the electrolyte. In practical applications, the interlayer should not affect Li-ion transport. Fig. 4.19A shows the capacity loss in terms of $\log (D_{1i}/D_{10})$ and $\log (D_{2i}/D_{20})$ for a fixed self-discharge time of $D_{10}t/L^2 = 150$ during resting. $D_{1i}$ is the diffusivity of $S_8$ in the interlayer, and $D_{2i}$ is the diffusivity of

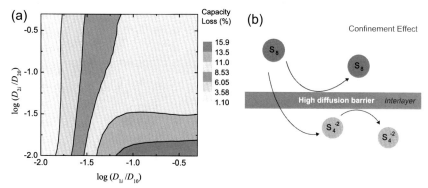

Fig. 4.19 Effect of interlayer on self-discharge kinetics. (A) For a fixed self-discharge time of $D_{10}t/L^2 = 150$, the capacity loss in terms of $\log(D_{1i}/D_{10})$ and $\log(D_{2i}/D_{20})$. $D_{1i}$ is the diffusivity of $S_8$ in the interlayer, and $D_{2i}$ is the diffusivity of $Li_2S_4$ in the interlayer. $D_{10}$ and $D_{20}$ are the reference diffusivities of $S_8$ and $Li_2S_4$ in the electrolyte, respectively. (B) Schematic of the confinement effect from the interlayer with high diffusion barriers for $S_8$ and $Li_2S_4$. (Adapted with permission from F. Hao, Z. Liu, P.B. Balbuena, P.P. Mukherjee, Mesoscale elucidation of self-discharge-induced performance decay in lithium–sulfur batteries, ACS Appl. Mater. Interfaces 11 (14) (2019) 13326–13333, Copyright (2019) American Chemical Society.)

$Li_2S_4$ in the interlayer. For dissolved $S_8$, when the interlayer has a much higher diffusion barrier than the electrolyte, the self-discharge rate primarily depends on $D_{1i}/D_{10}$. As the diffusion barrier of $S_8$ in the interlayer becomes close to that in the electrolyte, the diffusion kinetics of $Li_2S_4$ in the interlayer dominates the self-discharge behavior. For instance, the right side in Fig. 4.19A shows that the self-discharge rate increases with the increasing diffusion barrier of $Li_2S_4$ in the interlayer. In addition, for a fixed high diffusion barrier of $Li_2S_4$ in the interlayer, a slight change of self-discharge rate is caused by a decreasing diffusion barrier of $S_8$ in the interlayer, shown in the lower right corner in Fig. 4.19A.

For the above results, Fig. 4.19B illustrates the mechanism. For sulfur molecules, the interlayer with a high diffusion barrier will resist the diffusion of dissolved $S_8$ through it. Therefore, sulfur molecules are largely confined in the space above the interlayer. Consequently, the decreased $S_8$ concentration at the Li anode–electrolyte substantially reduces the self-discharge rate during cell resting. The result is consistent with the experiments [72], where excellent overall performance could be obtained using an ultrathin dense interlayer. In contrast, a special case is that the interlayer has a moderate diffusion barrier for $S_8$ but a high diffusion barrier for $Li_2S_4$, and thus, $Li_2S_4$ is largely confined in the space below the interlayer. Because of the confinement, the accumulated

$Li_2S_4$ could increase $Li_2S_4$ concentration and its gradient between the interlayer and the anode, which contributes to a relatively larger self-discharge rate. For a practical Li—S battery, the artificial interlayer needs to be designed to hinder the diffusion of sulfur and polysulfides, aimed at effectively suppressing the self-discharge during resting.

Many factors and their competition affect the self-discharge behavior: the binding of sulfur to the cathode, sulfur solubility and diffusivity in the electrolyte, and chemical stability of sulfur species against Li anode. These factors contribute to the self-discharge rate differences in experiments. For practical Li—S batteries, the use of interlayers alleviates self-discharge. Self-discharge rate could also be reduced by improving the separator and electrolyte, such as enhancing the resistance to the diffusion of sulfur species but without the compromise of Li-ion diffusion kinetics. In our model, a qualitative analysis is provided. However, a more rigorous model is further needed to clarify the mechanism of self-discharge. (1) In the model, we focus on the self-discharge caused by the reduction of the dissolved sulfur from the cathode and its produced polysulfides, and the reduction of pre-existing sulfur species in the electrolyte is not considered. (2) In addition to the reactions on the anode surface, parasitic reactions could also occur in the cathode and current collectors. (3) This model can only be applied to cell resting; however, self-discharge exists during the entire cell life.

## 4.4 Summary and outlook

Based on the understanding gained from the discussions in this chapter, a hierarchical modeling framework as illustrated in Fig. 4.20 can be developed as follows:

- **(i)** an atomistic scale investigation to unravel the mechanisms underlying the PS-substrate interactions
- **(ii)** calculation of the $Li_2S$ film growth rate at the interfacial level based on the interaction parameters
- **(iii)** 3D cathode microstructure reconstruction using SEM micrographs
- **(iv)** capturing the evolution of the microstructure based on the growth rate of the $Li_2S$ film
- **(v)** translation of the cathode microstructural transport properties to compute electrochemical performance based on a macroscale model.

A comprehensive understanding of the intricate nature of PS-substrate interactions is crucial toward regulating the $Li_2S$ film growth and improving the electrochemical performance of the Li—S battery. In this context,

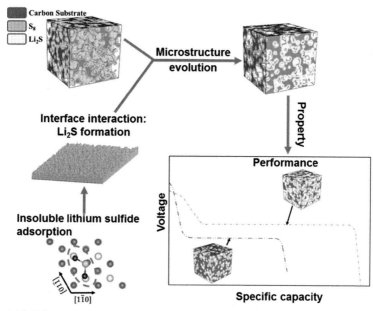

**Fig. 4.20** Schematic representation of the hierarchical modeling framework to understand the physicochemical interplays in a Li—S battery. *(Reproduced with permission from Z. Liu, A. Mistry, P.P. Mukherjee, Mesoscale physicochemical interactions in lithium–sulfur batteries: progress and perspective, J. Electrochem. Energy Convers. Storage 15 (1) (2018) 010802, https://doi.org/10.1115/1.4037785, Copyright (2018) by ASME.)*

macroscale modes have revealed the influence of sulfur loading, cathode microstructural arrangement, ion transport pathways, and reaction kinetics on the cell's electrochemical performance [73–75]. Owing to the insulating nature of the Li$_2$S precipitate, the passivating film formed on the substrate can increase the interfacial kinetic resistance and result in performance degradation. The microstructure of the cathode plays a pivotal role in cell performance. An ideal cathode microstructure is not limited by ion percolation and maintains sufficient effective Li$^+$ conductivity in the pore phase. Moreover, the pore size distribution of the cathode should be engineered taking into account the volume fluctuations and the associated mechanical degradation implications of the phase transitions from S$_8$ to Li$_2$S. In addition, the cathode microstructure should mitigate PS dissolution and surface passivation effects should be minimized by designing microstructures to provide sufficient interfacial area for Li$_2$S precipitation. In this regard, a large assortment of microstructures has been constructed toward attaining improved Li—S battery performance.

While macroscale models can connect with observable electrochemical signatures of the cell like the two plateau voltage profiles, DFT and AIMD simulations can accurately capture the dissociation of PS and estimate the conductivity of the discharge products. Macroscale models can additionally incorporate the effect of stochastic $Li_2S$ precipitation to shed light on the competing mechanisms of surface passivation and PS retention and quantify the deleterious implications of the polysulfide shuttle on cell performance. Hence, developing a hierarchical modeling methodology as depicted in Fig. 4.20 can bridge the gap between atomistic scale simulations and macroscale models to present novel inferences toward achieving an enhanced electrochemical performance of the Li—S battery.

## Acknowledgment

The authors acknowledge the American Chemical Society, the Royal Society of Chemistry, the Electrochemical Society, and the American Society of Mechanical Engineers for the figures reproduced or adapted in this chapter from the referenced publications of their respective journals.

## References

[1] J.M. Tarascon, Key challenges in future Li-battery research, Philos. Transact. A Math. Phys. Eng. Sci. 368 (1923) (2010) 3227–3241.

[2] J.M. Tarascon, M. Armand, Issues and challenges facing rechargeable lithium batteries, Nature 414 (6861) (2001) 359–367.

[3] V. Etacheri, R. Marom, R. Elazari, G. Salitra, D. Aurbach, Challenges in the development of advanced Li-ion batteries: a review, Energ. Environ. Sci. 4 (9) (2011) 3243–3262.

[4] M. Armand, J.M. Tarascon, Building better batteries, Nature 451 (7179) (2008) 652–657.

[5] E.J. Cairns, P. Albertus, Batteries for electric and hybrid-electric vehicles, Annu. Rev. Chem. Biomol. Eng. 1 (2010) 299–320.

[6] C. Barchasz, F. Mesguich, J. Dijon, J.-C. Leprêtre, S. Patoux, F. Alloin, Novel positive electrode architecture for rechargeable lithium/sulfur batteries, J. Power Sources 211 (0) (2012) 19–26.

[7] Y.V. Mikhaylik, J.R. Akridge, Polysulfide shuttle study in the Li/S battery system, J. Electrochem. Soc. 151 (11) (2004) A1969–A1976.

[8] Z. Liu, S. Bertolini, P.B. Balbuena, P.P. Mukherjee, $Li_2S$ film formation on lithium anode surface of Li–S batteries, ACS Appl. Mater. Interfaces 8 (7) (2016) 4700–4708.

[9] T. Ould Ely, D. Kamzabek, D. Chakraborty, M.F. Doherty, Lithium–sulfur batteries: state of the art and future directions, ACS Appl. Energy Mater. 1 (5) (2018) 1783–1814.

[10] S.-H. Chung, A. Manthiram, Lithium–sulfur batteries with the lowest self-discharge and the longest shelf life, ACS Energy Lett. 2 (5) (2017) 1056–1061.

[11] F.Y. Fan, M.S. Pan, K.C. Lau, R.S. Assary, W.H. Woodford, L.A. Curtiss, W.C. Carter, Y.-M. Chiang, Solvent effects on polysulfide redox kinetics and ionic conductivity in lithium-sulfur batteries, J. Electrochem. Soc. 163 (14) (2016) A3111–A3116.

[12] J. Chen, K.S. Han, W.A. Henderson, K.C. Lau, M. Vijayakumar, T. Dzwiniel, H. Pan, L.A. Curtiss, J. Xiao, K.T. Mueller, Restricting the solubility of polysulfides in Li-S batteries via electrolyte salt selection, Adv. Energy Mater. 6 (11) (2016) 1600160.

[13] L.E. Camacho-Forero, T.W. Smith, P.B. Balbuena, Effects of high and low salt concentration in electrolytes at lithium–metal anode surfaces, J. Phys. Chem. C 121 (1) (2016) 182–194.

[14] M.I. Nandasiri, L.E. Camacho-Forero, A.M. Schwarz, V. Shutthanandan, S. Thevuthasan, P.B. Balbuena, K.T. Mueller, V. Murugesan, In-situ chemical imaging of solid-electrolyte interphase layer evolution in Li-S batteries, Chem. Mater. 29 (2017) 4728–4737.

[15] Q. Zhang, Y. Wang, Z.W. Seh, Z. Fu, R. Zhang, Y. Cui, Understanding the anchoring effect of two-dimensional layered materials for lithium–sulfur batteries, Nano Lett. 15 (6) (2015) 3780–3786.

[16] Z. Li, L. Yin, Nitrogen-doped MOF-derived micropores carbon as immobilizer for small sulfur molecules as a cathode for lithium sulfur batteries with excellent electrochemical performance, ACS Appl. Mater. Interfaces 7 (7) (2015) 4029–4038.

[17] S. Xiao, S. Liu, J. Zhang, Y. Wang, Polyurethane-derived N-doped porous carbon with interconnected sheet-like structure as polysulfide reservoir for lithium–sulfur batteries, J. Power Sources 293 (2015) 119–126.

[18] Q. Pang, J. Tang, H. Huang, X. Liang, C. Hart, K.C. Tam, L.F. Nazar, A nitrogen and sulfur dual-doped carbon derived from polyrhodanine@cellulose for advanced lithium–sulfur batteries, Adv. Mater. 27 (39) (2015) 6021–6028.

[19] T.-Z. Hou, H.-J. Peng, J.-Q. Huang, Q. Zhang, B. Li, The formation of strong-couple interactions between nitrogen-doped graphene and sulfur/lithium (poly)sulfides in lithium-sulfur batteries, 2D Materials 2 (1) (2015), 014011.

[20] J. Yang, S. Wang, Z. Ma, Z. Du, C. Li, J. Song, G. Wang, G. Shao, Novel nitrogen-doped hierarchically porous coralloid carbon materials as host matrixes for lithium–sulfur batteries, Electrochim. Acta 159 (2015) 8–15.

[21] Y. Qiu, G. Rong, J. Yang, G. Li, S. Ma, X. Wang, Z. Pan, Y. Hou, M. Liu, F. Ye, W. Li, Z.W. Seh, X. Tao, H. Yao, N. Liu, R. Zhang, G. Zhou, J. Wang, S. Fan, Y. Cui, Y. Zhang, Highly nitridated graphene–$Li_2S$ cathodes with stable modulated cycles, Adv. Energy Mater. 5 (23) (2015) 1501369.

[22] S. Niu, W. Lv, C. Zhang, F. Li, L. Tang, Y. He, B. Li, Q.-H. Yang, F. Kang, A carbon sandwich electrode with graphene filling coated by N-doped porous carbon layers for lithium-sulfur batteries, J. Mater. Chem. A 3 (40) (2015) 20218–20224.

[23] K. Ding, Y. Bu, Q. Liu, T. Li, K. Meng, Y. Wang, Ternary-layered nitrogen-doped graphene/sulfur/ polyaniline nanoarchitecture for the high-performance of lithium-sulfur batteries, J. Mater. Chem. A 3 (15) (2015) 8022–8027.

[24] S. Zhang, M. Liu, F. Ma, F. Ye, H. Li, X. Zhang, Y. Hou, Y. Qiu, W. Li, J. Wang, J. Wang, Y. Zhang, A high energy density $Li_2S$@C nanocomposite cathode with a nitrogen-doped carbon nanotube top current collector, J. Mater. Chem. A 3 (37) (2015) 18913–18919.

[25] Y. Zhao, F. Yin, Y. Zhang, C. Zhang, A. Mentbayeva, N. Umirov, H. Xie, Z. Bakenov, A free-standing sulfur/nitrogen-doped carbon nanotube electrode for high-performance lithium/sulfur batteries, Nanoscale Res. Lett. 10 (1) (2015) 1–6.

[26] F. Sun, J. Wang, H. Chen, W. Li, W. Qiao, D. Long, L. Ling, High efficiency immobilization of sulfur on nitrogen-enriched mesoporous carbons for Li–S batteries, ACS Appl. Mater. Interfaces 5 (12) (2013) 5630–5638.

[27] X.G. Sun, X. Wang, R.T. Mayes, S. Dai, Lithium–sulfur batteries based on nitrogen-doped carbon and an ionic-liquid electrolyte, ChemSusChem 5 (10) (2012) 2079–2085.

[28] Y. Qiu, W. Li, W. Zhao, G. Li, Y. Hou, M. Liu, L. Zhou, F. Ye, H. Li, Z. Wei, High-rate, ultralong cycle-life lithium/sulfur batteries enabled by nitrogen-doped graphene, Nano Lett. 14 (8) (2014) 4821–4827.

[29] J. Song, T. Xu, M.L. Gordin, P. Zhu, D. Lv, Y.-B. Jiang, Y. Chen, Y. Duan, D. Wang, Nitrogen-doped mesoporous carbon promoted chemical adsorption of sulfur and fabrication of high-areal-capacity sulfur cathode with exceptional cycling stability for lithium-sulfur batteries, Adv. Funct. Mater. 24 (9) (2014) 1243–1250.

[30] Y. Qiu, W. Li, W. Zhao, G. Li, Y. Hou, M. Liu, L. Zhou, F. Ye, H. Li, Z. Wei, S. Yang, W. Duan, Y. Ye, J. Guo, Y. Zhang, High-rate, ultralong cycle-life lithium/sulfur batteries enabled by nitrogen-doped graphene, Nano Lett. 14 (8) (2014) 4821–4827.

[31] G. Zhou, E. Paek, G.S. Hwang, A. Manthiram, High-performance lithium-sulfur batteries with a self-supported, 3D $Li_2S$-doped graphene aerogel cathodes, Adv. Energy Mater. 6 (2015) 1501355.

[32] Y. Xie, Z. Meng, T. Cai, W.-Q. Han, Effect of boron-doping on the graphene aerogel used as cathode for the lithium–sulfur battery, ACS Appl. Mater. Interfaces 7 (45) (2015) 25202–25210.

[33] X. Liang, A. Garsuch, L.F. Nazar, Sulfur cathodes based on conductive MXene nanosheets for high-performance lithium–sulfur batteries, Angew. Chem. Int. Ed. 54 (13) (2015) 3907–3911.

[34] J. Song, M.L. Gordin, T. Xu, S. Chen, Z. Yu, H. Sohn, J. Lu, Y. Ren, Y. Duan, D. Wang, Strong lithium polysulfide chemisorption on electroactive sites of nitrogen-doped carbon composites for high-performance lithium–sulfur battery cathodes, Angew. Chem. Int. Ed. 54 (14) (2015) 4325–4329.

[35] P. Zhu, J. Song, D. Lv, D. Wang, C. Jaye, D.A. Fischer, T. Wu, Y. Chen, Mechanism of enhanced carbon cathode performance by nitrogen doping in lithium–sulfur battery: an X-ray absorption spectroscopic study, J. Phys. Chem. C 118 (15) (2014) 7765–7771.

[36] Y. Yang, G. Zheng, S. Misra, J. Nelson, M.F. Toney, Y. Cui, High-capacity micrometer-sized $Li_2S$ particles as cathode materials for advanced rechargeable lithium-ion batteries, J. Am. Chem. Soc. 134 (37) (2012) 15387–15394.

[37] R. Eithiraj, G. Jaiganesh, G. Kalpana, M. Rajagopalan, First-principles study of electronic structure and ground-state properties of alkali-metal sulfides–$Li_2S$, $Na_2S$, $K_2S$ and $Rb_2S$, Phys. Status Solidi B 244 (4) (2007) 1337–1346.

[38] V. Viswanathan, K.S. Thygesen, J. Hummelshøj, J.K. Nørskov, G. Girishkumar, B. McCloskey, A. Luntz, Electrical conductivity in $Li_2O_2$ and its role in determining capacity limitations in non-aqueous $Li-O_2$ batteries, J. Chem. Phys. 135 (21) (2011), 214704.

[39] Z. Liu, D. Hubble, P.B. Balbuena, P.P. Mukherjee, Adsorption of insoluble polysulfides $Li_2S_x$ (x = 1, 2) on $Li_2S$ surfaces, Phys. Chem. Chem. Phys. 17 (14) (2015) 9032–9039.

[40] J. Hummelshøj, A. Luntz, J. Nørskov, Theoretical evidence for low kinetic overpotentials in $Li-O_2$ electrochemistry, J. Chem. Phys. 138 (3) (2013), 034703.

[41] P.G. Bruce, S.A. Freunberger, L.J. Hardwick, J.-M. Tarascon, Li-O2 and Li-S batteries with high energy storage, Nat. Mater. 11 (1) (2012) 19–29.

[42] M. Nagao, A. Hayashi, M. Tatsumisago, High-capacity $Li_2S$–nanocarbon composite electrode for all-solid-state rechargeable lithium batteries, J. Mater. Chem. 22 (19) (2012) 10015–10020.

[43] K. Zhang, L. Wang, Z. Hu, F. Cheng, J. Chen, Ultrasmall $Li_2S$ nanoparticles anchored in graphene nanosheets for high-energy lithium-ion batteries, Sci. Rep. 4 (2014) 1–7.

[44] M. Barghamadi, A. Kapoor, C. Wen, A review on Li-S batteries as a high efficiency rechargeable lithium battery, J. Electrochem. Soc. 160 (8) (2013) A1256–A1263.

[45] Z. Feng, C. Kim, A. Vijh, M. Armand, K.H. Bevan, K. Zaghib, Unravelling the role of $Li_2S_2$ in lithium-sulfur batteries: a first principles study of its energetic and electronic properties, J. Power Sources 272 (2014) 518–521.

[46] X. Ji, L.F. Nazar, Advances in Li–S batteries, J. Mater. Chem. 20 (44) (2010) 9821–9826.

[47] K. Reuter, M. Scheffler, Composition, structure, and stability of $RuO_2(110)$ as a function of oxygen pressure, Phys. Rev. B 65 (3) (2001), 035406.

[48] M. Cuisinier, P.-E. Cabelguen, S. Evers, G. He, M. Kolbeck, A. Garsuch, T. Bolin, M. Balasubramanian, L.F. Nazar, Sulfur speciation in Li–S batteries determined by operando X-ray absorption spectroscopy, J. Phys. Chem. Lett. 4 (19) (2013) 3227–3232.

[49] M. Ghaznavi, P. Chen, Sensitivity analysis of a mathematical model of lithium–sulfur cells: part II: precipitation reaction kinetics and sulfur content, J. Power Sources 257 (2014) 402–411.

[50] J. Xiao, J.Z. Hu, H. Chen, M. Vijayakumar, J. Zheng, H. Pan, E.D. Walter, M. Hu, X. Deng, J. Feng, Following the transient reactions in lithium–sulfur batteries using an in situ nuclear magnetic resonance technique, Nano Lett. 15 (5) (2015) 3309–3316.

[51] P. Cunningham, S. Johnson, E. Cairns, Phase equilibria in lithium-chalcogen systems II. Lithium-sulfur, J. Electrochem. Soc. 119 (11) (1972) 1448–1450.

[52] L. Grande, E. Paillard, J. Hassoun, J.B. Park, Y.J. Lee, Y.K. Sun, S. Passerini, B. Scrosati, The lithium/air battery: still an emerging system or a practical reality? Adv. Mater. 27 (5) (2015) 784–800.

[53] R. Dominko, M.U. Patel, V. Lapornik, A. Vizintin, M. Koželj, N. Novak Tušar, I. Arcon, L. Stievano, G. Aquilanti, Analytical detection of polysulfides in the presence of adsorption additives by operando X-ray absorption spectroscopy, J. Phys. Chem. C 119 (2015) 19001–19010.

[54] N.A. Cañas, S. Wolf, N. Wagner, K.A. Friedrich, In-situ X-ray diffraction studies of lithium–sulfur batteries, J. Power Sources 226 (2013) 313–319.

[55] H. Park, H.S. Koh, D.J. Siegel, First-principles study of redox end members in lithium–sulfur batteries, J. Phys. Chem. C 119 (9) (2015) 4675–4683.

[56] Y.-X. Chen, P. Kaghazchi, Metalization of $Li_2S$ particle surfaces in Li-S batteries, Nanoscale 6 (22) (2014) 13391–13395.

[57] J.D. Weeks, G.H. Gilmer, Dynamics of crystal growth, Adv. Chem. Phys. 40 (489) (1979) 157–227.

[58] A.D. Dysart, J.C. Burgos, A. Mistry, C.-F. Chen, Z. Liu, C.N. Hong, P.B. Balbuena, P.P. Mukherjee, V.G. Pol, Towards next generation lithium-sulfur batteries: non-conventional carbon compartments/sulfur electrodes and multi-scale analysis, J. Electrochem. Soc. 163 (5) (2016) A730–A741.

[59] F.Y. Fan, W.C. Carter, Y.M. Chiang, Mechanism and kinetics of $Li_2S$ precipitation in lithium–sulfur batteries, Adv. Mater. 27 (35) (2015) 5203–5209.

[60] J. Song, Z. Yu, M.L. Gordin, D. Wang, Advanced sulfur cathode enabled by highly crumpled nitrogen-doped graphene sheets for high-energy-density lithium–sulfur batteries, Nano Lett. 16 (2) (2016) 864–870.

[61] T. Ould Ely, D. Kamzabek, D. Chakraborty, M.F. Doherty, Lithium–sulfur batteries: state of the art and future directions, ACS Appl. Energy Mater. 1 (5) (2018) 1783–1814.

[62] X.-B. Cheng, J.-Q. Huang, Q. Zhang, Review—Li metal anode in working lithium-sulfur batteries, J. Electrochem. Soc. 165 (1) (2018) A6058–A6072.

[63] Q. Long, Z. Chenxi, M. Arumugam, A high energy lithium-sulfur battery with ultrahigh-loading lithium polysulfide cathode and its failure mechanism, Adv. Energy Mater. 6 (7) (2016) 1502459.

[64] A. Dysart, A. Mistry, C.-F. Chen, Z. Liu, C. Hong, P.B. Balbuena, P.P. Mukherjee, V.G. POL, Towards next generation lithium-sulfur batteries: non-conventional carbon compartments/sulfur electrodes and multi-scale analysis, J. Electrochem. Soc 163 (2015) A730, https://iopscience.iop.org/article/10.1149/2.0481605jes.

[65] H. Ye, M. Li, T. Liu, Y. Li, J. Lu, Activating Li2S as the lithium-containing cathode in lithium-sulfur batteries, ACS Energy Lett. 5 (7) (2020) 2234–2245.

[66] A.F. Voter, Introduction to the kinetic Monte Carlo method, in: Radiation Effects in Solids, Springer, 2007, pp. 1–23.

[67] F. Hao, A. Verma, P.P. Mukherjee, Mesoscale complexations in lithium electrodeposition, ACS Appl. Mater. Interfaces 10 (31) (2018) 26320–26327.

[68] F. Hao, A. Verma, P.P. Mukherjee, Mechanistic insight into dendrite–SEI interactions for lithium metal electrodes, J. Mater. Chem. A 6 (40) (2018) 19664–19671.

[69] B.S. Vishnugopi, F. Hao, A. Verma, P.P. Mukherjee, Double-edged effect of temperature on lithium dendrites, ACS Appl. Mater. Interfaces 12 (2020) 23931–23938.

[70] B.S. Vishnugopi, F. Hao, A. Verma, P.P. Mukherjee, Surface diffusion manifestation in electrodeposition of metal anodes, Phys. Chem. Chem. Phys. 22 (20) (2020) 11286–11295.

[71] B.S. Vishnugopi, F. Hao, A. Verma, L.E. Marbella, V. Viswanathan, P.P. Mukherjee, Co-electrodeposition mechanism in rechargeable metal batteries, ACS Energy Lett. 6 (6) (2021) 2190–2197, https://doi.org/10.1021/acsenergylett.1c00677.

[72] Z. Li, Y. Han, J. Wei, W. Wang, T. Cao, S. Xu, Z. Xu, Suppressing shuttle effect using Janus cation exchange membrane for high-performance lithium–sulfur battery separator, ACS Appl. Mater. Interfaces 9 (51) (2017) 44776–44781.

[73] A. Mistry, P.P. Mukherjee, Precipitation–microstructure interactions in the Li-sulfur battery electrode, J. Phys. Chem. C 121 (47) (2017) 26256–26264.

[74] A.N. Mistry, P.P. Mukherjee, Electrolyte transport evolution dynamics in lithium–sulfur batteries, J. Phys. Chem. C 122 (32) (2018) 18329–18335.

[75] A.N. Mistry, P.P. Mukherjee, "Shuttle" in polysulfide shuttle: friend or foe? J. Phys. Chem. C 122 (42) (2018) 23845–23851.

# CHAPTER 5

# Reliable HPLC-MS method for the quantitative and qualitative analyses of dissolved polysulfide ions during the operation of Li-S batteries

**Dong Zheng, Tianyao Ding, and Deyang Qu**

Department of Mechanical Engineering, College of Engineering and Applied Science, University of Wisconsin Milwaukee, Milwaukee, WI, United States

## Contents

| | | |
|---|---|---|
| 5.1 | Introduction to HPLC-MS | 159 |
| | 5.1.1 High-performance liquid chromatography | 159 |
| | 5.1.2 Mass spectrometry and other detectors | 162 |
| 5.2 | Dissolved polysulfide ions and their behaviors in nonaqueous electrolytes | 164 |
| 5.3 | Advantages of HPLC-MS vs. other analytical techniques | 166 |
| 5.4 | One-step derivatization, separation, and determination of polysulfide ions | 171 |
| 5.5 | The mechanism of sulfur redox reaction determined in situ electrochemical-HPLC technique | 175 |
| | 5.5.1 Mechanism studies with other techniques | 175 |
| | 5.5.2 Investigation of sulfur redox mechanism using electrochemical HPLC techniques | 178 |
| 5.6 | Conclusions | 194 |
| | References | 195 |

## 5.1 Introduction to HPLC-MS

### 5.1.1 High-performance liquid chromatography

Over the past several decades, due to continuous progress in manufacturing, electronics, and material science, high–performance liquid chromatography (HPLC) has become a powerful tool for the quantitative and qualitative analyses of different types of complex samples in both industry and academia.

---

*Lithium-Sulfur Batteries*
https://doi.org/10.1016/B978-0-12-819676-2.00003-7

Copyright © 2022 Elsevier Inc.
All rights reserved.

The power and popularity of HPLC are mainly based on its superior separation capability and capacity. Under optimal conditions, baseline separation of dozens of analytes can be easily accomplished for a complex mixture within a 30-min HPLC run, whether analytes are organic or inorganic, or small molecules or polymers. The superior separation of HPLC systems is solely dependent on the interaction of the analyte, stationary phase, and mobile phase. Typically, stationary phases consist of fine particles (in the size of several μm, with modified surfaces) packed in HPLC columns, and mobile phases are solvents to elute analytes through HPLC columns (stationary phases).

According to the interaction between the stationary phases and the analytes, HPLC systems can roughly be classified into four different types based on different separation mechanisms: adsorption chromatography, partition chromatography, ion-exchange chromatography, and size-exclusion chromatography [1–3]. For adsorption chromatography (also called normal-phase chromatography), polar materials are used as stationary phases, and nonpolar solvents are used as mobile phases. The separation of analytes is achieved due to different strengths of analytes adsorbed on the polar stationary phases, the stronger adsorption, the later the elution-out. For a partition chromatography (also called reversed-phase chromatography), nonpolar materials are used as stationary phases and polar solvents are used as mobile phases. The separation of analytes is dependent on the partition of analytes between nonpolar stationary phases and polar mobile phases. Nonpolar analytes have a high partition with nonpolar stationary phases and thus will elute out later. For ion-exchange chromatography, the stationary phases normally have ionic groups. The mobile phases are buffer solutions, and the separation of analytes is based on the electronic static absorption between the ionic groups on stationary phases and the analytes. Analytes with higher ionic interaction with stationary phases will elute out later. For size-exclusion chromatography, porous materials with a particular averaged pore size are used as the stationary phases. The separation of analytes is according to their effective sizes in the mobile phases. The analytes with effective size larger than the averaged pore size of the stationary phases will elute out first and the analytes with effective size smaller than the averaged pore size of the stationary phases will elute out later. Reverse-phase HPLC (RP-HPLC) based on partition mechanism is now the most popular due to the ease of operation, the robustness of the system, the reproducibility of separation, and the versatility of application. In the following parts of this chapter, all discussions are based on RP-HPLC unless it is specified.

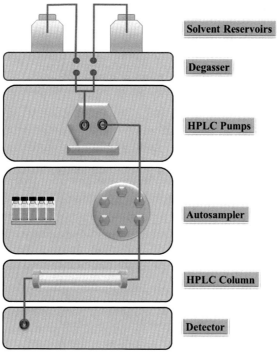

**Fig. 5.1** Typical scheme of HPLC system components. *(Source: Authors.)*

As shown in Fig. 5.1, a RP-HPLC system typically consists of a solvent delivery system for the mobile phase, a sample inlet, an HPLC column, and a detector. The solvent delivery system is essential to deliver the desired composition of mobile phases through an HPLC column at a preset flow rate, to flush the sample into the HPLC column, and to elute the samples out of the HPLC column to the detectors. The delivery system includes solvent reservoirs, degasser, pumps, mixers, valves, dampers, and flow/pressure controllers. At a preset flow rate, there are two types of elution for a solvent delivery system, isocratic elution, and gradient elution. In isocratic elution, the composition of the mobile phase is constant during the whole period of elution, while in gradient elution, and the composition of the mobile phase keeps changing during elution. Compared with the isocratic elution, the gradient elution has several advantages of enhanced detection sensitivity, enhanced chromatographic resolution and capacity, and alleviated column deterioration and residual carryover, especially for complex samples [1,2]. The sample inlets are dedicated to precisely and reproducibly introduce a specific amount (volume) of liquid samples into a pressurized and flowing mobile phase without any adverse effects on the column efficiency.

Although the automated sample introduction system is more sophisticated than the manual sample introduction system, the key components for both are the same. They both include an injection valve, a sample loop, and a sample syringe. A liquid sample is first drawn into the sample syringe, then the liquid sample in the syringe is injected into the sample loop, and at the end of the loop, the liquid sample is rapidly inserted into the mobile phase flow path by the injection valve. After this introduction (also called injection) process of a sample, the injected sample will be flushed into an HPLC column for separation. Normally, the sample inlet system will also send a signal to trigger the detector to be turned on.

An HPLC column is the core part of a HPLC system, where the state-of-the-art separation occurs. A typical HPLC column consists of a stainless tube with porous materials (stationary phase) packed inside. The mobile phase is then forced to flow through the stationary phase with high pressure (sometimes HPLC also is called high-pressure liquid chromatography), where the separation occurs. In a RP-HPLC column, the most common stationary phase is surface-modified hydrophobic silica particles. The partition or separation of analytes results from the interaction between the analytes and the hydrophobic surface. Most of the hydrophobic surface functionalities are the organosiloxane derivatives. Among them, the most common one is octadecylsiloxane, which has an octadecyl chain with 18 carbon atoms (referred to as C18).

## 5.1.2 Mass spectrometry and other detectors

After the separated analytes are eluted out of the HPLC column by mobile phases, a detector or several detectors connected to the end of a HPLC column will be used to identify each analyte. Over a century ago, the chromatographic pioneer Mikhail Semyonovich Tswett used his eye as a detector to identify his separated colorful analytes by chromatography. Today, many types of detectors with high sensitivity and fast response can be used in tandem with HPLC to ensure the adequate measurement of each analyte of interest from HPLC separation, such as UV–visible detector (UV-Vis), refractive index detector (RID), fluorescence detector, evaporative light scattering detector (ELSD), and mass spectrometric detector (MSD). Each type of detector has its advantages and limits [1,4]. For example, a UV–Vis detector has high sensitivity and wide linear range to analytes with high UV-Vis absorbance, but low sensitivity to analytes with low UV-Vis absorbance. The RID is a universal detector with good sensitivity and linear range independence to the properties of analytes, but is incompatible with a gradient elution condition. Most detectors tandem with HPLC have one major limitation for applications: they are incapable of performing untargeted analysis and identifying unknown analytes (or when standards of analytes are unavailable). In other words, the identification of analytes through HPLC separation is solely based on the retention time of each analyte. If coelution of analytes or an unknown chromatographic peak occurs, those optical instruments cannot offer any useful information for identification.

Only the MSD has the capability of identifying the analytes based on molecular properties of analytes. Mass spectrometer (MS) is a powerful analytical tool to obtain the molecular weight and structural information of analytes. A typical MS is composed of three core parts besides the vacuum system required: ion source, mass analyzer, and detector [5–7]. In an MS, the mass information (molecular weight) of a chemical is obtained through the measurement of the mass-to-charge ($m/z$) ratio of ions. The ion source of an MS is where chemicals form gas-phase ions, which can be negative or positive. Depending on the ionization methods, there are many different types of ion sources including (but not limited to) electron impact (EI) source, chemical ionization (CI) source, fast atom bombardment (FAB) source, inductively coupled plasma (ICP) source, matrix-assisted laser desorption ionization (MADLI), atmospheric pressure photoionization (APPI) source, atmospheric pressure chemical ionization (APCI) source, and electrospray ionization (ESI) source [7]. Once the ions are formed in an ionization source, they will enter a mass analyzer in which those ions are analyzed (or separated) according to their mass-to-charge ratios.

Just as there are many different types of ion sources, based on different mechanisms to differentiate ions of different $m/z$ ratios, different types of mass analyzers have been developed to work individually or tandem, such as electric/magnetic sectors (E/B) analyzers, quadrupole (Q) analyzer, ion trap (IT) analyzer, time-of-flight (TOF) analyzer, Fourier transform ion cyclotron resonance (FTICR), and Fourier transform Orbitrap (FT-OT) [7]. Through an MS analyzer, the separated ions of different $m/z$ ratios will reach a detector. The common MS detectors include photographic plate detectors, Faraday cup detectors, electron multiplier detectors, and electrooptical ion detectors [7]. To use MS to determine chromatographic separation, the choice of an adequate ion source is the key. For gas chromatography (GC), the EI source is perfect since the analytes from GC are already in a gas phase; however, for liquid chromatography (LC), the atmospheric pressure ionization (API) source is commonly used since the API sources (including ESI, APCI) can directly ionize the analytes in the pressurized mobile phases from LC [1,7].

The HPLC/(ESI)MS system has become one of the most popular and powerful platforms for separation, identification, and quantitation of really complex samples [1,2,7]. Due to the fast growth of interest in understanding the electrochemical processes for rechargeable lithium-based batteries, such as lithium-ion battery (LIB), lithium-oxygen battery (Li-O$_2$), and lithium-sulfur battery (Li-S), the HPLC/MS system has been applied in many studies of investigating the fundamental mechanism of the electrochemical process during charge/discharge of batteries [8–13]. In the following parts of this chapter, recent research works on studying the mechanism of a Li-S battery by the HPLC/MS method will be introduced, compared, and discussed.

## 5.2 Dissolved polysulfide ions and their behaviors in nonaqueous electrolytes

Dissolved polysulfide ions ($S_n^{2-}$, $n \geq 2$) are important discharge and recharge intermediates of Li-S batteries [14–20]. Polysulfide ions with $n > 2$ are highly soluble in nonaqueous electrolytes. The solubility of polysulfide ions can be several hundreds of mM in some electrolytes [21–23]. On the one hand, the high solubility of polysulfide ions can facilitate the sulfur redox reaction by preventing the formation of surface passivation on the sulfur cathode, which would reduce the reaction kinetics; on the other hand, the dissolved polysulfide ions are also the root cause of poor performance of Li-S batteries. The highly dissolved polysulfide ions diffuse to the metallic Li anode and become further reduced, which can passivate the anode and result in a polysulfide shuttle. A polysulfide shuttle or "shuttle effect" is responsible for the low coulombic efficiency, high self-discharge, and poor rechargeability [15,24]. Therefore, the analysis of dissolved polysulfide species has great importance not only in the fundamental mechanism study but also in the engineering design and the optimization of a Li-S battery.

It is well known that the sulfur redox reaction is one of the most complex reactions. The major obstacle for the investigation of the sulfur redox mechanism in a Li-S battery is the lack of analytical means for the qualitative and quantitative identification of the dissolved polysulfide ions during discharge and recharge. In solution phases (either in aqueous solution or in nonaqueous electrolytes), complicated equilibrium reactions of polysulfide ions always exist due to the low S-S bond dissociation energy (33 kcal/mol) and the close values of Gibbs free energy for each polysulfide anion [25]. In the early studies of polysulfide ions in aqueous conditions, Lev et al. reported that the Gibbs free energy of the polysulfide anions with sulfur chain length between 2 and 8 was from 66 to $77.4\,\mathrm{kJ\,mol^{-1}}$ [26]. In a recent study of polysulfide species in nonaqueous electrolytes, Vijayakumar et al. demonstrated that the energy barrier for dissociation reaction of $S_6^{2-}$ (into radical anions $2S_3^-$) was just about $0.25\,\mathrm{eV}$ in DMSO [27]. The equilibrium reactions of polysulfide ions can be generally categorized as one-molecule reaction (as shown in Eq. 5.1), two-molecule reaction (as shown in Eqs. 5.2 and 5.3), and multiple-molecule reaction (as shown in Eqs. 5.4, 5.5, and 5.6). Due to the multiple equilibrium reactions of fast kinetics, polysulfide ions of various sulfur chain lengths always coexist in a solution at ambient temperature, regardless of the starting materials. Therefore, it is impossible to make a pure standard solution of a single-chain-length polysulfide. For example, stoichiometric reaction (as shown in Eq. 5.7) [21] between $Li_2S$ and $S_8$ can only yield a solution with different polysulfide species (from $Li_2S_n$, $n$ is up to 10) instead of the pure stoichiometric polysulfide ions [26,27].

$$S_6^{2-} \leftrightarrow 2\,S_3^- \tag{5.1}$$

$$2\,S_6^{2-} \leftrightarrow S_5^{2-} + S_7^{2-} \tag{5.2}$$

$$S^{2-} + S_7^{2-} \leftrightarrow 2\,S_4^{2-} \tag{5.3}$$

$$3\,S_2^{2-} \leftrightarrow S_4^{2-} + 2\,S^{2-} \tag{5.4}$$

$$S_2^{2-} + S_3^{2-} + S_4^{2-} \leftrightarrow S_7^{2-} + 2\,S^{2-} \tag{5.5}$$

$$4\,S_6^{2-} \leftrightarrow 4\,S_5^{2-} + 1/2\,S_8 \tag{5.6}$$

$$Li_2S + (n-1)/8\,S_8 \rightarrow Li_2S_n \tag{5.7}$$

The equilibrium reactions, thus the distribution of polysulfide ions in a solution, also show the dependence on temperature, degree of solvation, and concentration. Vijayakumar et al. reported that the ESR signal intensity of polysulfide ions reversibly changed with temperature, indicating the influence of temperature on the distribution of $S_6^{2-}$ to $S_3^-$ radicals in the solution [27]. Seel et al. also observed that the UV–visible spectra of polysulfides in

EPD (electron-pair donor) solvents gradually changed as the temperature varied from 216 K to 293 K [28]. After synthesizing polysulfides in 12 organic solvents, Rauh et al. found that the UV-visible spectra for $Li_2S_{9.7}$ (in stoichiometric ratio) in pure tetrahydrofuran (THF) and in 3:2 THF/DMSO mixture were totally different [21]. While $Li_2S_{9.7}$ showed no absorption within 600 to 800 nm in THF solvent, a strong absorbance peak was observed in THF/DMSO mixture solvent.

The complexity of the behaviors of polysulfide ions makes it difficult to study them in nonaqueous electrolytes. To make things more complicated, polysulfide ions also are not stable against many chemicals, such as transition metals [29], acid [26], base [30], oxidizer [31], and even water [20]. When the polysulfide ions in a nonaqueous solution are in contact with those chemicals, disproportionation reactions could happen gradually or drastically, as shown in the following Eqs. (5.8)–(5.12).

$$S_n^{2-} + H_2O \rightarrow HS^- + OH^- + (n-1)/8\, S_8 \tag{5.8}$$

$$S_n^{2-} + H^+ \rightarrow HS^- + (n-1)/8\, S_8 \tag{5.9}$$

$$S_5^{2-} + 3\, OH^- \rightarrow S_2O_3^{2-} + 3\, HS^- \tag{5.10}$$

$$S_n^{2-} + O_2 \rightarrow S_2O_3^{2-} + (n-2)/8\, S_8 \tag{5.11}$$

$$S_n^{2-} + 2\, Cu \rightarrow Cu_2S + S_{(n-1)}^{2-} \tag{5.12}$$

## 5.3 Advantages of HPLC-MS vs. other analytical techniques

To fundamentally understand the redox mechanism of a sulfur electrode, numerous analytical techniques have been used, including advanced electrochemical methods, [32,33] UV-visible, [34,35] Raman spectroscopy, [36,37] X-ray diffraction (XRD), [38–40] X-ray absorption spectroscopy (XAS), [41,42] nuclear magnetic resonance (NMR), [43–45] electron paramagnetic resonance (EPR), [46] X-ray photoelectron (XPS), [47] atomic-force microscopy (AFM)/scanning electron microscopy (SEM)/transmission electron microscopy (TEM), [48–50] Fourier transform infrared (FT-IR), [51] resonant inelastic X-ray scattering (RIXS), [52] X-ray fluorescence microscopy (XRF), [53] transmission X-ray microscopy (TXM), [54] differential scanning calorimetry (DSC), [55] inductively coupled plasma–optical emission spectrometry (ICP-OES), [11] as well as HPLC (HPLC/UV, HPLC/RID, and HPLC-MS) [11,12,43,56–60].

Electrochemically produced sulfide chemicals from a sulfur cathode can be classified into two groups: low-solubility species, such as $Li_2S$, $Li_2S_2$, and high-solubility species, such as lithium polysulfides with sulfur chain length $\geq 3$. The low-solubility sulfides can be analyzed with techniques such as XRD, XAS, AFM, SEM, TEM, XRF, RIXS, and TXM. UV-Vis, HPLC, EPR, and DSC were reported for the analysis of high-solubility polysulfide species in an electrolyte. NMR, Raman, FT-IR, and ICP-OES were used for the investigation of both kinds. The advantages and limitations of those analytical techniques on the investigation of the redox mechanism of sulfur cathodes are summarized in Table 5.1.

Compared with other analytical methods, the HPLC tandem with MS (or UV/Vis, RID) has two obvious and unparalleled advantages that make it the only reliable analytical method for the qualitative and quantitative analysis of dissolved polysulfide ions:

1. Physical separation of each polysulfide ion from the mixtures. Due to the similarity of polysulfide ions of various chain lengths, most of the spectroscopic analytical methods, such as UV-Vis, Raman, FT-IR, and XPS, yield the overlap of peaks, which need to be deconvoluted. Mathematical deconvolution is always a major source of errors. However, before detection, HPLC does the baseline separation for elemental sulfur and each polysulfide species. Fig. 5.2 shows the difference of results from XPS analysis (mathematical deconvolution) and from HPLC analysis (experimental baseline separation).

2. Quantitative and qualitative identifications of each individual polysulfide species without pure standards. Almost all instrumental analysis techniques have difficulties for the measurement of polysulfides, because there is no standard solution of polysulfide anions with a precisely defined chain length for reference measurements. As described in the previous section, multiple chemical disproportionation and redox reactions will result in the distribution of polysulfide anions with various chain lengths regardless of the initial stoichiometry. Therefore, the assignment of the observed signals to polysulfide species becomes a challenge without pure standard solutions. Either theoretical calculations or controversial observations have been used to allocate the observed signals to each polysulfide. For example, in the literature reporting UV-Vis analysis of $S_4^{2-}$, it was assigned to 440 nm [63] and 425 nm [34,64] due to different electrolytes used. HPLC-MS is one of the very few techniques that do not need a pure chemical standard for analysis. The chromatographic peaks for each polysulfide species can be easily

**Table 5.1** Comparison of different analytical techniques used in understanding the sulfur redox in Li-S batteries.

| Method | Advantage | Limitation | Reference |
|---|---|---|---|
| HPLC | Fully speciation of all soluble polysulfide species ($S_n^{2-}$, $n \geq 3$); real-time and *in operando* analysis; qualitative and quantitative analyses of soluble polysulfide species and elemental sulfur; high sensitivity and high selectivity | Derivatization needed; not intact and in situ analysis; incompatible with insoluble species ($Li_2S_2$ and $Li_2S$) and radical species (such as $LiS_3$) | [12,43,57,59,60] |
| Electroanalytical | Intact and in situ analysis for charge transfer and system resistance; qualitative analysis | Indirect speciation of sulfur species; no quantitative analysis due to chemical process interference | [32,33,61,62] |
| UV–Vis | Intact and in situ analysis for sulfur species; qualitative and semiquantitative analyses; high sensitivity | Lack of pure standard for speciation of sulfur species; semiquantitative analysis fully based on subjective signal deconvolution; low selectivity | [34,35,63–66] |
| Raman | Intact and in situ analysis for sulfur species; qualitative and semiquantitative analyses | Lack of pure standard for speciation of sulfur species; semiquantitative analysis fully based on subjective signal deconvolution, and calculation; low selectivity and sensitivity | [36,37,67–70] |
| XRD | Intact and in situ analysis for sulfur species; qualitative and semiquantitative analyses | Only solid species observable; medium selectivity and sensitivity | [38–40] |
| XAS | Intact and in situ analysis for sulfur species; qualitative and semiquantitative analyses | Lack of pure standard for speciation of sulfur species; semiquantitative analysis fully based on subjective signal deconvolution; low selectivity and sensitivity | [41,42] |

| | | | |
|---|---|---|---|
| NMR | Intact and in situ qualitative analysis for $^7$Li and $^{33}$S NMR; qualitative and semiquantitative analyses with $^1$H NMR | For $^7$Li and $^{33}$S NMR, lack of pure standard for speciation of sulfur species, qualitative analysis fully based on subjective signal deconvolution, low selectivity and sensitivity<br>For $^1$H NMR, derivatization needed, semiquantitative ex situ analysis partially based on calculated and simulated signal assignment, medium selectivity, and sensitivity | [43–45,71] |
| EPR | Intact and in situ analysis for radical sulfur species; qualitative and semiquantitative analyses | Only radical species observable; lack of standards for radical species other than $S_3^-$; low sensitivity and medium selectivity | [27,46,72] |
| XPS | Qualitative analysis for different sulfur species | Lack of pure standard for speciation of sulfur species; ex situ analysis fully based on subjective signal deconvolution and calculation; low selectivity and sensitivity | [47,73] |
| Others including AFM/ SEM/TEM, FT-IR, RIXS, XRF, TXM, DSC, ICP-OES | Morphological, structural, and compositional analyses of sulfur species; qualitative (or semiqualitative) analyses | Incapable of speciation of individual sulfur species; incapable of quantitative analysis; medium to low sensitivity and selectivity | [11,48–55] |

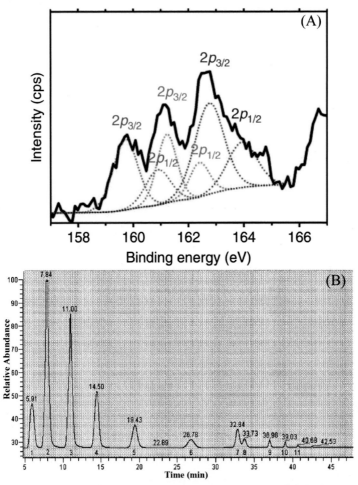

Fig. 5.2 Comparison of XPS analysis and HPLC analysis. (A) polysulfide XPS analysis with mathematical deconvolution; (B) HPLC chromatogram of organopolysulfides obtained by derivatization of inorganic polysulfides. *(Panel (A) reproduced with permission from Y.S. Su, Y. Fu, T. Cochell, A. Manthiram, A strategic approach to recharging lithium-sulphur batteries for long cycle life, Nat. Commun. 4 (2013), 2985, Copyright (2013) Springer Nature. Panel (B) reproduced with permission from D. Rizkov, O. Lev, J. Gun, B. Anisimov, I. Kuselman, Development of in-house reference materials for determination of inorganic polysulfides in water, Accred. Qual. Assur. 9 (2004) 399-403, Copyright (2004) Springer Nature.)*

identified by the different $m/z$ ratios of the polysulfide species of different chain lengths.

Due to these two advantages, the HPLC method provides the most reliable analysis of polysulfides in Li-S batteries quantitatively and qualitatively compared with other analytical methods.

## 5.4 One-step derivatization, separation, and determination of polysulfide ions

Diao et al. reported investigating polysulfides dissolved in the electrolyte of Li–S batteries during discharge and recharge by the HPLC/MS method [11]. Later, Manthiram and coworkers used HPLC/UV–Vis to identify the polysulfides in the catholyte of a lithium dissolved polysulfide battery (Li-PS) [74]. Those early HPLC analyses of polysulfides in Li-S and Li-PS batteries were problematic due to two aspects: first, the inorganic polysulfides lack of interaction with nonpolar $C_{18}$ stationary phases even with high water content in the mobile phase; thus, no retention and separation can be achieved; second, the instability of polysulfide ions may result in decomposition and disproportionation once in contact with the mobile phase, e.g., methanol or water/methanol mixture. Thus, the reequilibration and the redistribution of polysulfide ions will occur during the elution. Further investigation by Qu and coworkers [56] clearly demonstrated the infeasibility of direct analysis of inorganic polysulfides with HPLC. The major chromatographic peak reported as polysulfides in Diao's paper was proven to be elemental sulfur ($S_8$) formed once the polysulfide ions are in contact with the mobile phase [56]. Therefore, the key to apply HPLC in the analysis of polysulfide ions is to stabilize polysulfide ions and ensure that no subsequent chemical changes occur in a mobile phase during elution.

Lev and coworkers reported a one-step derivatization of polysulfide ions in an aqueous solution [26,75]. Through methylation, ionic polysulfide species can be correspondingly derivatized and form the covalent organopolysulfide species as shown in Eq. (5.13). Since the kinetics of the derivatization reaction are faster than that of polysulfide disproportionation, the original structure, and chain length of polysulfide ions are preserved [26]. The derivatized organopolysulfides have much higher stability than that of the original polysulfide ions [76]. It was demonstrated by Lev and coworkers that the decomposition and the reequilibration of organopolysulfides during the operation condition of HPLC were negligible [26].

$$S_n^{2-} + 2\,CF_3SO_3CH_3 \rightarrow (CH_3)_2S_n + 2\,CF_3SO_3^- \qquad (5.13)$$

Another benefit of derivatization is the polarity change of polysulfides. Through the derivatization, the highly polar ionic polysulfide ions are converted to the highly nonpolar organopolysulfides. Consequently, the partition of the organopolysulfides between nonpolar stationary phases and polar mobile phases can be satisfied. Therefore, good retention and separation of different organopolysulfides based on different sulfur chain lengths can be obtained by the RP-HPLC system [26]. The organopolysulfides with the longer sulfur chain length will have a longer retention time and elute out later in HPLC chromatogram.

Using different derivatization reagents as shown in Eqs. (5.14) and (5.15), Barchasz et al. [12], Kawase et al. [43], and Qu and coworkers [57] successfully applied the HPLC method with one-step derivatization to identify polysulfide ions in the electrolyte of Li-S batteries. The RID (refractive index detector) detector was used in Barchasz's study. They reported a linear dependency between the logarithm of the chromatographic retention time and the sulfur chain length of organopolysulfides ($RS_nR$, while $n \leq 8$) as shown in Fig. 5.3. Although the conclusion was consistent with that of Qu and coworkers [57], the assignment of the chromatographic peaks

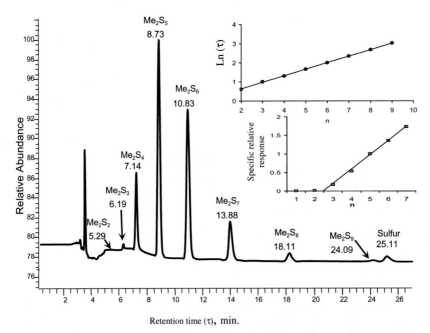

**Fig. 5.3** Typical HPLC chromatogram of organopolysulfide mixtures after derivatizing inorganic polysulfides with methyl triflate. Inserts indicate the dependency of the retention time and the sulfur chain length of organopolysulfides. *(Reproduced with permission from A. Kamyshny, A. Goifman, J. Gun, D. Rizkov, O. Lev, Equilibrium distribution of polysulfide ions in aqueous solutions at 25°C: a new approach for the study of polysulfides' equilibria, Environ. Sci. Technol. 38 (2004) 6633-6644, Copyright (2004) American Chemical Society.)*

was not based on direct experimental evidence [26]. Kawase et al. and Qu and coworkers identified the chromatographic peaks based on the $m/z$ ratio using MS detector, which is in tandem with the HPLC system (HPLC/MS). Depending on the derivatization reagents used (methyl triflate, benzyl chloride, or 4-(dimethylamino) benzoyl chloride), the organopolysulfides after HPLC separation can either form positive adduct ions in (+) ESI/MS or form negative fragment ions in (−) APCI/MS as shown in Table 5.2. MS can determine the $m/z$ ratio of the elutants of each chromatographic peak as shown in Fig. 5.4. Since the $m/z$ ratio of each of the organopolysulfides of different chain lengths is known, the chromatographic peaks of different organopolysulfides can be attributed easily and unambiguously. Although the retention time of organopolysulfides could be greatly influenced by the change of different derivatization reagents and different HPLC conditions, e.g., the type of column and mobile-phase composition, the retention time always has a linear relationship with the sulfur chain length of organopolysulfides [26] as shown in Figs. 5.3 and 5.4.

$$S_{(n)}^{2-} + 2 \left[ \text{Cl} \diagdown \bigcirc \right] \longrightarrow \bigcirc \diagdown S_{(n)} \diagdown \bigcirc + 2Cl^- \tag{5.14}$$

$$S_{(n)}^{2-} + 2 \left[ \text{Cl} \diagdown \overset{O}{\diagup} \bigcirc \diagdown N \diagdown \right] \longrightarrow N \diagdown \bigcirc \diagdown \overset{O}{\diagup} S_{(n)} \diagdown \overset{O}{\diagup} \bigcirc \diagdown N \diagdown + 2Cl^- \tag{5.15}$$

4-(dimethylamino) benzoyl chloride

While the organopolysulfides can be qualitatively identified based on the sequence of elution, the relative amount of organopolysulfides can be quantitatively determined by the integration of the corresponding chromatographic peaks. According to Beer's law, the signal intensity (the chromatographic peak area) of each organopolysulfide in a chromatogram should be proportional to the amount (concentration) of each organopolysulfide in the derivatized electrolyte; thus, the changes of chromatographic peak areas for different organopolysulfides can be correlated to the concentration changes of different lithium polysulfides in the electrolyte of Li-S battery. By observing the changes of chromatographic peak areas of different organopolysulfides derivatized at different stages of discharge and recharge, one can easily deduce how the concentrations of different polysulfide ions change at different stages of discharge/charge in Li-S batteries.

**Table 5.2** The summary of different *m/z* ions for different organopolysulfides observed at different HPLC/MS analyses after the inorganic polysulfides derivatized with different reagents [43,57,59,60].

| Organopolysulfide | Derivatization reagent: *Methyl triflate* [59,60] | | Derivatization reagent: *Benzyl chloride* [43] | | Derivatization reagent: *4-(Dimethylamino) benzoyl chloride* [57] | |
|---|---|---|---|---|---|---|
| | Fragment ion (−)APCI/MS | *m/z* ratio | Adduct ion (+) ESI/MS | *m/z* ratio | Adduct ion (+) ESI/MS | *m/z* ratio |
| $R_2S_2$ | $[CH_3S]^-$ | 47 | $[(C_7H_7)_2S_2 + NH_4]^+$ | 264 | $[(C_9H_{10}NO)_2S_2 + H]^+$ | 361 |
| $R_2S_3$ | $[CH_3S_2]^-$ | 79 | $[(C_7H_7)_2S_3 + NH_4]^+$ | 296 | $[(C_9H_{10}NO)_2S_3 + H]^+$ | 393 |
| $R_2S_4$ | $[CH_3S_3]^-$ | 111 | $[(C_7H_7)_2S_4 + NH_4]^+$ | 328 | $[(C_9H_{10}NO)_2S_4 + H]^+$ | 425 |
| $R_2S_5$ | $[CH_3S_4]^-$ | 143 | $[(C_7H_7)_2S_5 + NH_4]^+$ | 360 | $[(C_9H_{10}NO)_2S_5 + H]^+$ | 457 |
| $R_2S_6$ | $[CH_3S_5]^-$ | 175 | $[(C_7H_7)_2S_6 + NH_4]^+$ | 392 | $[(C_9H_{10}NO)_2S_6 + H]^+$ | 489 |
| $R_2S_7$ | $[CH_3S_6]^-$ | 207 | $[(C_7H_7)_2S_7 + NH_4]^+$ | 424 | $[(C_9H_{10}NO)_2S_7 + H]^+$ | 521 |
| $R_2S_8$ | $[CH_3S_7]^-$ | 239 | $[(C_7H_7)_2S_8 + NH_4]^+$ | 456 | $[(C_9H_{10}NO)_2S_8 + H]^+$ | 553 |

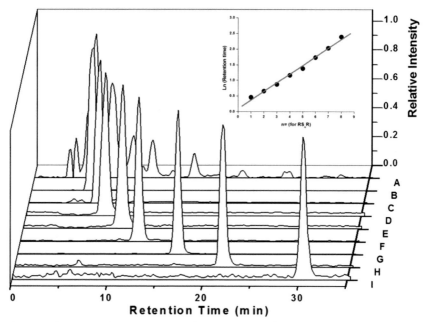

**Fig. 5.4** The HPLC chromatograms of organopolysulfide mixtures after derivatized by 4-(dimethylamino) benzoyl chloride. A=total ion current chromatogram; B=ion chromatogram for m/z 329; C=ion chromatogram for m/z 361; D=ion chromatogram for m/z 393; E=ion chromatogram for m/z 425; F=ion chromatogram for m/z 457; G=ion chromatogram for m/z 489; H=ion chromatogram for m/z 521; I=ion chromatogram for m/z 553. Inset: The dependencies of the retention time on polysulfide chain length for the derivatized $RS_nR$. (Reproduced with permission from D. Zheng, D. Qu, X. Yu, H. Lee, X. Yang, D. Qu, Quantitative and qualitative determination of polysulfide species in the electrolyte of a lithium-sulfur battery using HPLC ESI/MS with one-step derivatization, Adv. Energy Mater. 5 (2015) 1401888, Copyright (2015) John Wiley & Sons.)

## 5.5 The mechanism of sulfur redox reaction determined in situ electrochemical-HPLC technique

### 5.5.1 Mechanism studies with other techniques

Before introducing the progress of a mechanistic study by the HPLC technique for rechargeable Li-S batteries, it is essential to summarize the current status of mechanism studies by other techniques and to introduce the widely accepted mechanism based on those techniques. Fig. 5.5 shows typical cyclic voltammetry (CV) and galvanostatic discharge and recharge curves of sulfur electrodes in ether-based electrolytes. There are two discharge plateaus and two charge plateaus corresponding to two reduction peaks and two

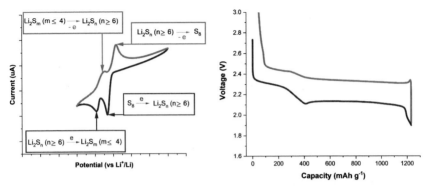

**Fig. 5.5** Typical CV and charge/discharge profiles for sulfur redox in Li-S batteries. *(Source: Authors.)*

oxidation peaks in CV, respectively [14–20]. The general and widely accepted redox mechanisms are [14–20]:

1. *The discharge or reduction of a sulfur electrode*: It includes three steps. The first discharge step is the first discharge plateau above 2.3 V corresponding to the first reduction peak in CV, in which elemental sulfur ($S_8$) gains electrons and is electrochemically reduced to long-chain linear polysulfides ($Li_2S_n$ with $n \geq 6$). The second discharge step is the transition period between 2.3 and 2.1 V, through which the first discharge plateau is connected with the second discharge plateau. In this step, the long-chain polysulfides gain electrons and are further electrochemically reduced to medium-length-chain linear polysulfides. The third and last step includes the second discharge plateau at around 2.1 V corresponding to the second reduction peak in CV, in which medium-length-chain linear polysulfides gain electrons and are eventually electrochemically reduced to insoluble lithium sulfide and lithium disulfide. The cathodic active materials during the discharge undergo a solid($S_8$)-liquid (polysulfides)-solid(sulfide) phase transfer.

2. *The recharge or oxidation of a sulfur electrode*: The recharge process also includes three steps. The first charge step is to the first charge plateau between 2.2 and 2.3 V corresponding to the first oxidation peak in CV. During this step, insoluble lithium sulfide and lithium disulfide lose electrons and are electrochemically oxidized to soluble medium-chain-length linear polysulfides. The second charge step is the transition period between 2.3 and 2.4 V, through which the first charge plateau is connected with the second charge plateau. The medium-chain-length polysulfides lose electrons and are further electrochemically oxidized to long-

chain linear polysulfides. The transition period in the recharge is not normally as evident as that in the discharge. The third step includes the second charge plateau at above 2.4 V corresponding to the second oxidation peak in CV. The long-chain linear polysulfides lose electrons and are eventually electrochemically oxidized to elemental sulfur ($S_8$). Coincidentally, the cathodic active material during charge also undergoes a solid(sulfide)-liquid(polysulfides)-solid($S_8$) phase transfer.

Although this mechanism was widely accepted, it lacks concrete experimental evidence to support the claims. The mechanism was mainly based on the stoichiometric electrochemical calculation [32]. With the assumption of a total 16-e redox reaction between $S_8$ and $Li_2S$, the theoretical specific capacity of the redox reaction would be $1672\,mAh\,g^{-1}$. Then, the polysulfides at various stages of discharge and recharge were determined based on the numbers of charges transferred. When the discharge capacity of a Li-S battery is increased from 0 to around $200\,mAh\,g^{-1}$ in the first discharge plateau, it roughly corresponds to 2-e transfer ($\approx(200/1672)*16$); thus, the main electrochemical process in this range was deduced to the formation of $S_8^{2-}$ (or long-chain polysulfides) from $S_8$; as the discharge capacity of a Li-S battery reaches $1000\,mAh\,g^{-1}$ (normally corresponding to the end of the second discharge plateau in galvanostatic discharge/charge curves), it roughly corresponds to 10-e transfer ($\approx(1000/1672)*16$); thus, the major electrochemical products at this point should be a mixture of $S_2^{2-}$ (8-e transfer) and $S^{2-}$ (16-e transfer). Despite the simplicity of the calculation based on stoichiometry, the disproportionation of polysulfide ions is overlooked. Therefore, in order to understand the mechanism of sulfur redox reactions, the composition of intermediate polysulfide ions must be quantitatively and qualitatively measured. Electrochemical techniques alone would not portray the whole picture. In situ or ex situ electrochemical spectroscopic techniques should be applied.

UV-visible is the most commonly used analytical technique to identify polysulfide ions in tandem with electrochemical methods [34,35,63,65]. Only $S_3^-$ was unanimously assigned to the peak at about 610 nm [34,63,65], which was also confirmed by EPR [27,46]. There has been no agreement regarding the assignment of UV-visible peaks to other polysulfides since the deconvolution of the broad UV-visible peaks always yielded different results [35,63–66]. In some UV-visible works [35,66], there was no assignment of polysulfide ions to wavelength. In fact, UV-visible spectroscopy is not much better than visual observation of color change during discharge and recharge of a Li-S battery.

By means of the calculation based on density functional theory (DFT), Raman spectroscopy can identify [36] elemental sulfur ($S_8$), $S^{2-}$ and $S_3^{-}$. $S_8$ can be assigned to the peaks at 150, 219, and $474 \, cm^{-1}$; $S^{2-}$ was assigned to the signal at $372 \, cm^{-1}$; while the peak around $535 \, cm^{-1}$ was attributed to $S_3^{-}$. The assignments of other polysulfide intermediates were inconsistent and sometimes subjective [67–70].

XRD is not feasible to investigate the dissolved polysulfide ions in electrolytes. Elemental sulfur ($S_8$) and $Li_2S$ in the solid phase were observed by XRD, while the XRD observation of $Li_2S_2$ was not conclusive [38,39]. XRD patterns were reported on the polysulfides absorbed on $SiO_2$, but the nature of the polysulfides was not identified [40]. Similar to Raman spectroscopy, XAS can determine elemental sulfur ($S_8$), $S^{2-}$ and $S_3^{-}$ [41,42]. The assignment of elemental sulfur was around 2472–2473 eV, $S^{2-}$ was around 2474 eV and $S_3^{-}$ was around 2468.5 eV, which can be only observed in EPD solvents. Other polysulfide species cannot be assigned. Due to the small observation window (from 2460 to 2490 eV), the signals overlapped and needed to be deconvoluted [41,42]. Sulfur binding energy can be measured using XPS, a value of 163.7 eV for $2p_{3/2}$ of elemental sulfur and 159.8 for $2p_{3/2}$ of the sulfur in $Li_2S$ [47,73]. The assignments of signals for other lithium polysulfide species are either deduced from the terminal to bridge ratios or assumed [73].

EPR was used to investigate the possible free radical formation during the sulfur redox reaction [27,46,72]. The formation of $S_3^{-}$ was reported, but no other radical species such as $S_2^{-}$ and $S_4^{-}$ were observed, although those were assumed to be the intermediate species in the sulfur redox reaction [27,72].

In summary, a substantial amount of research has been reported using traditional spectroscopic techniques. It has made a significant contribution to understand the redox mechanism of a sulfur electrode. However, owing to the lack of capability of quantitative and qualitative determinations of dissolved polysulfide ions, a detailed and trustworthy mechanism that should be based on solid experimental evidence was not revealed. The proposed mechanisms were all ambiguous, since only one or two polysulfide ions out of ten possible ones can be firmly identified.

## 5.5.2 Investigation of sulfur redox mechanism using electrochemical HPLC techniques

Barchasz et al. [12] derivatized polysulfide ions in the catholytes recovered from a battery polarized at 3 V, 1.95 V, and 1.5 V. Methyl triflate was used for the derivatization and HPLC/RID was used for the analysis. The chromatographic results are shown in Fig. 5.6. After being polarized at 3 V, the major broad chromatographic peak was the coeluted shoulder peak from 70 and 80 min. Those chromatographic peaks were assigned to the unreacted $S_8$ and the long-chain polysulfide $S_8^{2-}$ (derivatized into $(CH_3)_2S_8$). Medium-chain polysulfide $S_4^{2-}$ is only barely evident in the minor peak at 13 min as shown in Fig. 5.6A. After being polarized at 1.95 V, the chromatographic peaks of elemental sulfur $S_8$ and the long-chain polysulfide $S_8^{2-}$ disappear; instead, the chromatographic peaks that belong to the short-chain, e.g., $S_2^{2-}$ and $S_3^{2-}$, and medium-chain, e.g., $S_4^{2-}$ and $S_5^{2-}$,

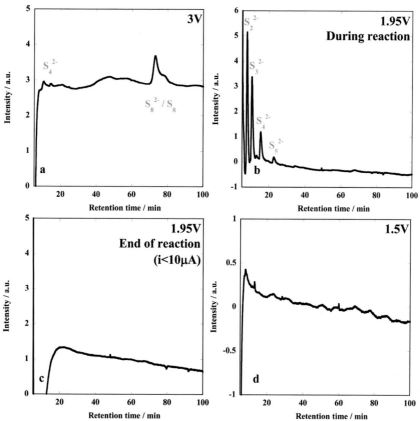

**Fig. 5.6** Chromatograms of catholyte samples obtained after the cell polarization at different potentials. *(Reproduced with permission from C. Barchasz, F. Molton, C. Duboc, J.C. Leprêtre, S. Patoux, F. Alloin, Lithium/sulfur cell discharge mechanism: an original approach for intermediate species identification, Anal. Chem. 84 (2012) 3973–3980, Copyright (2012) American Chemical Society.)*

polysulfides become abundant in a chromatogram from 6 to 22 min, as shown in Fig. 5.6B. At the end of the discharge, no elemental sulfur $S_8$ or polysulfides can be observed in the chromatograms as shown in Fig. 5.6C and 5.6D, indicating the formation of insoluble $Li_2S$. Although the HPLC results reported in Barchasz's work were only very limited, in combination with UV–Vis and ESR techniques, the reduction mechanism for the Li-S battery was proposed as a three-step process. The first step of reduction is the slow electrochemical reduction of $S_8$ to $S_8^{2-}$ in tandem with the fast disproportionation of $S_8^{2-}$ to $S_6^{2-}$ as well as the disproportionation of $S_6^{2-}$ to other polysulfides. The second step of reduction is the fast electrochemical reduction of $S_6^{2-}$ to $S_4^{2-}$ in tandem with the slow disproportionation of the polysulfides to $S_6^{2-}$. The third step of reduction corresponds to the electrochemical reduction of $S_4^{2-}$, $S_3^{2-}$, and $S_2^{2-}$ to $S_3^{2-}$, $S_2^{2-}$, and $S^{2-}$. One of the major issues of Barchasz's work was poor resolution and low sensitivity, because the isocratic HPLC condition and the low-sensitive RID detector were used in their qualitative work. Qu and coworkers later demonstrated that by simply using a different derivation reagent, e.g., 4-(dimethylamino) benzoyl chloride, all the polysulfide ions can be clearly separated and analyzed qualitatively and quantitatively by HPLC/MS [57].

Kawase et al. [43] reported the change of dissolved polysulfides and sulfide during discharge and recharge of a Li-S battery. The electrolyte samples were taken at different depths of discharge and at different depths of recharge as indicated in Fig. 5.7A. Benzyl chloride was used as the derivatization reagent to convert the ionic polysulfides and sulfide into organopolysulfides and organosulfides. The compositions of polysulfide ions at the different stages of discharge and recharge are summarized in Fig. 5.7B. The changes in the relative amount of the polysulfide ion at different depths of discharge and recharge are shown in Fig. 5.7C. Kawase et al. proposed a four-stage discharge mechanism for a sulfur electrode. During the first plateau, elemental sulfur was reduced either through 2-electron transfer to form $S_8^{2-}$ (Eq. 5.16) or through 4-electron transfer to form other polysulfides ($S_n^{2-}$, $n = 2$ to 7) and sulfide ($S^{2-}$) (Eqs. 5.17–5.20). The hypothesis was supported by the continuous decrease of the relative amount of elemental sulfur and the gradual increase of the relative amounts of other polysulfide/sulfide species as shown in Fig. 5.7. During the transition of the first discharge plateau to the second discharge plateau, the long- and medium-chain polysulfides were further reduced to short-chain polysulfides (such as $S_3^{2-}$ or $S_2^{2-}$) and sulfide through Eqs. (5.21)–(5.27). This is evidence of the decrease and eventually the complete disappearance of the long- and medium-chain polysulfides, while the short-chain polysulfides (especially $S_2^{2-}$) are significantly increased

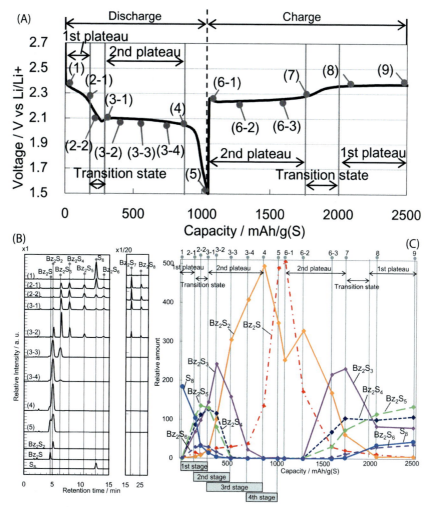

**Fig. 5.7** The discharge/charge profile of Li-S cell with different observation points (A); HPLC chromatograms for points 1 to 5, including three chemical standards (Bz₂S, Bz₂S₂, and S₈) (B); and the relative remaining quantity of Bz₂S_x (x=1 to 6) at different observation points as estimated from chromatographic data (C). (Reproduced with permission from A. Kawase, S. Shirai, Y. Yamoto, R. Arakawa, T. Takata, Electrochemical reactions of lithium-sulfur batteries: an analytical study using the organic conversion technique, Phys. Chem. Chem. Phys. 16 (2014) 9344–9350, Copyright (2014) Royal Society of Chemistry.)

as shown in Fig. 5.7. The third-stage covered most of the second discharge plateau. The short-chain polysulfide $S_3^{2-}$ became reduced into $S_2^{2-}$ and sulfide through Eq. (5.28). As shown in Fig. 5.7, a great decrease of benzylized (Bz = $C_6H_5CH_2$—) polysulfide $Bz_2S_3$ and the continuous increases of

Bz$_2$S$_2$ and Bz$_2$S are evident. The end of the second discharge plateau to the cutoff potential was the fourth stage, where the shortest polysulfide S$_2^{2-}$ was partially reduced to sulfide as shown in Eq. (5.29), which is supported by the decrease of Bz$_2$S$_2$ and the increase of Bz$_2$S, as shown in Fig. 5.7. The constant voltage of the second plateau can be attributed to the reduction of S$_3^{2-}$ to S$_2^{2-}$ and S$^{2-}$ with 2-electron transfer as shown in Eq. (5.28). As for the mechanism of charge, Kawase et al. believed that the reverse reactions occurred in the charge process. However, no detailed mechanism was proposed. Due to the lack of a shuttle inhibitor, e.g., LiNO$_3$ in the system, the charge process never reached the desired 3.0 V cutoff voltage, while the charge capacity was substantially higher than that of discharge, which was evident of the occurrence of a severe "shuttle effect."

$$S_8 + 2\,e^- \rightarrow S_8{}^{2-} \tag{5.16}$$

$$S_8 + 4\,e^- \rightarrow S_7{}^{2-} + S^{2-} \tag{5.17}$$

$$S_8 + 4\,e^- \rightarrow S_6{}^{2-} + S_2{}^{2-} \tag{5.18}$$

$$S_8 + 4\,e^- \rightarrow S_5{}^{2-} + S_3{}^{2-} \tag{5.19}$$

$$S_8 + 4\,e^- \rightarrow 2\,S_4{}^{2-} \tag{5.20}$$

$$S_8{}^{2-} + 2\,e^- \rightarrow 2\,S_4{}^{2-} \tag{5.21}$$

$$S_6{}^{2-} + 2\,e^- \rightarrow 2\,S_3{}^{2-} \tag{5.22}$$

$$S_8{}^{2-} + 2\,e^- \rightarrow S_6{}^{2-} + S_2{}^{2-} \tag{5.23}$$

$$S_8{}^{2-} + 2\,e^- \rightarrow S_5{}^{2-} + S_3{}^{2-} \tag{5.24}$$

$$S_7{}^{2-} + 2\,e^- \rightarrow S_4{}^{2-} + S_3{}^{2-} \tag{5.25}$$

$$S_5{}^{2-} + 2\,e^- \rightarrow S_3{}^{2-} + S_2{}^{2-} \tag{5.26}$$

$$S_4{}^{2-} + 2\,e^- \rightarrow 2\,S_2{}^{2-} \tag{5.27}$$

$$S_3{}^{2-} + 2\,e^- \rightarrow S_2{}^{2-} + S^{2-} \tag{5.28}$$

$$S_2{}^{2-} + 2\,e^- \rightarrow 2\,S^{2-} \tag{5.29}$$

Compared with Barchasz's work [12], one distinctive and meaningful attempt in Kawase's work [43] was to use a different derivatization procedure. Both the sulfur electrode and the electrolyte were exposed to a derivatization reagent for a long time, so the insoluble Li$_2$S and Li$_2$S$_2$, along with the dissolved polysulfide ions, were derivatized to soluble RSR and RS$_2$R, as shown in Eq. (5.14). In most of the other HPLC investigations, only the dissolved polysulfides ($Li_2S_n$, $n \geq 3$) in an electrolyte were derivatized and analyzed. The determination of insoluble $Li_2S$ and $Li_2S_2$ was conducted by other techniques, such as XRD [39]. A critical criterion for the mechanism study is the quick kinetics of the derivatization. Since polysulfides can go through rapid disproportionation reactions, the instant capture of the distribution of polysulfide ions at a certain stage of the redox reaction is the key for the reliable determination of reaction mechanism. We emphasized previously the importance of stabilizing the polysulfide ions with derivatization in order to prevent the further disproportionation of polysulfides in a mobile phase during HPLC separation. It is equally important that the derivatization reaction itself is faster than that of the disproportionation reactions so the electrochemically generated polysulfide ions can be preserved before they have time to become subsequently disproportionated. Therefore, the change of the polysulfides distribution during derivatization can be greatly minimized, if not totally eliminated.

Lev and coworkers demonstrated that by using methyl triflate, the total conversion of polysulfide ions into organopolysulfides was done in seconds [26,75]. In Kawase's work, the reactions between benzyl chloride and polysulfides took several hours (and even days). This made Kawase's mechanism less accurate. For example, $Li_2S_2$ and $Li_2S$ were already observable at the first discharge plateau while elemental sulfur still existed. This observation was not only contrary to the fact that the existing elemental sulfur could react with $Li_2S_2$ and $Li_2S$ to form soluble polysulfides but also contrary to the XRD results from other works in which the formation of $Li_2S$ is only observable at the second discharge plateau with the total consumption of elemental sulfur. Probably the slow and long derivatization process used by Kawase et al. [43] could be the reason for those abnormal and inconsistent observations.

In the work of both Barchasz et al. and Kawase et al., a series of Li-S coin cells were first cycled to different preset stages; then, the cells were dissembled and the cathodes (with the electrolytes) were recovered. The recovered cathodes were derivatized for further HPLC analysis. It is well known that the reproducibility of the coin cell type of Li-S batteries is poor. The variation between cells in their work was inevitable, not to mention that the recovery process itself could introduce some error to the final analysis. Because of these indispensable uncertainties associated with the ex situ method, in situ and *in operando* methods, such as in situ XRD, in situ UV-Vis, and in situ XAS, are more preferable.

Systematic investigation of the mechanism of a sulfur redox reaction by the qualitative and quantitative determination of dissolved polysulfide ions by HPLC/MS was reported in a series of papers by Qu and coworkers [57,59,60]. The quick and reliable identification of dissolved sulfur was reported using a novel derivation reagent of 4-(dimethylamino) benzoyl chloride. By means of RP-HPLC/MS, all the polysulfide ions in organic electrolytes with a high concentration of Li salt were separated and identified with confidence for the first time. The results are shown in Fig. 5.4. The results paved the way for using an electrochemical HPLC/MS assay as a reliable tool for the analysis of polysulfide species formed in the operation of a Li-S battery.

Armed with a reliable tool for the quantitative and qualitative analyses of the dissolved polysulfide species in the electrolyte, we now can embark on the investigation of a sulfur-redox reaction mechanism. The very first task to understand the mechanism was to determine the polysulfide ions formed during the first reduction wave of sulfur in a Li-S battery, in other words, the redox mechanism from elemental sulfur to polysulfide.

### 5.5.2.1 First reduction wave of sulfur: From elemental sulfur to polysulfide

Qu and coworkers investigated the CV scans of the catholytes including the catholytes saturated with elemental sulfur, the one with 20 mM methyl triflate (derivatization reagent), and the catholyte with both methyl triflate and elemental sulfur [59]. They found that methyl triflate was electrochemically inert between 1.5 and 3.2 V but could chemically react with the polysulfides rapidly as shown in Eq. (5.13). As shown in Fig. 5.8, reduction and oxidation peaks can be observed regardless of whether the starting catholyte was the dissolved elemental sulfur ($S_8$) or polysulfide ions (curve 4 and 3, respectively). However, the organopolysulfides did not participate in a redox reaction within the potential window of 1.4–3.4 V vs. Li (curve 2). Cyclic voltammograms of a sulfur-saturated LiTFSi/DME solution with the addition of 20 mM of methyl triflate are shown in curve 5 in Fig. 5.8. There was no oxidation peak in curve 5 and only one reduction peak instead of two as shown in curve 4, which was CV of the sulfur-saturated catholyte without methyl triflate. It was the firm evidence that all the electrochemically generated polysulfide ions were immediately converted into organopolysulfides, which were electrochemically inert—no second reduction peak of polysulfides and no oxidation peak. However, the addition of methyl triflate had no impact on the reduction of elemental sulfur.

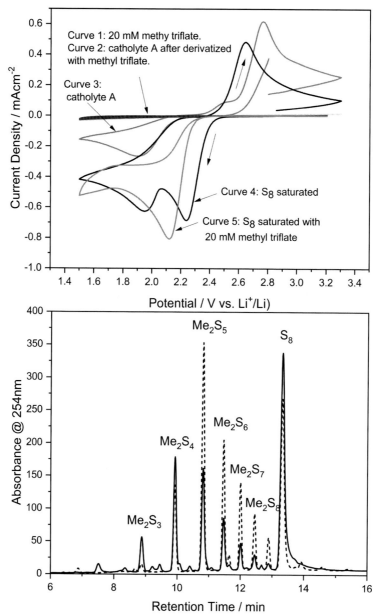

**Fig. 5.8** CV of different electrolytes on glassy carbon electrode *(top)*, *curve 1*: 20 mM methyl triflate; *curve 2*: simulated electrolyte A after derivatization by methyl triflate; *curve 3*: simulated electrolyte A; *curve 4*: $S_8$ saturated electrolyte; *curve 5*: $S_8$ saturated electrolyte with 20 mM methyl triflate; and chromatograms for the derivatized electrolytes from Li-S batteries polarized at 2.3 V (vs Li/Li$^+$) *(bottom)*, *dash line*: ex situ derivatization method; *solid line*: in situ derivatization method. (Reproduced with permission from D. Zheng, X. Zhang, J. Wang, D. Qu, X. Yang, D. Qu, Reduction mechanism of sulfur in lithium-sulfur battery: from elemental sulfur to polysulfide, J. Power Sources 301 (2016) 312–316, Copyright (2015) Elsevier.)

Taking into consideration the fast kinetics for the derivatization reaction and electrochemical inertia of organopolysulfides, a "snapshot" of the distribution of the electrochemically produced polysulfides can be done at a certain potential or a stage of operation of a Li-S battery. Any electrochemically produced polysulfides can be captured by methyl triflate and the resulting organopolysulfides are immune to both subsequent chemical and electrochemical reactions.

It was widely accepted that the first reduction peak in CV or during the first discharge plateau was a 2-e transfer process in tandem with chemical reactions [14–19]. Elemental sulfur gained two electrons to form $S_8^{2-}$ polysulfide, and then, the $S_8^{2-}$ quickly disproportionated into $S_6^{2-}$ through a chemical reaction as shown in Eq. (5.30). However, whether it was a simultaneous 2-e transfer or a continuous two 1-e transfer was not confirmed. EIS results [33,61,62] revealed that the charge transfer resistance was first decreased and increased in the first discharge plateau while the electrolyte resistance was first increased and then decreased. The reverse trends were observed during the recharge. Qu and coworkers [59] compared the polysulfide distribution in the electrolyte polarized at 2.3 V by ex situ and in situ methods of derivatization. In an ex situ derivatization, the electrolyte was first taken out of the cell before methyl triflate was added, while in an in situ derivatization, methyl triflate was added to the catholyte. The electrochemically generated polysulfide ions became stable organopolysulfides before they had a chance to participate in further disproportionation reactions in an in situ process. The polysulfides formed due to the first electron transfer process have enough time to become disproportionated before derivatization in an ex situ process. Therefore, in situ derivatization can provide a "snapshot" for the distribution of polysulfides before they further reacted chemically.

$$S_8{}^{2-} \rightarrow S_6{}^{2-} + 1/4 S_8 \tag{5.30}$$

The reduction mechanism from elemental sulfur to polysulfides that corresponds to the first reduction wave in CV and the first discharge plateau in discharge and recharge curve was investigated [59]. Two experimental pieces of evidence were reported that contradicted all previous assumptions that elemental sulfur gained 2-e forming $S_8^{2-}$ linear polysulfide. As shown in Fig. 5.8, in an ex situ HPLC experiment, more polysulfides with a long chain ($Me_2S_n$, with $n=5$, 6, 7, 8, and most abundant peak is $Me_2S_5$) were observed, whereas in the in situ HPLC experiment, more short- and middle-chain polysulfides ($Me_2S_n$, with $n=3$, 4, 5, most abundant peak is $Me_2S_4$) were observed. Second, in the ex situ experiment, the sulfur consumption was higher than that of the electrochemical reduction as tabulated in Table 5.3. The additional sulfur was consumed by the reaction with dissolved polysulfides. In the in situ derivatization, the consumption of sulfur almost matched the 2-e stoichiometry electrochemical reduction, since methyl triflate was added to the catholyte, the polysulfide ions were immediately derivatized and subsequent chemical reactions with sulfur were avoided. Apparently, neither $Me_2S_8$ nor $Me_2S_6$ was the dominant polysulfide species at 2.3 V, and the widely adopted reduction mechanism for the first discharge plateau mentioned earlier was not supported.

Qu and coworkers proposed that the very first step of sulfur electrochemical reduction involved a multiple -S-S- bond cleavage to form multiple shorter chain-length polysulfides that were chemically more stable, instead of the less stable long-chain $S_8^{2-}$ or $S_6^{2-}$ species [59]. The electrochemical reduction mechanism of sulfur with multiple parallel pathways into different polysulfide species was consistent with that of Kawase et al. [43], while they did provide concrete experimental evidence. The higher than expected sulfur consumption in the ex situ experiment also demonstrated the severity of the "shuttle reaction" as shown in Eq. (5.31), bearing in mind that an ex situ derivatization only took less than 1 min. Since elemental sulfur cannot be derivatized, the investigation of the first-step electron transfer in the sulfur reduction was very successful by means of in situ derivatization. But the in situ technique has a major downside. It permanently alters the subsequent reduction reaction. Therefore, this in situ HPLC method cannot be applied to the study of the whole redox mechanism of Li–S batteries.

$$S_8 + 2\,S_4{}^{2-} \rightarrow 2\,S_8{}^{2-} \tag{5.31}$$

**Table 5.3** Summary of discharge capacities and elemental sulfur remained in the Li-S catholyte after ex situ and in situ derivatization. The $S_8$ concentration in the electrolytes before discharge is 4.01 mM.)

| Derivatization method | Discharge capacity from polarization at 2.3 V (mAh) | Percentage of elemental sulfur left from theoretical calculation based on 2-electron transfer | Percentage of elemental sulfur left from HPLC/UV measurement |
|---|---|---|---|
| In situ | 1.42 | 34.0% | 37.9% |
| Ex situ | 1.43 | 33.3% | 24.9% |

Reproduced with permission from D. Zheng, X. Zhang, J. Wang, D. Qu, X. Yang, D. Qu, Reduction mechanism of sulfur in lithium–sulfur battery: from elemental sulfur to polysulfide, J. Power Sources 301 (2016) 312–316, Copyright (2015), Elsevier B.V.

### 5.5.2.2 The change of dissolved polysulfide distribution during sulfur redox reaction

Qu and coworkers established an *in operando* HPLC method for the systematic and quantitative study of the sulfur redox reaction mechanism [60]. A homemade Li-S cell was cycled at a low C rate. A large volume of catholyte containing dissolved elemental sulfur was used as the supporting electrolyte and the source of sulfur instead of the common sulfur/carbon electrode. A carbon felt and Li metal were used as working electrode and counter electrode, respectively. A very small (neglectable) amount of catholyte was taken out and derivatized with methyl trifle at different stages of discharge and recharge. The derivatized organopolysulfides were analyzed by HPLC. Through the real-time (*in operando*) HPLC analysis, the qualitative and quantitative changes of both elemental sulfur ($S_8$) and lithium polysulfides ($Li_2S_n$, $8 \geq n \geq 3$) were monitored during discharge (12 steps) and recharge (11 steps). A total of seven sulfur and polysulfide species were observed.

Fig. 5.9 shows the discharge curve of the Li-S cell, the real-time sampling points, and the comparison of the HPLC chromatograms taken at corresponding sampling points. Fig. 5.10 shows the recharge profile of the same Li-S cell, the real-time sampling points, and the corresponding chromatograms taken at different sampling points. As shown in Fig. 5.9, the first discharge plateau between 2.3 V and 2.5 V can be attributed to the reduction of elemental sulfur to polysulfides. Evidentially, elemental sulfur almost disappears in the chromatogram D4 as shown in Fig. 5.9, whereas polysulfide species become abundant. The long-chain polysulfide ions $S_n^{2-}$ ($n \geq 5$) were reduced to $S_n^{2-}$ ($n \leq 4$) between 2.1 V and 2.3 V, since the chromatographic peaks of $Me_2S_n$ ($n \geq 5$) species are gradually decreased from D4 to D7, whereas the chromatographic peak of $Me_2S_4$ gradually increased as shown in Fig. 5.9. Although the formation of solid $Li_2S$ and $Li_2S_2$ was confirmed by in situ XRD [39], XANES [41], and NMR [43] when the cell was discharged to around 1.7 V, the chromatographic peak ascribed to $Me_2S_4$ is still abundant in D12 as shown in Fig. 5.9. It appears that not all the dissolved polysulfides can be electrochemically reduced. At 1.7 V, the discharge capacity was about 70% of the theoretical discharge capacity.

Fig. 5.10 shows the oxidation of polysulfide ions during recharge and the change of the concentrations of polysulfide ions, respectively. Clearly, the recharge reaction started to oxidize insoluble $S_2^{2-}$ and $S^{2-}$ into almost all soluble polysulfides in sequence. It was supported by the in situ XRD work [39]. $S_3^{2-}$, $S_4^{2-}$, $S_5^{2-}$, $S_6^{2-}$, $S_7^{2-}$, and $S_8^{2-}$ coexisted in the solution throughout

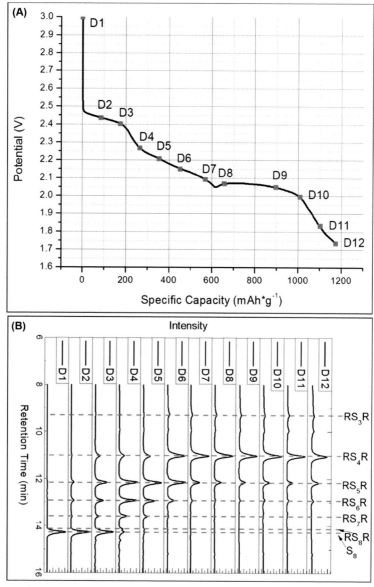

**Fig. 5.9** The discharge profile of Li-S cell with sampling points (*red, dark gray in print versions, squares from D1 to D13*) during operation (A) and the corresponding HPLC chromatograms (B). *(Reproduced with permission from D. Zheng, D. Liu, J. Harris, T. Ding, J. Si, S. Andrew, D. Qu, X. Yang, D. Qu, Investigation of the Li-S battery mechanism by real-time monitoring of the changes of sulfur and polysulfide species during the discharge and charge, ACS Appl. Mater. Interfaces 9 (2017) 4326–4332, Copyright (2017) American Chemical Society.)*

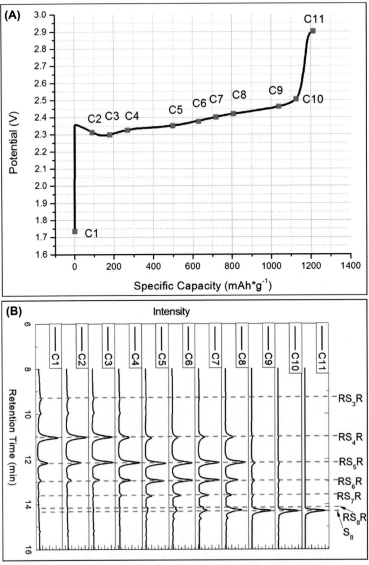

**Fig. 5.10** The charge profile of Li-S cell with sampling points (*red, dark gray in print versions, squares from C1 to C11*) during operation (A) and the corresponding HPLC chromatograms (B). *(Reproduced with permission from D. Zheng, D. Liu, J. Harris, T. Ding, J. Si, S. Andrew, D. Qu, X. Yang, D. Qu, Investigation of the Li-S battery mechanism by real-time monitoring of the changes of sulfur and polysulfide species during the discharge and charge, ACS Appl. Mater. Interfaces 9 (2017) 4326–4332, Copyright (2017) American Chemical Society.)*

the recharge process. Except for $S_3^{2-}$ and $S_4^{2-}$, the concentrations of all other polysulfides started to decrease at 2.4 V.

### 5.5.2.3 Chemical equilibrium

The chromatographic peak areas in each chromatogram in Figs. 5.9 and 5.10 were normalized and plotted in Fig. 5.11, which portrays a picture of the quantitative change of each polysulfide ion in the catholyte. The mechanism of the sulfur redox reaction, especially the auxiliary homogenous chemical reactions, becomes quite obvious. It is worth pointing out that if the change rates of the concentrations of the polysulfide ions are the same, as shown by the parallel curves in Fig. 5.11, these polysulfides remain in a chemical equilibrium. Although the total amount of those dissolved polysulfide species changed, their ratio remained the same. Such parallel curves can be seen from the increases of the abundances of $S_8^{2-}$, $S_7^{2-}$, and $S_6^{2-}$, indicating the existence of a chemical equilibrium among $S_7^{2-}$, $S_6^{2-}$, and $S_5^{2-}$. The equilibrium was maintained until the end of first discharge plateau at 2.3, where the consumption of $S_7^{2-}$ and $S_6^{2-}$ became substantial and the corresponding abundance starts to drop, as shown in Fig. 5.11. During the slope discharge between 2.3 V and 2.1 V, elemental sulfur was almost totally consumed, while the decrease of the relative abundance of long–chain polysulfides $S_8^{2-}$, $S_7^{2-}$, $S_6^{2-}$, and $S_5^{2-}$ and the increase of the relative abundance of medium–chain–length polysulfides including $S_4^{2-}$ and $S_3^{2-}$ indicated that the main electrochemical processes in this stage were the reduction of the long–chain–length polysulfides to medium–chain–length and short–chain–length polysulfides. Apparently, the concentrations of all the polysulfide ions changed independently. There is no evidence of the parallel curves of relative abundance in Fig. 5.11. Therefore, we can safely assume there was no chemical equilibrium reaction at this stage. It was reported that solid $Li_2S_2$ and $Li_2S$ may be formed at this stage with in situ XRD [39], XAS [41], and NMR [43]. During the second plateau of the discharge at about 2.1 V, almost all of the polysulfide species with chain length $\geq 6$ were completely consumed. The concentrations of $S_5^{2-}$, $S_4^{2-}$, and $S_3^{2-}$ decrease nicely along three parallel lines, as shown in Fig. 5.11, revealing the existence of chemical equilibrium reactions among them. The corresponding discharge mechanisms are shown in Scheme 5.1.

As shown in Fig. 5.11, the mechanism of the oxidation of polysulfides can be divided into two regions before and after 2.4 V. Chemical equilibrium reactions among $S_5^{2-}$, $S_6^{2-}$, $S_7^{2-}$, and $S_8^{2-}$ remain throughout the entire

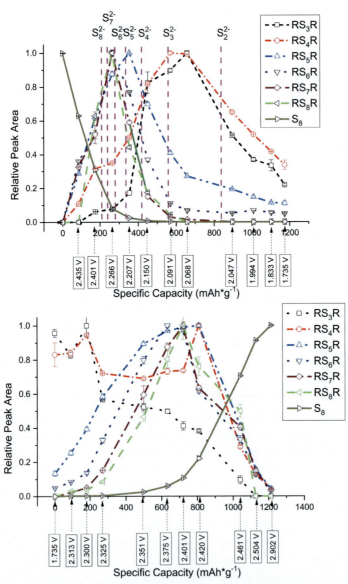

**Fig. 5.11** The normalized chromatographic peak for each derivatized polysulfide species (R=CH$_3$) from real-time HPLC results during discharge and charge of Li-S cell. *(Reproduced with permission from D. Zheng, D. Liu, J. Harris, T. Ding, J. Si, S. Andrew, D. Qu, X. Yang, D. Qu, Investigation of the Li-S battery mechanism by real-time monitoring of the changes of sulfur and polysulfide species during the discharge and charge, ACS Appl. Mater. Interfaces 9 (2017) 4326–4332, Copyright (2016) American Chemical Society.)*

**>2.3 V, first plateau, major species are bold**

$$S_8 \xrightarrow{ne} \left( S_8^{2-} + S_7^{2-} + S_6^{2-} + S_5^{2-} + S_4^{2-} + S_3^{2-} \right)$$

$$S_7^{2-} \rightleftarrows S_6^{2-} \rightleftarrows S_5^{2-}$$

**Between 2.3 and 2.1 V, major species are bold**

$$S_8^{2-} + S_7^{2-} + S_6^{2-} + S_5^{2-} \xrightarrow{ne} S_4^{2-} + S_3^{2-} + Li_2S_2 \downarrow + Li_2S \downarrow$$

**<2.1 plateau, major species are bold**

$$S_5^{2-} + S_4^{2-} + S_3^{2-} \xrightarrow{ne} Li_2S_2 \downarrow + Li_2S \downarrow$$

$$S_5^{2-} \rightleftarrows S_4^{2-} \rightleftarrows S_3^{2-}$$

**Scheme 5.1** The proposed discharge mechanism of sulfur in Li-S batter. *(Reproduced with permission from D. Zheng, D. Liu, J. Harris, T. Ding, J. Si, S. Andrew, D. Qu, X. Yang, D. Qu, Investigation of the Li-S battery mechanism by real-time monitoring of the changes of sulfur and polysulfide species during the discharge and charge, ACS Appl. Mater. Interfaces 9 (2017) 4326–4332, Copyright (2016) American Chemical Society.)*

duration of oxidation. It was evident that the changes of concentration of the four polysulfide ions stayed parallel in the whole process. 2.4 V was the turning point when elemental sulfur started to form, which drove the equilibrium to the right (Le Châtelier's Principle). Unlike in the discharge process, $S_3^{2-}$ and $S_4^{2-}$ species did not seem to participate in any equilibrium reaction since the concentrations of the species stayed independent. Unlike in the discharge process, there was a distinct transition period at 2.4 V. The mechanism of the recharge processes is listed in Scheme 5.2.

**Cathode potential <2.4 V**

$$Li_2S_2 \downarrow + Li_2S \downarrow \xrightarrow{-ne} S_8^{2-} + S_7^{2-} + S_6^{2-} + S_5^{2-}$$

$$S_8^{2-} \rightleftarrows S_7^{2-} \rightleftarrows S_6^{2-} \rightleftarrows S_5^{2-}$$

**Cathode potential > 2.4 V**

$$S_8^{2-} + S_7^{2-} + S_6^{2-} + S_5^{2-} + S_4^{2-} + S_3^{2-} \xrightarrow{-ne} S_8$$

$$S_8^{2-} \rightleftarrows S_7^{2-} \rightleftarrows S_6^{2-} \rightleftarrows S_5^{2-}$$

**Scheme 5.2** The proposed charge mechanism on the cathode in a Li-S battery. *(Reproduced with permission from D. Zheng, D. Liu, J. Harris, T. Ding, J. Si, S. Andrew, D. Qu, X. Yang, D. Qu, Investigation of the Li-S battery mechanism by real-time monitoring of the changes of sulfur and polysulfide species during the discharge and charge, ACS Appl. Mater. Interfaces 9 (2017) 4326–4332, Copyright (2016) American Chemical Society.)*

## 5.6 Conclusions

The discharge and recharge profiles of a sulfur cathode can be explained with the fast-reaching chemical equilibrium reactions among the polysulfide ions. At a low rate, the Nernst equation may be applied. According to it, the potential is roughly related to the ratio of redox couples. If there remains a chemical equilibrium, the ratio stays the same, as does the potential. The two separated chemical equilibrium reactions shown in Scheme 5.1 correlate with the two plateaus of discharge; the less discrete equilibrium during recharge is shown in Scheme 5.2 results in less distinctive plateaus in the recharge curve.

Table 5.4 tabulates the comparison of the potential of the most abundant polysulfide species during discharge and recharge processes. It is clearly shown that the mechanism of the oxidation is not simply the reverse of that of the discharge. The sulfur cathode may be rechargeable, but definitely not thermodynamically reversible.

The total number of dissolved polysulfide ions can be estimated with the integration of all the chromatographic peaks. The total number of polysulfides topped at around 2.3 V and 2.4 V during discharge and recharge, respectively. The two potentials (2.3 and 2.4 V) indicated the start of the precipitation of solid $Li_2S_2$, $Li_2S$ and the formation of sulfur, respectively. The observation was consistent with the reports of investigating solid sulfur phases [39,41].

The most interesting observation was that almost all the solid $Li_2S_2$ and $Li_2S$ precipitates and dissolved polysulfide ions can be electrochemically

**Table 5.4** The highest chromatographic peak areas and the corresponding cell potentials for different polysulfide species during charge and discharge of a Li-S cell.

| | Charge | | Discharge | | Ratio of peak area of charge/ discharge |
|---|---|---|---|---|---|
| | Potential (V) | Peak area | Potential (V) | Peak area | |
| $RS_3R$ | 2.300 | 49,904 | 2.068 | 219,102 | 0.23 |
| $RS_4R$ | 2.420 | 754,517 | 2.091 | 1,865,579 | 0.40 |
| $RS_5R$ | 2.420 | 1,603,892 | 2.207 | 1,965,162 | 0.82 |
| $RS_6R$ | 2.375 | 1,387,329 | 2.266 | 1,400,488 | 0.99 |
| $RS_7R$ | 2.401 | 537,688 | 2.266 | 550,688 | 0.98 |

Reproduced with permission from D. Zheng, D. Liu, J. Harris, T. Ding, J. Si, S. Andrew, D. Qu, X. Yang, D. Qu, Investigation of the Li-S battery mechanism by real-time monitoring of the changes of sulfur and polysulfide species during the discharge and charge, ACS Appl. Mater. Interfaces 9 (2017) 4326–4332, Copyright (2016) American Chemical Society.

oxidized back to elemental sulfur at 2.9 V. In Fig. 5.11, only the chromatographic peak of elemental sulfur can be observed in the chromatogram C11. By comparing the initial chromatographic peak area for $S_8$ before discharge and that of recharge with 2.9 V, over 95% of the sulfur was recovered. This is the first time that the coulombic efficiency for a sulfur redox reaction was determined, not with electrochemical analysis but with quantitative chemical analysis, which are more accurate and reliable. However, if "shuttle effect" inhibitors, e.g., $LiNO_3$, were not used, the complete recovery of elemental sulfur was never achieved.

## References

[1] L.R. Snyder, J.J. Kirkland, J.W. Dolan, Introduction to Modern Liquid Chromatography, John Wiley & Sons, 2010.

[2] L.R. Snyder, J.J. Kirkland, J.L. Glajch, Practical HPLC Method Development, second ed., John Wiley & Sons, 1997.

[3] C.F. Poole, The Essence of Chromatography, Elsevier, 2003.

[4] D. Harvey, Modern Analytic Chemistry, McGraw-Hill, 2000.

[5] G.L. Glish, R.W. Vachet, The basics of mass spectrometry in the twenty-first century, Nat. Rev. Drug Discov. 2 (2003) 140–150.

[6] A. Doerr, J. Finkelstein, I. Jarchum, C. Goodman, B. Dekker, Nature Milestones: Mass Spectrometry, 2015. https://www.nature.com/milestones/milemassspec/pdf/milemassspec_all.pdf.

[7] E. Hoffman, V. Stroobant, Mass Spectrometry: Principles and Applications, third ed., Wiley, 2007.

[8] G. Gourdin, J. Collins, D. Zheng, M. Foster, D. Qu, Spectroscopic compositional analysis of electrolyte during initial SEI layer formation, J. Phys. Chem. C 118 (2014) 17383–17394.

[9] C. Schultz, S. Vedder, M. Winter, S. Nowak, Qualitative investigation of the decomposition of organic solvent based lithium ion battery electrolytes with LC-IT-TOF-MS, Anal. Chem. 88 (2016) 11160–11168.

[10] J. Henschel, J.L. Schwarz, F. Glorius, M. Winter, S. Nowak, Further insights into structural diversity of phosphorus-based decomposition products in lithium ion battery electrolytes via liquid chromatographic techniques hyphenated to ion trap-time-of-flight mass spectrometry, Anal. Chem. 91 (2019) 3980–3988.

[11] Y. Diao, K. Xie, S. Xiong, X. Hong, Analysis of polysulfide dissolved in electrolyte in discharge-charge process of Li-S battery, J. Electrochem. Soc. 159 (2012) A421–A425.

[12] C. Barchasz, F. Molton, C. Duboc, J.C. Leprêtre, S. Patoux, F. Alloin, Lithium/sulfur cell discharge mechanism: an original approach for intermediate species identification, Anal. Chem. 84 (2012) 3973–3980.

[13] N. Mahne, B. Schafzahl, C. Leypold, M. Leypold, S. Grumm, A. Leitgeb, G.A. Strohmeier, M. Wilkening, O. Fontaine, D. Kramer, C. Slugovc, S.M. Borisov, S.A. Freunberger, Singlet oxygen generation as a major cause for parasitic reactions during cycling of aprotic lithium-oxygen batteries, Nat. Energy 2 (2017) 17036.

[14] Y. Yang, G. Zheng, Y. Cui, Nanostructured sulfur cathodes, Chem. Soc. Rev. 42 (2013) 3018–3032.

[15] S.S. Zhang, Liquid electrolyte lithium/sulfur battery: fundamental chemistry, problems, and solutions, J. Power Sources 231 (2013) 153–162.

[16] D. Bresser, S. Passerini, B. Scrosati, Recent progress and remaining challenges in sulfur-based lithium secondary batteries—a review, Chem. Commun. 49 (2013) 10545–10562.

[17] A. Manthiram, Y. Fu, S.H. Chung, C. Zu, Y.S. Su, Rechargeable lithium-sulfur batteries, Chem. Rev. 114 (2014) 11751–11787.

[18] M. Wild, L. O'Neill, T. Zhang, R. Purkayastha, G. Minton, M. Marinescu, G.J. Offer, Lithium sulfur batteries, a mechanistic review, Energ. Environ. Sci. 8 (2015) 3477–3494.

[19] R. Xu, J. Lu, K. Amine, Progress in mechanistic understanding and characterization techniques of Li-S batteries, Adv. Energy Mater. 5 (2015) 1500408.

[20] D. Zheng, G. Wang, D. Liu, J. Si, T. Ding, D. Qu, X. Yang, D. Qu, The progress of Li-S batteries—understanding of the sulfur redox mechanism: dissolved polysulfide ions in the electrolytes, Adv. Mater. Technol. 3 (2018) 1700233.

[21] R.D. Rauh, F.S. Shuker, J.M. Marston, S. Brummer, Formation of lithium polysulfides in aprotic media, J. Inorg. Nucl. Chem. 39 (1977) 1761–1766.

[22] Y. Yang, G. Zheng, Y. Cui, A membrane-free lithium/polysulfide semi-liquid battery for large-scale energy storage, Energ. Environ. Sci. 6 (2013) 1552–1558.

[23] X.W. Yu, A. Manthiram, A class of polysulfide catholytes for lithium-sulfur batteries: energy density, cyclability, and voltage enhancement, Phys. Chem. Chem. Phys. 17 (2015) 2127–2136.

[24] J. Chen, K.S. Han, W.A. Henderson, K.C. Lau, M. Vijayakumar, T. Dzwiniel, H. Pan, L.A. Curtiss, J. Xiao, K.T. Mueller, Y. Shao, J. Liu, Restricting the solubility of polysulfides in Li-S batteries via electrolyte salt selection, Adv. Energy Mater. 6 (2016) 1600160.

[25] B. Meyer, Elemental sulfur, Chem. Rev. 76 (1976) 367–388.

[26] A. Kamyshny, A. Goifman, J. Gun, D. Rizkov, O. Lev, Equilibrium distribution of polysulfide ions in aqueous solutions at 25 °C: a new approach for the study of polysulfides' equilibria, Environ. Sci. Technol. 38 (2004) 6633–6644.

[27] M. Vijayakumar, N. Govind, E. Walter, S.D. Burton, A. Shukla, A. Devaraj, J. Xiao, J. Liu, C.M. Wang, A. Karim, S. Thevuthasan, Molecular structure and stability of dissolved lithium polysulfide species, Phys. Chem. Chem. Phys. 16 (2014) 10923–10932.

[28] F. Seel, J.H. Guttler, G. Simon, A. Wieckowski, Colored sulfur species in EPD-solvents, Pure Appl. Chem. 49 (1977) 45–54.

[29] L. Jia, T. Wu, J. Lu, L. Ma, W. Zhu, X. Qiu, Polysulfides capture-copper additive for long cycle life lithium sulfur batteries, ACS Appl. Mater. Interfaces 8 (2016) 30248–30255.

[30] J. Gun, A. Modestov, A. Kamyshny, D. Ryzkov, V. Gitis, A. Goifman, O. Lev, V. Hultsch, T. Grischek, E. Worch, Electrospray ionization mass spectrometric analysis of aqueous polysulfide solutions, Microchim. Acta 146 (2004) 229–237.

[31] R. Steudel, G. Holdt, T. Göbel, Ion-pair chromatographic separation of inorganic sulphur anions including polysulphide, J. Chromatogr. A 475 (1989) 442–446.

[32] Y. Lu, Q. He, H.A. Gasteiger, Probing the lithium-sulfur redox reactions: a rotating-ring disk electrode study, J. Phys. Chem. C 118 (2014) 5733–5741.

[33] Z. Deng, Z. Zhang, Y. Lai, J. Liu, J. Li, Y. Liu, Electrochemical impedance spectroscopy study of a lithium/sulfur battery: modeling and analysis of capacity fading, J. Electrochem. Soc. 160 (2013) A553–A558.

[34] D. Han, B. Kim, S. Choi, Y. Jung, J. Kwak, S. Park, Time-resolved in situ spectroelectrochemical study on reduction of sulfur in N,N'-dimethylformamide, J. Electrochem. Soc. 151 (2004) E283–E290.

[35] M. Patel, R. Dominko, Application of in operando UV/Vis spectroscopy in lithium-sulfur batteries, ChemSusChem 7 (2014) 2167–2175.

Reliable HPLC-MS method for the quantitative and qualitative analyses **197**

[36] M. Hagen, P. Schiffels, M. Hammer, S. Doerfler, J. Tuebke, M. Hoffmann, H. Althues, S. Kaskel, In-situ Raman investigation of polysulfide formation in Li-S cells, J. Electrochem. Soc. 160 (2013) A1205–A1214.

[37] H. Wu, L. Huff, A. Gewirth, In situ Raman spectroscopy of sulfur speciation in lithium-sulfur batteries, ACS Appl. Mater. Interfaces 7 (2015) 1709–1719.

[38] J. Nelson, S. Misra, Y. Yang, A. Jackson, Y. Liu, H. Wang, H. Dai, J.C. Andrews, Y. Cui, M. Toney, In operando X-ray diffraction and transmission X-ray microscopy of lithium Sulfur batteries, J. Am. Chem. Soc. 134 (2012) 6337–6343.

[39] S. Walus, C. Barchasz, R. Bouchet, J. Lepretre, J. Colin, J. Martin, E. Elkaim, C. Baehtz, F. Alloin, Lithium/sulfur batteries upon cycling: structural modifications and species quantification by in situ and operando X-ray diffraction spectroscopy, Adv. Energy Mater. 5 (2015) 1500165.

[40] J. Conder, R. Bouchet, S. Trabesinger, C. Marino, L. Gubler, C. Villevieille, Direct observation of lithium polysulfides in lithium-sulfur batteries using operando X-ray diffraction, Nat. Energy 2 (2017) 17069.

[41] M. Cuisinier, P. Cabelguen, S. Evers, G. He, M. Kolbeck, A. Garsuch, T. Bolin, M. Balasubramanian, L. Nazar, Sulfur speciation in Li-S batteries determined by operando X-ray absorption spectroscopy, J. Phys. Chem. Lett. 4 (2013) 3227–3232.

[42] M. Lowe, J. Gao, H. Abruna, Mechanistic insights into operational lithium-sulfur batteries by in situ X-ray diffraction and absorption spectroscopy, RSC Adv. 4 (2014) 18347–18353.

[43] A. Kawase, S. Shirai, Y. Yamoto, R. Arakawa, T. Takata, Electrochemical reactions of lithium-sulfur batteries: an analytical study using the organic conversion technique, Phys. Chem. Chem. Phys. 16 (2014) 9344–9350.

[44] L. Huff, J. Rapp, J. Baughman, P. Rinaldi, A. Gewirth, Identification of lithium-Sulfur battery discharge products through $^6$Li and $^{33}$S solid-state MAS and $^7$Li solution NMR spectroscopy, Surf. Sci. 631 (2015) 295–300.

[45] H. Wang, N. Sa, M. He, X. Liang, L. Nazar, M. Balasubramanian, K. Gallagher, B. Key, In situ NMR observation of the temporal speciation of lithium sulfur batteries during electrochemical cycling, J. Phys. Chem. C 121 (2017) 6011–6017.

[46] Q. Wang, J. Zheng, E. Walter, H. Pan, D. Lv, P. Zuo, H. Chen, Z.D. Deng, B. Liaw, X. Yu, X. Yang, J. Zhang, J. Liu, J. Xiao, Direct observation of sulfur radicals as reaction media in lithium sulfur batteries, J. Electrochem. Soc. 162 (2015) A474–A478.

[47] Y.S. Su, Y. Fu, T. Cochell, A. Manthiram, A strategic approach to recharging lithium-sulphur batteries for long cycle life, Nat. Commun. 4 (2013) 2985.

[48] R. Xu, I. Belharouak, X. Zhang, R. Chamoun, C. Yu, Y. Ren, A. Nie, R. Shahbazian-Yassar, J. Lu, J. Li, K. Amine, Insight into sulfur reactions in Li-S batteries, ACS Appl. Mater. Interfaces 6 (2014) 21938–21945.

[49] F. Fan, W. Carter, Y. Chiang, Mechanism and kinetics of $Li_2S$ precipitation in lithium-sulfur batteries, Adv. Mater. 27 (2015) 5203–5209.

[50] S. Lang, Y. Shi, Y. Guo, D. Wang, R. Wen, L. Wan, Insight into the interfacial process and mechanism in lithium–sulfur batteries: an in situ AFM study, Angew. Chem. Int. Ed. 55 (2016) 15835–15839.

[51] N. Saqib, C. Silva, C. Maupin, J. Porter, A novel optical diagnostic for in situ measurements of lithium polysulfides in battery electrolytes, Appl. Spectrosc. 71 (2017) 1593–1599.

[52] M. Kavčič, K. Bučar, M. Petric, M. Žitnik, I. Arčon, R. Dominko, A. Vizintin, Operando resonant inelastic X-ray scattering: an appropriate tool to characterize sulfur in Li-S batteries, J. Phys. Chem. C 120 (2016) 24568–24576.

[53] X.Q. Yu, H. Pan, Y. Zhou, P. Northrup, J. Xiao, S. Bak, M. Liu, K. Nam, D. Qu, J. Liu, T. Wu, X. Yang, Direct observation of the redistribution of sulfur and polysulfides

in Li-S batteries during the first cycle by in situ X-ray fluorescence microscopy, Adv. Energy Mater. 5 (2015) 1500072.

[54] C. Lin, W. Chen, Y. Song, C. Wang, L. Tsai, N. Wu, Understanding dynamics of polysulfide dissolution and *re*-deposition in working lithium-sulfur battery by in-operando transmission X-ray microscopy, J. Power Sources 263 (2014) 98–103.

[55] Y. Wang, W. Wang, C. Huang, Z. Yu, H. Zhang, J. Sun, A. Wang, K. Yuan, Structural change of the porous sulfur cathode using gelatin as a binder during discharge and charge, Electrochim. Acta 54 (2009) 4062–4066.

[56] D. Zheng, D. Qu, Chromatographic separation of polysulfide species in non-aqueous electrolytes-revisted, J. Electrochem. Soc. 161 (2014) A1164–A1166.

[57] D. Zheng, D. Qu, X. Yu, H. Lee, X. Yang, D. Qu, Quantitative and qualitative determination of polysulfide species in the electrolyte of a lithium-sulfur battery using HPLC ESI/MS with one-step derivatization, Adv. Energy Mater. 5 (2015) 1401888.

[58] D. Zheng, X. Zhang, C. Li, M. Mckinnon, R. Sadok, D. Qu, X. Yu, H. Lee, X. Yang, D. Qu, Quantitative chromatographic determination of dissolved elemental sulfur in the non-aqueous electrolyte for lithium-sulfur batteries, J. Electrochem. Soc. 162 (2015) 203–206.

[59] D. Zheng, X. Zhang, J. Wang, D. Qu, X. Yang, D. Qu, Reduction mechanism of sulfur in lithium-sulfur battery: from elemental sulfur to polysulfide, J. Power Sources 301 (2016) 312–316.

[60] D. Zheng, D. Liu, J. Harris, T. Ding, J. Si, S. Andrew, D. Qu, X. Yang, D. Qu, Investigation of the Li-S battery mechanism by real-time monitoring of the changes of sulfur and polysulfide species during the discharge and charge, ACS Appl. Mater. Interfaces 9 (2017) 4326–4332.

[61] V. Kolosnitsyn, E. Kuzmina, E. Karaseva, S. Mochalov, A study of the electrochemical processes in lithium-sulphur cells by impedance spectroscopy, J. Power Sources 196 (2011) 1478–1482.

[62] N. Canas, K. Hirose, B. Pascucci, N. Wagner, K. Friedrich, R. Hiesgen, Investigations of lithium-sulfur batteries using electrochemical impedance spectroscopy, Electrochim. Acta 97 (2013) 42–51.

[63] N. Manan, L. Aldous, Y. Alias, P. Murray, L. Yellowlees, C. Lagunas, C. Hardacre, Electrochemistry of sulfur and polysulfides in ionic liquids, J. Phys. Chem. B 115 (2011) 13873–13879.

[64] N. Canas, D. Fronczek, N. Wagner, A. Latz, K. Friedrich, Experimental and theoretical analysis of products and reaction intermediates of lithium-sulfur batteries, J. Phys. Chem. C 118 (2014) 12106–12114.

[65] Q. Zou, Y. Lu, Solvent-dictated lithium sulfur redox reactions: an operando UV–vis spectroscopic study, J. Phys. Chem. Lett. 7 (2016) 1518–1525.

[66] M. Patel, M. Demir-Cakan, M. Morcrette, J. Tarscon, M. Gaberscek, R. Dominko, Li-S battery analyzed by UV/Vis in operando mode, ChemSusChem 6 (2013) 1177–1181.

[67] J. Hannauer, J. Scheers, J. Fullenwarth, B. Fraisse, L. Stievano, P. Johansson, The quest for polysulfides in lithium-sulfur battery electrolytes: an operando confocal Raman spectroscopy study, ChemPhysChem 16 (2015) 2755–2759.

[68] B. Vinayan, T. Diemant, X. Lin, M. Cambaz, U. Golla-Schindler, U. Kaiser, R. Jürgen Behm, M. Fichtner, Nitrogen rich hierarchically organized porous carbon/sulfur composite cathode electrode for high performance Li/S battery: a mechanistic investigation by operando spectroscopic studies, Adv. Mater. Interfaces 3 (2016) 1600372.

[69] J. Chen, R. Yuan, J. Feng, Q. Zhang, J. Huang, G. Fu, M. Zheng, B. Ren, Q. Dong, Conductive Lewis base matrix to recover the missing link of $Li_2S_8$ during the sulfur redox cycle in Li-S battery, Chem. Mater. 27 (2015) 2048–2055.

[70] W. Zhu, A. Paolella, C. Kim, D. Liu, Z. Feng, C. Gagnon, J. Trottier, A. Vijh, A. Guerfi, A. Mauger, C. Julien, M. Armand, K. Zaghib, Investigation of the reaction mechanism of lithium sulfur batteries in different electrolyte systems by: in situ Raman spectroscopy and in situ X-ray diffraction, Sustain. Energy Fuels 1 (2017) 737–747.

[71] M. Patel, I. Arcon, G. Aquilanti, L. Stievano, G. Mali, R. Dominko, X-ray absorption near-edge structure and nuclear magnetic resonance study of the lithium-sulfur battery and its components, ChemPhysChem 15 (2014) 894–904.

[72] K. Wujcik, D. Wang, A. Raghunathan, M. Drake, T. Pascal, D. Prendergast, N. Balsara, Lithium polysulfide radical anions in ether-based solvents, J. Phys. Chem. C 120 (2016) 18403–18410.

[73] X. Liang, C. Hart, Q. Pang, A. Garsuch, T. Weiss, L. Nazar, A highly efficient polysulfide mediator for lithium-sulfur batteries, Nat. Commun. 6 (2015) 5682.

[74] Y. Fu, Y. Su, A. Manthiram, Highly reversible lithium/dissolved polysulfide batteries with carbon nanotube electrodes, Angew. Chem. Int. Ed. 52 (2013) 6930–6935.

[75] A. Kamyshny, I. Ekeltchik, J. Gun, O. Lev, Method for the determination of inorganic polysulfide distribution in aquatic systems, Anal. Chem. 78 (2006) 2631–2639.

[76] D. Rizkov, O. Lev, J. Gun, B. Anisimov, I. Kuselman, Development of in-house reference materials for determination of inorganic polysulfides in water, Accred. Qual. Assur. 9 (2004) 399–403.

# CHAPTER 6

# Modeling of electrode, electrolyte, and interfaces of lithium-sulfur batteries

## Venkat Srinivasan[a,b] and Aashutosh Mistry[a]

[a]Chemical Sciences and Engineering Division, Argonne National Laboratory, Lemont, IL, United States
[b]Argonne Collaborative Center for Energy Storage Science, Argonne National Laboratory, Lemont, IL, United States

## Contents

| | | |
|---|---|---|
| **6.1** Introduction | | 202 |
| **6.2** Mathematical description of porous electrode performance | | 206 |
| **6.3** Evolution of cathode porous electrode structure | | 213 |
| **6.4** Concentrated electrolyte transport effects | | 218 |
| **6.5** Dynamics of the polysulfide shuttle effect | | 221 |
| **6.6** Sources of variability: Mechanisms and properties | | 225 |
| **6.7** Summary and outlook | | 226 |
| Acknowledgments | | 227 |
| References | | 227 |

## Nomenclature

| Symbol | Description | Units |
|---|---|---|
| $a$ | specific surface area | $m^2/m^3$ |
| $i, i_{app}$ | current density | $A/m^2$ |
| $i_q^0$ | exchange current density for reaction q | $A/m^2$ |
| $j$ | volume-specific reaction current | $A/m^3$ |
| $c$ | concentration of ionic species | $mol/m^3$ |
| $D$ | self-diffusivity of ionic species | $m^2/s$ |
| $\mathfrak{D}$ | Stefan–Maxwell diffusivity | $m^2/s$ |
| $F$ | Faraday's constant | $C/mol$ |
| $f$ | identity of sulfur species — expressed as a fraction; as $f_p \to 1$, sulfur species p is present most dominantly | – |
| $K_q^{sp}$ | solubility product for reaction q (multiple units) | |
| $k_q^{rxn}$ | reaction rate for reaction q (multiple units) | |

*Continued*

*Lithium-Sulfur Batteries*
https://doi.org/10.1016/B978-0-12-819676-2.00012-8

Copyright © 2022 Elsevier Inc.
All rights reserved.

201

## Nomenclatur—cont'd

| Symbol | Description | Units |
|---|---|---|
| $L_{cat}$, $L_{sep}$ | cathode and separator thicknesses | m |
| $M$ | dimensionless mechanism number, as $M \to 1$, the corresponding mechanism is more dominant | – |
| $N_p$ | molar flux for species p | $mol/m^2$ /s |
| $Q_\Delta$ | capacity lost due to shuttle effect over one discharge | $mAh/g$ $S_8$ |
| $R$ | Universal gas constant | $J/mol \cdot K$ |
| $\dot{r}$ | reaction rate (volumetric rate in $mol/m^3$ /s; surface rate in $mol/m^2$ /s) | $mol/m^3$ /s $mol/m^2$ /s |
| $T$ | temperature | K |
| $t$ | time | s |
| $U_q^0$ | open circuit potential for reaction q | V |
| $\overline{V}$ | molar volume | $m^3/mol$ |
| $z$ | coordinate along cell thickness | m |
| $z_p$ | charge on ionic species p | – |

### Greek symbols

| | | |
|---|---|---|
| $\varepsilon$ | volume fractions for pores (=porosity) and solid phases | – |
| $\kappa$ | ionic conductivity of the electrolyte | S/m |
| $\mu_p$ | (electro-) chemical potential of species p | J/mol |
| $\rho$ | density | $kg/m^3$ |
| $\Sigma_S$ | total sulfur content in the cell | $mol/m^2$ |
| $\sigma$ | (effective) electronic conductivity of the porous cathode | S/m |
| $\tau$ | tortuosity of the pore network | m/m |
| $\phi$ | electric potentials for electrolyte and electrode phases | V |
| $\omega$ | precipitate morphology | – |

### Subscript/superscript

| | |
|---|---|
| 0 | property of the pristine (i.e., without $S_{8(s)}$ or $Li_2S_{(s)}$) cathode microstructure |
| c | porous cathode (or 3D current collector) |
| e | electrolyte pore network |
| p, p′ | indices for ionic species |
| q | index for the elemental reaction |

## 6.1 Introduction

A lithium-sulfur (Li-S) cell operates due to two half-cell reactions; the anode reaction is: $Li \Leftrightarrow Li^+ + e^-$; the cathode reaction is: $1/2S^{2-} \Leftrightarrow 1/16 S_8 + e^-$. In Li-sulfur cells with a liquid organic electrolyte, it has been found that the $S^{2-}$ ions precipitate as $Li_2S$ in the presence of $Li^+$ ions since $Li_2S$ is a sparingly soluble ionic solid in organic solvents. Thus, during cell discharge, the net half-cell reactions are $Li \rightarrow Li^+ + e^-$ and $Li^+ + \frac{1}{16}S_8 + e^- \rightarrow \frac{1}{2}Li_2S_{(s)}$. Depending on the choice of the electrolyte and sulfur loading, at the start of discharge, sulfur can be in a dissolved state, $S_{8(l)}$, or it can exist as a solid, $S_{8(s)}$, that dissolves in the electrolyte as electrochemical reactions take place. In either scenario, the amount of dissolved sulfur in the electrolyte is related to solubility—an equilibrium (thermodynamic) property [1]. The dissolution and precipitation reactions at the cathode side constantly evolve the pore geometry and the porous electrode structure changes during each operation [2]. Thus, the chemical changes (macroscopically identified by the state of charge) are closely related to the physical (geometrical) evolution of the cathode. These effects exist even in a very slow (close to equilibrium) operation. Hence, an ideally behaving Li-sulfur cell is one where only the two net half-cell reactions shown above take place, all $S_8$ is converted to $Li_2S$ and vice versa, and geometrically the electrode resumes its physical arrangement at the end of every discharge-charge operation. When all $S_8$ converts to $Li_2S$, ~1675 mAh/g $S_8$ charge is passed (Faraday's second law). This metric is central to the promise of Li-sulfur cell chemistry.

Most Li-sulfur cells, however, do not exhibit such reversibility. Instead, less than 1675 mAh/g $S_8$ capacity and rapid deterioration of cell performance over a few (tens-hundreds of) cycles are observed. Historically, three mechanisms have been recurringly proposed to justify such limited behavior [1,3–6], and in turn, solutions have been developed to counter either of these effects:

(1) **Polysulfide shuttle effect**: Since the same electrolyte is in contact with both the electrodes (Fig. 6.1), just as $Li^+$ can transport to cathode, sulfur species can travel to lithium anode and chemically react. Lithium reduces sulfur species. This reduction is a chemical reaction (i.e., electrons are exchanged between lithium and sulfur), unlike the electrochemical reaction where electrons are exchanged with the external circuit (Fig. 6.1C). In turn, the observed electrochemical capacity is less than the theoretical 1675 mAh/g $S_8$ value.

(2) **Poor electronic conductivity**: The solid phases at the cathodes, $S_8$ and $Li_2S$, are electronically insulating which can hinder the electron exchange with the external circuit, in turn, terminating the battery operation before the cell reactions proceed to completion.

(3) **High volume expansion**: Every mole of $S_8$ produces 8 moles of $Li_2S$ and occupies ~80% more volume than $S_8$ (calculated using material densities and molecular weights). Equivalently, the pore volume needs to shrink to accommodate the new phase. Since the electrolyte is an

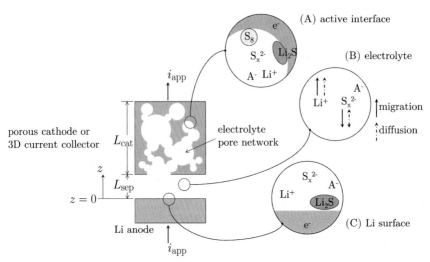

**Fig. 6.1** A schematic diagram showing the essential components of a Li-sulfur cell: Li metal anode, electrolyte, and porous cathode. Relevant interactions contributing to electrochemical signatures are illustrated in insets. $S_x^{2-}$ ($x=1,2,\ldots,8$) represents various polysulfide species present in the electrolyte; $A^-$ is the anion from the lithium salt used to prepare the electrolyte, such as $TFSI^-$ for LiTFSI salt. *Source: Authors.*

incompressible liquid, the volumetric variations of the solid phase can generate mechanical stresses in the backbone porous structure leading to fracture and detachment of the electrochemically active solids from the external electronic contact. Accordingly, the observed performance falls short of the theoretical promise.

While each of these is a logical argument and at play during the electrochemical operation of Li-sulfur cells, it is unclear if they can completely account for the observed response. For example, $LiNO_3$ has been suggested as an electrolyte additive and shown to mitigate the polysulfide shuttle effect [7,8]; but the corresponding gain in the cell performance is not enough to recover the theoretical response (e.g., Fig. 2(b) vs (c) in [9]). On the other hand, carbon variants—Ketjenblack, graphene, acetylene black, etc., — are a typical choice for the backbone structure given their high electronic conductivity [10]. However, even with a sufficiently high amount of carbon, the electrochemical performance is not ideal (e.g., Fig. 5(a) in [11]). Having to accommodate new $Li_2S$ volume is a thermodynamic requirement, but it can happen in multiple ways. The incompressible liquid electrolyte can displace to make room for the new phase (there is always extra room in the coin cells). Also, 3D tomography studies [12] reveal a minimal fracture of the backbone and rather a rearrangement of solid sulfur as the predominant geometrical change [2]. Hence, the high volume expansion is likely not the root cause of the observed nonideal response.

When examined collectively, these studies [2,9–12] and many others show that the electrochemical behavior of the Li-sulfur system is poorly understood. The fact that the cell response does not approach ideal expectations even for a very slow discharge (e.g., Fig. 3(c–f) in [9]) makes it more fascinating (for well-behaved Li-ion materials, the thermodynamic performance is recovered at very slow rates [13,14]). Such observations allude to more involved fundamental interactions, and consequently, many other mechanisms have been proposed to explain the observations. These coupled interactions represent a unique set of complexities as compared to Li-ion batteries (where electrolyte transport [15], diffusion inside active particles [16], and electrode microstructure properties [17] can be studied separately to piece together the cell response [18,19]) and to a certain extent do not allow us to borrow tools and techniques from the Li-ion research. Unlike Li-ion where electrochemical energy is stored in active particles and electrolyte behaves as a conduit to transfer energy from one electrode to another using ions as charge carriers, the Li-sulfur chemistry (Fig. 6.1) is fundamentally different. Herein, the energy is stored at the electrochemical interface in terms of $S_8$ and $Li_2S$ (Fig. 6.1A) and as polysulfide ions, $S_x^{2-}$ ($x = 1, 2, ..., 8$), in the electrolyte (Fig. 6.1B). The porous electrode structure forms the active interface and is responsible for suppling electrons to the reaction spots. In addition to transporting ions from one electrode to another, the electrolyte also stores the polysulfide ions (Fig. 6.1B).

Since multiple interactions are simultaneously active and likely combine synergistically, it is difficult and somewhat misleading to study them separately. In this context, continuum-scale modeling becomes an attractive tool to help demystify the observed behavior. The modeling allows one to examine *What-If* questions to assess the relevance of proposed mechanisms. The following discussion starts with the general description of the Li-sulfur porous electrode electrochemical response—originally proposed by Kumaresan, Mikhaylik, and White [20] in 2008. Subsequent modeling studies build on this description by explicitly resolving various mechanisms such as morphology-dependent precipitate growth of the 3D electrode volumes [21], nucleation dynamics of the precipitate phase [22], concentrated electrolyte transport [23], and polysulfide shuttle effect [24,25]. These mechanisms are grouped as those related to the porous sulfur cathode (Fig. 6.1A), the electrolyte (Fig. 6.1B), and the lithium anode (Fig. 6.1C). For a poorly understood system such as the Li-sulfur cell, not all the details in the electrochemical performance model are known to

## 206 Lithium-sulfur batteries

the same certainty and give rise to variabilities in the predictions. The sources of such variabilities are discussed. In closing, the unanswered questions are highlighted for future studies.

## 6.2 Mathematical description of porous electrode performance

Fig. 6.1 shows a schematic for the typical Li–sulfur cell where lithium is the anode of choice (to take advantage of high capacity sulfur, the anode's weight-specific capacity should be as high as possible; hence most traditional anode materials [26] are not suitable for Li-sulfur cells). During discharge, in the external circuit, the current, $i_{app}$, flows from cathode (positive electrode) to anode (negative electrode). Equivalently, the ionic current in the electrolyte flows from anode to cathode. Here, $i_{app}$ is the current density. The global charge neutrality ensures that identical currents pass through each cross-section of the cell. Hence, the electronic current density in the lithium anode is $i_{app}$. An equivalent amount of $Li^+$ ions are generated at the lithium/electrolyte interface ($z=0$), and the ionic current through the separator is $i_{app}$ at every location along the thickness ($0 \leq z < L_{sep}$). Throughout the porous cathode, the electrolyte current gradually gets converted to the electronic current such that at every $z$ location ($L_{sep} \leq z < L_{sep} + L_{cat}$) the total electronic + ionic current is $i_{app}$. This creates a distribution of the reaction current throughout the porous cathode such that the total reaction current matches the current passing through the cell, i.e.,

$$\int_{L_{sep}}^{L_{sep}+L_{cat}} -j\mathrm{d}z = i_{app} \tag{6.1}$$

The volume-specific reaction current, $j$, is positive when cations are generated (or anions are consumed) and electrons are released to the external circuit. Accordingly, $j$ is negative during the discharge of the Li-sulfur cells. The electrical response of the Li-sulfur cell is described by prescribing two current conservation equations:

$$\text{Electrolyte pore network}: \quad \frac{\partial i_e}{\partial z} = j \tag{6.2}$$

$$\text{Porous electrode (or 3D current collector)}: \quad \frac{\partial i_c}{\partial z} = -j \tag{6.3}$$

Eq. (6.2) is valid in both separator and cathode, i.e., $0 \leq z < L_{sep} + L_{cat}$ (note that $j=0$ in the separator). The ionic and electronic current densities are related to the respective potential fields and concentrations as per the following constitutive relations:

$$\text{Ionic current density}: \quad i_e = -\kappa \frac{\varepsilon}{\tau} \frac{\partial \phi_e}{\partial z} - \sum_p z_p F D_p \frac{\varepsilon}{\tau} \frac{\partial c_p}{\partial z} \tag{6.4}$$

$$\text{Electronic current density}: \quad i_c = -\sigma \frac{\partial \phi_c}{\partial z} \tag{6.5}$$

The ionic current is driven by the electric field established in the electrolyte. This tendency is countered by the concentration overpotential of the ionic species and quantified using the second term in Eq. (6.4). The effect of the electrolyte pore network is accounted for using the ratio of porosity, $\varepsilon$, and tortuosity, $\tau$. Similarly, the electronic conductivity, $\sigma$, implicitly contains the effects of the 3D geometry of the porous backbone (e.g., Fig. S3 in [21]) and approaches zero as the amount of background phase decreases. Such behavior of $\sigma$ consistently describes the role of the backbone structure in ensuring sufficient electronic conduction throughout the porous cathode.

The concentration for each of the species, p, is separately tracked using individual species conservation relation

$$\frac{\partial (\varepsilon c_p)}{\partial t} = -\frac{\partial N_p}{\partial z} + \dot{r}_p \tag{6.6}$$

where the species flux, $N_p$, consistent with the ionic current (Eq. 6.4) is described by

$$N_p = -D_p \frac{\varepsilon}{\tau} \frac{\partial c_p}{\partial z} - z_p F \frac{D_p}{RT} \frac{\varepsilon}{\tau} c_p \frac{\partial \phi_e}{\partial z} \tag{6.7}$$

Since the electrolyte is both locally and globally charge-neutral, Eq. (6.6) is solved for all but one ionic species (often the anion, $A^-$ of the lithium salt is not solved for explicitly, and its concentration is obtained by invoking the local charge neutrality). Solvent is considered as a background species and its evolution is typically ignored. Note that the pore network effects are also included in Eq. (6.7). One can further show that the ionic conductivity, $\kappa$, consistent with Eqs. (6.4), (6.7) is

$$\kappa = \sum_p z_p^2 F^2 \frac{D_p}{RT} c_p \tag{6.8}$$

Eqs. (6.4), (6.7), and (6.8) represent the dilute electrolyte theory based on the Nernst-Planck constitutive relation [18]. As discussed later, an alternative electrolyte description based on the Stefan-Maxwell relation does exist, which leads to alternative constitutive relations. The molar species generation rates, $\dot{r}_p$, are composed of elementary reaction steps that jointly describe the net half-cell reaction $Li^+ + \frac{1}{16}S_{8(s)} + e^- \rightarrow \frac{1}{2}Li_2S_{(s)}$. The reaction scheme also prescribes the identity of polysulfide ions. For example, according to Kumaesan, Mikhaylik, and White [20], the reaction scheme and intermediate ionic species are:

chemical reaction (precipitation—dissolution):

$$\text{I} \qquad S_{8(l)} \Longleftrightarrow S_{8(s)}$$

electrochemical reactions:

$$\text{II} \qquad \frac{1}{2}S_8^{2-} \Longleftrightarrow \frac{1}{2}S_{8(l)} + e^-$$

$$\text{III} \qquad 2S_6^{2-} \Longleftrightarrow \frac{3}{2}S_8^{2-} + e^-$$

$$\text{IV} \qquad \frac{3}{2}S_4^{2-} \Longleftrightarrow S_6^{2-} + e^- \qquad (6.9)$$

$$\text{V} \qquad S_2^{2-} \Longleftrightarrow \frac{1}{2}S_4^{2-} + e^-$$

$$\text{VI} \qquad S^{2-} \Longleftrightarrow \frac{1}{2}S_2^{2-} + e^-$$

chemical reaction (precipitation—dissolution):

$$\text{VII} \qquad 2Li^+ + S^{2-} \Longleftrightarrow Li_2S_{(s)}$$

Since reactions I and VII are chemical reactions, they do not contribute to the volume-specific reaction current, $j$, but they do contribute to the molar species generation rates, $\dot{r}_p$. Using the reaction scheme in Eq. (6.9), it can be shown that

$$j = \sum_{q=\text{II},\dots,\text{VI}} j_q \qquad (6.10)$$

and the molar species generation rates for the ionic species as

$$\dot{r}_{Li^+} = -2\dot{r}_{VII}$$

$$\dot{r}_{S_{8(l)}} = -\dot{r}_I + \frac{j_{II}}{2F}$$

$$\dot{r}_{S_8^{2-}} = \frac{-j_{II}}{2F} + \frac{3j_{III}}{2F}$$

$$\dot{r}_{S_6^{2-}} = \frac{-2j_{III}}{F} + \frac{j_{IV}}{F} \qquad (6.11)$$

$$\dot{r}_{S_4^{2-}} = \frac{-3j_{IV}}{2F} + \frac{j_V}{2F}$$

$$\dot{r}_{S_2^{2-}} = \frac{-j_V}{F} + \frac{j_{VI}}{2F}$$

$$\dot{r}_{S^{2-}} = \frac{-j_{VI}}{F} - \dot{r}_{VII}$$

where the coefficients in Eq. (6.11) correspond to stoichiometries in Eq. (6.9). These coefficients also reflect in the rate expressions for each of the reactions:

$$\dot{r}_I = k_I^{rxn}\left(c_{S_{8(l)}} - K_I^{sp}\right)$$

$$j_{II} = ai_{II}^0\left\{\left(\frac{c_{S_8^{2-}}}{c_{S_8^{2-}}^0}\right)^{\frac{1}{2}} e^{\frac{F}{2RT}(\phi_c - \phi_e - U_{II}^0)} - \left(\frac{c_{S_{8(l)}}}{c_{S_{8(l)}}^0}\right)^{\frac{1}{2}} e^{\frac{-F}{2RT}(\phi_c - \phi_e - U_{II}^0)}\right\}$$

$$j_{III} = ai_{III}^0\left\{\left(\frac{c_{S_6^{2-}}}{c_{S_6^{2-}}^0}\right)^{2} e^{\frac{F}{2RT}(\phi_c - \phi_e - U_{III}^0)} - \left(\frac{c_{S_8^{2-}}}{c_{S_8^{2-}}^0}\right)^{\frac{3}{2}} e^{\frac{-F}{2RT}(\phi_c - \phi_e - U_{III}^0)}\right\}$$

$$j_{IV} = ai_{IV}^0\left\{\left(\frac{c_{S_4^{2-}}}{c_{S_4^{2-}}^0}\right)^{\frac{3}{2}} e^{\frac{F}{2RT}(\phi_c - \phi_e - U_{IV}^0)} - \left(\frac{c_{S_6^{2-}}}{c_{S_6^{2-}}^0}\right) e^{\frac{-F}{2RT}(\phi_c - \phi_e - U_{IV}^0)}\right\} \qquad (6.12)$$

$$j_{V} = ai_{V}^0\left\{\left(\frac{c_{S_2^{2-}}}{c_{S_2^{2-}}^0}\right) e^{\frac{F}{2RT}(\phi_c - \phi_e - U_{V}^0)} - \left(\frac{c_{S_4^{2-}}}{c_{S_4^{2-}}^0}\right)^{\frac{1}{2}} e^{\frac{-F}{2RT}(\phi_c - \phi_e - U_{V}^0)}\right\}$$

$$j_{VI} = ai_{VI}^0\left\{\left(\frac{c_{S^{2-}}}{c_{S^{2-}}^0}\right) e^{\frac{F}{2RT}(\phi_c - \phi_e - U_{VI}^0)} - \left(\frac{c_{S_2^{2-}}}{c_{S_2^{2-}}^0}\right)^{\frac{1}{2}} e^{\frac{-F}{2RT}(\phi_c - \phi_e - U_{VI}^0)}\right\}$$

$$\dot{r}_{VII} = k_{VII}^{rxn}\left(c_{Li^+}^2 c_{S^{2-}} - K_{VII}^{sp}\right)$$

Here, $U_q^0$ is an open circuit potential for reaction q (q = II to VI in Eq.(6.9)) at a reference concentration, $c_p^0$ for the species p involved in the reaction. Note that $\dot{r}_I$ and $\dot{r}_{VII}$ are volumetric rates and expressed in mol/m$^3$/s . $a$ is electrochemically active area. In the absence of any precipitates, i.e., $S_{8(s)}$ or $Li_2S_{(s)}$, it is the interfacial area for the porous backbone (e.g., carbon) and electrolyte contact. This microstructure state can be characterized using the corresponding porosity, $\varepsilon^0$, and interfacial area, $a^0$. In the presence of either of the solid species, actual porosity is smaller compared to the pristine value, $\varepsilon^0$. Since the solid species are electronically insulating, the presence of solid species decreases the area available for the electrochemical reactions (II to VI) and $a \leq a^0$. As the cell operation takes place, the amount of solid species keeps changing, and leads to a concurrent change in the porosity, active area, and tortuosity. The variations in the amount of solid

Lithium-sulfur batteries

species are tracked in terms of their local volume fractions, governed by the following conservation relations:

$$\frac{\partial \varepsilon_{S_{8(s)}}}{\partial t} = \overline{V}_{S_{8(s)}} \dot{r}_{I} \tag{6.13}$$

$$\frac{\partial \varepsilon_{Li_2S_{(s)}}}{\partial t} = \overline{V}_{Li_2S_{(s)}} \dot{r}_{VII} \tag{6.14}$$

$$\frac{\partial}{\partial t}\left(\varepsilon + \varepsilon_{S_{8(s)}} + \varepsilon_{Li_2S_{(s)}}\right) = 0 \tag{6.15}$$

These equations describe the microstructure evolution of the cathode, while the separator microstructure is assumed to remain unchanged. Each cathode microstructure state is described by any two of $\varepsilon$, $\varepsilon_{S_{8(s)}}$, $\varepsilon_{Li_2S_{(s)}}$ volume fractions. At each of these states, the active area and tortuosity are estimated using microstructure relations

$$\frac{a}{a^0} = \left(\frac{\varepsilon}{\varepsilon^0}\right)^{1.5} \tag{6.16}$$

$$\tau = {}^1\!/\sqrt{\varepsilon} \tag{6.17}$$

Any structural differences, for example, a porous cathode made up of carbon particles versus carbon fibers, or different interface growth morphologies are captured by specifying appropriate area and tortuosity relations (Eqs. 6.16, 6.17), while the conservation relations (Eqs. 6.13–6.15) remain invariant.

The governing equations are solved for potentials (Eqs. 6.2, 6.3), concentrations (Eq. 6.6) and volume fractions (Eqs. 6.13–6.15). In addition to the programming challenge of keeping track of multiple species and simultaneously solving a system of elliptic partial differential equations (Eqs. 6.2, 6.3), parabolic partial differential equations (Eq. 6.6) and ordinary differential equations (Eqs. 6.13–6.15), the most difficult aspect of any meaningful Li-sulfur solution algorithm is to conserve species [21]. The conservation is quite difficult to ensure because intermediate polysulfide species approach very low (ideally zero) concentrations and any improper discretization leads to unphysical behavior, e.g., negative species concentrations. These difficulties are indirectly reflected in the Li-sulfur modeling literature where more than half of the published works [27–31] average along the thickness, $z$, direction to convert all partial differential equations to ordinary differential equations that are straightforward to solve. Physically, such an

Modeling of electrode, electrolyte of lithium-sulfur batteries **211**

averaging is only sensible if the concentration profiles exhibit negligible gradients. This simplification implicitly assumes that any ion generated at the cathode due to a local reaction is immediately available at the anode to participate in the side reactions, which is difficult to justify for thick or low porosity cathodes operated at faster rates—all of which are the design targets for practically useful Li-sulfur batteries. A quantitative indicator for the numerical accuracy of sulfur conservation is to keep track of the total sulfur content:

$$\Sigma_S = \int_0^{L_{sep}+L_{cat}} \left\{ \varepsilon \left( c_{S_{8(l)}} + c_{S_8^{2-}} + \frac{3}{4}c_{S_6^{2-}} + \frac{1}{2}c_{S_4^{2-}} + \frac{1}{4}c_{S_2^{2-}} + \frac{1}{8}c_{S^{2-}} \right) + \frac{\varepsilon_{S_{8(s)}}}{\overline{V}_{S_{8(s)}}} + \frac{\varepsilon_{Li_2S_{(s)}}}{8\overline{V}_{Li_2S_{(s)}}} \right\} dz \quad (6.18)$$

Theoretically, $d\Sigma_S/dt = 0$; and any reliable numerical implementation must have a very small $d\Sigma_S/dt$. Accurate numerical solutions exhibit $d\Sigma_S/dt \ll i_{app}/F$.

This modeling discussion is nearly identical to that reported in Kumaresan, Mikhaylik, and White [20], except for two differences

- For precipitation–dissolution reactions, their rate expressions for $\dot{r}_I$ and $\dot{r}_{VII}$ include $\varepsilon_{S_{8(s)}}$ and $\varepsilon_{Li_2S_{(s)}}$ as proportionality factors. This is purely for mathematical convenience to guarantee that the reaction rates approach zero when the corresponding solid phase disappears. This, however, causes numerical issues while trying to precipitate a new phase during a subsequent charge.

- They also include the precipitation of lithium polysulfides, $Li_2S_x$, where $x = 2, 3, \ldots, 8$. Most experiments only observe $Li_2S$ as the precipitated phase, and it is unnecessary to include precipitation reactions for these intermediate species.

Fig. 6.2A shows a simulated electrochemical response from Kumaresan, Mikhaylik, and White [20] for a C/60 discharge rate. The C–rate sets the applied current based on initial sulfur loading as

$$i_{app} = C - rate \times \frac{16F\varepsilon_{S_{8(s)}}^0 L_{cat}}{3600\overline{V}_{S_{8(s)}}} \left( A/m^2 \right) \quad (6.19)$$

As shown in Fig. 6.2A, the simulated performance exhibits the experimentally observed two discharge plateaus. Origins of the plateaus become

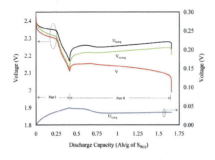
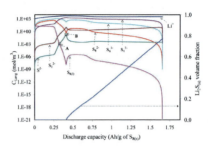

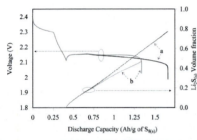

a: without Li$_2$S$_2$ precipitation
b: with Li$_2$S$_2$ precipitation

**Fig. 6.2** Results from [20]: (A) electrochemical performance and (B) species evolution at C/60; (C) performance comparison with and without Li$_2$S$_2$ precipitation *(Reproduced with permission from K. Kumaresan, Y. Mikhaylik, R.E. White, A Mathematical model for a lithium–sulfur cell, J. Electrochem. Soc. 155 (8) (2008), A576–A582, Copyright (2008) The Electrochemical Society.)*

clearer when the corresponding electrode averaged species concentrations (Fig. 6.2B are examined. The first (∼2.3 V) discharge plateau arises when the dissolved sulfur concentration is nearly constant as the rate of dissolution of solid sulfur matches the electrochemical reduction of the dissolved sulfur, such that $\dot{r}_{S_{8(l)}} \approx 0$. Similarly, when the concentration of sulfide ions, $S^{2-}$, becomes nearly time invariant, the cell voltage exhibits the lower voltage plateau. The sudden voltage drop at the end of discharge results from the consumption of all sulfur species in the electrolyte such that the electrochemical reactions, II to VI, cannot be sustained further.

Since the first plateau is related to sulfur dissolution, the rate constant, $k_I^{rxn}$, affects the shape of the first plateau (Fig. 6 in [20]). A smaller $k_I^{rxn}$ leads to a slower sulfur dissolution such that less dissolved sulfur is available to sustain the electrochemical reaction II. Consequently, subsequent reduction reactions activate early to sustain the current demand, $i_{app}$. Since these

Modeling of electrode, electrolyte of lithium-sulfur batteries 213

subsequent reactions have different open circuit potentials, the cell voltage changes faster while sulfur has not dissolved completely. Qualitatively similar behavior is also seen in the dissolved sulfur concentration.

This model [20] does not include side reactions at the anode surface that can reduce cell capacity, i.e., the polysulfide shuttle reactions. Accordingly, the model predicts complete theoretical capacity except when intermediate lithium polysulfides precipitate. For example, if $Li_2S_2$ precipitates, $S_2^{2-}$ ions are removed from the electrolyte and cannot reduce further to $S^{2-}$ to deliver the remaining capacity. Hence, when $Li_2S_2$ precipitation is allowed (Fig. 6.2C), the cell capacity decreases appreciably.

The partial differential equations (Eqs. 6.2, 6.3, 6.6) require appropriate boundary conditions to solve them. The boundary conditions for the current equations are straightforward: $i_e = i_{app}$ at $z = 0$, $i_c = 0$ at $z = L_{sep}$, $\phi_e = 0$ at $z = L_{sep} + L_{cat}$ and $i_c = i_{app}$ at $z = L_{sep} + L_{cat}$. In the absence of side reaction at anode, the current entering electrolyte is carried by $Li^+$ ions, i.e., $N_{Li^+} = i_{app}/F$ and $N_p = 0$ for $p \neq Li^+$. Similarly, no ions exit at $z = L_{sep} + L_{cat}$; hence, $N_p = 0$ for all p. In the presence of anode side reactions, only $N_p$ boundary conditions at $z = 0$ are to be updated.

Ghaznavi and Chen [32–34] have shown detailed results using this description [20] and systematically varying different properties appearing in the model equations. Other similar studies are [35,36]. We do not explicitly discuss the thickness averaged performance calculations [27–31] as they are a subset of the thickness-resolved models highlighted herein [20]. Subsequent examples add various mechanisms to this description [20]. The conservation statements (Eqs. 6.2, 6.3, 6.6, 6.13–6.15) describe the essence of a continuum-scale description for Li-sulfur performance. Note that given the fundamental similarities between Li-sulfur and Li-air performance [37], the governing equations and the numerical solution can be repurposed for predicting the electrochemical response of a Li-air cell [38].

## 6.3 Evolution of cathode porous electrode structure

The microstructure evolution relations (Eqs. 6.16, 6.17) qualitatively capture the expected porous electrode evolution wherein more precipitation decreases the active area available for electrochemical reactions (II to VI in Eq. 6.9) as well as increases the transport resistance by decreasing the pore space. The functional forms of Eqs. (6.16), (6.17) are such that both these limitations of microstructure evolution take place concurrently, which is overly simplistic. For example, when cells with excess electrolyte are tested for typical cyclic voltammetry experiments [39] or cells with flat substrates [40], the porosity change is minimal but the cell shuts down when the entire surface is covered with insulating $Li_2S$ (also the polysulfide shuttle is arguably negligible as the polysulfide concentration is very low with the excess electrolyte). The active area diminishes primarily due to precipitation and less so because of the porosity changes. Mathematically, a passivation metric can be defined to quantitatively identify this limitation:

$$M_{passivation} = 1 - {}^a/_{a^0} \qquad (6.20)$$

As $M_{passivation}$ approaches 1, the active area decreases and the reaction overpotentials terminate the electrochemical operation. Equivalently, another indicator can be defined to track the evolution of the electrolyte pore network:

$$M_{blockage} = 1 - \frac{\varepsilon/\tau}{\varepsilon^0/\tau^0} \qquad (6.21)$$

Herein, $a^0$, $\varepsilon^0$, and $\tau^0$ are properties of the pristine surface and structure before sulfur addition.

To construct more representative microstructure relations, one must study growth in porous structures belonging to the same class. Here, the microstructure class identifies similar-looking structures, e.g., all microstructures with spherical pores and different extents of precipitation at the interface as shown in Fig. 6.3A. Other common microstructure classes (related to electrochemical systems) include structures composed of spherical or platelet particles (e.g., typical Li–ion microstructures) and fibrous structures (e.g., fuel cell gas diffusion layers). Growth in each of these microstructure classes can be examined to characterize active area and tortuosity as a function of porosity and precipitate volume [21]. Each of the 3D structures, e.g., Fig. 6.3A, is analyzed separately to estimate active area and tortuosity. Once enough such structures are examined, the resultant dataset is analyzed to develop accurate microstructure relations to replace Eqs. (6.16), (6.17).

For the same backbone structure (e.g., Fig. 6.3A and B), the precipitate growth morphology can vary as substrate and/or electrolyte materials are changed [40]. To account for such morphological variation, the precipitate growth description in [21] includes $\omega$—a growth characteristic that directly influences the precipitate morphology. $\omega$ quantifies the likelihood of new precipitate growing on itself as compared to the remaining active surface. As $\omega \to 0$, the precipitate phase prefers depositing on the active area and the active area diminishes quickly as shown in Fig. 6.3D—the curve

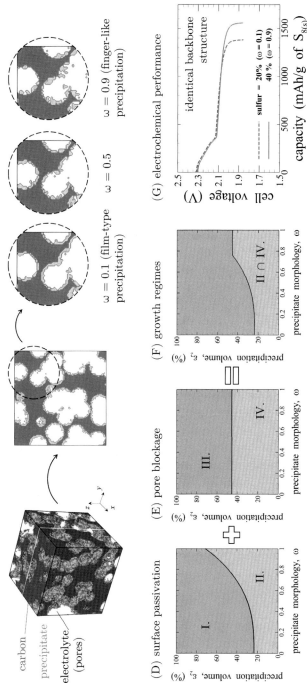

**Fig. 6.3** Accounting for detailed electrode microstructure evolution: (A) electrode subvolume with relevant phases (B) corresponding 2D slice (C) differences in precipitate morphology become apparent when the interface is examined closely; the precipitate growth leads to two different microstructure limitations (D) where insulating precipitates cover the active interface (E) when precipitate growth decreases the pore volume—either of them becomes dominant for a specific type of precipitate morphology (F); (G) accordingly the cell performance varies strongly with the precipitate morphology. *(Reproduced with permission from C.A. Mistry, P.P. Mukherjee, Precipitation-microstructure interactions in the Li-sulfur battery electrode, J. Phys. Chem. C 121 (47) (2017) 26256–26264, Copyright (2017) American Chemical Society.)*

separating the two regime marks $M_{passivation} = 0.9$. As $\omega$ increases, the active area coverage is delayed. Representative growth patterns for different $\omega$ values are shown in Fig. 6.3C. As the tortuosity is a bulk characteristic of the pore network, the morphology has a relatively subdued effect on tortuosity evolution as shown in Fig. 6.3E using $M_{blockage}$ (Eq. 6.21). Since both of these effects jointly exist, the relative severity of each mode is directly related to the morphology descriptor, $\omega$ (Fig. 6.3F).

Fig. 6.3G uses the electrochemical performance description discussed previously in Eqs. (6.2), (6.3), (6.6), (6.13)–(6.15) and substitutes class-specific microstructure growth relations for Eqs. (6.16), (6.17). Accordingly, even for the same pristine structure, a high sulfur loading can be made to perform better by tailoring the precipitate morphology as shown in Fig. 6.3G. Essentially more three-dimensional (finger-like) morphology delays the surface coverage, thus allowing for better discharge performance. These predictions match results from controlled experiments [41]. Note that depending on the precipitate morphology and how quickly the active area is passivated, the observed performance can be less than ideal. The surface passivation accounts for a nonideal electrochemical response even in the absence of the polysulfide shuttle effect. By developing microstructure growth relations for different microstructure classes, one can explore the relative merits of different porous structures to delay passivation [38].

The discussion in Fig. 6.3 assumes that there is no barrier to nucleating a new precipitate phase ($Li_2S$ during discharge or $S_8$ during charge). Depending on the surface energies involved at the active interface, energy may have to be spent nucleating a new phase, which gives rise to a slight voltage undershoot as the cell transitions from the upper plateau to the lower plateau and the $Li_2S$ nucleation is imminent [39,41]. Mathematically, such a response is captured by explicitly bringing in nucleation and growth dynamics of individual $Li_2S$ particles [22,42–44]. Since the nucleation events are strongly rate-dependent, it also causes rate dependent precipitate morphology (Fig. 6.4). The continuum-scale description in Eqs. (6.2), (6.3), (6.6), (6.13)–(6.15) is modified by resolving $Li_2S$ precipitation reaction VII (Eq. 6.9) into distinct nucleation (of individual $Li_2S$ particles) and growth steps. Supersaturation is required to nucleate new $Li_2S$ particles; hence, $S^{2-}$ concentration builds up before it starts nucleating. This concentration buildup is responsible for the voltage undershoot just before the nucleation onset in Fig. 6.4A. At higher rates, $S^{2-}$ is generated more rapidly and leads to a greater supersaturation and in turn a smaller $Li_2S$ particle size (Fig. 6.4B). At higher rates, more nuclei are activated simultaneously which leads to a

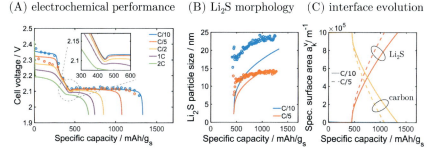

*symbols are experimental measurements from Yu et al. (2018) Energy Environ. Sci. 11(1) 202-210

**Fig. 6.4** Incorporating dynamics of nucleating new precipitate phase in electrochemical description: (A) a voltage undershoot appears just before the onset of the lower voltage plateau; (B) Li$_2$S particles grow smaller at faster rates and (C) the active area is passivated more quickly. *(Reproduced with permission from T. Danner, A. Latz, On the influence of nucleation and growth of S$_8$ and Li$_2$S in Lithium-sulfur batteries, Electrochim. Acta 322 (2019) 134719.)*

faster surface coverage and the active interface is passivated quickly as the rates increase (Fig. 6.4C). Note that even without including the nucleation physics, simulations can predict the voltage undershoot for a particular choice of exchange current densities for the intermediate reactions III to V (Eq. 6.9) [31]. However, this will exhibit different C-rate dependence than reflected by the nucleation physics, which in turn helps underpin the root cause. The nucleation dynamics is a candidate mechanism for current-dependent electrochemical performance. The current-dependent morphology is also observed in other similar electrochemical systems, for example, Li-air [45,46], and quantitatively explains the observed capacity trend with discharge rates [38,47].

Explicitly resolving nucleation and growth of Li$_2$S particles captures the interface evolution at the early stages of precipitation. Each nucleus cannot grow larger than its distance from its neighboring particles. Such individual growth ends there and the particles likely coalesce at this stage [39] which needs to be treated separately. In principle, these two different growth regimes would exhibit different microstructure evolution and lead to different functional forms of area and tortuosity evolution relations. In the preceding discussion (Figs. 6.2–6.4), the electrolyte transport is not rate-limiting and the observed performance characteristics contain signatures of reaction limitations. The dilute electrolyte treatment in Eqs. (6.4),

(6.7), (6.8) is conceptually insufficient to describe the transport in concentration electrolytes. When the transport becomes rate-limiting, the cell response is dominated by how quickly ions can reach reaction sites, and not by how quickly the ions can react.

## 6.4 Concentrated electrolyte transport effects

The dilute electrolyte treatment in Eqs. (6.4), (6.7), (6.8) predicts higher conductivity at higher salt concentrations. This is only true if the electrolyte is dilute, i.e., low ion concentrations. Even for a comparatively simple electrolyte containing one salt in solvents such as used for Li-ion batteries, at higher salt concentrations, the ionic conduction is found to be inferior [15]. Another trait of concentrated solution is more sluggish ionic diffusion at higher concentrations. Both these effects can limit the electrochemical performance of Li-sulfur cells, especially for states of charge around the transition from the upper voltage plateau to the lower one, since most species are in the electrolyte phase at this stage (e.g., Fig. 6.2B). Given our interest in building Li-sulfur cells with high sulfur-to-electrolyte ratio [6,48,49], the concentrated solution effects are practically relevant. To treat these concentrated electrolyte effects [18], one has to resort to the Stefan-Maxwell relation for ionic fluxes:

$$\frac{-c_p c_T}{RT} \nabla \mu_p = \sum_{p' \neq p} \frac{c_{p'} N_p - c_p N_{p'}}{\mathfrak{D}_{pp'}} \tag{6.22}$$

where $c_T = \sum_p c_p$ is the total species concentration in the electrolyte. $\mu_p$ is the (electro-) chemical potential of species p. $\mathfrak{D}$'s are Stefan-Maxwell diffusivities and differ from the self-diffusivity, $D$, used earlier. Using Eq. (6.22) and charge neutrality, one can reexpress species fluxes, $N_p$, as driven by concentration and potential gradients. This treatment substitutes Eqs. (6.4), (6.7), (6.8) with equivalent concentrated theory forms, while retaining the same conservation statements (Eqs. 6.2, 6.3, 6.6, 6.13–6.15) in the continuum description [23].

Fig. 6.5A shows a simulated electrochemical performance with concentrated electrolyte transport [23]. The performance also exhibits a voltage undershoot while transitioning from the first to the second discharge plateau. This voltage undershoot results from low electrolyte conductivity when there are too many ions in the electrolyte. To quantitatively describe the identity of the dominant ionic species, sulfur fractions, $f_p$. Eq. (6.18) for $\Sigma_S$ provides a consistent definition for $f_p$, for example,

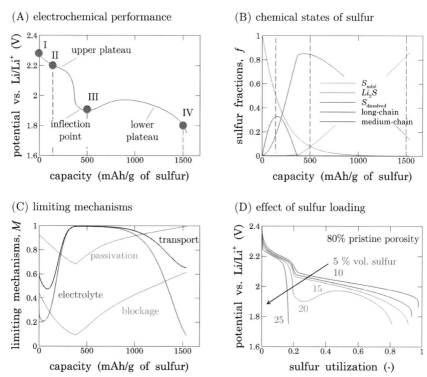

**Fig. 6.5** When multiple ionic species are present in the electrolyte, the interspecies interactions strongly affect the transport: (A) such effects are strong in between the two voltage plateaus since (B) polysulfide concentration is very high around these states of charge; (C) limitations arising from microstructure growth and concentrated electrolyte transport become active over specific time intervals; (B) and (C) correspond to the electrochemical performance as shown in (A); (D) as sulfur loading is increased, the concentrated electrolyte effect becomes stronger. *(Reproduced with permission from C.A.N. Mistry, P.P. Mukherjee, Electrolyte transport evolution dynamics in lithium-sulfur batteries, J. Phys. Chem. C 122 (32) (2018) 18329–18335, Copyright (2018) American Chemical Society.)*

$$f_{S_4^{2-}} = \frac{1}{\Sigma_S} \int_0^{L_{sep}+L_{cat}} \frac{1}{2}\left(\varepsilon c_{S_4^{2-}}\right) dz \qquad (6.23)$$

Essentially, the species with the highest $f_p$ is present is the largest concentration at a given time.

As shown in Fig. 6.5B, the medium-chain polysulfide, $S_4^{2-}$, is the dominant sulfur species when the voltage undershoot appears in Fig. 6.5A. The concentration of each of the ionic species is correlated with the solid sulfur loading at the onset of discharge. It can be shown that,

$$\max\left(c_{S_4^{2-}}\right) \approx \frac{2\varepsilon_{S_{8(s)}}^0}{\varepsilon^0 \overline{V}_{S_{8(s)}}} \tag{6.24}$$

which is $\sim$2 M for 10% volume sulfur loading and pristine porosity of 80%.

To track the severity of ionic conductivity decline for the concentrated solution, additional limitation metrics can be defined (in a similar manner as defined in Eqs. 6.20, 6.21):

$$M_{\text{electrolyte}} = 1 - {}^{\kappa}/\max(\kappa) \tag{6.25}$$

Here, $\max(\kappa)$ is a property of the electrolyte. The conductivities in Eq. (6.25) are bulk electrolyte transport properties. Accounting for the pore network variations with precipitation, an additional limitation metric can be defined that quantifies the transport limitation in the porous electrode:

$$M_{\text{transport}} = 1 - \frac{\kappa\varepsilon/\tau}{\max(\kappa)\varepsilon^0/\tau^0} \tag{6.26}$$

These four metrics are tracked along the discharge process (Fig. 6.5A) in Fig. 6.5C. It reveals that the performance is limited by surface passivation during the two plateaus, while the transition between the two plateaus is limited by poor electrolyte transport. For thin electrodes (e.g., Fig. 6.5 [23]), the sluggish diffusion does not create appreciable electrolyte concentration gradients. This would be detrimental for thicker electrodes wherein sluggish diffusion of ions leads to a strong distribution of reaction rates along the electrode thickness and equivalently smaller capacities stemming from nonuniform utilization of the electrode. This effect is similar to mass transport limitations in Li-ion electrodes [14,50]. As the sulfur-to-electrolyte ratio, S/E, is varied, the ionic concentrations vary in the electrolyte. The concentrated electrolyte effects become worse at higher S/E ratios as shown in Fig. 6.5D.

$$S/E = \frac{\rho_{S_8}\varepsilon_{S_{8(s)}}^0}{\varepsilon^0 - \varepsilon_{S_{8(s)}}^0} \tag{6.27}$$

To maximize the energy density of Li-sulfur cells, a key design requirement is to use high S/E ratios [6,48,49], which makes it necessary to understand the concentrated electrolyte response.

## 6.5 Dynamics of the polysulfide shuttle effect

The transport of ionic species through the electrolyte is essential to understand the polysulfide shuttle effect. As shown in Fig. 6.1C, if the polysulfide ions reach the anode, they can be chemically reduced by lithium. Every electrochemical reaction at the cathode (II to VI in Eq. 6.9) can chemically occur at the anode surface where the electron is provided by lithium oxidizing to Li+ and a higher-order polysulfide undergoing reduction to a lower-order polysulfide in the immediate vicinity. Specific reactions are shown below:

chemical reactions at anode surface (unwanted side reactions):

$$
\begin{aligned}
&\text{VIII} \quad \text{Li}^+ + \frac{1}{2}\text{S}_8^{2-} \Leftrightarrow \frac{1}{2}\text{S}_{8(l)} + \text{Li} \\
&\text{IX} \quad \text{Li}^+ + 2\text{S}_6^{2-} \Leftrightarrow \frac{3}{2}\text{S}_8^{2-} + \text{Li} \\
&\text{X} \quad \text{Li}^+ + \frac{3}{2}\text{S}_4^{2-} \Leftrightarrow \text{S}_6^{2-} + \text{Li} \\
&\text{XI} \quad \text{Li}^+ + \text{S}_2^{2-} \Leftrightarrow \frac{1}{2}\text{S}_4^{2-} + \text{Li} \\
&\text{XII} \quad \text{Li}^+ + \text{S}^{2-} \Leftrightarrow \frac{1}{2}\text{S}_2^{2-} + \text{Li} \\
&\text{XIII} \quad 2\text{Li}^+ + \text{S}^{2-} \Leftrightarrow \text{Li}_2\text{S}_{(s)}
\end{aligned}
\tag{6.28}
$$

The corresponding reaction rates, $\dot{r}_{\text{VIII}}$ to $\dot{r}_{\text{XIII}}$, are area-specific reaction rates at the anode surface (units of mol/m$^2$/s). Note that the newly formed lithium and sulfide ions in reaction XII combine and precipitate at the anode surface via reaction XIII. Each of these rates contributes to the species flux boundary conditions, $N_p$, at the anode surface. Each of VIII to XII reactions chemically reduces sulfur species and represents a lost capacity. If the reduced ions remain in the electrolyte, they remain electrochemically active and can be oxidized back (in principle). However, Li$_2$S precipitating at the anode surface is irreversibly lost. Thus, the total capacity lost due to the shuttle effect, $Q_\Delta$, is related to reactions VIII–XIII (Eq. 6.28)—most of which can be recovered electrochemically (e.g., by overcharging [9]) except when Li$_2$S precipitates at anode. Fig. 6.6A compares electrochemical performance with and without these side reactions [24], and the components of the lost capacity, $Q_\Delta$, are shown in Fig. 6.6B.

The anode side reactions VIII–XIII require sufficient ionic concentrations in the vicinity of the anode to react. If the ionic transport is sluggish due to poor electrolyte transport properties, these ions do not reach the anode in ample quantities and the shuttle effect capacity loss becomes transport-limited. Such transport limitations do not hamper the electrochemical functioning of the anode since it generates lithium cations and the side reactions are starving because polysulfide species do not reach the

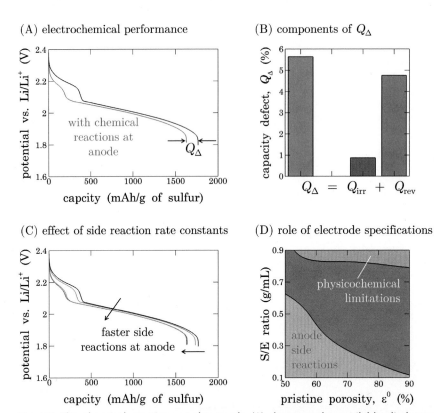

**Fig. 6.6** The chemical reactions at the anode (A) decrease the available discharge capacity by chemically reducing some of the ionic species at the anode; (B) this capacity defect is composed of capacity that can be recovered by overcharging (reversible) and the rest is in terms of Li$_2$S precipitated on the anode which cannot be oxidized electrochemically (irreversible). (C) As these side reactions become faster, more capacity is lost at the same C-rate; beyond a limit, faster reactions do not cause greater capacity defect since the reactions now become transport-limited. (D) The anode side reactions are dominant for a low S/E ratio, while cathode microstructure evolution and concentrated electrolyte transport are dominant limitations at a high S/E ratio. *(Reproduced with permission from C.A.N. Mistry, P.P. Mukherjee, "Shuttle" in polysulfide shuttle: friend or foe? J. Phys. Chem. C 122 (42) (2018) 23845–23851, Copyright (2018) American Chemical Society.)*

anode in sufficient quantities. The ionic fluxes of the polysulfides are composed of migration and diffusive components—the negatively charged polysulfides are driven toward an anode because of migration but its diffusion can counter this tendency such that the net polysulfide flux reaching the anode is not as high. Fig. 6.6C compares the cell response by gradually increasing the rate constants for the side reactions. Initially, as the rate constant is increased, the performance worsens as more capacity is lost in faster side reactions, but subsequently, faster kinetics of the side reactions do not change the performance as the side reactions become transport-limited.

The transport limitations dominate for concentrated electrolytes. Hence, higher S/E ratios diminish the severity of the shuttle effect. Similarly, for an identical S/E ratio, a higher pristine porosity houses more sulfur and leads to more concentrated electrolytes. Fig. 6.6D exhibits both these trends where high S/E ratio and/or high pristine porosity ameliorates the severity of the anode side reactions. Note that very high S/E ratio and/or pristine porosity are also detrimental as they trigger microstructure growth limitations (Fig. 6.3) or poor electrolyte conductivity (Fig. 6.5). For dilute electrolytes, ionic conductivity and species diffusion are positively correlated; but the concentrated electrolytes can exhibit sluggish species diffusion and reasonable conductivity [18], which is the desirable mode of electrolyte transport to decrease polysulfide transport to anode without generating an appreciable ohmic drop in the cell.

As the deleterious consequence of the polysulfide shuttle is associated with the chemical side reactions at the anode, one can decrease the total amount of chemical reduction by operating at higher current densities (at higher current densities, the cell operation spans a shorter time period, which leads to a smaller amount of chemical reaction). However, most experiments show a monotonic capacity decline with C-rate, further implying that the polysulfide shuttle is not the only limitation. Alternatively, the cathode (Figs. 6.3 and 6.4) and electrolyte (Fig. 6.5) limitations qualitatively explain this observed monotonic trend.

Since the anode side reactions are active even when the cell is not being operated electrochemically, theycause self-discharge. The self-discharge measurements can provide meaningful information related to the severity of the shuttle effect when combined with suitable models [24,51].

Most literature studies identifying the polysulfide shuttle effect as the dominant limitation either use thickness averaged models, e.g., Mikhaylik and Akridge [52] and Busche et al. [53], or tune modeling parameters by assuming that all the capacity loss is due to the polysulfide shuttle, e.g., Hofmann, Fronczek, and Bessler [25]—Fig. 6.7. Fig. 6.7 adds an additional mechanism to the original continuum description in Eqs. (6.2), (6.3), (6.6), (6.13)–(6.15) [20] in the form of anode passivation by $Li_2S$ deposition. For each of the experimental curves in Fig. 6.7, the maximum amount of $Li_2S$ precipitation at anode for complete passivation is adjusted to match the measured capacity. Herein, the cell shuts down because the lithium anode surface is covered with insulating $Li_2S$ and the anode overpotential

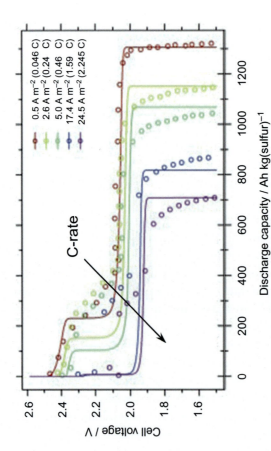

**Fig. 6.7** Predicted cell performance is matched to experiments by adjusting the amount of Li$_2$S precipitation to passivate reaction area at the anode; Li$_2$S precipitates at anode due to a chemical side reaction; symbols are experiments from Cheon et al. (2003) J. Electrochem. Soc. 150(6) A800–A805 *(Reproduced with permission from A.F. Hofmann, D.N. Fronczek, W.G. Bessler, Mechanistic modeling of polysulfide shuttle and capacity loss in lithium-sulfur batteries. J. Power Sources 259 (2014) 300–310, Copyright (2014) Elsevier.)*

drives the cell voltage drop. These studies are older compared to the recent investigations identifying cathode passivation as a dominant limitation [39]. Also, the systematic experiments [9] with $LiNO_3$ eliminate the shuttle effect as the dominant capacity-limiting mechanism.

## 6.6 Sources of variability: Mechanisms and properties

The fundamental interactions in Li–sulfur cells are neither well understood (i.e., all mechanisms are not known) nor well characterized (i.e., corresponding properties have not all been measured). Accordingly, the predicted electrochemical behavior, e.g., Figs. 6.2–6.7, intrinsically contains variabilities. The elegance of physics-based continuum modeling lies in that the simulated response does not violate physical laws such as species conservation and charge neutrality. It allows a general framework to test hypotheses about different mechanisms. However, one must be careful in trying to fit model predictions to experiments by varying the properties because the models described herein contains many property values, which can be varied arbitrarily to match limited experimental data. Ideally, one should be able to measure relevant properties affecting performance by simpler experiments but the uniqueness of Li–sulfur chemistry may partially negate such systematic studies.

To make matters worse, multiple fundamental mechanisms exhibit similar macroscopic signatures, e.g., the less than theoretical capacity can arise from polysulfide shuttle [25] or cathode passivation [21,22] or electrolyte transport limitations [23,35]; in the same spirit, the voltage undershoot during discharge can arise due to a particular sequencing of exchange current densities [31], or nucleation dynamics [22], or concentrated electrolyte effects [23]. Each of these varies differently with C-rates or electrode composition [54], and if appropriate experimental data are available, it is possible to further pinpoint the root cause. However, one discharge profile is insufficient to unambiguously identify the dominant limiting mechanisms.

With this background, the possible sources of variability can be categorized as:

- **Mechanisms selected to describe the electrochemical performance**: For example, thickness averaged models [29,52] make polysulfides available to react at the anode as soon as they are generated at the cathode, which overemphasizes the role of a polysulfide shuttle effect; to safeguard against misinterpretation, the model should at least have two or more capacity-limiting mechanisms such that their relative importance can be isolated by varying operating conditions or electrode composition.

- **Variability in properties**: The fundamental transport and reaction interactions have not been well characterized for the materials relevant to Li-sulfur, e.g., the solubility of sulfur in the electrolyte is not reported; limited experimental data from the organic solvent literature suggest that its solubility strongly varies with salt concentration [55], which can appreciably change computational predictions. Similarly, electrolyte transport [56] and reaction rates [57] are incompletely characterized.
- **Uncertainty of reaction pathways**: The reaction steps in Eq. (6.9) preset the redox-active species and how many properties are to be characterized (or assumed). Various studies report different intermediate steps [58–60]. Another ambiguity relates to the precipitation and dissolution reactions—whether they are chemical steps [20] or electrochemical [39,41] will potentially change the avenues to counter their limitations, e.g., redox mediators can be used to modify the growth if the $Li_2S$ formation is an electrochemical step [41,61]. Alternatively, if these end reactions are chemical, they can also precipitate the solid phase in separator and accordingly require a different suit of solutions.

## 6.7 Summary and outlook

The Li-sulfur electrochemistry holds a great theoretical promise as the next-generation energy storage system, but its implementation is limited by our incomplete understanding of its electrochemical response. Early investigations pointed to side reactions at the anode as the root cause of all nonidealities; however, controlled experimental and modeling studies in the last five years have identified many other mechanisms that potentially alter the macroscopically measured response. Still, the mechanisms contributing to the subsequent electrochemical steps and repeated cycling have not been conclusively identified.

In this context, continuum–scale modeling is well suited to explore various candidate mechanisms by interpreting suitably controlled experiments. While experimental techniques that provide sufficiently resolved information for operating cells are in a nascent stage (in part due to the uniqueness of Li-sulfur chemistry), the continuum–scale modeling inherently describes the concentration and potential fields. If used synergistically, the continuum–scale modeling and controlled experiments can explore the underlying mechanisms that lead to the peculiarly measured response.

One should be cautious to acknowledge the variabilities associated with such an approach while combining modeling and experiments to unambiguously isolate mechanisms.

Key unanswered questions are the reaction pathways, morphological variations of the lithium anode, mechanical stresses, associated structural changes at the cathode, and cell degradation mechanisms for repeated operations. Once corresponding fundamental mechanisms have been characterized, we can borrow tools from Li-ion research to design thick electrodes and explore the role of microstructure inhomogeneities [62] and thermoelectrochemical response [63]. The continuum description also identifies the properties directly contributing to the electrochemical performance. Accordingly, the relevance of molecular and atomic modifications to the battery materials can be quantitatively assessed by using the material property changes for each structural modifications in the continuum modeling.

## Acknowledgments

The authors gratefully acknowledge the support from the U.S. Department of Energy (DOE), Vehicle Technologies Office. Argonne National Laboratory is operated for DOE Office of Science by UChicago Argonne, LLC, under contract number DE-AC02-06CH11357.

## References

[1] S. Zhang, K. Ueno, K. Dokko, M. Watanabe, Recent advances in electrolytes for Lithium-sulfur batteries, Adv. Energy Mater. 5 (16) (2015) 1500117, https://doi.org/10.1002/aenm.201500117.

[2] A. Yermukhambetova, C. Tan, S.R. Daemi, Z. Bakenov, J.A. Darr, D.J.L. Brett, P.R. Shearing, Exploring 3D microstructural evolution in Li-sulfur battery electrodes using in-situ X-ray tomography, Sci. Rep. 6 (2016) 35291, https://doi.org/10.1038/srep35291.

[3] D. Bresser, S. Passerini, B. Scrosati, Recent progress and remaining challenges in sulfur-based lithium secondary batteries—a review, Chem. Commun. 49 (90) (2013) 10545–10562, https://doi.org/10.1039/c3cc46131a.

[4] A. Rosenman, E. Markevich, G. Salitra, D. Aurbach, A. Garsuch, F.F. Chesneau, Review on Li-sulfur battery systems: an integral perspective, Adv. Energy Mater. 5 (16) (2015), https://doi.org/10.1002/aenm.201500212.

[5] H. Zhao, N. Deng, J. Yan, W. Kang, J. Ju, Y. Ruan, X. Wang, X. Zhuang, Q. Li, B. Cheng, A review on anode for lithium-sulfur batteries: progress and prospects, Chem. Eng. J. (2018) 343–365, https://doi.org/10.1016/j.cej.2018.04.112.

[6] Z.W. Seh, Y. Sun, Q. Zhang, Y. Cui, Designing high-energy lithium-sulfur batteries, Chem. Soc. Rev. (2016) 5605–5634, https://doi.org/10.1039/c5cs00410a.

[7] A. Rosenman, R. Elazari, G. Salitra, E. Markevich, D. Aurbach, A. Garsuch, The effect of interactions and reduction products of $LiNO_3$, the anti-shuttle agent, in Li-S battery

systems, J. Electrochem. Soc. 162 (3) (2015) A470–A473, https://doi.org/10.1149/2.0861503jes.

[8] D. Aurbach, E. Pollak, R. Elazari, G. Salitra, C.S. Kelley, J. Affinito, On the surface chemical aspects of very high energy density, rechargeable Li–sulfur batteries, J. Electrochem. Soc. 156 (8) (2009) A694, https://doi.org/10.1149/1.3148721.

[9] T. Poux, P. Novák, S. Trabesinger, Pitfalls in Li–S rate-capability evaluation, J. Electrochem. Soc. 163 (7) (2016) A1139–A1145.

[10] J. Zheng, M. Gu, M.J. Wagner, K.A. Hays, X. Li, P. Zuo, C. Wang, J.-G. Zhang, J. Liu, J. Xiao, Revisit carbon/sulfur composite for Li-S batteries, J. Electrochem. Soc. 160 (10) (2013) A1624–A1628.

[11] J. Zheng, D. Lv, M. Gu, C. Wang, J.-G. Zhang, J. Liu, J. Xiao, How to obtain reproducible results for lithium sulfur batteries? J. Electrochem. Soc. 160 (11) (2013) A2288–A2292, https://doi.org/10.1149/2.106311jes.

[12] C. Tan, T.M.M. Heenan, R.F. Ziesche, S.R. Daemi, J. Hack, M. Maier, S. Marathe, C. Rau, D.J.L. Brett, P.R. Shearing, Four-dimensional studies of morphology evolution in lithium-sulfur batteries, ACS Appl. Energy Mater. 1 (9) (2018) 5090–5100, https://doi.org/10.1021/acsaem.8b01148.

[13] S.-L. Wu, W. Zhang, X. Song, A.K. Shukla, G. Liu, V. Battaglia, V. Srinivasan, High rate capability of $Li(Ni_{1/3}Mn_{1/3}Co_{1/3})O_2$ electrode for Li-ion batteries, J. Electrochem. Soc. 159 (4) (2012) A438–A444, https://doi.org/10.1149/2.062204jes.

[14] H. Zheng, J. Li, X. Song, G. Liu, V.S. Battaglia, A comprehensive understanding of electrode thickness effects on the electrochemical performances of Li-ion battery cathodes, Electrochim. Acta 71 (2012) 258–265, https://doi.org/10.1016/j.electacta.2012.03.161.

[15] L.O. Valøen, J.N. Reimers, Transport properties of LiPF[sub 6]-based Li-ion battery electrolytes, J. Electrochem. Soc. 152 (5) (2005) A882–A891, https://doi.org/10.1149/1.1872737.

[16] R. Amin, Y.-M. Chiang, Characterization of electronic and ionic transport in $Li_{1-x-}Ni_{0.33}Mn_{0.33}Co_{0.33}O_2$ ($NMC_{333}$) and $Li_{1-x}Ni_{0.50}Mn_{0.20}Co_{0.30}O_2$ ($NMC_{523}$) as a function of Li content, J. Electrochem. Soc. 163 (8) (2016) A1512–A1517, https://doi.org/10.1149/2.0131608jes.

[17] J. Landesfeind, J. Hattendorff, A. Ehrl, W.A. Wall, H.A. Gasteiger, Tortuosity determination of battery electrodes and separators by impedance spectroscopy, J. Electrochem. Soc. 163 (7) (2016) A1373–A1387, https://doi.org/10.1149/2.1141607jes.

[18] J. Newman, K.E. Thomas-Alyea, Electrochemical Systems, John Wiley & Sons, 2012.

[19] M. Doyle, T.F. Fuller, J. Newman, Modeling of Galvanostatic charge and discharge of the lithium/polymer/insertion cell, J. Electrochem. Soc. 140 (6) (1993) 1526–1533, https://doi.org/10.1149/1.2221597.

[20] K. Kumaresan, Y. Mikhaylik, R.E. White, A Mathematical model for a lithium–sulfur cell, J. Electrochem. Soc. 155 (8) (2008) A576–A582.

[21] A. Mistry, P.P. Mukherjee, Precipitation-microstructure interactions in the Li-sulfur battery electrode, J. Phys. Chem. C 121 (47) (2017) 26256–26264, https://doi.org/10.1021/acs.jpcc.7b09997.

[22] T. Danner, A. Latz, On the influence of nucleation and growth of $S_8$ and $Li_2S$ in Lithium-sulfur batteries, Electrochim. Acta 322 (2019) 134719, https://doi.org/10.1016/j.electacta.2019.134719.

[23] A.N. Mistry, P.P. Mukherjee, Electrolyte transport evolution dynamics in lithium-sulfur batteries, J. Phys. Chem. C 122 (32) (2018) 18329–18335, https://doi.org/10.1021/acs.jpcc.8b05442.

[24] A.N. Mistry, P.P. Mukherjee, "Shuttle" in polysulfide shuttle: friend or foe? J. Phys. Chem. C 122 (42) (2018) 23845–23851, https://doi.org/10.1021/acs.jpcc.8b06077.

[25] A.F. Hofmann, D.N. Fronczek, W.G. Bessler, Mechanistic modeling of polysulfide shuttle and capacity loss in lithium-sulfur batteries, J. Power Sources 259 (2014) 300–310, https://doi.org/10.1016/j.jpowsour.2014.02.082.

[26] R. Marom, S.F. Amalraj, N. Leifer, D. Jacob, D. Aurbach, A review of advanced and practical lithium battery materials, J. Mater. Chem. 21 (27) (2011) 9938–9954, https://doi.org/10.1039/c0jm04225k.

[27] K. Yoo, M.K. Song, E.J. Cairns, P. Dutta, Numerical and experimental investigation of performance characteristics of lithium/sulfur cells, Electrochim. Acta 213 (2016) 174–185, https://doi.org/10.1016/j.electacta.2016.07.110.

[28] M. Marinescu, T. Zhang, G.J. Offer, A zero dimensional model of lithium-sulfur batteries during charge and discharge, Phys. Chem. Chem. Phys. 18 (1) (2016) 584–593, https://doi.org/10.1039/c5cp05755h.

[29] M. Marinescu, L. O'Neill, T. Zhang, S. Walus, T.E. Wilson, G.J. Offer, Irreversible vs reversible capacity fade of lithium-sulfur batteries during cycling: the effects of precipitation and shuttle, J. Electrochem. Soc. 165 (1) (2018) A6107–A6118, https://doi.org/10.1149/2.0171801jes.

[30] Y. Yin, A.A. Franco, Unraveling the operation mechanisms of lithium sulfur batteries with ultramicroporous carbons, ACS Appl. Energy Mater. 1 (11) (2018) 5816–5821, https://doi.org/10.1021/acsaem.8b01159.

[31] D.N. Fronczek, W.G. Bessler, Insight into lithium-sulfur batteries: elementary kinetic modeling and impedance simulation, J. Power Sources 244 (2013) 183–188, https://doi.org/10.1016/j.jpowsour.2013.02.018.

[32] M. Ghaznavi, P. Chen, Sensitivity analysis of a mathematical model of lithium-sulfur cells: part II: precipitation reaction kinetics and sulfur content, J. Power Sources 257 (2014) 402–411, https://doi.org/10.1016/j.jpowsour.2013.12.145.

[33] M. Ghaznavi, P. Chen, Sensitivity analysis of a mathematical model of lithium-sulfur cells part I: applied discharge current and cathode conductivity, J. Power Sources 257 (2014) 394–401, https://doi.org/10.1016/j.jpowsour.2013.10.135.

[34] M. Ghaznavi, P. Chen, Analysis of a mathematical model of lithium-sulfur cells part III: electrochemical reaction kinetics, transport properties and charging, Electrochim. Acta 137 (2014) 575–585, https://doi.org/10.1016/j.electacta.2014.06.033.

[35] T. Zhang, M. Marinescu, S. Walus, G.J. Offer, Modelling transport-limited discharge capacity of lithium-sulfur cells, Electrochim. Acta 219 (2016) 502–508, https://doi.org/10.1016/j.electacta.2016.10.032.

[36] N. Erisen, D. Eroglu, Modeling the discharge behavior of a lithium-sulfur battery, Int. J. Energy Res. 44 (13) (2020) 10599–10611, https://doi.org/10.1002/er.5701.

[37] M. Mehta, V. Bevara, P. Andrei, Maximum theoretical power density of lithium–air batteries with mixed electrolyte, J. Power Sources 286 (2015) 299–308, https://doi.org/10.1016/j.jpowsour.2015.03.158.

[38] A.N. Mistry, F. Cano-Banda, D. Law, A. Hernandez-Guerrero, P.P. Mukherjee, Non-equilibrium thermodynamics in electrochemical complexation of Li-oxygen porous electrodes, J. Mater. Chem. A 7 (15) (2019) 8882–8888, https://doi.org/10.1039/c9ta01339c.

[39] F.Y. Fan, W.C. Carter, Y.M. Chiang, Mechanism and kinetics of $Li_2S$ precipitation in lithium-sulfur batteries, Adv. Mater. 27 (35) (2015) 5203–5209, https://doi.org/10.1002/adma.201501559.

[40] G. Babu, K. Ababtain, K.Y.S. Ng, L.M.R. Arava, Electrocatalysis of lithium polysulfides: current collectors as electrodes in Li/S battery configuration, Sci. Rep. 5 (2015) 8763, https://doi.org/10.1038/srep08763.

[41] L.C.H. Gerber, P.D. Frischmann, F.Y. Fan, S.E. Doris, X. Qu, A.M. Scheuermann, K. Persson, Y.M. Chiang, B.A. Helms, Three-dimensional growth of $Li_2S$ in lithium-sulfur batteries promoted by a redox mediator, Nano Lett. 16 (1) (2016) 549–554, https://doi.org/10.1021/acs.nanolett.5b04189.

[42] Y.X. Ren, T.S. Zhao, M. Liu, P. Tan, Y.K. Zeng, Modeling of lithium-sulfur batteries incorporating the effect of $Li_2S$ precipitation, J. Power Sources 336 (2016) 115–125, https://doi.org/10.1016/j.jpowsour.2016.10.063.

[43] P. Andrei, C. Shen, J.P. Zheng, Theoretical and experimental analysis of precipitation and solubility effects in lithium-sulfur batteries, Electrochim. Acta 284 (2018) 469–484, https://doi.org/10.1016/j.electacta.2018.07.045.

[44] C. Xiong, T.S. Zhao, Y.X. Ren, H.R. Jiang, X.L. Zhou, Mathematical modeling of the charging process of Li-S batteries by incorporating the size-dependent Li2S dissolution, Electrochim. Acta 296 (2019) 954–963, https://doi.org/10.1016/j.electacta.2018.11.159.

[45] L.D. Griffith, A.E.S. Sleightholme, J.F. Mansfield, D.J. Siegel, C.W. Monroe, Correlating $Li/O_2$ cell capacity and product morphology with discharge current, ACS Appl. Mater. Interfaces 7 (14) (2015) 7670–7678, https://doi.org/10.1021/acsami.5b00574.

[46] R.R. Mitchell, B.M. Gallant, Y. Shao-Horn, C.V. Thompson, Mechanisms of morphological evolution of $Li_2O_2$ particles during electrochemical growth, J. Phys. Chem. Lett. 4 (7) (2013) 1060–1064, https://doi.org/10.1021/jz4003586.

[47] B. Horstmann, T. Danner, W.G. Bessler, Precipitation in aqueous lithium-oxygen batteries: a model-based analysis, Energ. Environ. Sci. 6 (4) (2013) 1299–1314, https://doi.org/10.1039/c3ee24299d.

[48] L. Cheng, L.A. Curtiss, K.R. Zavadil, A.A. Gewirth, Y. Shao, K.G. Gallagher, Sparingly solvating electrolytes for high energy density lithium-sulfur batteries, ACS Energy Lett. (2016) 503–509, https://doi.org/10.1021/acsenergylett.6b00194.

[49] S.S. Zhang, Improved cyclability of liquid electrolyte lithium/sulfur batteries by optimizing electrolyte/sulfur ratio, Energies 5 (12) (2012) 5190–5197, https://doi.org/10.3390/en5125190.

[50] Aashutosh Mistry, et al., Quantifying Negative Effects of Carbon-Binder Networks from Electrochemical Performance of Porous Li-Ion Electrodes, Journal of the Electrochemical Society 168 (7) (2021) 070536, https://doi.org/10.1149/1945-7111/ac1033.

[51] R. Richter, J. Häcker, Z. Zhao-Karger, T. Danner, N. Wagner, M. Fichtner, K.A. Friedrich, A. Latz, Insights into self-discharge of lithium– and magnesium–sulfur batteries, ACS Appl. Energy Mater. 3 (9) (2020) 8457–8474, https://doi.org/10.1021/acsaem.0c01114.

[52] Y.V. Mikhaylik, J.R. Akridge, Polysulfide shuttle study in the Li/S battery system, J. Electrochem. Soc. 151 (11) (2004) A1969, https://doi.org/10.1149/1.1806394.

[53] M.R. Busche, P. Adelhelm, H. Sommer, H. Schneider, K. Leitner, J. Janek, Systematical electrochemical study on the parasitic shuttle-effect in lithium-sulfur-cells at different temperatures and different rates, J. Power Sources 259 (2014) 289–299, https://doi.org/10.1016/j.jpowsour.2014.02.075.

[54] S. Urbonaite, P. Novák, Importance of "unimportant" experimental parameters in Li-S battery development, J. Power Sources 249 (2014) 497–502, https://doi.org/10.1016/j.jpowsour.2013.10.095.

[55] Y. Ren, H. Shui, C. Peng, H. Liu, Y. Hu, Solubility of elemental sulfur in pure organic solvents and organic solvent-ionic liquid mixtures from 293.15 to 353.15K, Fluid Phase Equilib. 312 (1) (2011) 31–36, https://doi.org/10.1016/j.fluid.2011.09.012.

[56] M. Safari, C.Y. Kwok, L.F. Nazar, Transport properties of polysulfide species in lithium-sulfur battery electrolytes: coupling of experiment and theory, ACS Cent. Sci. 2 (8) (2016) 560–568, https://doi.org/10.1021/acscentsci.6b00169.

[57] V. Thangavel, A. Mastouri, C. Guéry, M. Morcrette, A.A. Franco, Understanding the reaction steps involving polysulfides in 1 M LiTFSI in TEGDME : DOL using cyclic voltammetry experiments and modelling, Batteries Supercaps (2020), https://doi.org/10.1002/batt.202000175.

[58] Y.C. Lu, Q. He, H.A. Gasteiger, Probing the lithium-sulfur redox reactions: a rotating-ring disk electrode study, J. Phys. Chem. C 118 (11) (2014) 5733–5741, https://doi.org/10.1021/jp500382s.

[59] C.W. Lee, Q. Pang, S. Ha, L. Cheng, S.D. Han, K.R. Zavadil, K.G. Gallagher, L.F. Nazar, M. Balasubramanian, Directing the lithium-sulfur reaction pathway via sparingly solvating electrolytes for high energy density batteries, ACS Cent. Sci. 3 (6) (2017) 605–613, https://doi.org/10.1021/acscentsci.7b00123.

[60] K.H. Wujcik, J. Velasco-Velez, C.H. Wu, T. Pascal, A.A. Teran, M.A. Marcus, J. Cabana, J. Guo, D. Prendergast, M. Salmeron, et al., Fingerprinting lithium-sulfur battery reaction products by X-ray absorption spectroscopy, J. Electrochem. Soc. 161 (6) (2014) A1100–A1106, https://doi.org/10.1149/2.078406jes.

[61] A. Mistry, V. Srinivasan, On our limited understanding of electrodeposition, MRS Adv. 4 (51–52) (2019) 2843–2861, https://doi.org/10.1557/adv.2019.443.

[62] A.N. Mistry, F.L.E. Usseglio-Viretta, A.M. Colclasure, K. Smith, P.P. Mukherjee, Fingerprinting redox heterogeneity in electrodes during extreme fast charging, J. Electrochem. Soc. 167 (9) (2020) 090542, https://doi.org/10.1149/1945-7111/ab8fd7.

[63] C.Y. Wang, V. Srinivasan, Computational battery dynamics (CBD)—electrochemical/thermal coupled modeling and multi-scale modeling, J. Power Sources 110 (2002) 364–376, https://doi.org/10.1016/S0378-7753(02)00199-4.

**PART III**

# Performance improvement

CHAPTER 7

# Recent progress in fundamental understanding of selenium-doped sulfur cathodes during charging and discharging with various electrolytes

**Chen Zhao[a,b], Gui-Liang Xu[a], Tianshou Zhao[b], and Khalil Amine[a,c]**
[a]Chemical Science and Engineering Division, Argonne National Laboratory, Lemont, IL, United States
[b]Department of Mechanical and Aerospace Engineering, The Hong Kong University of Science and Technology, Kowloon, Hong Kong, China
[c]Materials Science and Engineering, Stanford University, Stanford, CA, United States

## Contents

**7.1** Introduction 235
**7.2** Overview of $Se_xS_y$ cathode composition and electrochemistry 236
**7.3** Progress on Li-$Se_xS_y$ batteries with liquid electrolytes 238
    **7.3.1** Carbonate-based electrolytes 238
    **7.3.2** Ether-based electrolytes 242
    **7.3.3** Highly concentrated electrolytes 247
    **7.3.4** Fluorinated electrolytes 251
**7.4** All-solid-state Li-$Se_xS_y$ batteries 253
**7.5** Concluding remarks and future design strategies for $Se_xS_y$-based battery systems 256
Acknowledgments 257
References 258

## 7.1 Introduction

The successful commercialization of lithium–ion batteries (LIBs) in the 1990s by Akira Yoshino triggered numerous transformations in various industries, for which he was awarded the 2019 Nobel Prize in chemistry. Today, LIBs are widely used to power not only portable electronics but also electric vehicles. However, the ever-growing demand for high-energy and low-cost batteries has boosted the exploration of novel electrochemistry beyond conventional LIBs, whose energy density is quickly approaching

*Lithium-Sulfur Batteries*
https://doi.org/10.1016/B978-0-12-819676-2.00015-3

Copyright © 2022 Elsevier Inc.
All rights reserved.

their ceilings based on conventional ion intercalation chemistry [1]. Emerging technologies such as zinc-air batteries (ZABs) [2], lithium-air batteries (LABs) [3], and lithium-sulfur (Li-S) batteries [4] have been proposed to improve the energy density of the battery system.

The lithium (Li) metal-sulfur (S) battery system is considered to be one of the most promising technologies to reach the goal of 500 Wh kg$^{-1}$ due to the low cost and high specific capacity of sulfur (1675 mAh g$^{-1}$) [5], as well as the low potential ($-3.04$ V vs. standard hydrogen electrode) and high specific capacity (3860 mAh g$^{-1}$) of Li metal [6]. However, the shuttling of highly soluble lithium polysulfides (LiPs) between the cathode and the anode, referred to as "shuttle effect," leads to a rapid capacity decay, dramatically deteriorating the cycling stability of Li-S batteries [7]. Moreover, due to the electronic/ionic insulating nature of S/Li$_2$S, the electrochemical cathode redox kinetic is sluggish [8]. Meanwhile, the high reactivity of Li metal with most electrolytes (e.g., carbonate, ether, and dimethylsulfoxide) leads to inhomogeneous Li deposition and eventually to dendrite growth, resulting in low coulombic efficiency and raising serious safety concerns [9].

Recently, selenium-doped S (Se/S) cathodes have been proposed to replace sulfur as the cathode material due to the much higher electronic conductivity of Se (25 magnitude of order higher than S), which can enhance the cathode redox kinetics [10]. More importantly, the introduction of the Se—S bond can modify the electrochemical behavior of the cathode [11]. By carefully choosing the doping ratio and the electrolyte composition, one can modify the cathode chemistry to improve the cycling stability.

To date, the electrochemical behavior and capacity fading mechanism of the Se/S cathodes have been revealed for some common electrolyte systems with the assistance of advanced in situ experimental studies and theoretical simulations [12,13]. The electrochemical performance of Li-Se/S batteries has also been improved under the guidance of these fundamental studies. In this chapter, recent progress in understanding the Se$_x$S$_y$ electrochemical behavior and the corresponding cathode design strategies in different electrolytes will be discussed. Also presented will be concluding remarks and perspectives for future research development of Li-Se/S batteries.

## 7.2 Overview of Se$_x$S$_y$ cathode composition and electrochemistry

Since Se (atomic number 34, [Ar]4s$^2$3d$^{10}$4p$^4$) and S (atomic number 16, [Ne]3s$^2$3p$^4$) are members of the VIA group, they both have two electrons in the outer s orbital and four electrons in the outer p orbital. Thus, S and Se share very similar chemical and electrochemical properties [14]. However, due to the difference in atom size and the number of electrons per atom for these two elements, several properties are different and follow a trend based on the location of each element in the periodic table. The metallic property, for example, increases as the atomic number rises in the same group. As a result, although Se is still classified as a nonmetal, it has much higher electron conductivity than S, and it is often used as the semiconductor in the manufacture of rectifiers.

Because of the much higher electron conductivity of Se ($1 \times 10^{-3}\,Sm^{-1}$) compared with that of S ($5 \times 10^{-28}\,Sm^{-1}$), Li-Se batteries exhibit a much better rate and cycling performance than Li-S batteries. When coupling the Se cathode with a Li metal anode, similar to Li-S batteries, the theoretical energy density is determined by the overall cell reaction:

$$Se + 2\,Li \leftrightarrow Li_2Se$$

Based on the density of Se ($4.8\,g\,cm^{-3}$) and S ($2.07\,g\,cm^{-3}$), a comparable theoretical volumetric capacity of Se ($3268\,mAh\,cm^{-3}$) vs. S ($3467\,mAh\,cm^{-3}$) makes Se a promising electrode material for Li storage. However, the inferior gravimetric specific capacity of Se ($675\,mAh\,g^{-1}$) limits the energy density of the Li-Se battery [15]. To address this, Amine's group conducted a pioneering study to develop a Se-doped S ($Se_xS_y$) cathode, which has a higher specific capacity than a pure Se cathode and superior electrochemical redox kinetics than a pure S cathode [10]. Afterward, the concept of applying $Se_xS_y$ as the cathode material for Li storage has attracted global research interest in developing Li metal batteries (LMBs) with higher energy densities, greater rate capabilities, and better cycling stabilities [16].

The ratio of Se to S ($x/y$) in the $Se_xS_y$ cathode material plays a key role in determining the cathode electrochemical behavior as well as the energy densities of the Li-$Se_xS_y$ battery [15]. Typically, the preparation of $Se_xS_y$ was carried out by heating a homogeneous mixture of the calculated amounts of Se and S powder. As evidenced by high-energy X-ray diffraction (HEXRD), Xu et al. found that in most cases, the as-collected $Se_xS_y$ is crystalline. However, in the case of $Se_5S_3$, $Se_1S_1$, and $Se_4S_5$, the cathode materials are amorphous. Meanwhile, the pore structure of the host material also influences the state of the $Se_xS_y$ crystallization [15]. As a result, the Se doping ratio and the host material structure are important factors in optimizing the electrochemical behavior of the $Se_xS_y$ cathode and the energy densities of the as-assembled Li-$Se_xS_y$ batteries.

Apart from the electrode itself, since the electrochemical reactions take place at the interphase between the electrode and the electrolyte, the choice of electrolyte is also crucial in optimizing the chemistries of both cathode and anode. Currently, most electrolytes in Li-Se$_x$S$_y$ cells are carbonate- and ether-based, fluorinated, high-concentration, and solid-state electrolytes. The following sections discuss the recent progress in understanding the electrochemical behavior of the Se$_x$S$_y$ cathode in these electrolytes.

## 7.3 Progress on Li-Se$_x$S$_y$ batteries with liquid electrolytes

### 7.3.1 Carbonate-based electrolytes

Carbonate-based electrolytes have been widely used in commercial LIBs due to their high stabilities and safety, as well as wide operating temperature window [17]. After around 30 years of development, many different carbonate-based electrolytes with various solvents, salts, and additives have been successfully developed and applied in commercial LIBs [18]. Therefore, it is natural to directly introduce carbonate-based electrolytes into the Li-Se$_x$S$_y$ system. The pioneering studies of the electrochemical behavior of the Se$_x$S$_y$ cathode in carbonate-based electrolytes were conducted by Amine's group in 2012 [10]. They used SeS$_2$/multiwalled carbon nanotubes (MWCNTs) as the cathode composite in a carbonate-based electrolyte composed of 1.2 M lithium hexafluorophosphate (LiPF$_6$) in 3:7 (v/v ratio) of ethylene carbonate (EC)/ethyl methyl carbonate (EMC) solvents to demonstrate that the Se$_x$S$_y$ cathode can offer higher theoretical capacities than Se alone. A single-phase transition in the Se$_x$S$_y$ cathode during (de)lithiation was also found in the carbonate-based electrolyte.

Fig. 7.1A shows the pair distribution function (PDF) plots at different charge/discharge stages of the Li-Se cell in the carbonate-based electrolyte. Both Li$_2$Se and Se can be identified at the long discharge plateau (2.0 V), in accordance with a single-phase transition. To further confirm this first-order phase transition, in situ investigations with high-energy X-ray diffraction (HEXRD) and X-ray absorption near-edge structure (XANES) were conducted with a Li-Se cell in the carbonate-based electrolyte [19]. As shown in Fig. 7.1B, during cell discharging, the intensity of the Se peak keeps decreasing, and the Li$_2$Se peak starts to appear at 1.86 V and keeps increasing to the end of discharge. The XRD patterns proved that there are no intermediate phases during the reduction of Se to Li$_2$Se and the oxidation of Li$_2$Se to Se. The 2D contour plot of the in situ Se XANES spectra (Fig. 7.1C and D) does not exhibit edge position shifts during cell cycling, indicating the absence of

Selenium-doped sulfur cathodes 239

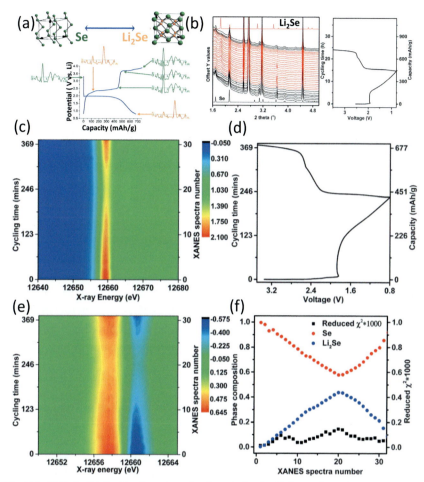

Fig. 7.1 (A) Pair distribution functions for the pristine Se-C electrode and upon recovery from various states of discharge/charge. Structural representations of the Se and Li$_2$Se (antifluorite) phases are shown. (B) In situ HEXRD characterization of the Li-Se cell in GenII electrolyte cycled between 0.8 and 3.5 V with a charging rate of 30 mA g$^{-1}$ and the voltage profile. In situ XANES measurement for Li-Se pouch cell in GenII electrolyte: (C) normalized XANES spectra of the cycling cell, (D) voltage profile, (E) derivative of normalized XANES spectra, and (F) linear combination fitting of residue values and corresponding phase compositions in different states of charge-discharge. *(Reprinted with permission from: (A) A. Abouimrane, D. Dambournet, K.W. Chapman, P.J. Chupas, W. Weng, K. Amine, A new class of lithium and sodium rechargeable batteries based on selenium and selenium-sulfur as a positive electrode, J. Am. Chem. Soc. 134 (2012) 4505–4508, Copyright (2012) American Chemical Society and (B–F) Y. Cui, A. Abouimrane, C. Sun, Y. Ren, K. Amine, Li–Se battery: Absence of lithium polyselenides in carbonate based electrolyte, Chem. Commun. 50 (2014) 5576–5579, Copyright (2014) Royal Society of Chemistry.)*

lithium polyselenides (LiPSes) intermediates in the carbonate-based electrolyte. The fitting results (Fig. 7.1E and F) further prove that no observable intermediate phases formed in the carbonate-based electrolytes. These results support the conclusion that Se is directly transformed to Li$_2$Se, corresponding well with the in situ XRD results.

The pore structure of the cathode host material also plays an important role in determining the electrochemical behavior of the Se$_x$S$_y$ cathode in the carbonate-based electrolyte. Microporous materials, because of the large specific surface area, have been widely applied as the host materials of Se$_x$S$_y$ to enhance the electrochemical reaction kinetics and improve the electronic conductivity. Goodenough's group revealed that Se is confined as single-chain molecules, rather than 8-atom ring molecules, in carbon host material with microporous structure [20]. They also performed theoretical calculations to study the electrochemistry of Se$_x$ chain molecules confined in microporous carbon materials. As shown in Fig. 7.2A, the energy released by the formation of two Li—Se bonds (5.909 eV) is much higher than that absorbed for breaking a Se—Se bond (3.447 eV). In addition, as shown in Fig. 7.2B, free space for Li$^+$ transportation still exists in the micropores after Se insertion. Therefore, Se atoms in the Se$_x$ chain could simultaneously react with Li$^+$, corresponding well with the first-order phase transition.

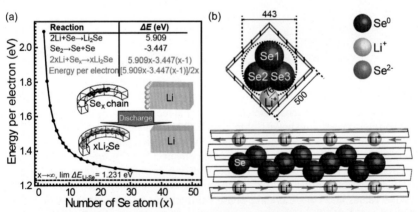

Fig. 7.2 Electrochemistry of a Li-Se$_x$ battery: (A) energy (per electron) released in one-step lithiation of Se$_x$ and (B) sectional and front views showing Li$^+$ migration in a Se$_x$ preoccupied slit pore. (Reproduced with permission from S. Xin, L. Yu, Y. You, H. Cong, Y. Yin, X. Du, Y. Guo, S. Yu, Y. Cui, J.B. Goodenough, The electrochemistry with lithium versus sodium of selenium confined to slit micropores in carbon, Nano Lett. 16 (2016) 4560–4568, Copyright (2016) American Chemical Society.)

All these features seem to make the Li-Se$_x$S$_y$ an ideal high-energy-density system for energy storage with the carbonate-based electrolyte. However, the nucleophilic attack of soluble LiPs and Li$_2$S toward carbonate solvents such as EC and EMC raises a concern on directly applying Se$_x$S$_y$ as the cathode composite in the carbonate-based electrolyte [13,21]. Although Guo et al. found that applying microporous materials with the extremely small pore size (0.5 nm) could relieve this nucleophilic attack, the low active material content ($\sim$40%) will greatly affect the energy density of the cell [22]. Another serious issue is the relatively sluggish electrochemical reaction kinetics due to the solid-solid (de)lithiation process. Although the highly soluble LiPs intermediates that form during charge/discharge will cause a severe shuttle effect, the dissolved LiPs will enhance the subsequent Li$_2$S/S precipitation process along the triple-phase boundaries to boost the electrochemical reaction kinetics. The absence of highly soluble intermediates will result in a sluggish electrochemical reaction between the solid-solid interphase.

As a result, selenium doping combined with rational cathode structure design to boost the electrochemical redox kinetics and stabilize the Se$_x$S$_y$ in the carbonate-based electrolyte can enhance the electrochemical performance. Qian et al. encapsulated S$_{0.94}$Se$_{0.06}$ in the porous carbon host material to get the amorphous S-rich Se$_x$S$_y$ composite. Based on the electrochemical performance characterizations, they found that S cathode without selenium doping can only deliver a reversible specific capacity of $\sim$500 mAh g$^{-1}$. After Se doping, the S$_{0.94}$Se$_{0.06}$ cathode composite delivered a higher reversible specific capacity of $\sim$1100 mAh g$^{-1}$. Moreover, due to the protection of porous carbon host, the cycling stability of the Li-Se$_x$S$_y$ cell was also greatly improved, corresponding to a 98.6% capacity retention rate for over 200 cycles [23]. Other host materials, such as porous carbon nanofibers [24], 3D network SeS$_x$/NCPAN [25], and SeS/PAN [26], have been developed to enhance the redox kinetics and Se$_x$S$_y$ cathode stability in the carbonate-based electrolyte. However, current success mainly relies on the use of low loading cathode, thick Li metal anode, and flooded electrolytes in a coin cell. More importantly, the continuously parasitic reaction between carbonate solvent molecules and Li metal will lead to the formation of inhomogeneous solid-electrolyte interphase (SEI) film, resulting in a low coulombic efficiency and severe dendrite growth [27]. Thus, to realize the practical application of Li-Se$_x$S$_y$ batteries in carbonate-based electrolytes, rational cathode structure design combined with electrolyte engineering and Li metal protection should be taken into consideration

## 7.3.2 Ether-based electrolytes

to achieve the high electrochemical performance under high cathode loading, thin Li metal anode, and lean electrolyte.

## 7.3.2 Ether-based electrolytes

Because of the poor cycling stability and efficiency associated with conventional carbonate-based electrolytes, in the 1980s, researchers turned to ether-based electrolytes as the alternative for LMBs because of their low viscosity and resultant high ionic conductivity [17]. More importantly, the high reductive stability of ethers with Li metal leads to much higher coulombic efficiency and less dendritic Li than the conventional carbonate-based electrolytes. At present, the ether-based electrolyte is the most frequently used electrolyte for Li-S batteries [28], which is typically composed of 1.0 M lithium bis-trifluoromethanesulfonimide (LiTFSI) salt in 1,2-dimethoxyethane (DME)/1,3-dioxolane (DOL) solvents. Although DME is reactive with Li metal, the higher LiPs solubility in DME can enhance the reaction kinetics of the S cathode. Furthermore, DOL has relatively lower LiPs solubility and can assist the formation of a more stable SEI layer on the Li metal anode. The synergistic effect of DME and DOL solvents helps to achieve higher energy densities and cycling stabilities in Li-S batteries [28]. Due to the similar properties of Se and S, ether-based electrolytes are also frequently applied in Li-Se$_x$S$_y$ cells.

To reveal the electrochemical process of the Se$_x$S$_y$ cathode in DME-based electrolyte, pioneering work was conducted by Amine's group in 2013 [29]. In situ HEXRD and XANES characterizations were conducted during charge/discharge to reveal the electrochemical process of the Se$_x$S$_y$ cathode in DME-based electrolyte. Fig. 7.3 shows the in situ HEXRD patterns of Se and SeS$_2$ cathodes in lithium coin cells with DME-based electrolyte during charge/discharge. As shown in Fig. 7.3A and B, the peak intensity of Se decreases, while the Li$_2$Se peaks grow during discharge, and the process is reversed during charge. From the enlarged portion of HEXRD patterns (Fig. 7.3C), the intensities of the Se peaks (1.64 degrees and 2.06 degrees) decrease slowly with the lithiation process, while no new phase appears until the cell is discharged to $\sim$1.69 V. Meanwhile, the Li$_2$Se phase starts to form when the Se peaks have completely vanished. During the charging process, the Li$_2$Se peak remains until the cell potential reaches $\sim$2.22 V. When further charging to 2.24 V, the Se peak steadily grows, while Li$_2$Se disappears. More importantly, during discharge, the Li$_2$Se peak does not form until a capacity of $\sim$500 mAh g$^{-1}$ is reached

Selenium-doped sulfur cathodes 243

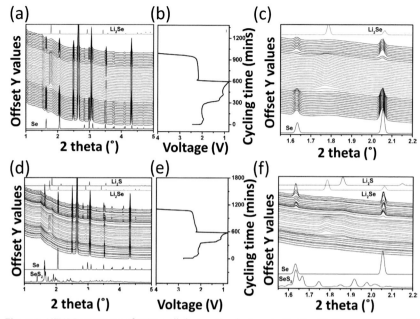

**Fig. 7.3** HEXRD patterns of Li/Se cell during the first cycle (A, C) with the voltage profile (B). HEXRD patterns of Li/SeS$_2$ cell in the first cycle (D, F) with the voltage profile (E). (Reproduced with permission from Y. Cui, A. Abouimrane, J. Lu, T. Bolin, Y. Ren, W. Weng, C. Sun, V.A. Maroni, S.M. Heald, K. Amine, (De)lithiation mechanism of Li/SeS$_x$ (x = 0-7) batteries determined by in situ synchrotron X-ray diffraction and X-ray absorption spectroscopy, J. Am. Chem. Soc. 135 (2013) 8047–8056, Copyright (2013) American Chemical Society.)

without any other phase detected. In the case of the SeS$_2$ cathode (Fig. 7.3D–F), no crystalline phase is present at the beginning of the discharge. When the cell potential reaches 2.08 V, the Li$_2$S phase is formed and followed by the Li$_2$Se crystallization at ∼2.04 V. During the charging process, Li$_2$Se and Li$_2$S remain until the cell is charged to 2.19 and 2.30 V, respectively. After the cell potential reaches 2.25 V, the Se phase appears and keeps increasing to the end of charge. Since the HEXRD characterization method can only detect crystalline phases, to reveal where the Li$^+$ goes and the exact electrochemical process during cycling, in situ XANES characterizations were conducted.

The 2D contour plots of Se XANES spectra during cycling are shown in Fig. 7.4A and B. The shift of the XANES spectrum to lower energy occurs at 2.04 V during discharge, corresponding well with previous in situ HEXRD characterizations. The first derivatives of the XANES spectra are shown in Fig. 7.4C, indicating that the Se edges shift significantly with (de)lithiation.

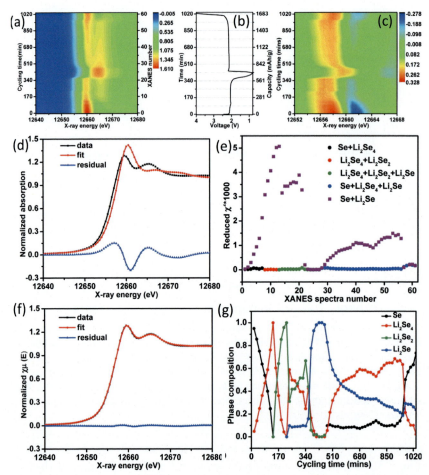

**Fig. 7.4** (A) Normalized XANES spectra of Li/Se cell during cycling, (B) battery voltage profile, and (C) derivative of normalized XANES spectra of Li/Se cell during cycling. Selenium XANES spectra linear combination fitting: (D) 19th spectrum fitted with two components (Se and Li$_2$Se) showing huge residue, (E) reduced $\chi^2 \times 1000$ of the linear combination fitting of spectra using different component combinations, and (F) 19th spectrum fitted with three components (Li$_2$Se$_4$, Li$_2$Se$_2$, and Li$_2$Se). (G) Relative composition evolution of possible phases during cycling. *(Reproduced with permission from Y. Cui, A. Abouimrane, J. Lu, T. Bolin, Y. Ren, W. Weng, C. Sun, V.A. Maroni, S.M. Heald, K. Amine, (De)lithiation mechanism of Li/SeS$_x$ (x = 0-7) batteries determined by in situ synchrotron X-ray diffraction and X-ray absorption spectroscopy, J. Am. Chem. Soc. 135 (2013) 8047–8056, Copyright (2013) American Chemical Society.)*

When the cell potential reaches 2.04 V, the edge shifts from 12,657.9 to 12,656.7 eV, and no change can be identified in the following 3 h of the electrochemical process (more than $300 \, \mathrm{mAh \, g^{-1}}$ capacity). When the cell is further discharged to 1.67 V, the Se K-edge jumps to 12,660 eV, corresponding to the formation of $Li_2Se$. During charge, the Se edges slowly shift back to lower energy. The Se K-edge shift during charge/discharge indicates the formation of LiPSes during cycling but in a noncrystalline form. At the very beginning of the discharge, Se reacts with $Li^+$ and yields the LiPSes products. The consumption of Se leads to the decrease of peak intensity as observed in the in situ HEXRD results. Because of the reduction of Se to LiPSes, the Se K-edge shifts to lower energy. As lithiation increases, the elemental Se is depleted, and LiPSes are reduced to crystallized $Li_2Se$, where the $Se^{2-}$ is surrounded by 8 $Li^+$ ions. The strong coulombic interaction reduces the screening effect and shifts the Se edge to higher energy.

To determine the components in the mixture of the cell from the in situ XANES results, Cui et al. [29] conducted a linear combination analysis based on the hypothesis that a two-phase one-step transition occurs from Se directly to $Li_2Se$. However, as shown in Fig. 7.4D and E, most of the spectrum is poorly fitted. When a third component, $Li_2Se_n$, was introduced to the electrochemical process for linear combination, however, a good fit was obtained (Fig. 7.4F).

Fig. 7.4G shows the composition of the three phases vs. cycling time. Based on the in situ characterizations and the fitting results, an electrochemical process for the Se cathode in DME-based electrolyte was proposed and confirmed (Fig. 7.5). During discharge, the Se is first reduced to LiPSes ($Li_2Se_n$, $n \geq 4$), then $Li_2Se_2$, and finally $Li_2Se$. During charge, $Li_2Se$ is directly oxidized to LiPSes ($Li_2Se_n$, $n \geq 4$) and Se without the formation of $Li_2Se_2$.

Since highly soluble LiPSes will be formed during cycling in the DME-based electrolyte, the $Se_xS_y$ cathode delivers higher specific capacity and exhibits better electrochemical kinetics in this electrolyte as opposed to the carbonate-based electrolyte. However, since the $Se_xS_y$ cathode involves the highly soluble intermediates $Li_2Se_n/Li_2S_n$, the dissolution of these intermediates and their subsequent reaction with Li metal will lead to a severe shuttle effect, resulting in a sharp capacity decay. Many groups have applied advanced nanotechnology to construct the host materials to relieve the shuttle effect of $Li-Se_xS_y$ batteries in the ether-based electrolyte. Lou et al. synthesized a mesoporous carbon@TiN hollow sphere as the host materials to immobilize the dissolved the LiPs/LiPSes inside the cathode material and thus enhance the cycling stability of $Li-Se_xS_y$ cell in the ether-based electrolyte.

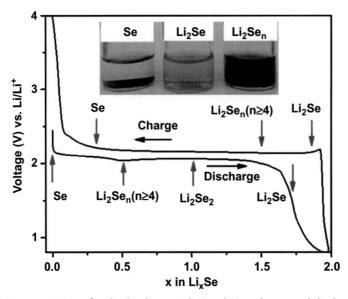

**Fig. 7.5** Representation of cathode-phase evolution during charge and discharge of a Li/Se cell in an ether-based electrolyte. *(Reproduced with permission from Y. Cui, A. Abouimrane, J. Lu, T. Bolin, Y. Ren, W. Weng, C. Sun, V.A. Maroni, S.M. Heald, K. Amine, (De)lithiation mechanism of Li/SeS$_x$ (x = 0-7) batteries determined by in situ synchrotron X-ray diffraction and X-ray absorption spectroscopy, J. Am. Chem. Soc. 135 (2013) 8047–8056, Copyright (2013) American Chemical Society.)*

The SeS$_2$ cathode delivered a reversible specific capacity of 987 mAh g$^{-1}$ at 0.2 C due to the selenium doping and rational cathode design. Moreover, the cycling stability of the cell was also greatly improved because of the introduction of binding sites, enabling a stable cycle life of 200 cycles with a ~70% capacity retention rate [30]. Other host materials, such as CoS$_2$-decorated multichannel carbon fiber [31], CMK-3 [32], and N-doped mesoporous carbon [33], have also been proven to be effective in relieving the shuttle effect of LiPs/LiPSes in ether-based electrolytes. However, flooded electrolytes, thin cathode, and thick lithium metal foil were still used, which will significantly affect the realistic energy density of the cell. Future development of Li-Se$_x$S$_y$ batteries in ether-based electrolytes should focus on lowering the electrolyte-to-sulfur (E/S) ratio and further improving the cathode loading to improve the realistic energy density of the cell. Another issue is the high reactivity of DME with Li metal, which will not only consume a large amount of solvent but also lead to an inhomogeneous Li stripping/platting process during cycling, lowering the electrochemical performance and resulting in harmful

Li dendrite growth [27]. As a result, Li metal protection should also be taken into consideration for future practical applications of Li-Se$_x$S$_y$ batteries in ether-based electrolytes. Since the electrolyte plays a key role in determining the chemistries of the cathode and anode, developing an ideal electrolyte system that can simultaneously enable robust cathode and anode chemistries is a promising strategy to truly solve the above-mentioned issues.

## 7.3.3 Highly concentrated electrolytes

Because of the troublesome cathode and anode chemistries of Li-Se$_x$S$_y$ batteries in conventional carbonate- and ether-based electrolytes, researchers worldwide are searching for new electrolyte systems that will enable robust Se$_x$S$_y$ cathode and Li metal anode chemistries and thereby enhance battery cycling stabilities, electrochemical reaction kinetics, and efficiencies required for practical use. Among the newly proposed electrolyte systems, highly concentrated electrolytes (HCEs) have received increasing attention recently [34].

Different from conventional electrolytes with normal salt concentration ($\sim$1.0 M), the HCE contains much higher salt concentration ($>$3.0 M). This higher concentration enhances the interactions between cations and anions/solvents and decreases the content of free-state solvent molecules. When the concentration is above a threshold ($>$3–5 M, depending on the salt-solvent combinations), free solvent molecules disappear, and a new class of electrolytes with a peculiar three-dimensional (3D) solution structure forms [34]. In 2014, Yamada et al. applied a quantum mechanical density functional theory based on molecular dynamics (DFT-MD) to simulate the location of the lowest unoccupied molecular orbital (LUMO) shift from the solvent to the salt, resulting in the reductive decomposition of the salt before the solvent at low potential [35]. Moreover, due to the disappearance of free solvent, their simulation showed that the HCE exhibits less solubilization (corrosion) of the substance inside the cell, such as Se$_x$S$_y$, LiPSes/LiPs, and aluminum current collector [34]. The change of solution structure will not only influence the physical properties of the electrolyte but also modify the electrochemical behavior of the Se$_x$S$_y$ cathode and Li metal anode in the HCE. These effects provide an opportunity to simultaneously achieve the robust cathode and anode chemistries in Li-Se$_x$S$_y$ batteries necessary to realize the practical application.

For the cathode, tuning the electrolyte structure by increasing the salt concentration can relieve the capacity decay caused by the shuttle effect.

In 2018, Nazar et al. used a diglyme (G2) electrolyte with saturated lithium bis(trifluoromethanesulfonyl)imide (LiTFSI) salt (G2:LiTFSI = 0.8:1) to demonstrate the transition of the S electrochemical pathway from the conventional solid-liquid-solid process to a quasi-solid-state reaction [36]. Ab initio molecular dynamics (AIMD) simulations were conducted to understand the solvation structure around $Li^+$ in the electrolyte. Fig. 7.6A shows the atomic coordination number around $Li^+$ as a function of the separation distance, and the curves show that $Li^+$ prefers tetrahedral coordination with the O atoms in $TFSI^-$ or G2. As shown in Fig. 7.6B, the salt concentration in the electrolyte will influence the relative contribution of the O atoms that belong to $TFSI^-$ or G2 in the first solvation shell. As shown in Fig. 7.6C, the increased concentration of lithium salt in the electrolyte will make the $TFSI^-$ bind with multiple $Li^+$ ions to form a reinforced 3D network. The change of solvation structure will thus lead to the transformation of cathode redox chemistry. As shown in Fig. 7.6D, the discharge curves of S cathodes exhibit a transition from the two- to single-plateau profile as the concentration of lithium salt increases. Fig. 7.6E exhibits the galvanostatic intermittent titration (GITT) results for the S cathode in electrolytes with different salt concentrations to study thermodynamics. As shown in Fig. 7.6F, before the knee point ($\sim 400\,mAh\,g^{-1}$), the average voltage decreases as the salt concentration increases. However, after the knee point, the average voltage shows a reverse trend. A higher average voltage corresponds to higher-order LiPs, and the decreasing of the first-stage voltage indicates an earlier transformation of higher- to lower-order LiPs. The long-term cycling performance of the S cathode was evaluated in various electrolytes with different salt concentrations (Fig. 7.6G). The cell with the higher salt concentration shows better stability and higher coulombic efficiency, proving that tuning the solvation structure of the electrolyte can modify the S cathode chemistry to a quasi-solid-solid pathway that eliminates the capacity decay caused by the LiPs shuttle effect.

The Li metal anode will also benefit from the high salt concentration. Because of the high concentration of $Li^+$ in the electrolyte, the Li plating/stripping process will take place in a more homogeneous way to avoid the formation of Li dendrite and dead Li. In addition, the shift of the LUMO of the electrolyte from solvent to salts due to the increase of salt concentration leads to a robust anion-derived SEI layer on the surface of the Li metal, which will relieve any parasitic reactions as well as dendrite growth [34]. As a result, the cycling stability and coulombic efficiency of the Li metal anode are greatly enhanced in the HCE electrolyte system. Until now, many

Selenium-doped sulfur cathodes 249

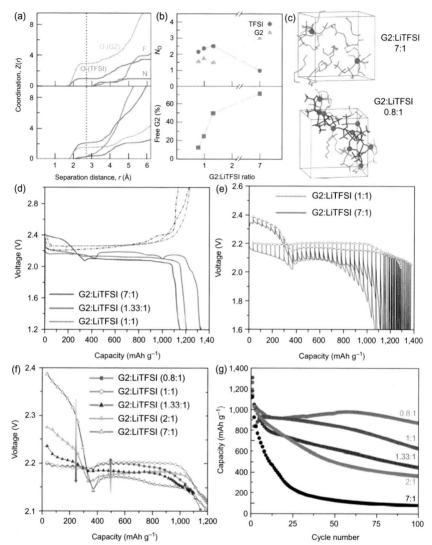

**Fig. 7.6** (A) AIMD simulations showing Z(r) for the two end-members, G2:LiTFSI (7:1, top) and G2:LiTFSI (0.8:1, bottom). The *dotted vertical line* represents the cutoff distance (2.7 Å) for the first solvation shell. (B) AIMD simulations showing the number of oxygen atoms (NO) that belong to TFSI$^-$ or G2 within the first solvation shell (top), and the fraction of free G2 molecules (bottom) as a function of the G2:LiTFSI ratio. (C) Typical snapshots from the AIMD trajectories for the two end-member electrolytes. Li$^+$ cations, *purple* (*dark gray* in print version); free and coordinated G2 molecules, *cyan* (*light gray* in print version) and *gray*, respectively; (contact-ion-pairs) CIPs and aggregates of TFSI$^-$ anions, *gold* (*dark gray* in print version) and *blue* (*light gray* in print version), respectively. Hydrogen atoms are not displayed for clarity. The *cubic box* represents the supercell for the calculated system. *a.u.*, arbitrary units.

*(Continued)*

**Fig. 7.6, Cont'd** (D) Voltage profiles of bulk sulfur electrodes in different G2:LiTFSI electrolytes at a rate of C/30 and at 55°C. (E) GITT voltage profiles in the two representative G2:LiTFSI electrolytes, with 20-min discharge at C/15 followed by 2h at rest; the equilibrium voltage is shown by *circles*. (F) Equilibrium voltages as a function of the state of discharge for different G2:LiTFSI electrolytes; the *arrows* indicate the descending G2:LiTFSI. (G) Capacity retention of the sulfur cells using different G2:LiTFSI electrolytes over 100 cycles at C/5; the first cycle is at C/15. The G2:LiTFSI (1:1) electrolyte shows an inferior capacity retention, which indicates that polysulfide shuttling occurs even at such low polysulfide concentrations. Note that some wetting and activation cycles are necessary for the low G2:LiTFSI ratio electrolytes. *(Reproduced with permission from Q. Pang, A. Shyamsunder, B. Narayanan, C.Y. Kwok, L.A. Curtiss, L.F. Nazar, Tuning the electrolyte network structure to invoke quasi-solid state sulfur conversion and suppress lithium dendrite formation in Li-S batteries, Nat. Energy 3 (2018) 783–791, Copyright (2018) Nature Springer.)*

groups around the world have reported the improved cycling life and coulombic efficiency in Li|Cu and Li|Li cells by utilizing different kinds of HCE electrolyte, demonstrating the effectiveness of HCE in improving the stability of the Li metal anode [34,37,38].

The robust chemistries of the S cathode and Li metal anode enabled by HCE greatly enhance the cycling stability and efficiency of the Li-S battery. However, the poor wettability, high viscosity, and low ionic conductivity of the HCE still limit the further practical applications of Li-S batteries with this electrolyte. Thus, Wang's group proposed a new class of localized high–concentration electrolyte (LHCE) enabled by adding 1,1,2,2-tetrafluoroethyl ether into HCE as the dilution to improve the wettability, lower the viscosity, and enhance the ionic conductivity [39]. Although stable cycling performance and high specific capacity can be achieved, current success mainly relies on the use of a thin cathode ($1-2\,mg\,cm^{-2}$) and flooded electrolyte (electrolyte-to-sulfur ratio $>30\,\mu L\,mg^{-1}$), which greatly affect the real energy densities of the as-assembled cells. The thin cathode and flooded electrolytes are often used in the HCE system because of the sluggish electrochemical reaction kinetics of the S cathode caused by the insulating nature of elemental sulfur. As a result, Se doping and cathode structure design could be solutions to enhance the reaction kinetics. Yushin et al. have successfully applied Se cathode in HCE, and because of the higher electronic conductivity of Se than that of S, the Se cathode delivered a reversible specific capacity of $\sim560\,mAh\,g^{-1}$ at 45°C, which is very close the theoretical specific capacity of Se cathode [40]. However, due to the limited specific capacity of Se, the energy density of the Li-Se cell in HCE is still low. Until now, research based on the $Se_xS_y$ cathode in HCE/LHCE is still lacking. An in-depth understanding of the electrochemical behavior of $Se_xS_y$ in HCE/LHCE as well as the cathode structure design is needed to realize practical Li-$Se_xS_y$ batteries with high-energy densities.

### 7.3.4 Fluorinated electrolytes

As mentioned above, the robust anode and cathode chemistries enabled by HCE can greatly enhance the electrochemical performance of Li-$Se_xS_y$ batteries. However, the high cost, extreme viscosity, and sluggish cathode kinetics in HCE still hinder its adoption. As a result, a new combination of solvent and salt with normal concentration is another pathway to consider in order to simultaneously achieve robust anode and cathode chemistries.

Recently, a fluorinated electrolyte has received increasing attention for LMBs because it induces a stable Li plating/stripping process, and it also enables the formation of a strong and uniform SEI layer on the surface of the Li metal anode. For example, Wang and Amine's group proposed a highly fluorinated electrolyte composed of 1.0 M $LiPF_6$ in a mixture of fluoroethylene carbonate/3,3,3-fluoroethylmethyl carbonate/1,1,2,2,-tetrafluoroethyl-2′,2′,2′-trifluoroethyl ether (2:6:2 by weight) for LMBs [27]. In a Li|Cu cell test, a cell with this fluorinated electrolyte exhibited a low overpotential of 130 mV at a current density of $5\,mA\,cm^{-2}$, which is comparable to the case with HCE. This group also successfully set up a full cell test coupling the fluorinated electrolyte with a NCM811 cathode. The cell maintained a high coulombic efficiency of 99.81% over 1000 cycles and formed a robust cathode-electrolyte interphase (CEI) on the surface of the NCM811 cathode.

Later in 2019, Xu et al. extended the research on the fluorinated electrolyte to Li-$Se_xS_y$ batteries to gain a fundamental understanding of the electrochemical behavior of $Se_xS_y$ in the fluorinated electrolyte [15]. As shown in Fig. 7.7A and C, the $Se_xS_y$ cathode exhibited a similar single discharge plateau in both carbonate- and 1,1,2,2-tetrafluoroethyl-2,2,3,3-tetrafluoropropyl ether (HFE)-based electrolytes, but the multiple discharge plateaus in the DME-based electrolyte (Fig. 7.7B) suggest that the $Se_xS_y$ cathode undergoes a solid-solid lithiation process in the HFE-based electrolyte. To unravel the reaction mechanism of the $Se_xS_y$ cathode in HFE-based electrolyte, in operando synchrotron HEXRD and in situ $^7Li$ NMR studies were carried out. As shown in Fig. 7.7D, $Li_2S$ and $Li_2Se$ can still be identified at the end of the charging process, indicating a chemical reaction between the LiPs/LiPSes

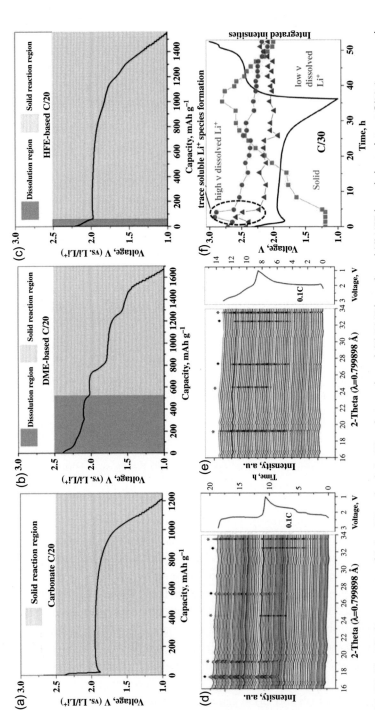

**Fig. 7.7** First discharge curve of S$_5$Se$_2$/KB cathode in (A) carbonate-based, (B) DME-based, and (C) HFE-based electrolyte at C/20. In operando HEXRD patterns collected during the first cycle of space-confined S$_5$Se$_2$/KB cathode at C/10 in (D) DME-based and (E) HFE-based electrolyte. The *diamond* represents Li$_2$S, and the *star* represents Li$_2$Se in panels (D) and (E). (F) Integrated areas of in operando $^7$Li NMR of the S$_5$Se$_2$/KB cathode in the HFE-based electrolyte as a function of charge/discharge process. *(Reproduced with permission from G. Xu, H. Sun, C. Luo, L. Estevez, M. Zhuang, H. Gao, R. Amine, H. Wang, X. Zhang, C. Sun, Solid-state lithium/selenium-sulfur chemistry enabled via a robust solid-electrolyte interphase, Adv. Energy Mater. 9 (2019) 1802235, Copyright (2019) John Wiley & Sons.)*

and Li metal to form the deposited $Li_2S/Li_2Se$. However, in HFE-based electrolyte, only a small amount of crystalline $Li_2Se$ is evident at the end of the charge (Fig. 7.7E). This amount of $Li_2Se$ is mainly attributed to its sluggish electrochemical oxidation kinetics in the HFE-based electrolyte due to the solid–solid (de)lithiation process. To confirm this hypothesis, in situ $^7Li$ NMR spectroscopy, which is sensitive to the solid and liquid $Li^+$ species inside the cells, was conducted. Fig. 7.7F shows the integrated areas of the high frequency, low frequency, and broad resonance along with the charge/discharge process. It reveals very little soluble polysulfide/polyselenide formation in the HFE-based electrolyte during operation. Most of the $Se_xS_y$ was converted directly to $Li_2S$ and $Li_2Se$, confirming a solid–solid (de)lithiation process.

Since the HFE-based electrolyte can simultaneously enable robust cathode and anode chemistries for $Li-Se_xS_y$ batteries with normal salt concentration, the application of HFE-based electrolyte can greatly enhance the stabilities of $Li-Se_xS_y$ batteries. However, due to the absence of LiPs/LiPSes, the solid–solid (de)lithiation process will result in electrochemical redox kinetics that is relatively sluggish during operation, affecting the performance of the cell. As a result, rational cathode structure design should be combined with the HFE-based electrolyte to achieve a stable high-energy-density $Li-Se_xS_y$ battery.

## 7.4 All-solid-state $Li-Se_xS_y$ batteries

All-solid-state lithium metal batteries have attracted increasing attention because of their high energy density and safety. To couple with Li metal anode in these batteries, S is one of the most promising cathode candidates because of the low cost, low toxicity, and high theoretical capacity. Meanwhile, the absence of highly soluble LiPs intermediates in the solid-state electrolyte will fundamentally eliminate the capacity decay caused by the shuttle effect in a liquid electrolyte. Several solid-state electrolytes (SSEs) with high ionic conductivities ($\sim 10^{-2}\,S\,cm^{-1}$), such as $Li_{10}GeP_2S_{12}$ [41], $Li_7P_3S_{11}$ [42], $Li_{9.54}Si_{1.74}P_{1.44}S_{11.7}Cl_{0.3}$ [43], and poly(ethylene oxide)-based solid polymer [44], have been developed for all-solid-state Li-S batteries (ASSLSBs). However, because of the solid–solid (de)lithiation process in these batteries, the electrochemical reaction kinetics of the S cathode is more sluggish than that of liquid-based electrolytes. Thus, the ionic and electronic conductivities of the S cathode need to be improved for ASSLSBs. Selenium doping has been widely proven to be an effective strategy to enhance the electronic conductivity of the S cathode. Moreover, the

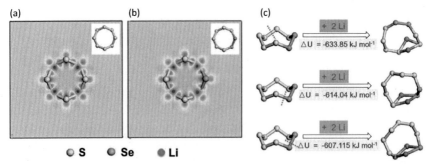

**Fig. 7.8** Electron densities of (A) $S_8$ and (B) $Se_2S_6$. (C) DFT calculations for the bond breaking and lithiation reaction of the $Se_2S_6$ ring and proposed intermediates. *Orange (gray in print version), yellow (light gray in print version), and purple (dark gray in print version) stand for Se, S, and Li atoms, respectively. (Reproduced by permisison from X. Li, J. Liang, J. Luo, C. Wang, X. Li, Q. Sun, R. Li, L. Zhang, R. Yang, S. Lu, H. Huang, X. Sun, High-performance Li–SeSx all-solid-state lithium batteries, Adv. Mater. 31 (2019) 1808100, Copyright (2019) John Wiley & Sons.)*

interfacial ionic conductivity between the electrolyte and active materials can also be improved. Thus, the all-solid-state Li-$Se_xS_y$ battery is a promising candidate for a high-energy-density battery system that is both stable and safe.

Sun's group first achieved high utilization of the S cathode in ASSLBs by Se doping. They used $Se_xS_y$ and commercial $Li_3PS_4$ electrolyte as the composite, and a bilayer SSE with $Li_3PS_4$ (on the anode side) and $Li_{10}GeP_2S_{12}$ (on the cathode side) was used as the separator to achieve a $SeS_2$/$Li_{10}GeP_2S_{12}$-$Li_3PS_4$/Li solid-state cell [11].

The electron density distributions of $S_8$ and $Se_2S_6$ are shown in Fig. 7.8A and B. Because of the higher p orbital and lower electronegativity of the Se atom than those of S, the Se can induce more states in the electronic structure of S atoms. Thus, the more electron densities in the $Se_xS_y$ can result in higher electronic conductivity. DFT calculations on the bond breaking and lithiation energies ($\Delta U$) of S—S, S—Se, and Se—Se bonds in the ortho-$Se_2S_6$ revealed that the Se doping can lower the activation energy of the lithiation process.

Due to the improved electronic conductivity, the $SeS_2$-LPS-C cathode in the ASSLBs delivers a high and stable specific capacity of 930 mAh g$^{-1}$ with minimum capacity decay (Fig. 7.9A and B). Meanwhile, the improved conductivity of the cathode also enhances the electrochemical redox kinetics

Selenium-doped sulfur cathodes    255

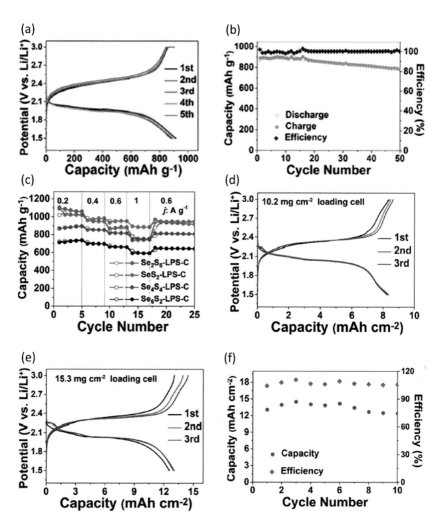

Fig. 7.9 (A) Discharge-charge curves of cells with SeS$_2$-Li$_3$PS$_4$(LPS)-C cathode at 50 mA g$^{-1}$. (B) Cycling performance and coulombic efficiencies of the SeS$_2$-LPS-C cathode at 25°C. (C) Rate capabilities of SeS$_x$-LPS-C cathodes at different current densities from 0.2 to 1 A g$^{-1}$. Electrochemical performance of the SeS$_2$-LPS-C cathodes with high loading at 60°C. (D, E) Discharge-charge curves of cells with loading of: (D) 10.2 mg cm$^{-2}$ and (E) 15.3 mg cm$^{-2}$ at 30 mA g$^{-1}$. (F) Cycling performance of the SeS$_2$-LPS-C cathodes with SeS$_2$ loading of 15.3 mg cm$^{-2}$. (Reproduced by permisison from X. Li, J. Liang, J. Luo, C. Wang, X. Li, Q. Sun, R. Li, L. Zhang, R. Yang, S. Lu, H. Huang, X. Sun, High-performance Li–SeS$_x$ all-solid-state lithium batteries, Adv. Mater. 31 (2019) 1808100, Copyright (2019) John Wiley & Sons.)

to ensure that ASSLBs with the Se–doped S cathode can deliver a high specific capacity of $887\,mAh\,g^{-1}$ under a current density of $1\,A\,g^{-1}$ (Fig. 7.9C). Moreover, even under high cathode loading, the $SeS_2$-LPS-C still can be stably operated. As shown in Fig. 7.9D and E, under the areal loading of $10.2\,mg\,cm^{-2}$ and $15.3\,mg\,cm^{-2}$, the $SeS_2$-LPS-C cathode can still deliver a high areal capacity of 8.5 and $12.6\,mAh\,cm^{-2}$, respectively. Upon cycling, the capacity of the $15.3\,mg\,cm^{-2}$ loading cell maintained a high areal capacity of $11.8\,mAh\,cm^{-2}$ at the tenth cycle, corresponding to a 94% capacity retention rate of the initial cycle. The high areal capacity and stable cycling stability of the ASSLBs based on the $Se_xS_y$ cathode make it a promising candidate for a stable and safe next-generation battery.

As a result, Se doping has been proven to be effective in improving the ionic/electronic conductivities of the cathode to enhance the electrochemical performance of ASSLBs. However, research of this battery system based on the $Se_xS_y$ cathode is still in a very initial phase, and many factors need to be considered for future practical applications, such as the Se doping ratio, the choice of SSEs, and the thickness/weight of the cell.

## 7.5 Concluding remarks and future design strategies for $Se_xS_y$-based battery systems

Due to the enhanced electrochemical redox kinetics and the high theoretical capacity of $Se_xS_y$, Li-$Se_xS_y$ batteries exhibit great promise in achieving high-energy-density battery systems that are stable and safe. The past decades have witnessed significant progress in both fundamental understanding and electrochemical performance of Li-$Se_xS_y$ batteries. The following summarizes knowledge gained to date and future research on the $Se_xS_y$ cathode in various electrolytes:

1. In the carbonate-based electrolyte, the $Se_xS_y$ cathode undergoes a solid-solid (de)lithiation process without the formation of highly soluble LiPs/LiPSes intermediates. Although there is no shuttling effect concern for the $Se_xS_y$ cathode in the carbonate-based electrolyte, the sluggish cathode redox kinetics, the continuous parasitic reaction between Li metal and electrolyte molecules, and the nucleophilic attack of soluble LiPs/$Li_2S$ toward carbonate solvents should be addressed to enable a stable Li-$Se_xS_y$ battery in the carbonate-based electrolyte.

2. In the ether-based electrolyte, the $Se_xS_y$ cathode reaction involves the formation of highly soluble LiPs/LiPSes during operation, which will cause a severe shuttle effect. Moreover, the weak SEI formed on the surface of the Li metal in the ether-based electrolyte leads to harmful

dendrite growth. Thus, immobilizing of LiPs/LiPSes and Li metal protection are needed for Li-Se$_x$S$_y$ batteries in the ether-based electrolyte.

3. In the highly concentrated electrolyte, both the Se$_x$S$_y$ cathode and Li metal anode exhibit robust chemistries with no shuttling and Li dendrite concern. However, the sluggish redox kinetics, the high viscosity, the high cost, and the high electrolyte-to-sulfur ratio still hinder their practical applications. Thus, the modification of the electrolyte components is still needed to address the above issues.

4. In fluorinated electrolyte, robust shuttle- and dendrite-free cathode and anode chemistries can be simultaneously enabled for Li-Se$_x$S$_y$ batteries. Moreover, compared with highly concentrated electrolyte, the normal salt concentration of fluorinated electrolyte makes it more promising in future practical application. However, since the Se$_x$S$_y$ cathode undergoes a solid–solid (de)lithiation process in the fluorinated electrolyte, rational cathode structure design should be conducted to enhance the redox kinetics for Li-Se$_x$S$_y$ batteries in the fluorinated electrolyte.

5. The research on the all-solid-state Li-Se$_x$S$_y$ battery is still in a very initial stage. The ionic conductivities of SSEs should be further improved. Moreover, the thickness/weight of the cell should also be taken into consideration to achieve a practical high-energy-density battery. While choosing the SSEs, their interfacial compatibilities to both the Se$_x$S$_y$ cathode and Li metal anode should be considered.

Although great progress has been made for Li-Se$_x$S$_y$ batteries, many issues need to be addressed for future practical applications. Based on the studies reviewed in this chapter, suggestions for the development of practical Li-Se$_x$S$_y$ batteries include the following: (1) combination of electrolyte engineering and rational cathode structure design to enable robust chemistries of both the cathode and anode while enhancing the electrochemical redox kinetics; (2) modification of the Se doping ratio and the porosity of the host material to optimize the electrochemical performance of the Se$_x$S$_y$ cathode; and (3) full cell configuration design to enable high energy density and long cycling life at the cell level, especially under high cathode loading, lean electrolyte, and thin metal foil conditions.

## Acknowledgments

Research at the Argonne National Laboratory was funded by the U.S. Department of Energy (DOE), Vehicle Technologies Office under Contract No. DE–AC02-06CH11357. Support from Tien Duong of the U.S. DOE's Office of Vehicle Technologies Program is gratefully acknowledged.

# References

[1] Y. Cao, M. Li, J. Lu, J. Liu, K. Amine, Bridging the academic and industrial metrics for next-generation practical batteries, Nat. Nanotechnol. 14 (2019) 200–207.

[2] Y. Deng, Y. Jiang, R. Liang, S. Zhang, D. Luo, Y. Hu, X. Wang, J. Li, A. Yu, Z. Chen, Dynamic electrocatalyst with current-driven oxyhydroxide shell for rechargeable zinc-air battery, Nat. Commun. 11 (2020) 1–10.

[3] M. Wang, Y. Yao, X. Bi, T. Zhao, G. Zhang, F. Wu, K. Amine, J. Lu, Optimization of oxygen electrode combined with soluble catalyst to enhance the performance of lithium–oxygen battery, Energy Storage Mater. 28 (2020) 73–81.

[4] M. Zhang, W. Chen, L. Xue, Y. Jiao, T. Lei, J. Chu, J. Huang, C. Gong, C. Yan, Y. Yan, Y. Hu, X. Wang, J. Xiong, Adsorption-catalysis design in the lithium-sulfur battery, Adv. Energy Mater. 10 (2020) 1903008.

[5] S. Chung, A. Manthiram, Rational design of statically and dynamically stable lithium-sulfur batteries with high sulfur loading and low electrolyte/sulfur ratio, Adv. Mater. 30 (2018) 1705951.

[6] J. Liu, Z. Bao, Y. Cui, E.J. Dufek, J.B. Goodenough, P. Khalifah, Q. Li, B.Y. Liaw, P. Liu, A. Manthiram, Pathways for practical high-energy long-cycling lithium metal batteries, Nat. Energy 4 (2019) 180–186.

[7] X. Tao, J. Wang, C. Liu, H. Wang, H. Yao, G. Zheng, Z.W. Seh, Q. Cai, W. Li, G. Zhou, Balancing surface adsorption and diffusion of lithium-polysulfides on nonconductive oxides for lithium-sulfur battery design, Nat. Commun. 7 (2016) 1–9.

[8] H. Yuan, H. Peng, B. Li, J. Xie, L. Kong, M. Zhao, X. Chen, J. Huang, Q. Zhang, Conductive and catalytic triple-phase interfaces enabling uniform nucleation in high-rate lithium-sulfur batteries, Adv. Energy Mater. 9 (2019) 1802768.

[9] X. Cao, X. Ren, L. Zou, M.H. Engelhard, W. Huang, H. Wang, B.E. Matthews, H. Lee, C. Niu, B.W. Arey, Y. Cui, C. Wang, J. Xiao, J. Liu, W. Xu, J. Zhang, Monolithic solid–electrolyte interphases formed in fluorinated orthoformate-based electrolytes minimize Li depletion and pulverization, Nat. Energy 4 (2019) 796–805.

[10] A. Abouimrane, D. Dambournet, K.W. Chapman, P.J. Chupas, W. Weng, K. Amine, A new class of lithium and sodium rechargeable batteries based on selenium and selenium-sulfur as a positive electrode, J. Am. Chem. Soc. 134 (2012) 4505–4508.

[11] X. Li, J. Liang, J. Luo, C. Wang, X. Li, Q. Sun, R. Li, L. Zhang, R. Yang, S. Lu, H. Huang, X. Sun, High-performance Li–SeS$_x$ all-solid-state lithium batteries, Adv. Mater. 31 (2019) 1808100.

[12] G. Xu, T. Ma, C. Sun, C. Luo, L. Cheng, Y. Ren, S.M. Heald, C. Wang, L. Curtiss, J. Wen, Insight into the capacity fading mechanism of amorphous Se$_2$S$_5$ confined in micro/mesoporous carbon matrix in ether-based electrolytes, Nano Lett. 16 (2016) 2663–2673.

[13] G. Xu, J. Liu, R. Amine, Z. Chen, K. Amine, Selenium and selenium-sulfur chemistry for rechargeable lithium batteries: interplay of cathode structures, electrolytes, and interfaces, ACS Energy Lett. 2 (2017) 605–614.

[14] R.O. Jones, D. Hohl, Structure, bonding, and dynamics in heterocyclic sulfur-selenium molecules, Se$_x$S$_y$, J. Am. Chem. Soc. 112 (1990) 2590–2596.

[15] G. Xu, H. Sun, C. Luo, L. Estevez, M. Zhuang, H. Gao, R. Amine, H. Wang, X. Zhang, C. Sun, Solid-state lithium/selenium-sulfur chemistry enabled via a robust solid-electrolyte interphase, Adv. Energy Mater. 9 (2019) 1802235.

[16] H. Du, S. Feng, W. Luo, L. Zhou, L. Mai, Advanced Li–Se$_x$S$_y$ battery system: electrodes and electrolytes, J. Mater. Sci. Technol. 55 (2020) 1–15.

[17] K. Xu, Nonaqueous liquid electrolytes for lithium-based rechargeable batteries, Chem. Rev. 104 (2004) 4303–4418.

[18] K. Xu, Electrolytes and interphases in Li-ion batteries and beyond, Chem. Rev. 114 (2014) 11503–11618.

[19] Y. Cui, A. Abouimrane, C. Sun, Y. Ren, K. Amine, Li–Se battery: absence of lithium polyselenides in carbonate based electrolyte, Chem. Commun. 50 (2014) 5576–5579.

[20] S. Xin, L. Yu, L. You, H. Cong, Y. Yin, X. Du, Y. Guo, S. Yu, Y. Cui, J.B. Goodenough, The electrochemistry with lithium versus sodium of selenium confined to slit micropores in carbon, Nano Lett. 16 (2016) 4560–4568.

[21] C. Yang, Y. Yin, Y. Guo, Elemental selenium for electrochemical energy storage, J. Phy. Chem. Lett. 6 (2015) 256–266.

[22] S. Xin, L. Gu, N. Zhao, Y. Yin, L. Zhou, Y. Guo, L. Wan, Smaller sulfur molecules promise better lithium-sulfur batteries, J. Am. Chem. Soc. 134 (2012) 18510–18513.

[23] X. Li, J. Liang, K. Zhang, Z. Hou, W. Zhang, Y. Zhu, Y. Qian, Amorphous S-rich $S_{1-x}Se_x/C$ ($x \leq 0.1$) composites promise better lithium-sulfur batteries in a carbonate-based electrolyte, Energ. Environ. Sci. 8 (2015) 3181–3186.

[24] L. Zeng, Y. Yao, J. Shi, Y. Jiang, W. Li, L. Gu, Y. Yu, A flexible $S_{1-x}Se_x@$ porous carbon nanofibers ($x \leq 0.1$) thin film with high performance for Li-S batteries and room-temperature Na-S batteries, Energy Storage Mater. 5 (2016) 50–57.

[25] S. Guo, C. Li, Y. Chi, Z. Ma, H. Xue, Novel 3-D network $SeS_x$/NCPAN composites prepared by one-pot in-situ solid-state method and its electrochemical performance as cathode material for lithium-ion battery, J. Alloys Compd. 664 (2016) 92–98.

[26] V.H. Pham, J.A. Boscoboinik, D.J. Stacchiola, E.C. Self, P. Manikandan, S. Nagarajan, Y. Wang, V.G. Pol, J. Nanda, E. Paek, Selenium-sulfur (SeS) fast charging cathode for sodium and lithium metal batteries, Energy Storage Mater. 20 (2019) 71–79.

[27] X. Fan, L. Chen, O. Borodin, X. Ji, J. Chen, S. Hou, T. Deng, J. Zheng, C. Yang, S. Liou, Non-flammable electrolyte enables Li-metal batteries with aggressive cathode chemistries, Nat. Nanotechnol. 13 (2018) 715–722.

[28] G. Li, Z. Li, B. Zhang, Z. Lin, Developments of electrolyte systems for lithium-sulfur batteries: a review, Front. Energy Res. 3 (2015) 5.

[29] Y. Cui, A. Abouimrane, J. Lu, T. Bolin, Y. Ren, W. Weng, C. Sun, V.A. Maroni, S.M. Heald, K. Amine, (De)lithiation mechanism of Li/SeS$_x$ ($x = 0$-7) batteries determined by in situ synchrotron X-ray diffraction and X-ray absorption spectroscopy, J. Am. Chem. Soc. 135 (2013) 8047–8056.

[30] Z. Li, J. Zhang, B.Y. Guan, X.W. Lou, Mesoporous carbon@ titanium nitride hollow spheres as an efficient $SeS_2$ host for advanced Li-SeS$_2$ batteries, Angew. Chem. Int. Ed. 56 (2017) 16003–16007.

[31] J. Zhang, Z. Li, X.W. Lou, A freestanding selenium disulfide cathode based on cobalt disulfide-decorated multichannel carbon fibers with enhanced lithium storage performance, Angew. Chem. Int. Ed. 129 (2017) 14295–14300.

[32] Z. Li, J. Zhang, H.B. Wu, X.W. Lou, An improved Li-SeS$_2$ battery with high energy density and long cycle life, Adv. Energy Mater. 7 (2017) 1700281.

[33] F. Sun, H. Cheng, J. Chen, N. Zheng, Y. Li, J. Shi, Heteroatomic $Se_nS_{8-n}$ molecules confined in nitrogen-doped mesoporous carbons as reversible cathode materials for high-performance lithium batteries, ACS Nano 10 (2016) 8289–8298.

[34] Y. Yamada, J. Wang, S. Ko, E. Watanabe, A. Yamada, Advances and issues in developing salt-concentrated battery electrolytes, Nat. Energy 4 (2019) 269–280.

[35] Y. Yamada, K. Furukawa, K. Sodeyama, K. Kikuchi, M. Yaegashi, Y. Tateyama, A. Yamada, Unusual stability of acetonitrile-based superconcentrated electrolytes for fast-charging lithium-ion batteries, J. Am. Chem. Soc. 136 (2014) 5039–5046.

[36] Q. Pang, A. Shyamsunder, B. Narayanan, C.Y. Kwok, L.A. Curtiss, L.F. Nazar, Tuning the electrolyte network structure to invoke quasi-solid state sulfur conversion and suppress lithium dendrite formation in Li-S batteries, Nat. Energy 3 (2018) 783–791.

[37] C. Yang, J. Chen, T. Qing, X. Fan, W. Sun, A. von Cresce, M.S. Ding, O. Borodin, J. Vatamanu, M.A. Schroeder, 4.0 V aqueous Li-ion batteries, Joule 1 (2017) 122–132.

[38] Z. Zeng, V. Murugesan, K.S. Han, X. Jiang, Y. Cao, L. Xiao, X. Ai, H. Yang, J. Zhang, M.L. Sushko, Non-flammable electrolytes with high salt-to-solvent ratios for Li-ion and Li-metal batteries, Nat. Energy 3 (2018) 674–681.

[39] J. Zheng, G. Ji, X. Fan, J. Chen, Q. Li, H. Wang, Y. Yang, K.C. DeMella, S.R. Raghavan, C. Wang, High-fluorinated electrolytes for Li-S batteries, Adv. Energy Mater. 9 (2019) 1803774.

[40] J.T. Lee, H. Kim, M. Oschatz, D. Lee, F. Wu, H. Lin, B. Zdyrko, W.I. Cho, S. Kaskel, G. Yushin, Micro- and mesoporous carbide-derived carbon–selenium cathodes for high-performance lithium selenium batteries, Adv. Energy Mater. 5 (2015) 1400981.

[41] N. Kamaya, K. Homma, Y. Yamakawa, M. Hirayama, R. Kanno, M. Yonemura, T. Kamiyama, Y. Kato, S. Hama, K. Kawamoto, A lithium superionic conductor, Nat. Mater. 10 (2011) 682–686.

[42] Y. Seino, T. Ota, K. Takada, A. Hayashi, M. Tatsumisago, A sulphide lithium super ion conductor is superior to liquid ion conductors for use in rechargeable batteries, Energ. Environ. Sci. 7 (2014) 627–631.

[43] Y. Kato, S. Hori, T. Saito, K. Suzuki, M. Hirayama, A. Mitsui, M. Yonemura, H. Iba, R. Kanno, High-power all-solid-state batteries using sulfide superionic conductors, Nat. Energy 1 (2016) 1–7.

[44] R. Fang, H. Xu, B. Xu, X. Li, Y. Li, J.B. Goodenough, Reaction mechanism optimization of solid-state Li-S batteries with a PEO-based electrolyte, Adv. Funct. Mater. (2020) 2001812.

# CHAPTER 8

# Suppression of lithium dendrite growth in lithium-sulfur batteries

**XiaoLong Xu[a,b] and Hao Wang[a]**

[a]The College of Materials Science and Engineering, Beijing University of Technology, Beijing, China
[b]School of Materials Science and Engineering, Qilu University of Technology (Shandong Academy of Sciences), Jinan, Shandong Province, China

## Contents

| | |
|---|---|
| 8.1 Introduction | 261 |
| 8.2 Dendritic growth mechanism | 263 |
|     8.2.1 Thermodynamics | 263 |
|     8.2.2 Kinetics | 264 |
|     8.2.3 Crystallography | 264 |
| 8.3 Effect of Li dendrite growth on Li-S batteries | 265 |
| 8.4 Suppression method | 266 |
|     8.4.1 Separator | 267 |
|     8.4.2 Anode | 269 |
|     8.4.3 Electrolyte | 278 |
| 8.5 Conclusions | 287 |
| References | 289 |

## 8.1 Introduction

Lithium-sulfur (Li-S) battery technology is a very promising high-energy candidate compared to state-of-the-art Li-ion batteries due to high theoretical capacities of sulfur cathode ($1672\,mAh\,g^{-1}$) and Li anode ($3862\,mAh\,g^{-1}$), which leads to a high specific energy of $2600\,Wh\,kg^{-1}$ [1]. Similar to a typical Li-ion battery with Li metal anode, Li dendrite growth is unavoidable [2]. In fact, Li dendrite is a kind of needle-like and tree-like morphology, which forms in the condition of deviation from balance (nonuniform distribution of charges) at the electrode/electrolyte interface as shown in Fig. 8.1. This is the typical dendritic morphology of electrodeposited Li, which was first reported by Tatsuma et al. [3].

Unlike standard Li-ion batteries, Li polysulfides form at the cathode and dissolve in the electrolyte during charge/discharge cycle, which is in direct

---

*Lithium-Sulfur Batteries*
https://doi.org/10.1016/B978-0-12-819676-2.00004-9

Copyright © 2022 Elsevier Inc.
All rights reserved.

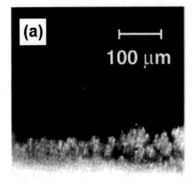

**Fig. 8.1** Micrographs of lithium surfaces after electrochemical deposition of Li (3 mA cm$^{-1}$ × 1 h) from the PMMA-based electrolytes containing TGD as a cross-linking agent. The TGD-MMA molar ratio is (A) 4% and (B) 8%. *(Reproduced with permission from T. Tatsuma, M. Taguchi, N. Oyama, Inhibition effect of covalently cross-linked gel electrolytes on lithium dendrite formation, Electrochim. Acta 46 (2001) 1201–1205, Copyright (2001) Elsevier.)*

contact with the Li anode and forms a complicated electrode/electrolyte interface system with the coexistence of Li dendrites and Li polysulfides [4,5]. Generally, Li dendrite formation is mainly caused by the following: (1) inhomogeneous distribution of the current density on the electrode surface, and (2) the high concentration gradient of Li-ions at the electrolyte/electrode interface [6–9]. Li dendrites with sharp tip may pierce the polymer separator, resulting in short circuit and subsequent thermal runaway of the cell. Meanwhile, the root of the Li dendrites prefers to lose Li first because the dendrite Li possesses higher reactivity than planar front Li [10,11]. It rapidly dissolves at the local region and breaks away from the Li anode during the delithiation, leading to the formation of "dead Li" that is detached from the current collector and contributes no capacity. This also decreases the efficiency and stability of Li anode in Li-S batteries [11].

This chapter describes the suppression of Li dendrite growth in Li-S batteries. First, dendrite growth mechanisms (i.e., driving force for interface perturbation) are discussed. Then, the strategies for delaying and suppressing dendrite growth by the development of novel battery components are described, including separator, anode, and electrolyte. Finally, the advantages and disadvantages of these strategies are compared.

## 8.2 Dendritic growth mechanism

### 8.2.1 Thermodynamics

A series of theoretical models based on mass and charge transfer of Li-ions toward the electrode/electrolyte interface under thermodynamic driving force (electrochemical potential gradient) have been developed that can explain the thermodynamics of Li dendrite initiation and growth, such as the solid electrolyte interphase (SEI) model [12–14], charge-induced growth model [15], constitutional underpotential plating model [16], and phase-field model [17]. The classical theoretical model was proposed by Yamaki et al. [18] in 1998, and it was the deposition/dissolution model (Fig. 8.2). This model is described as follows: Li metal was deposited under the SEI film and supplied with an external power, and Li-ions in the electrolyte transported to Li metal surface through the protective SEI film. The deposition sites on the protective film exhibited higher Li-ion conductivity; thus, crystal defects and grain

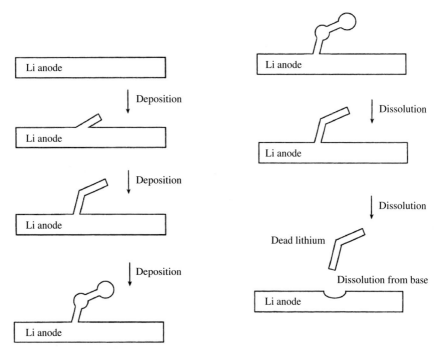

**Fig. 8.2** The model of Li deposition and dissolution. *(Reproduced with permission from J. Yamaki, S. Tobishima, K. Hayashi, K. Saito, Y. Nemoto, M. Arakawa, A consideration of the morphology of electrochemically deposited lithium in an organic electrolyte, J. Power Sources 74 (1998) 219–227, Copyright (1998) Elsevier.)*

boundaries in the SEI initiated the continuous Li deposition. The mechanical stress within the Li anode induced an asymmetrical Li deposition, resulting in the formation of Li dendrites. Recently, Kumta et al. [16] have developed the concept of constitutional underpotential plating (CUP) and interface perturbation theory to understand the fundamental mechanism of dendrite growth during Li metal deposition. The CUP condition, a measure of the driving force for interface perturbation, occurs when the true potential gradient at the interface is smaller than the equilibrium electrodeposition potential gradient at the Li/electrolyte interface, resulting in the electrolyte ahead of the Li-electrolyte interface to be in an underpotential state. Detailed CUP criterion calculations establish the relationship of battery operating conditions (e.g., current density, true potential gradient) with the electrolyte (e.g., Li-ion transference number and diffusion coefficient) and current collector properties (e.g., Gibbs Thomson parameter) predicting the morphological stability condition at the interface.

## 8.2.2 Kinetics

Over the past century, a rigorous experimental and theoretical foundation was developed to explain and quantify the nature of the planar front interface instabilities occurring during crystal growth from solidification of liquid metal, metal generation via metal electrodeposition, as well as crystal growth from solution and vapor condensation techniques. The kinetics research of Chang et al. [19] produced a fundamental understanding of the growth mechanism of Li dendrites by the in situ $^7$Li magnetic resonance (Fig. 8.3). The chemical shift imaging shows that mossy types of microstructure grow close to the surface of the anode from the beginning of charge, while Li dendritic growth is triggered much later. Comparison of a series of cells charged at different current densities demonstrates that at high charge rates, there was a strong correlation between the onset time of Li dendrite growth and the local depletion of the electrolyte (i.e., Li-ion) at the surface of the electrode both observed experimentally and predicted theoretically. Therefore, high charge rates, low Li-ion transference number, and low Li-ion diffusivity at the electrode/electrolyte interface could cause the growth of Li dendrites.

## 8.2.3 Crystallography

From crystallographic studies, it is well known that metallic Li of bcc structure has two characteristic X-ray diffraction (XRD) peaks at 36 degrees and

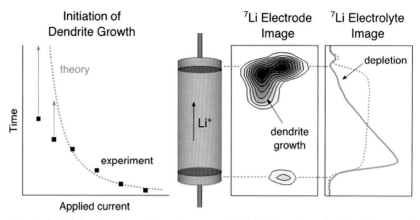

**Fig. 8.3** Plot of the theoretical Sand's time and the initiation time of Li dendrite growth measured experimentally from the discontinuity in the 270 ppm metal peak. Schematic of the cell used for in situ $^7$Li magnetic resonance imaging (MRI). MRI time series showing the evolution of the $^7$Li electrolyte concentration profile (top) and the $^7$Li chemical shift image of the metal (bottom) for the cell charged at 0.76 mA cm$^{-2}$. *(Reproduced with permission from H.J. Chang, A.J. Ilott, N.M. Trease, M. Mohammadi, A. Jerschow, C.P. Grey, Correlating microstructural lithium metal growth with electrolyte salt depletion in lithium batteries using $^7$Li MRI, J. Am. Chem. Soc. 137 (48) (2015) 15209–15216, Copyright (2015) American Chemical Society.)*

52 degrees, corresponding to (110) and (200) crystal faces, respectively; the peak at 52 degrees is stronger than the peak at 36 degrees of pristine Li [2]. In 2014, Zhang et al. [20] studied the change of these two peaks before and after cycling, showing that the peak at 36 degrees is significantly stronger than the peak at 52 degrees after Li deposition. This transformation demonstrates the orientational crystallization of Li metal along (001) during deposition, which can be the underlying reason of Li dendrite formation. Zu and Manthiram [5] reported a stabilized Li metal surface in a polysulfide-rich environment of Li-S batteries (Fig. 8.4), indicating the Li deposition with a preferential orientation in contrast to the bulk Li metal.

## 8.3 Effect of Li dendrite growth on Li-S batteries

When employing Li metal as the anode, two major failure mechanisms are typically associated with the application of Li-S batteries. On the one hand, the problem of continuous Li erosion is compounded with the dissolved polysulfides that get involved in the passivation film formation during

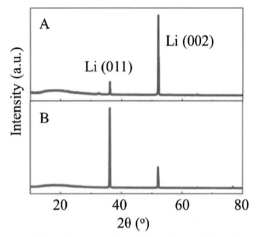

**Fig. 8.4** XRD patterns of the Li anode before (A) and after (B) the discharge and charge cycles. (Reproduced with permission from C. Zu, A. Manthiram, Stabilized lithium-metal surface in a polysulfide-rich environment of lithium-sulfur batteries, J. Phys. Chem. Lett. 5 (15) (2014) 2522–2527, Copyright (2015) American Chemical Society.)

discharge [21,22]. On the other hand, uncontrolled high-surface-area Li dendrite formation during charge is a main reason for the failure of Li-S batteries [23,24]. The former causes serious attenuations of cycle performance and specific capacity [25], while the latter leads to lower Coulombic efficiency and even serious safety issues caused by cell short circuit [26].

Although there are two main failure mechanisms, they have different effects on performances of Li-S batteries. The Li dendrite growth on the surface of Li anode is particularly important due to its relevance to the safety issue. In the following, we summarize and compare the methods of suppressing the growth of Li dendrites based on above failure mechanisms.

## 8.4 Suppression method

Recently, compared with cathodes and electrolytes, the safety of the Li-S batteries' anode has become one of the more urgent challenges to overcome in order to produce practical Li-S batteries for high energy applications. Since the electrode/electrolyte interface microstructures are major links connecting the performance and electrodeposition conditions, it is important to understand how the Li metal deposition conditions and the battery component properties influence the resulting microstructures. This is important to predict, modify, and control the microstructure of the deposited Li metal by meticulously designing the appropriate battery components. In this case, three main battery-component-targeted approaches exist for suppressing the growth of Li dendrite: separator [27,28], anode [29–34], and electrolyte [15,35–40]. All of these methods are able to suppress the growth of Li dendrites to a certain extent.

## 8.4.1 Separator

The separators of Li-S batteries are the inactive fundamental components, which are necessarily sandwiched between the anode and the cathode [41] and have to possess two functions of (1) resting between the anode and the cathode to prevent internal short circuit and (2) providing a path for ionic conduction in the liquid electrolyte throughout the interconnected porous structure. For liquid electrolyte, batteries are currently engineered as porous membranes, nonwoven mats, or multilayers consisting of porous membranes and/or nonwoven mats [42,43]. The safety problems caused by the growth of Li dendrites may be solved by taking advantages of the special properties of novel separators. Currently, the novel separators mainly involve ceramic, polymer, ceramic-coated polymer separator, cellulose, and solid state electrolyte (SSE) separator [44,45].

As a kind of Li-ion conductor, the inorganic solid state electrolyte separator is able to prevent the safety problems caused by Li dendrite growth and hence turns out to be an ideal electrolyte material for Li-S batteries because of the sufficient hardness and elastic modulus of SSE to prevent Li dendrites piercing the SSE separator [46,47]. Wang et al. [48] exploited an inorganic solid electrolyte, $Li_{1.5}Al_{0.5}Ge_{1.5}(PO_4)_3$ (LAGP), as a separator for fabricating a hybrid electrolyte (HE) Li-S battery formed by using a Li metal anode. The 1 M $LiN(CF_3SO_2)_2$ in 1,3-dioxolane and 1,2-dimethoxyethane (1:1, v/v) (LiTFSI/DOL/DME) as liquid electrolyte was used to connect the electrode and the separator. Such solid electrolyte separator is able to avoid the Li anode corrosion and prevent Li dendrite initiation due to its excellent mechanical strength.

Polymers and cellulose membranes are promising separators for Li-S batteries due to their ability to be prepared as thin films [49]. Their advantages are that they are safer with increased mechanical stability and more shape conforming [50]. Rao et al. [51] reported a Li-S battery consisting of carbon nanofibers-sulfur cathode and polymer separator. The poly(acrylonitrile)/poly(methyl methacrylate) (PAN/PMMA) membrane as the polymer separator were prepared by electrospinning technique, possessing a

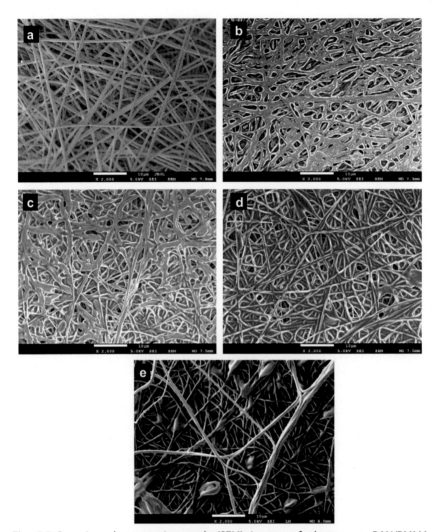

**Fig. 8.5** Scanning electron micrograph (SEM) images of electrospun PAN/PMMA membrane before (A) and after electrolyte uptake: (B) PPR14TFSI; (C) PPR14TFSI: PEGDME (2:1, by weight); (D) PPR14TFSI:PEGDME (1:1, by weight); and (E) PPR14TFSI: PEGDME (1:2, by weight). *(Reproduced with permission from M. Rao, X. Geng, X. Li, S. Hu, W. Li, Lithium-sulfur cell with combining carbon nanofibers–sulfur cathode and gel polymer electrolyte, J. Power Sources 212 (2012) 179–185, Copyright (2012) Elsevier.)*

three-dimensional (3D) network of interlaid fibers and imparting sufficient mechanical strength (Fig. 8.5A). About $1\,mol\,kg^{-1}$ lithium bis(trifluoromethylsulfonyl)imide in *N*-methyl-*N*-butylpiperidinium bis (trifluoromethanesulfonyl) imide (PPR14TFSI) and poly (ethylene glycol) dimethyl ether (PEGDME) were used as the liquid electrolytes. After electrolyte absorption, the separator swelled and became more entangled. The fibers were in close contact with each other, and the fibrous structure was retained (Fig. 8.5B–E), thus enhancing the advantages of Li-S batteries as well as suppressing Li dendrite growth and the dissolution of the intermediate products generated during the cycle process. Yu et al. [52] used a 30-µm-thick cellulose-based membrane with a pore size between 50 and 100 nm as the separator in Li-S batteries, which is low cost, holds an organic-liquid electrolyte longer than Celgard 2500, has better thermal stability, and is wet by metallic lithium. Cells with this separator show excellent electrochemical performance for up to 1000 cycles without dendrite formation [52].

The safety problems caused by the growths of Li dendrites can be solved by using novel separators. The mechanical strength and 3D structure of the separator prevent the lithium dendrites from piercing the separators. However, recent studies indicated that Li metal dendrites can pierce the solid state separator (e.g., $Li_7La_3Zr_2O_{12}$) causing the short circuit of the cell [53]. Sudo et al. [54] reported that Li|SSE|Li cell with the 0.5 wt% $Al_2O_3$-doped $Li_7La_3Zr_2O_{12}$ separator shows a short circuit after a limited polarization period, and short circuiting is considered to be due to Li dendrite formation. It is necessary to seek low-cost and more efficient separators with 3D structure for practical applications.

## 8.4.2 Anode

### 8.4.2.1 3D anode

To suppress the growth of Li dendrite, replacement of the Li anode with other Li host anode materials provides a feasible way to improve the battery's safety [55–57]. In this strategy, special 3D structures of anodes can effectively suppress growth of Li dendrite, because the 3D structure controls the deposition and dissolution of Li crystal and limits growth kinetics of dendritic nuclei by promoting heterogeneous nucleation rate of metallic lithium [58,59]. Generally, the 3D structure has hierarchical pores and abundant electrode/electrolyte interfaces for lowering the absolute local current density, and the interconnected hierarchical pores among the 3D scaffolds benefit the full penetration of electrolytes and fast ionic diffusion [11].

Yang et al. [59] reported a Li-S battery consisting of a $Li_2S$/mesoporous carbon composite cathode and a silicon nanowire anode (Fig. 8.6), delivering a theoretical specific energy of 1550 Wh kg$^{-1}$ and an initial discharge-specific energy of 630 Wh kg$^{-1}$. The silicon nanowire anode in Li-S battery effectively solved the safety problem arising from the growth of Li dendrites

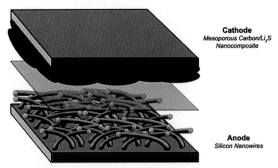

**Fig. 8.6** Schematic diagram of battery structure: The cathode contains lithium sulfide (Li$_2$S) encapsulated within ordered mesoporous carbon, and the anode consists of silicon nanowires. *(Reproduced with permission from Y. Yang, M.T. McDowell, A. Jackson, J.J. Cha, S.S. Hong, Y. Cui, New nanostructured Li$_2$S/silicon rechargeable battery with high specific energy, Nano Lett. 10 (4) (2010) 1486–1491, Copyright (2010) American Chemical Society.)*

during cycling because the silicon nanowires are able to control the uniform deposition of Li metal by its 3D structure.

In order to modify the silicon anode, Hagen et al. [60] performed a prelithiation treatment of silicon microwire array anodes by an electrochemical process. The Li-S battery with sulfur-infiltrated carbon nanotube (CNT-S) cathode, prelithiated Si anode, and various polysulfide or Li$_2$S-containing electrolytes exhibited stable cycle performance more than 200 cycles without Li dendrite growth problems.

In addition to silicon-based anodes, the Li-B alloy of two phase mixture (Li + Li-B intermetallic) also attracts researchers' interest due to their special 3D structure. In general, the Li metal anode shows significant Li dendrite growth on the surface after 200 cycles in the Li-S battery, while the Li-B alloy anode can retain a smooth surface without Li dendrite growth under the same test conditions [11]. Therefore, the cycle performance and Coulombic efficiency of Li-S batteries comprising of Li-B alloy as an anode are greatly improved.

Zhang et al. [20] enhanced cycle performance and component safety of Li-S batteries by employing a Li-B alloy anode with abundant-free Li embedded in a stable Li$_7$B$_6$ framework. In Fig. 8.7, the peak intensity of (200) peak of pristine Li was stronger than the (110) peak, yet at the end of Li deposition, (110) peak was significantly stronger than (200) peak, demonstrating oriented crystallization of Li during deposition. However, Li-B alloy anode did not show oriented Li crystallization. From the above results, the intermetallic Li$_7$B$_6$ constituent demonstrated three important advantages: (1) eliminated orientational crystallization of free Li; (2) reduced effective current density and promoted the formation of the stable solid state electrolyte interface (SEI); and (3) protected host alloy bulk materials from deformation, volume expansion, or collapse when cycling. This marked contrast further verified that Li-B alloy (Li$_7$B$_6$) has several key advantages in restraining the formation of Li dendrites (Fig. 8.7). The investigations of silicon-based and Li-B alloy anodes give the mechanism of the 3D structure suppressing the growth of Li dendrites, providing a basis for the developments of other novel 3D anodes.

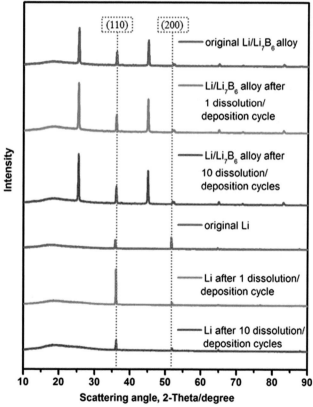

**Fig. 8.7** XRD patterns of Li-B alloy (40 wt% B) and metallic Li electrodes before and after dissolution/deposition cycles. *(Reproduced with permission from X. Zhang, W. Wang, A. Wang, Y. Huang, K. Yuan, Z. Yu, J. Qiu, Y. Yang, Improved cycle stability and high security of Li-B alloy anode for lithium–sulfur battery, J. Mater. Chem. A 2 (30) (2014) 11660–11665, Copyright (2014) Royal Society of Chemistry.)*

Liang et al. [30] designed a 3D oxidized polyacrylonitrile nanofiber network on top of the current collector, delivering a Coulombic efficiency of more than 97.4% at a current density of $3\,mA/cm^2$ and a capacity of $1\,mAh/cm^2$ after 120 cycles. The polar functional groups on the polymer nanofiber surface serve as adhesion/adsorption sites to bind with Li-ions in the electrolyte. As a result, the movement of Li-ions toward the "hot spots" was retarded by the attraction force from polymer fiber containing element oxygen functional groups (i.e., O—H and C=O) during electrochemical cycling, facilitating a relatively homogeneous Li-ionic flux and reducing the Li-ion concentration gradient at the electrode/electrolyte interface (Fig. 8.8). In addition, the electrically insulating polymer was not involved in the Li deposition although it could play the role of a 3D scaffold to guide and confine the growth of Li along the fiber, because the polar surface functional groups have a strong affinity with Li metal.

Similar to the surface 3D structure on the current collector, Chi et al. [61] employed nickel (Ni) foam as a host for prestoring Li via a thermal infusion strategy to form a dendrite-free Li-Ni composite anode, demonstrating a low electrode dimensional change and dendrite-free behavior after 100 cycles at a current density of $5\,mA\,cm^{-2}$. The Ni foam host not only acts as a cage entrapment for Li metal, but also accommodates the surface energy of the Li-Ni composite anode during electrochemical cycling, thus preventing dendritic growth of Li metal and restricting dimensional variation of Li electrode.

### 8.4.2.2 Surface treatment

The surface treatment of Li metallic anode also presents a strategy for suppressing the growth of Li in Li-S batteries [62,63], which is able to fundamentally limit the growth of Li dendrite due to the electrochemical deposition of Li metal along the surface treatment layers [64,65].

For example, Huang et al. [7] designed a Li-S battery using electrically connected graphite and Li metal as a hybrid anode to control undesirable Li surface reactions. A lithiated graphite electrode, connected in parallel with the Li electrode, placed in front of the Li metal functions as an artificial, self-regulating solid electrolyte interface layer to actively control the electrochemical reactions and suppress the Li dendrites growth, leading to significant performance improvements. Such Li-S battery delivered a specific discharge capacity of $800\,mAh\,g^{-1}$ after 400 cycles at a current density of $1737\,mA\,g^{-1}$ with a capacity fade of 11% and a Coulombic efficiency of 99% (Fig. 8.9).

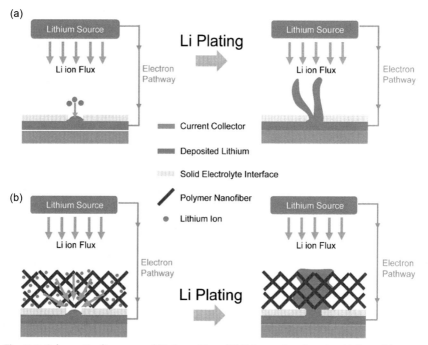

**Fig. 8.8** Schematic diagrams of Li deposition. (A) Schematics showing Li deposition on bare Cu electrode. Li-ions concentrate in the vicinity of the dendrite tip, which led to a mossy lithium growth and severe electrolyte decomposition. (B) Modifying the Cu electrode with a layer of polymer nanofiber network results in a homogeneous Li deposition and less electrolyte consumption. *(Reproduced with permission from Z. Liang, G. Zheng, C. Liu, N. Liu, W. Li, K. Yan, H. Yao, P.C. Hsu, S. Chu, Y. Cui, Polymer nanofiber-guided uniform lithium deposition for battery electrodes, Nano Lett. 15 (5) (2015) 2910–2916, Copyright (2015) American Chemical Society.)*

Employing a more obvious surface treatment via a simple process, Ma et al. [66] fabricated a Li$_3$N protection layer on the surface of a Li anode to suppress the Li dendrite growth (Fig. 8.10). The Li$_3$N protective layer was in situ fabricated on the surface of a Li anode by the direct reaction between the Li and nitrogen gas at room temperature. First, a Li$_3$N layer has a high ionic conductivity, which did not hinder Li-ion migration. Second, a Li$_3$N layer could prevent the side reaction between the Li anode and the electrolyte, forming a stable SEI layer. Finally, contact between the Li polysulfides and the Li anode was thus prevented, and the undesired corrosive reaction was suppressed. As a result, the Li$_3$N layer could suppress the Li dendrite originating from a nonuniform deposition of Li for improving the component (or material or layer) integrity of the battery.

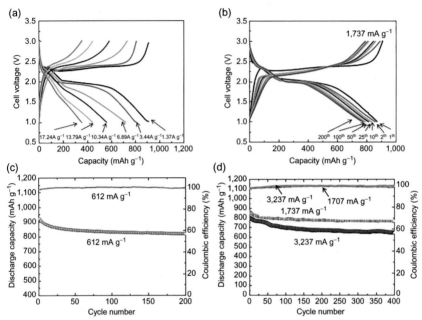

**Fig. 8.9** Electrochemical behaviors of Li-S battery using the proposed hybrid anode. (A) Charge and discharge curves of Li-S battery at different rates. (B) Cycling ability of the Li-S battery at a current density of 1737 mA g$^{-1}$. (C, D) Long-term cycling behavior and corresponding Coulombic efficiencies of Li-S battery at different current densities. *(Reproduced with permission from R. Tao, X. Bi, S. Li, Y. Yao, F. Wu, Q. Wang, C. Zhang, J. Lu, Kinetics tuning the electrochemistry of lithium dendrites formation in lithium batteries through electrolytes, ACS Appl. Mater. Interfaces 9 (8) (2017) 7003–7008, Copyright (2017) American Chemical Society.)*

Dong et al. [15] prepared a flexible and tensile polymer coating on Li metal surface (Fig. 8.11), which induces the uniform and dendrite-free deposition of Li metal during cycling. The combination of 18-crown-6 and polyvinylidene fluoride synergistically leads to the robust flexibility and the high conductivity of Li-ions of the prepared coating film, suppressing Li dendrites obviously, and demonstrating a Coulombic efficiency of 97.38% and a good cycle stability after 200 cycles in Li-S battery.

In a large-scale study, Wang et al. [14] established a dense and homogenous lithium phosphorus oxynitride (LiPON) coating on the Li anode surface via nitrogen-plasma-assisted deposition of electron-beam reaction evaporation (Fig. 8.12A). The LiPON coating serves as a highly ionic conductive, chemically stable, and mechanically robust protective layer, suppresses the

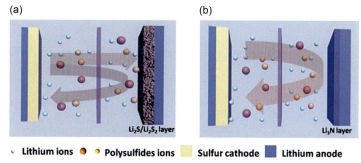

**Fig. 8.10** Schematic of the design of a Li metal electrode in Li-S battery configurations. (A) A battery without the Li₃N layer. (B) A battery with the Li₃N layer. *(Reproduced with permission from G. Ma, Z. Wen, M. Wu, C. Shen, Q. Wang, J. Jin, X. Wu, A lithium anode protection guided highly-stable lithium-sulfur battery, Chem. Commun. 50 (91) (2014) 14209–14212, Copyright (2014) Royal Society of Chemistry.)*

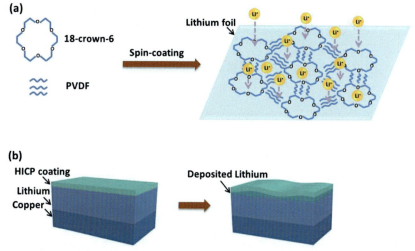

**Fig. 8.11** (A) Schematic diagram of the formation of the demonstrated HICP and (B) electrodes with HICP. The electrode surface is relatively flat after cycling, suggesting homogeneous Li metal deposition. *(Reproduced with permission from H. Dong, X. Xiao, C. Jin, X. Wang, P. Tang, C. Wang, Y. Yin, D. Wang, S. Yang, C. Wu, High lithium-ion conductivity polymer film to suppress dendrites in Li metal batteries, J. Power Sources 423 (2019) 72–79, Copyright (2019) Elsevier.)*

corrosion reaction with organic electrolytes, and promotes uniform Li plating/stripping (Fig. 8.12B), thus enabling a stable and dendrite-free cycling of the symmetric Li metal cells for 900 cycles under a current density of 3 mA cm$^{-2}$. This Li-S battery displays a specific energy density of 300 Wh kg$^{-1}$, a Coulombic efficiency of 91%, and a life span of 120 cycles.

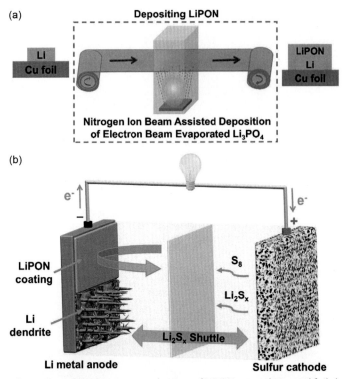

**Fig. 8.12** Illustration of (A) the mass production of LiPON-coated Li metal foils by using a continuous roll-to-roll system and (B) the performance of a LiPON-coated Li metal anode in Li-S battery configurations. *(Reproduced with permission from X.B. Cheng, R. Zhang, C.Z. Zhao, F. Wei, J.G. Zhang, Q. Zhang, A review of solid electrolyte interphases on lithium metal anode, Adv. Sci. 3 (3) (2016) 1500213, Copyright (2015) John Wiley & Sons.)*

To further exploit the effect of various coating structures, Xu et al. [67] constructed an artificial protective layer (denoted as APL) with synergistic soft-rigid feature on the surface of Li anode to offer superior interfacial stability suppressing random Li deposition and enabled a dendrite-free morphology of Li metal anode. Furthermore, three cases for dendrite growth and suppression were considered: (1) The fragility and heterogeneity of SEI are origins of nonuniform Li deposition and formation of Li dendrites (Fig. 8.13A), and during Li plating, SEI, normally possessing a low modulus, is easily broken by stresses induced by dendritic Li growth and electrode volume change; (2) low ionic conductivity, insufficient mechanical robustness of polymer, and poor interfacial contact of ceramics remain formidable challenges to render efficient and dendrite-free Li metal anodes for practical

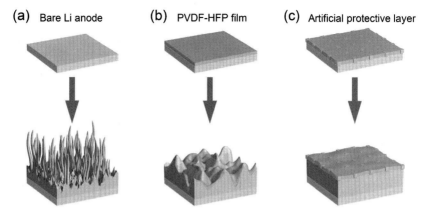

**Fig. 8.13** Schematic illustrations of Li deposition: (A) without protection, lithium metal dendrites and dead Li form after cycling; (B) with a pure PVDF-HFP layer that is of poor mechanical modulus, interfacial fluctuation with dendrites piercing the PVDF-HFP layer occurs after cycling; and (C) with APL composed of organic PVDF-HFP and inorganic LiF that is conformal and mechanically strong to suppress Li dendrites penetration and stabilize Li metal surface. *(Reproduced with permission from R. Xu, X.Q. Zhang, X. B. Cheng, H.J. Peng, C.Z. Zhao, C. Yan, J.Q. Huang, Artificial soft-rigid protective layer for dendrite-free lithium metal anode, Adv. Funct. Mater. 28 (8) (2018) 1705838, Copyright (2018) John Wiley & Sons.)*

applications (Fig. 8.13B); (3) poly(vinylidene-*co*-hexafluoropropylene) (PVDF-HFP) and LiF are rationally hybridized into a composite film, which serves as an APL on Li metal anode. The APL possesses favorable attributes including a high mechanical strength, a high Li-ion conductivity, superb shape compliance, and good compatibility with Li metal anode. All these merits are inherited from and synergistically enhanced by PVDF-HFP matrix that is soft and sticky and embedded tough LiF particles to realize uniform Li deposition (Fig. 8.13C).

### 8.4.2.3 Li powder anode

In addition to the above approaches, Li powder anodes also play an important role in the investigation of Li dendrite suppression. Dating back to the early 21st century, Yoon et al. [68,69] reported that the Li dendritic growth was suppressed and the cycling efficiency was improved by about 10% in cells with a Li powder anode. Seong et al. [70] found that the Li dendritic growth was accelerated on the Li foil anode, but suppressed on the Li powder anode after repeated cycling. Moreover, the volume changes and electrolyte loss during cycling were observed to be very small in the cell with Li powder anode [71,72].

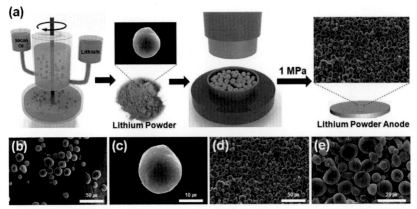

**Fig. 8.14** (A) Schematic illustration of preparation of Li powder anode by droplet emulsion technique (DET). SEM images of (B, C) Li powder (B: X1K, C: X5K) and (D, E) surface of Li powder anode (D: X1K, E: X3K). *(Reproduced with permission from J. Shim, J.W. Lee, K.Y. Bae, H.J. Kim, W.Y. Yoon, J.C. Lee, Dendrite suppression by synergistic combination of solid polymer electrolyte cross-linked with natural terpenes and lithium-powder anode for lithium-metal batteries, ChemSusChem 10 (10) (2017) 2274–2283, Copyright (2017) John Wiley & Sons.)*

Shim et al. [73] suppressed the formation and growth of Li dendrites by utilizing Li powder as the anode materials, prepared by a compression procedure as shown in Fig. 8.14A. The lithium powder anode was obtained by pressing the spherical Li powder particle with average diameter of about 10 nm (Fig. 8.14B and C) on a stainless steel mesh. The spherical shape of the Li powder particles was maintained (Fig. 8.14D and E), producing a porous structure with a larger surface area than that of the conventional Li foil. The increased surface area of the Li powder anode contributes to decreasing the effective current density applied to each Li powder particle compared to Li foil anode; thus, the formation and growth of Li dendrites can be suppressed.

3D anode, Li surface treatment, and Li powder anode are all able to suppress the Li dendrite growth to some extent. Different suppression strategies have their own advantages, while the control of electrochemical deposition of lithium metal is the most important in the suppression of lithium dendrites.

### 8.4.3 Electrolyte
#### 8.4.3.1 Ionic liquid electrolyte
Compared with the enhanced-anode approach, the investigation of improved electrolytes may provide a better protection of Li anodes from

Li dendrite growth in Li-S batteries. Ionic liquid electrolytes (ILEs) possess the ability to tune the formation of SEI layer and consequently improve stability and safety [74–76]. For Li-S batteries, conventional ionic liquid electrolytes mainly include the following: 1-ethyl-3-methylimidazolium bis(trifluoromethanesulfonyl)imide [77], lithium bis(trifluoromethanesulfonyl)imide (LiTFSI) [78], $N$-methyl-$N$-allylpyrrolidinium bis(trifluoromethanesulfonyl)imide (RTIL P1A3TFSI) [79], etc. Such electrolytes are able to suppress the formation of Li dendrites and have further advantageous properties such as nonflammability, negligible vapor pressure, high Li-ion conductivity, and wide electrochemical windows [80,81].

Ma and co-workers [82] reported the application of the ionic liquid electrolyte $N$-methyl-($n$-butyl) pyrrolidinium bis(trifluoromethanesulfonyl) imide (PYR14TFSI) in Li-S batteries for suppressing the growth of Li dendrites. It found that the Li anode cycled in the electrolyte without the PYR14TFSI ionic liquid electrolyte showed loose and dendritic morphology, indicating severe corrosion of Li anode by the shuttled polysulfide species during charge/discharge processes, while PYR14TFSI ionic liquid electrolyte offered better protection of Li anode surface and suppression of Li dendrite formation. Therefore, the higher capacity and the capacity retention were obtained when using the ILE PYR14TFSI in Li-S batteries.

To modify the SEI layer on the Li metal surface, Zheng et al. [21] not only investigated the effects of Pyr14TFSI ion liquid electrolyte on the performance of Li-S battery but also analyzed the properties and morphologies of the SEI layers formed on the Li metal surface in Li-S batteries. When the 75% Pyr14TFSI ionic liquid electrolyte containing electrolyte was used, the electrochemical performances were significantly improved, exhibiting high Coulombic efficiency and very stable cycling stability with high capacity retention of 94.3% capacity fade after 120 cycles. Integrated electrochemical impedance spectroscopy and SEM studies (Fig. 8.15) revealed that the ionic liquid facilitated the formation of a more stable SEI layer with improved quality on the Li metal surface, which wove some $Py14^+$ cations within the reduction products of $TFSI^-$ anion and solvents. This dense SEI layer effectively prevented the continuous penetration of soluble polysulfides into bulk Li, which slowed down the otherwise rapid corrosion of Li and the increase of the impedance. The intensive side reactions between polysulfides and Li metal were largely reduced; therefore, Li dendritic growth was effectively suppressed.

For other case, Yang et al. [83] reported that the LiF-rich SEI can guide the uniform Li deposition due to its low diffusion energy barrier and high

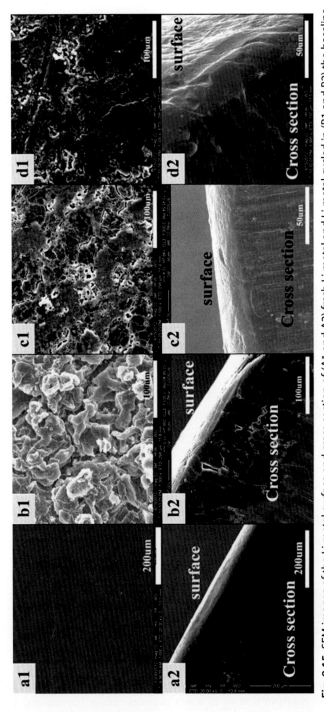

**Fig. 8.15** SEM images of the Li metal surface and cross-section of (A1 and A2) fresh Li metal and Li metal cycled in (B1 and B2) the baseline electrolyte, (C1 and C2) 50% ion liquid electrolyte containing electrolyte, and (D1 and D2) 75% ion liquid electrolyte containing electrolyte for 100 cycles at 0.2C. *(Reproduced with permission from J. Zheng, M. Gu, H. Chen, P. Meduri, M.H. Engelhard, J.-G. Zhang, J. Liu, J. Xiao, Ionic liquid-enhanced solid state electrolyte interface (SEI) for lithium–sulfur batteries, J. Mater. Chem. A 1 (29) (2013) 8464–8470, Copyright (2013) Royal Society of Chemistry.)*

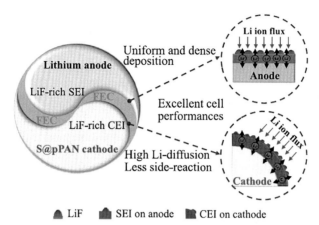

**Fig. 8.16** Schematic illustration of the function of FEC-based electrolyte for robust and elastic interfaces for electrodes. *(Reproduced with permission from J.-H. Yang, A. Naveed, Q. Li, C. Guo, J. Chen, J. Lei, J. Yang, Y. Nuli, J. Wang, Lithium sulfur batteries with compatible electrolyte both for stable cathode and dendrite-free anode, Energy Storage Mater. 15 (2018) 299–307, Copyright (2018) Elsevier.)*

surface diffusivity. The correlation of fluoroethylene carbonate (FEC)-induced SEI both on sulfur composite cathode and on Li metal anode was summarized in Fig. 8.16. Unlike the directed LiF-coated anode and HF pretreatment, the bulky FEC solvent can continuously produce dispersive cross-linked polymers and LiF-rich coating to repair the Li anode after infinite volume change.

### 8.4.3.2 Electrolyte additives

Additives to organic fluid electrolytes (i.e., $LiPF_6$ in ethylene carbonate/dimethyl carbonate/methyl ethyl carbonate and $LiPF_6$ propylene carbonate) may be the most effective method for suppressing the growth of Li dendrites. The organic fluid electrolyte is an important component in the Li-S battery due to their easy availability, wide array of structural variations, excellent wettability to sulfur cathodes and Li anodes, low viscosity to fill in the micropores of the sulfur cathode, relatively high ionic conductivity, low interfacial resistance, reasonable polysulfides solubility, good chemical stability with Li metal anode, and the good electrochemical window [84–87].

For Li-S batteries, carbonate-based solvents are only harnessed when molecules with 2–4 sulfur atoms serve as cathode materials [88,89]. Carbonate-based solvents react with intermediate products like polysulfides

[90]; therefore, most studies focus on seeking other solvents such as ether-based solvents. Hu et al. [91] reported that the ether-based LiTFSI electrolyte greatly enhances the reversibility of the electrochemical reaction by stabilizing the surface/interface of the electrodes. Assembled Li/$FeS_2$ batteries show a discharge capacity of $680\,mAh\,g^{-1}$ at $100\,mA\,g^{-1}$ and a capacity retention of 90% after 100 cycles at $1000\,mA\,g^{-1}$ without lithium dendrite growth.

The ether-based solvents are the most common organic solvents in Li–S batteries because of their stability with the Li metal and their compatibility with elemental sulfur cathodes. The Li salts commonly used in Li–S batteries include LiTFSI, lithium trifluoromethanesulfonate ($LiSO_3CF_3$ or LiOTf), etc. [87,90,92,93].

There are several effects of additives to organic fluid electrolytes on the performance of the Li–S batteries. First, they are able to modify the interface between the electrolyte and the Li anode to protect the Li surface from soluble polysulfides [94,95]. They also can act as catalytic carriers to collect insoluble Li sulfide and disulfide [96]. If additives can improve the stability and uniformity of the interfaces between electrolyte and Li electrode, the dendrite formation and growth would be suppressed accordingly [97,98]. Therefore, various electrolyte additives (such as carbon dioxide/sulfur dioxide, hydrogen fluoride, 2-methylfuran, $LiNO_3$, and $CsNO_3$) [47,99,100] with higher reduction voltages than solvents and salts are often employed to reinforce the interfaces on the surface of Li metal anode. They react with Li anodes quickly, and form a dense and protective interphase; this further minimizes severe parasitic reactions between Li and electrolytes.

Zu et al. [5] demonstrated that the Li surface in the corrosive polysulfide-rich environment could be stabilized by the addition of copper acetate (Fig. 8.17A and B), promising for effective Li protection and a safe, long-term operation of high specific energy Li–S batteries, especially with high sulfur content for practical application. Copper acetate addition improved the physical/chemical characteristics of the Li metal anode and the performance of Li–S batteries. The improved surface morphology and chemistry could be attributed to the in situ chemical formation of a stable passivation film consisting of a good combination of a sulfide matrix and electrolyte decomposition products. The passivation film suppresses Li dendrite initiation by controlling the Li deposition sites.

Li et al. [101] demonstrated that the parasitic reaction between Li polysulfide and Li could be used to effectively suppress the growth of Li dendrites (Fig. 8.18). Both $Li_2S_8$ and $LiNO_3$ were used as additives to the ether-based

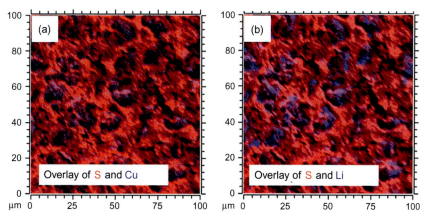

**Fig. 8.17** Time-of-flight secondary ion mass spectrometer (TOFSIMS) chemical mapping images showing (A) the overlay of sulfur and copper (the sulfur signal is in *red* and the copper signal is in *blue*) and (B) the overlay of sulfur and Li (the sulfur signal is in *red* and the Li signal is in *blue*) at the surface of the passivation film after the first charge in the experimental cell. *(Reproduced with permission from C. Zu, A. Manthiram, Stabilized lithium-metal surface in a polysulfide-rich environment of lithium-sulfur batteries, J. Phys. Chem. Lett. 5 (15) (2014) 2522–2527, Copyright (2014) Royal Society of Chemistry.)*

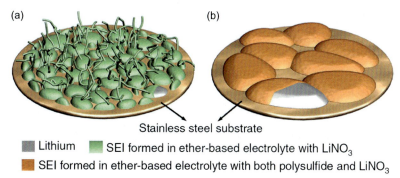

**Fig. 8.18** Schematic illustration showing the morphology difference of Li deposited on the stainless steel substrate in the two electrolytes (both contain Li nitrate), but (A) without Li polysulfide and (B) containing Li polysulfide. *(Reproduced with permission from W. Li, H. Yao, K. Yan, G. Zheng, Z. Liang, Y.M. Chiang, Y. Cui, The synergetic effect of lithium polysulfide and lithium nitrate to prevent lithium dendrite growth, Nature Commun. 6 (2015) 7436, Copyright (2015) Springer Nature.)*

electrolyte. They enabled a synergetic effect leading to the formation of a stable and uniform SEI layer on Li anode surface, which could greatly minimize the electrolyte decomposition and prevent dendrites from shooting out. By simply manipulating the concentrations of $Li_2S_8$ and $LiNO_3$, the

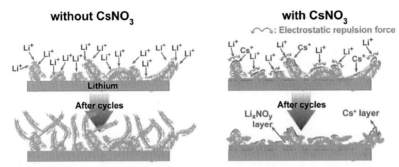

**Fig. 8.19** Addition of CsNO₃ in the electrolyte to suppress Li dendrite growth by electrostatic repulsion between Cs⁺ ion and incoming Li-ion. NO₃⁻ also contributes to formation of stable interfacial layer on the Li metal anode. *(Reproduced with permission from J.S. Kim, T.H. Hwang, B.G. Kim, J. Min, J.W. Choi, A lithium-sulfur battery with a high areal energy density, Adv. Funct. Mater. 24 (34) (2014) 5359–5367, Copyright (2014) John Wiley & Sons.)*

formation of Li dendrites was prevented at a current density of $2\,mA\,cm^{-2}$ up to a deposited areal capacity of $6\,mAh\,cm^{-2}$. Moreover, the Coulombic efficiency was maintained at 99% after 300 cycles at $2\,mA\,cm^{-2}$ with a deposited capacity of $1\,mAh\,cm^{-2}$.

In an effort to mitigate the Li dendrite growth further, Kim et al. [102] added CsNO₃ in the electrolyte; the addition of CsNO₃ was made for a dual purpose (Fig. 8.19). First, Cs ions were expected to remain positively charged even at the voltages where Li-ions were reduced to zero-valent Li. Also, the presence of positively charged Cs ions electrostatically repelled Li-ions and thus prevented deposition of Li metal, especially at sharp tips on the Li surface where electron densities were higher than other locations with flat morphology [102]. On the other hand, $NO_3^-$ was chosen as a counter ion because $NO_3^-$ has been known to assist formation of stable interfacial layers on Li metal surface in Li-S batteries [101].

Li et al. [103] reported a thionyl chloride (SOCl₂) additive to in situ build a stable interfacial protective layer on Li anode surface to prevent Li dendrite growth. Furthermore, the decomposition of SOCl₂ can produce active sulfur to offer extra capacity for cathode in Li-S battery. As-assembled Li-S battery delivered a discharge capacity of $2202.3\,mAh\,g^{-1}$ at $400\,mA\,g^{-1}$ and a capacity of $1348.6\,mAh\,g^{-1}$ at $3000\,mA\,g^{-1}$.

More details of the formation of the stable interfacial protective layer and the related functions are described below. SOCl₂ is easily dissolved in nonpolar solvents. After the redox reaction taking place with Li metal, a solid LiCl-rich film is formed on the metal surface. This reaction occures before the reactions between Li metal and electrolyte, and thus, the stable interfacial protective layer can significantly diminish the decomposition of electrolyte as well as suppress the formation of dendrites and dead Li.

### 8.4.3.3 Novel electrolytes

In addition to the ionic liquid electrolyte and organic electrolyte additive(s), seeking out novel electrolytes with optimized solvents, Li salts, and concentration is an active area of research [39,104]. It indicates that highly concentrated electrolytes, dual-salt electrolytes, and ether-based electrolytes have important role in suppressing Li dendrite growth.

Recently, electrolytes that have high $Li^+$ concentrations (>1 M) have been proposed for the design of electrolytes that promote stable anode cycling, because high concentrations of Li-ions in these electrolytes can facilitate the fast Li deposition/stripping even at high current density conditions [39,105]. Cells that incorporate this kind of electrolytes show unique features, such as enhanced electrolyte stability when in contact with Li metal and decreased dendritic morphology of the deposited Li [106,107]. Togasaki et al. [108] reported a novel electrolyte design for enhancing the cycling performance of Li anodes by using a highly concentrated dimethyl sulfoxide (DMSO)-based electrolyte with the Li salt of $LiNO_3$. The 4.0 M $LiNO_3$/DMSO electrolyte displays enhanced Coulombic efficiency (Fig. 8.20) and stable cycling performance without Li dendrite formation and

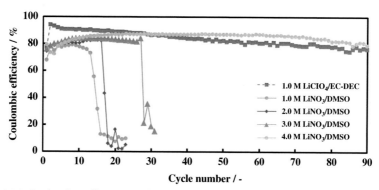

**Fig. 8.20** Coulombic efficiencies of the Li metal anode during cycling at 0.2 mA cm$^{-2}$ in the DMSO-based electrolytes with different $LiNO_3$ concentrations and in the carbonate-based electrolyte 1.0 M $LiClO_4$/EC-DEC. (Reproduced with permission from N. Togasaki, T. Momma, T. Osaka, Enhanced cycling performance of a Li metal anode in a dimethyl sulfoxide-based electrolyte using highly concentrated lithium salt for a lithium − oxygen battery, J. Power Sources 307 (2016) 98–104, Copyright (2016) Elsevier.)

comparable to that of carbonate-based electrolytes. This enhancement is due to the absence of free DMSO solvent in the electrolyte and the promotion of deposition of Li-ions on the solid electrolyte interphase surface, both being consequences of the unique structure of the electrolyte.

Miao et al. [104] developed a dual-salt electrolyte with $Li[N(SO_2F)_2]$ and $Li[N(SO_2CF_3)_2]$, exhibiting the dendrite-free Li deposition at a current density of $10\,mA\,cm^{-2}$. Xiang et al. [109] demonstrated the rapid charging capability of Li-S battery which can be enabled by a dual-salt electrolyte with LiTFSI and lithium bis(oxalato)borate (LiBOB) in a carbonate solvent mixture. At the charging current density of $1.50\,mA\,cm^{-2}$, a Li-S battery with the dual-salt electrolyte delivered a discharge capacity of $131\,mAh\,g^{-1}$ and a capacity retention of 80% after 100 cycles. It is attributed to the good film-formation ability of the electrolyte on the Li metal anode and the highly conductive nature of the sulfur-rich interphase layer.

Pang et al. [110] demonstrated that by tuning the electrolyte bonding structure of organic molecules, the challenges at both electrodes can be tackled simultaneously. Specifically, the sulfur speciation pathway transforms from a dissolution-precipitation route to a quasi-solid state conversion in the presence of a lowered solvent activity and an extended electrolyte network, reducing the need for high electrolyte volumes. With such an optimized structure, the electrolyte allows dendrite-free Li plating and shows a 20-fold reduction in parasitic reactions with Li, which avoids electrolyte consumption and greatly extends the lifetime of a low electrolyte/sulfur ($5\,\mu L\,mg^{-1}$) battery.

### 8.4.3.4 Solid polymer electrolyte

Replacement of solid polymer electrolyte (SPE) for routine liquid electrolyte is highly desirable for high-energy-density Li-S batteries due to its low volatility, high safety, and the ability to suppress shuttle effect and Li dendrite growth. Chen et al. [111] combined cellulose acetate (CA) matrix with bigrafted polysiloxane (BPSO) copolymer, lithium bis(trifluoromethanesulfonyl)imide (LiTFSI), and poly(vinylidene fluoride) (PVDF) to prepare SPE that possesses an ionic conductivity of $4.0 \times 10^{-4}\,S\,cm^{-1}$, a mechanical strength of $6.8\,MPa$, an electrochemical stability window of $4.7\,V$, and an ion transference number of $0.52$ at room temperature. The SPE effectively suppresses the growth of Li dendrites and the diffusion of polysulfides, which is attributed to a synergistic effect of the uniform and stable SEI and the enhanced mechanical strength of SPE (Fig. 8.21). The high-content anions of polymer-in-salt electrolytes easily react with Li metal to produce a stable

**Fig. 8.21** Schematic illustrations of Li plating/stripping behaviors of Li symmetrical cells using 90% (BPSO-150% LiTFSI)-10% PVDF+CA. *(Reproduced with permission from L. Chen, L.Z. Fan, Dendrite-free Li metal deposition in all-solid-state lithium sulfur batteries with polymer-in-salt polysiloxane electrolyte, Energy Storage Mater. 15 (2018) 37–45, Copyright (2018) Elsevier.)*

SEI layer, which contains inorganic components ($Li_2S$, LiF, and $Li_3N$) and has a low resistance to Li-ion transport. The assembled Li-S battery with SPE exhibits good cycling performance at 1C and room temperature.

The four electrolyte strategies considered above have different approaches and advantages. For instance, the ionic liquid electrolyte can tune the SEI layer because of its property advantages, such as nonflammability, negligible vapor pressure, high Li-ion conductivity, and wide electrochemical window. The electrolyte additives are able to improve the stability and uniformity of the interfaces between electrolytes and Li anodes, and their significant advantages are simple and convenient. The novel electrolytes with ether-based solvents, highly concentrated Li salts, and dual salts have the functions to form film on the surface of anode and reduce the side reactions between electrolyte and polysulfides. For solid polymer electrolyte, the synergistic effect of the uniform/stable SEI and the good mechanical strength of SPE play an important role in the suppression of shuttle effect and Li dendrite growth. Therefore, the Li dendrite formation and growth are suppressed effectively by each electrolyte modification.

## 8.5 Conclusions

With the continuous development of energy technology, enhancing the specific energy of energy storage devices has attracted increasing attention. Li-S batteries represented one of the most promising high energy candidates due to the high theoretical capacities of sulfur cathode (1672 mAh g$^{-1}$) and Li anode (3862 mAh g$^{-1}$), which leads to a specific energy of approximately

$2600 \, \mathrm{Wh \, kg^{-1}}$. However, similar to Li–ion batteries, Li dendrite growth is unavoidable when Li metal is used as the anode, which also leads to lower Coulombic efficiency and safety issue. Therefore, the failure mechanisms of Li-S batteries are investigated to search suppression methods of Li dendrite growth.

On this basis, various strategies are employed to suppress the Li dendrite growth, which could be summarized via three routes: separator, anode, and electrolyte. For separator, it could prevent the growths of Li dendrites by using the 3D structure of the separators. However, it needs more effort to seek separators for practical applications. For anode strategies, the 3D anodes suppress the growth of Li dendrite due to their 3D structures. Similar to the 3D structure of the separator, it mainly slows down the orientation of free Li crystals. Moreover, Li anode surface treatment is the effective and simple strategy for fundamental limiting the growth of Li dendrite due to the electrochemical deposition of Li metal at the surface treatment layer. Li powder anode suppresses the formation and growth of Li dendrites, because its increased surface area decreases the effective current density. Compared with the anode, electrolyte modification is simpler and more convenient for solving the safety issues caused by the Li dendrite growth. Use of an ionic liquid electrolyte is able to prevent the formation of Li dendrite because it can modify the SEI layer for controlling the Li deposition. The electrolyte additives can improve the stability and uniformity of the interfaces between electrolyte and Li anode by acting as catalytic carriers and modifying the interface; therefore, the dendrite formation and growth are suppressed effectively. The novel electrolytes have the functions to form film on the surface of anode and reduce side reactions between electrolyte and polysulfides. To summarize, electrolyte modification mainly promotes the in situ formation of film on the Li anode surface by regulating their own components. Therefore, it is easier to suppress the Li dendrite growth compared with anode modification. The comparisons of typical examples indicate that different strategies possess their own advantages in performance enhancement. Hence, finding effective processing approaches also needs to take into account the cost and simplicity of the processes, and these will be practical concerns for Li-S battery researchers.

In summary, recent studies have demonstrated that Li-S batteries are promising potential candidates for energy storage due to their high specific energy. The consensus of the literature is that Li dendrite growth is unavoidable when Li metal was used as the anode. In this chapter, we summarized various strategies to address failure mechanisms of Li-S batteries. However, there are still some challenges that must be addressed. First, theoretical modeling of dendrite growth on the surface of the Li metal in the Li–S batteries needs to be improved to better understand the effects of battery cycling parameters on the morphology and size of Li microstructures (e.g., dendrite tip radius, dendrite arm spacing) and develop mitigation strategies/approaches to suppress the growth of Li dendrites. Second, there is an urgent need to find cost-effective and efficient surface treatments to modify the surface of Li metal anode. Third, it is necessary to investigate electrolytes with abilities to catalyze Li dendrite dissolution, thus to fundamentally suppress the dendrite growth. Also, the application of SPEs in suppressing Li dendrites needs to be further studied. Finally, future materials investigations must be focused on approaches to improve the performance, longevity, and hence practicality of Li–S batteries. With further and more comprehensive materials research, we believe that Li–S batteries have great potential for numerous practical applications for high-energy devices.

## References

[1] S. Urbonaite, T. Poux, P. Novák, Progress towards commercially viable Li-S battery cells, Adv. Energy Mater. 5 (16) (2015) 1500118.

[2] X. Xu, S. Wang, H. Wang, C. Hu, Y. Jin, J. Liu, H. Yan, Recent progresses in the suppression method based on the growth mechanism of lithium dendrite, J. Energy Chem. 27 (2) (2018) 513–527.

[3] T. Tatsuma, M. Taguchi, N. Oyama, Inhibition effect of covalently cross-linked gel electrolytes on lithium dendrite formation, Electrochim. Acta 46 (2001) 1201–1205.

[4] T.A. Pascal, K.H. Wujcik, D.R. Wang, N.P. Balsara, D. Prendergast, Thermodynamic origins of the solvent-dependent stability of lithium polysulfides from first principles, Phys. Chem. Chem. Phys. 19 (2) (2017) 1441–1448.

[5] C. Zu, A. Manthiram, Stabilized lithium-metal surface in a polysulfide-rich environment of lithium-sulfur batteries, J. Phys. Chem. Lett. 5 (15) (2014) 2522–2527.

[6] H.J. Peng, J.Y. Liang, L. Zhu, J.Q. Huang, X.B. Cheng, X.F. Guo, W.P. Ding, W.C. Zhu, Q. Zhang, Catalytic self-limited assembly at hard templates: a mesoscale approach to graphene nanoshells for lithium-sulfur batteries, ACS Nano 8 (11) (2014) 11280–11289.

[7] C. Huang, J. Xiao, Y. Shao, J. Zheng, W.D. Bennett, D. Lu, L.V. Saraf, M. Engelhard, L. Ji, J. Zhang, X. Li, G.L. Graff, J. Liu, Manipulating surface reactions in lithium-sulphur batteries using hybrid anode structures, Nat. Commun. 5 (2014) 3015.

[8] R. Tao, X. Bi, S. Li, Y. Yao, F. Wu, Q. Wang, C. Zhang, J. Lu, Kinetics tuning the electrochemistry of lithium dendrites formation in lithium batteries through electrolytes, ACS Appl. Mater. Interfaces 9 (8) (2017) 7003–7008.

[9] X.B. Cheng, Q. Zhang, Dendrite-free lithium metal anodes: stable solid electrolyte interphases for high-efficiency batteries, J. Mater. Chem. A 3 (14) (2015) 7207–7209.

[10] X.B. Cheng, H.J. Peng, J.Q. Huang, F. Wei, Q. Zhang, Dendrite-free nanostructured anode: entrapment of lithium in a 3D fibrous matrix for ultra-stable lithium-sulfur batteries, Small 10 (21) (2014) 4257–4263.

[11] M. Arakawa, S. Tobishima, Y. Nemoto, M. Ichimura, Lithium electrode cycleability and morphology dependence on current density, J. Power Sources 43–44 (1993) 27–35.

[12] X.B. Cheng, T.Z. Hou, R. Zhang, H.J. Peng, C.Z. Zhao, J.Q. Huang, Q. Zhang, Dendrite-free lithium deposition induced by uniformly distributed lithium ions for efficient lithium metal batteries, Adv. Mater. 28 (15) (2016) 2888–2895.

[13] E. Paled, The electrochemical behavior of alkali and alkaline earth metals in nonaqueous battery systems—the solid electrolyte interphase model, J. Electrochem. Soc. 126 (12) (1979) 2047–2051.

[14] X.B. Cheng, R. Zhang, C.Z. Zhao, F. Wei, J.G. Zhang, Q. Zhang, A review of solid electrolyte interphases on lithium metal anode, Adv. Sci. 3 (3) (2016) 1500213.

[15] H. Dong, X. Xiao, C. Jin, X. Wang, P. Tang, C. Wang, Y. Yin, D. Wang, S. Yang, C. Wu, High lithium-ion conductivity polymer film to suppress dendrites in Li metal batteries, J. Power Sources 423 (2019) 72–79.

[16] M.K. Datta, B. Gattu, R. Kuruba, P.M. Shanthi, P.N. Kumta, Constitutional underpotential plating (CUP)—new insights for predicting the morphological stability of deposited lithium anodes in lithium metal batteries, J. Power Sources 467 (2020) 228243.

[17] Y. Okajima, Y. Shibuta, T. Suzuki, A phase-field model for electrode reactions with Butler–Volmer kinetics, Comput. Mater. Sci. 50 (1) (2010) 118–124.

[18] J. Yamaki, S. Tobishima, K. Hayashi, K. Saito, Y. Nemoto, M. Arakawa, A consideration of the morphology of electrochemically deposited lithium in an organic electrolyte, J. Power Sources 74 (1998) 219–227.

[19] H.J. Chang, A.J. Ilott, N.M. Trease, M. Mohammadi, A. Jerschow, C.P. Grey, Correlating microstructural lithium metal growth with electrolyte salt depletion in lithium batteries using $^{7}$Li MRI, J. Am. Chem. Soc. 137 (48) (2015) 15209–15216.

[20] X. Zhang, W. Wang, A. Wang, Y. Huang, K. Yuan, Z. Yu, J. Qiu, Y. Yang, Improved cycle stability and high security of Li–B alloy anode for lithium–sulfur battery, J. Mater. Chem. A 2 (30) (2014) 11660–11665.

[21] J. Zheng, M. Gu, H. Chen, P. Meduri, M.H. Engelhard, J.-G. Zhang, J. Liu, J. Xiao, Ionic liquid-enhanced solid state electrolyte interface (SEI) for lithium–sulfur batteries, J. Mater. Chem. A 1 (29) (2013) 8464–8470.

[22] S.S. Zhang, J.A. Read, A new direction for the performance improvement of rechargeable lithium/sulfur batteries, J. Power Sources 200 (2012) 77–82.

[23] D. Aurbach, E. Zinigrad, H. Teller, P. Dan, Factors which limit the cycle life of rechargeable lithium (metal) batteries, J. Electrochem. Soc. 147 (4) (2000) 1274–1279.

[24] R. Elazari, G. Salitra, A. Garsuch, A. Panchenko, D. Aurbach, Sulfur-impregnated activated carbon fiber cloth as a binder-free cathode for rechargeable Li-S batteries, Adv. Mater. 23 (47) (2011) 5641–5644.

[25] S.S. Zhang, Role of $LiNO_3$ in rechargeable lithium/sulfur battery, Electrochim. Acta 70 (2012) 344–348.

[26] N. Jayaprakash, J. Shen, S.S. Moganty, A. Corona, L.A. Archer, Porous hollow carbon@sulfur composites for high-power lithium-sulfur batteries, Angew. Chem. 50 (26) (2011) 5904–5908.

[27] N. Deng, W. Kang, Y. Liu, J. Ju, D. Wu, L. Li, B.S. Hassan, B. Cheng, A review on separators for lithium sulfur battery: progress and prospects, J. Power Sources 331 (2016) 132–155.

[28] X. Qian, L. Jin, D. Zhao, X. Yang, S. Wang, X. Shen, D. Rao, S. Yao, Y. Zhou, X. Xi, Ketjen black-MnO composite coated separator for high performance rechargeable lithium–sulfur battery, Electrochim. Acta 192 (2016) 346–356.

[29] M.H. Ryou, Y.M. Lee, Y. Lee, M. Winter, P. Bieker, Mechanical surface modification of lithium metal: towards improved Li metal anode performance by directed Li plating, Adv. Funct. Mater. 25 (6) (2015) 834–841.

[30] Z. Liang, G. Zheng, C. Liu, N. Liu, W. Li, K. Yan, H. Yao, P.C. Hsu, S. Chu, Y. Cui, Polymer nanofiber-guided uniform lithium deposition for battery electrodes, Nano Lett. 15 (5) (2015) 2910–2916.

[31] M.S. Whittingham, Lithium batteries and cathode materials, Chem. Rev. 104 (2004) 4271–4301.

[32] T.J. Richardson, G. Chen, Solid solution lithium alloy cermet anodes, J. Power Sources 174 (2) (2007) 810–812.

[33] Y. Yang, G.H. Yu, J.J. Cha, H. Wu, M. Vosgueritchian, Y. Yao, Z.N. Bao, Y. Cui, Improving the performance of lithium-sulfur batteries by conductive polymer coating, ACS Nano 5 (11) (2011) 9187–9193.

[34] W. Liu, W. Li, D. Zhuo, G. Zheng, Z. Lu, K. Liu, Y. Cui, Core-shell nanoparticle coating as an interfacial layer for dendrite-free lithium metal anodes, ACS Cent. Sci. 3 (2) (2017) 135–140.

[35] K. Xu, Electrolytes and interphases in Li-ion batteries and beyond, Chem. Rev. 114 (23) (2014) 11503–11618.

[36] Z. Tu, Y. Kambe, Y. Lu, L.A. Archer, Nanoporous polymer-ceramic composite electrolytes for lithium metal batteries, Adv. Energy Mater. 4 (2) (2014) 1300654.

[37] G. Zhou, L. Li, D.W. Wang, X.Y. Shan, S. Pei, F. Li, H.M. Cheng, A flexible sulfurgraphene-polypropylene separator integrated electrode for advanced Li-S batteries, Adv. Mater. 27 (4) (2015) 641–647.

[38] K. Yan, H.W. Lee, T. Gao, G. Zheng, H. Yao, H. Wang, Z. Lu, Y. Zhou, Z. Liang, Z. Liu, S. Chu, Y. Cui, Ultrathin two-dimensional atomic crystals as stable interfacial layer for improvement of lithium metal anode, Nano Lett. 14 (10) (2014) 6016–6022.

[39] J. Qian, W.A. Henderson, W. Xu, P. Bhattacharya, M. Engelhard, O. Borodin, J.G. Zhang, High rate and stable cycling of lithium metal anode, Nat. Commun. 6 (2015) 6362.

[40] R. Khurana, J.L. Schaefer, L.A. Archer, G.W. Coates, Suppression of lithium dendrite growth using cross-linked polyethylene/poly(ethylene oxide) electrolytes: a new approach for practical lithium-metal polymer batteries, J. Am. Chem. Soc. 136 (20) (2014) 7395–7402.

[41] X. Huang, Separator technologies for lithium-ion batteries, J. Solid State Electrochem. 15 (4) (2010) 649–662.

[42] P. Arora, Z.M. (John) Zhang, Battery separators, Chem. Rev. 104 (2004) 4419–4462.

[43] S.S. Zhang, A review on the separators of liquid electrolyte Li-ion batteries, J. Power Sources 164 (1) (2007) 351–364.

[44] K. Xie, W. Wei, K. Yuan, W. Lu, M. Guo, Z. Li, Q. Song, X. Liu, J.G. Wang, C. Shen, Toward dendrite-free lithium deposition via structural and interfacial synergistic effects of 3D graphene@Ni scaffold, ACS Appl. Mater. Interfaces 8 (39) (2016) 26091–26097.

[45] M. Nagao, Y. Imade, H. Narisawa, T. Kobayashi, R. Watanabe, T. Yokoi, T. Tatsumi, R. Kanno, All-solid-state Li–sulfur batteries with mesoporous electrode and thio-LISICON solid electrolyte, J. Power Sources 222 (2013) 237–242.

[46] K. Takada, Progress and prospective of solid-state lithium batteries, Acta Mater. 61 (3) (2013) 759–770.

[47] W. Xu, J. Wang, F. Ding, X. Chen, E. Nasybulin, Y. Zhang, J.-G. Zhang, Lithium metal anodes for rechargeable batteries, Energ. Environ. Sci. 7 (2) (2014) 513–537.

[48] Q. Wang, J. Jin, X. Wu, G. Ma, J. Yang, Z. Wen, A shuttle effect free lithium sulfur battery based on a hybrid electrolyte, Phys. Chem. Chem. Phys. 16 (39) (2014) 21225–21229.

[49] W.G. Chong, J.-Q. Huang, Z.-L. Xu, X. Qin, X. Wang, J.-K. Kim, Lithium-sulfur battery cable made from ultralight, flexible graphene/carbon nanotube/sulfur composite fibers, Adv. Funct. Mater. 27 (4) (2017) 1604815.

[50] J.W. Fergus, Ceramic and polymeric solid electrolytes for lithium-ion batteries, J. Power Sources 195 (15) (2010) 4554–4569.

[51] M. Rao, X. Geng, X. Li, S. Hu, W. Li, Lithium-sulfur cell with combining carbon nanofibers–sulfur cathode and gel polymer electrolyte, J. Power Sources 212 (2012) 179–185.

[52] B.C. Yu, K. Park, J.-H. Jang, J.B. Goodenough, Cellulose-based porous membrane for suppressing Li dendrite formation in lithium–sulfur battery, ACS Energy Lett. 1 (3) (2016) 633–637.

[53] J.A. Lewis, J. Tippens, F.J.Q. Cortes, M.T. McDowell, Chemo-mechanical challenges in solid-state batteries, Trends Chem. 1 (9) (2019) 845–857.

[54] R. Sudo, Y. Nakata, K. Ishiguro, M. Matsui, A. Hirano, Y. Takeda, O. Yamamoto, N. Imanishi, Interface behavior between garnet-type lithium-conducting solid electrolyte and lithium metal, Solid State Ion. 262 (2014) 151–154.

[55] X. Pu, G. Yang, C. Yu, Safe and reliable operation of sulfur batteries with lithiated silicon, Nano Energy 9 (2014) 318–324.

[56] H. Jha, I. Buchberger, X. Cui, S. Meini, H.A. Gasteiger, Li-S batteries with $Li_2S$ cathodes and Si/C anodes, J. Electrochem. Soc. 162 (9) (2015) A1829–A1835.

[57] H.S. Kim, T.-G. Jeong, Y.-T. Kim, Electrochemical properties of lithium sulfur battery with silicon anodes lithiated by direct contact method, J. Electrochem. Sci. Technol. 7 (3) (2016) 228–233.

[58] J. Hassoun, J. Kim, D.-J. Lee, H.-G. Jung, S.-M. Lee, Y.-K. Sun, B. Scrosati, A contribution to the progress of high energy batteries: a metal-free, lithium-ion, silicon–sulfur battery, J. Power Sources 202 (2012) 308–313.

[59] Y. Yang, M.T. McDowell, A. Jackson, J.J. Cha, S.S. Hong, Y. Cui, New nanostructured $Li_2S$/silicon rechargeable battery with high specific energy, Nano Lett. 10 (4) (2010) 1486–1491.

[60] M. Hagen, E. Quiroga-González, S. Dörfler, G. Fahrer, J. Tübke, M.J. Hoffmann, H. Althues, R. Speck, M. Krampfert, S. Kaskel, H. Föll, Studies on preventing Li dendrite formation in Li–S batteries by using pre-lithiated Si microwire anodes, J. Power Sources 248 (2014) 1058–1066.

[61] S.S. Chi, Y. Liu, W.-L. Song, L.-Z. Fan, Q. Zhang, Prestoring lithium into stable 3D nickel foam host as dendrite-free lithium metal anode, Adv. Funct. Mater. 27 (24) (2017) 1700348.

[62] Z. Liang, D.C. Lin, J. Zhao, Z.D. Lu, Y.Y. Liu, C. Liu, Y.Y. Lu, H.T. Wang, K. Yan, X.Y. Tao, Y. Cui, Composite lithium metal anode by melt infusion of lithium into a 3D conducting scaffold with lithiophilic coating, Proc. Natl. Acad. Sci. U. S. A. 113 (11) (2016) 201518188.

[63] A. Zhang, X. Fang, C. Shen, Y. Liu, C. Zhou, A carbon nanofiber network for stable lithium metal anodes with high coulombic efficiency and long cycle life, Nano Res. 9 (11) (2016) 3428–3436.

[64] Y. Liu, D. Lin, Z. Liang, J. Zhao, K. Yan, Y. Cui, Lithium-coated polymeric matrix as a minimum volume-change and dendrite-free lithium metal anode, Nat. Commun. 7 (2016) 10992.

[65] H. Lee, D.J. Lee, Y.-J. Kim, J.-K. Park, H.-T. Kim, A simple composite protective layer coating that enhances the cycling stability of lithium metal batteries, J. Power Sources 284 (2015) 103–108.

[66] G. Ma, Z. Wen, M. Wu, C. Shen, Q. Wang, J. Jin, X. Wu, A lithium anode protection guided highly-stable lithium-sulfur battery, Chem. Commun. 50 (91) (2014) 14209–14212.

[67] R. Xu, X.Q. Zhang, X.B. Cheng, H.J. Peng, C.Z. Zhao, C. Yan, J.Q. Huang, Artificial soft-rigid protective layer for dendrite-free lithium metal anode, Adv. Funct. Mater. 28 (8) (2018) 1705838.

[68] J.S. Kim, W.Y. Yoon, Improvement in lithium cycling efficiency by using lithium powder anode, Electrochim. Acta 50 (2–3) (2004) 531–534.

Suppression of lithium dendrite growth **293**

[69] W.S. Kim, W.Y. Yoon, Observation of dendritic growth on Li powder anode using optical cell, Electrochim. Acta 50 (2–3) (2004) 541–545.

[70] I.W. Seong, C.H. Hong, B.K. Kim, W.Y. Yoon, The effects of current density and amount of discharge on dendrite formation in the lithium powder anode electrode, J. Power Sources 178 (2) (2008) 769–773.

[71] J.H. Chung, W.S. Kim, W.Y. Yoon, S.W. Min, B.W. Cho, Electrolyte loss and dimensional change of the negative electrode in Li powder secondary cell, J. Power Sources 163 (1) (2006) 191–195.

[72] J.S. Kim, W.Y. Yoon, K.Y. Yi, B.K. Kim, B.W. Cho, The dissolution and deposition behavior in lithium powder electrode, J. Power Sources 165 (2) (2007) 620–624.

[73] J. Shim, J.W. Lee, K.Y. Bae, H.J. Kim, W.Y. Yoon, J.C. Lee, Dendrite suppression by synergistic combination of solid polymer electrolyte cross-linked with natural terpenes and lithium-powder anode for lithium-metal batteries, ChemSusChem 10 (10) (2017) 2274–2283.

[74] J.W. Park, K. Ueno, N. Tachikawa, K. Dokko, M. Watanabe, Ionic liquid electrolytes for lithium–sulfur batteries, J. Phys. Chem. C 117 (40) (2013) 20531–20541.

[75] K. Ueno, J.W. Park, A. Yamazaki, T. Mandai, N. Tachikawa, K. Dokko, M. Watanabe, Anionic effects on solvate ionic liquid electrolytes in rechargeable lithium–sulfur batteries, J. Phys. Chem. C 117 (40) (2013) 20509–20516.

[76] N.W. Li, Y.X. Yin, J.Y. Li, C.H. Zhang, Y.G. Guo, Passivation of lithium metal anode via hybrid ionic liquid electrolyte toward stable Li plating/stripping, Adv. Sci. 4 (2) (2017) 1600400.

[77] J. Wang, S.Y. Chew, Z.W. Zhao, S. Ashraf, D. Wexler, J. Chen, S.H. Ng, S.L. Chou, H.K. Liu, Sulfur–mesoporous carbon composites in conjunction with a novel ionic liquid electrolyte for lithium rechargeable batteries, Carbon 46 (2) (2008) 229–235.

[78] X.G. Sun, X. Wang, R.T. Mayes, S. Dai, Lithium-sulfur batteries based on nitrogen-doped carbon and an ionic-liquid electrolyte, ChemSusChem 5 (10) (2012) 2079–2085.

[79] Y. Yan, Y.X. Yin, S. Xin, J. Su, Y.G. Guo, L.J. Wan, High-safety lithium-sulfur battery with prelithiated Si/C anode and ionic liquid electrolyte, Electrochim. Acta 91 (2013) 58–61.

[80] J. Jin, H.H. Li, J.P. Wei, X.K. Bian, Z. Zhou, J. Yan, Li/LiFePO$_4$ batteries with room temperature ionic liquid as electrolyte, Electrochem. Commun. 11 (7) (2009) 1500–1503.

[81] Y.X. Yin, S. Xin, L.J. Wan, C.J. Li, Y.G. Guo, Electrospray synthesis of silicon/carbon nanoporous microspheres as improved anode materials for lithium-ion batteries, J. Phys. Chem. C 115 (29) (2011) 14148–14154.

[82] G. Ma, Z. Wen, J. Jin, M. Wu, G. Zhang, X. Wu, J. Zhang, The enhanced performance of Li–S battery with P14YRTFSI-modified electrolyte, Solid State Ion. 262 (2014) 174–178.

[83] H. Yang, A. Naveed, Q. Li, C. Guo, J. Chen, J. Lei, J. Yang, Y. Nuli, J. Wang, Lithium sulfur batteries with compatible electrolyte both for stable cathode and dendrite-free anode, Energy Storage Mater. 15 (2018) 299–307.

[84] M. Barghamadi, A. Kapoor, C. Wen, A review on Li-S batteries as a high efficiency rechargeable lithium battery, J. Electrochem. Soc. 160 (8) (2013) A1256–A1263.

[85] L. Chen, L.L. Shaw, Recent advances in lithium–sulfur batteries, J. Power Sources 267 (2014) 770–783.

[86] G. Xu, B. Ding, J. Pan, P. Nie, L. Shen, X. Zhang, High performance lithium–sulfur batteries: advances and challenges, J. Mater. Chem. A 2 (32) (2014) 12662–12676.

[87] J. Scheers, S. Fantini, P. Johansson, A review of electrolytes for lithium–sulphur batteries, J. Power Sources 255 (2014) 204–218.

[88] S. Xin, L. Gu, N.H. Zhao, Y.X. Yin, L.J. Zhou, Y.G. Guo, L.J. Wan, Smaller sulfur molecules promise better lithium-sulfur batteries, J. Am. Chem. Soc. 134 (45) (2012) 18510–18513.

[89] K. Zhang, G.H. Lee, M. Park, W. Li, Y.M. Kang, Recent developments of the lithium metal anode for rechargeable non-aqueous batteries, Adv. Energy Mater. 6 (20) (2016) 1600811.

[90] T. Yim, M.S. Park, J.S. Yu, K.J. Kim, K.Y. Im, J.H. Kim, G. Jeong, Y.N. Jo, S.G. Woo, K.S. Kang, I. Lee, Y.J. Kim, Effect of chemical reactivity of polysulfide toward carbonate-based electrolyte on the electrochemical performance of Li–S batteries, Electrochim. Acta 107 (2013) 454–460.

[91] Z. Hu, K. Zhang, Z. Zhu, Z. Tao, J. Chen, $FeS_2$ microspheres with an ether-based electrolyte for high-performance rechargeable lithium batteries, J. Mater. Chem. A 3 (24) (2015) 12898–12904.

[92] J. Gao, M.A. Lowe, Y. Kiya, H.D. Abruña, Effects of liquid electrolytes on the charge–discharge performance of rechargeable lithium/sulfur batteries: electrochemical and in-situ X-ray absorption spectroscopic studies, J. Phys. Chem. C 115 (50) (2011) 25132–25137.

[93] S.S. Zhang, Liquid electrolyte lithium/sulfur battery: fundamental chemistry, problems, and solutions, J. Power Sources 231 (2013) 153–162.

[94] S. Xiong, X. Kai, X. Hong, Y. Diao, Effect of LiBOB as additive on electrochemical properties of lithium–sulfur batteries, Ionics 18 (3) (2011) 249–254.

[95] X. Liang, Z. Wen, Y. Liu, M. Wu, J. Jin, H. Zhang, X. Wu, Improved cycling performances of lithium sulfur batteries with $LiNO_3$-modified electrolyte, J. Power Sources 196 (22) (2011) 9839–9843.

[96] B.A. Trofimov, M.V. Markova, L.V. Morozova, G.F. Prozorova, S.A. Korzhova, M.-D. Cho, V.V. Annenkov, A.B.I. Mikhaleva, Protected bis(hydroxyorganyl) polysulfides as modifiers of Li/S battery electrolyte, Electrochim. Acta 56 (5) (2011) 2458–2463.

[97] T.T. Tung, M.J. Nine, M. Krebsz, T. Pasinszki, C.J. Coghlan, D.N.H. Tran, D. Losic, Recent advances in sensing applications of graphene assemblies and their composites, Adv. Funct. Mater. 27 (46) (2017) 1702891.

[98] Z. Zeng, W.I. Liang, H.G. Liao, H.L. Xin, Y.H. Chu, H. Zheng, Visualization of electrode-electrolyte interfaces in $LiPF_6$/EC/DEC electrolyte for lithium ion batteries via in situ TEM, Nano Lett. 14 (4) (2014) 1745–1750.

[99] J.S. Kim, D.J. Yoo, J. Min, R.A. Shakoor, R. Kahraman, J.W. Choi, Poreless separator and electrolyte additive for lithium-sulfur batteries with high areal energy densities, ChemNanoMat 1 (4) (2015) 240–245.

[100] T.Y. Ding, In-Situ Optical Microscopic Investigation of the Dendrite Formation on Lithium Anode Under Different Electrolyte Conditions in Li-S Battery (Theses and dissertations), 2016, p. 1359.

[101] W. Li, H. Yao, K. Yan, G. Zheng, Z. Liang, Y.M. Chiang, Y. Cui, The synergetic effect of lithium polysulfide and lithium nitrate to prevent lithium dendrite growth, Nat. Commun. 6 (2015) 7436.

[102] J.S. Kim, T.H. Hwang, B.G. Kim, J. Min, J.W. Choi, A lithium-sulfur battery with a high areal energy density, Adv. Funct. Mater. 24 (34) (2014) 5359–5367.

[103] S. Li, H. Dai, Y. Li, C. Lai, J. Wang, F. Huo, C. Wang, Designing Li-protective layer via $SOCl_2$ additive for stabilizing lithium-sulfur battery, Energy Storage Mater. 18 (2019) 222–228.

[104] R. Miao, J. Yang, X. Feng, H. Jia, J. Wang, Y. Nuli, Novel dual-salts electrolyte solution for dendrite-free lithium-metal based rechargeable batteries with high cycle reversibility, J. Power Sources 271 (2014) 291–297.

[105] Y. Yamada, K. Furukawa, K. Sodeyama, K. Kikuchi, M. Yaegashi, Y. Tateyama, A. Yamada, Unusual stability of acetonitrile-based superconcentrated electrolytes for fast-charging lithium-ion batteries, J. Am. Chem. Soc. 136 (13) (2014) 5039–5046.

[106] S.K. Jeong, H.Y. Seo, D.H. Kim, H.K. Han, J.G. Kim, Y.B. Lee, Y. Iriyama, T. Abe, Z. Ogumi, Suppression of dendritic lithium formation by using concentrated electrolyte solutions, Electrochem. Commun. 10 (4) (2008) 635–638.

[107] L. Suo, Y.S. Hu, H. Li, M. Armand, L. Chen, A new class of solvent-in-salt electrolyte for high-energy rechargeable metallic lithium batteries, Nat. Commun. 4 (2013) 1481.

[108] N. Togasaki, T. Momma, T. Osaka, Enhanced cycling performance of a Li metal anode in a dimethyl sulfoxide-based electrolyte using highly concentrated lithium salt for a lithium − oxygen battery, J. Power Sources 307 (2016) 98–104.

[109] H. Xiang, P. Shi, P. Bhattacharya, X. Chen, D. Mei, M.E. Bowden, J. Zheng, J.G. Zhang, W. Xu, Enhanced charging capability of lithium metal batteries based on lithium bis(trifluoromethanesulfonyl)imide-lithium bis(oxalato)borate dual-salt electrolytes, J. Power Sources 318 (2016) 170–177.

[110] Q. Pang, A. Shyamsunder, B. Narayanan, C.Y. Kwok, L.A. Curtiss, L.F. Nazar, Tuning the electrolyte network structure to invoke quasi-solid state sulfur conversion and suppress lithium dendrite formation in Li–S batteries, Nat. Energy 3 (9) (2018) 783–791.

[111] L. Chen, L.Z. Fan, Dendrite-free Li metal deposition in all-solid-state lithium sulfur batteries with polymer-in-salt polysiloxane electrolyte, Energy Storage Mater. 15 (2018) 37–45.

# CHAPTER 9

# The role of advanced host materials and binders for improving lithium-sulfur battery performance

**Shahid Hussain[a,\*], Naseem Akhtar[b,\*], Awais Ahmad[c], Muhammad Khurram Tufail[d], Muhammad Kashif Aslam[e], Muhammad Sufyan Javed[a], and Xiangzhao Zhang[a]**

[a]School of Materials Science and Engineering, Jiangsu University, Zhenjiang, China
[b]College of Materials Science and Engineering, Beijing University of Chemical Technology, Beijing, China
[c]Department of Chemistry, The University of Lahore, Lahore, Pakistan
[d]Key Laboratory of Cluster Science of Ministry of Education Beijing Key Laboratory of Photoelectronic/ Electrophotonic Conversion Materials, School of Chemistry and Chemical Engineering, Beijing Institute of Technology, Beijing, China
[e]Faculty of Materials and Energy, Southwest University, Chongqing, China

## Contents

| | |
|---|---|
| 9.1 Introduction to energy sources and rechargeable batteries | 298 |
| 9.2 Complex energy storage challenges and solutions | 298 |
| 9.3 Host materials | 300 |
|    9.3.1 Three-dimensional graphene hollow spheres | 301 |
|    9.3.2 Reduced graphene oxide nanocomposite/nitrogen-doped carbon framework | 302 |
|    9.3.3 Three-dimensional porous carbon composites | 303 |
|    9.3.4 Micro-mesoporous graphitic carbon spheres | 304 |
|    9.3.5 Carbon nanotube cathodes | 305 |
|    9.3.6 Hierarchical network macrostructure | 309 |
|    9.3.7 In situ wrapping process | 313 |
| 9.4 Binders | 314 |
|    9.4.1 Multifunctional polar binder | 314 |
|    9.4.2 Polyamidoamine dendrimer-based binders | 317 |
|    9.4.3 PAA/PEDOT: PSS as a functional binder | 319 |
| 9.5 Conclusions and future directions | 322 |
| References | 322 |

[\*]Authors contributed equally.

*Lithium-Sulfur Batteries*
https://doi.org/10.1016/B978-0-12-819676-2.00013-X

Copyright © 2022 Elsevier Inc.
All rights reserved.

## 9.1 Introduction to energy sources and rechargeable batteries

The growing concern of global warming and air pollution as a result of the use of nonrenewable fossil fuels requires the development of alternative sustainable and renewable energy sources, such as wind and solar power [1]. The function of advanced energy storage technologies including renewable energy generation and hybrid automobile systems becomes increasingly vital as a way of mitigating global warming. An energy storage device is used to store and release or generate electrical energy when needed [2]. To generate intermittent power, those technologies require reliable, low-value, environmentally friendly, and large-scale energy storage infrastructures.

There is no question that the development of superior power storage devices with higher energy density is vital for improved quality of life in the future [3], as shown in Fig. 9.1. Among many prospective candidates for high-energy storage systems of the following decade, metal-sulfur energy storage devices with lithium-sulfur (Li–S), magnesium-sulfur (Mg–S), and sodium-sulfur (Na–S) cell chemistries carry high theoretical power and energy density, making them specifically appropriate [4]. Among these, Li—S battery has the very best theoretical energy density of $2567\,W\,h\,kg^{-1}$, calculated on the concept of the Li anode (nearly $3860\,mAh\,g^{-1}$) and the S electrode (almost $1675\,mAh\,g^{-1}$), rendering it a promising choice for rechargeable high-energy batteries to be produced afterward (Fig. 9.2).

## 9.2 Complex energy storage challenges and solutions

The Li—S battery mechanism is demonstrated in Fig. 9.3. In order to achieve practical implementation, many challenging problems need to be tackled and overcome. The main issues preventing lithium-sulfur batteries from wide commercialization are (i) insulating nature of active material [5,6], which inhibits the transportation of electrons in the sulfur cathode and leads to low active material utilization; (ii) polysulfide dissolution [7], which aggravates the shuttle phenomenon when active PS species commute between the cathode and anode, resulting in low coulombic efficiency and substantial mass loss of the cathode material and inducing continuous side reactions on the surface of electrodes causing a drastic decay of the electric potential [8]; (iii) active material's volume fluctuations during electrochemical cycling leading to the integrity failure of the composite electrode [9]. To date, significant research efforts have been undertaken to resolve the previously discussed issues and fulfill our future society's energy needs by developing the

Advanced host materials and binders for improved perfomance 299

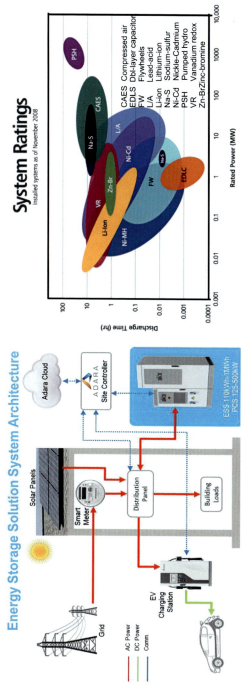

Fig. 9.1 Energy consumption and rechargeable storage system. (Reprinted with permission from S. Zhang, K. Ueno, K. Dokko, M. Watanabe, Recent advances in electrolytes for lithium–sulfur batteries, Adv. Energy Mater. 5 (2015) 1500117, Copyright (2015) John Wiley & Sons.)

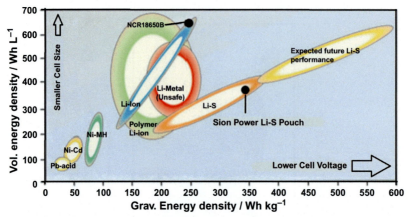

**Fig. 9.2** Practical performance of Lithium-sulfur batteries. *(Reprinted with permission from G. Benveniste, H. Rallo, L.C. Casals, A. Merino, B. Amante, Comparison of the state of Lithium-Sulphur and lithium-ion batteries applied to electromobility, J. Environ. Manage. 226 (2018) 1–12, Copyright (2018) Elsevier.)*

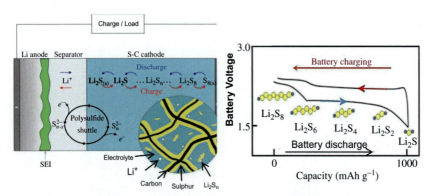

**Fig. 9.3** Redox mechanism of lithium-sulfur battery. *(Reprinted with permission from N. Akhtar, et al., A gelatin-based artificial SEI for lithium deposition regulation and polysulfide shuttle suppression in lithium-sulfur batteries, J. Energy Chem. 52 (2020) 310–317, Copyright (2020) Elsevier.)*

following advanced components: host materials and binders [10,11]. These two aspects will be discussed in the upcoming sections of the chapter.

## 9.3 Host materials

Many types of host materials have been developed to improve the electrochemical performance of lithium-sulfur batteries. Among them, the following few modifications will be discussed in this section:
- Three-dimensional hollow graphene spheres
- Reduced graphene oxide nanocomposite/N-doped carbon framework

- Three-dimensional porous carbon composites
- Micro-mesoporous graphitic carbon spheres
- Carbon nanotubes cathode
- Hierarchical network macrostructure
- In situ wrapping process

## 9.3.1 Three-dimensional graphene hollow spheres

A novel three-dimensional (3D) hollow graphene spheres (HGs) structure was reported by Wu. et al as an effective sulfur host (as shown in Fig. 9.4), resulting in enhanced Li—S battery performance [10]. By tailoring the ionic concentration in solutions, the $SiO_2$ nanospheres were effectively wrapped with a graphene oxide (GO) film, followed by $SiO_2$ carbonization and etching, leading to the formation of homogeneous hollow graphene nanospheres (HGs) with 3D conductive interlinked network. The innovative nanostructure of 3D HGs along with pure sulfur is utilized to form the middle-shell HGs/S electrode via the melt-diffusion process. Nanostructured HGs/S is rationally designed and controlled and has important functional benefits for Li—S cells: (i) for high-rate capability (please check) lithium-sulfur batteries, the HGs offer sufficient space for high S loading (up to 90 wt%); (ii) in the HGs/S, a uniform 3D conductive network can construct very efficient 3D electron transfer pathways and ion diffusion networks; and (iii) the HGs/S hole nanospheres can efficiently contain polysulfides inside the porous structure of HGs/S, effectively controlling the volume expansion of active material and shuttling effect during the

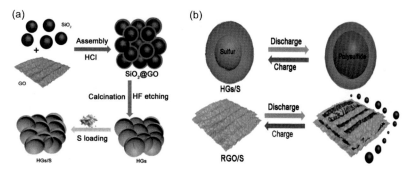

**Fig. 9.4** (A) Schematic illustration of the artificial system of the HGs/S complex, (B) Advantage of the HGs/S composite vs. RGO/S. *(Reprinted with permission from Z. Wu et al., Three-dimensional graphene hollow spheres with high sulfur loading for high-performance lithium-sulfur batteries, Electrochim. Acta 224 (2017) 527–533, https://doi.org/10.1016/j.electacta.2016.12.072, Copyright (2017) Elsevier.)*

lithiation. Moreover, such a design will maintain the structural integrity of the electrode and eliminate the loss of electric contact with the current collector. The HGs/S cathode material remains a high discharge capacity of 810 mAh g$^{-1}$ after 200 cycles at 0.5C rate. Furthermore, it demonstrates a low capacity decay rate of 0.083% per cycle after 600 cycles at 1C rate. With respect to pristine graphene oxide@S-composites cathode, HGs/S cathode shows dramatically improved electrochemical performance in terms of high specific capacity, excellent rate capability, and cycling stability [10].

### 9.3.2 Reduced graphene oxide nanocomposite/nitrogen-doped carbon framework

Graphene-wrapped nitrogen-doped carbon framework (NCF) is an excellent reservoir of S for increasing the S content in the cathode materials as shown in Fig. 9.5 [12]. The composite NCF-G@S showed significantly improved functional capability and a more robust cyclic performance. The enhanced electrochemical efficiency is ascribed to the adequate NCF-G@S chemical and physical properties, such as (i) N-doping polysulfides capping effect and (ii) fast ion transfer and highly conductive graphene/NCF composite. The NCF-G@S cathode consequently demonstrates an excessive 629.3 mAh g$^{-1}$ discharge capacity at 0.5C rate over 150 cycles and excellent (close to 100%) coulombic efficiency [12].

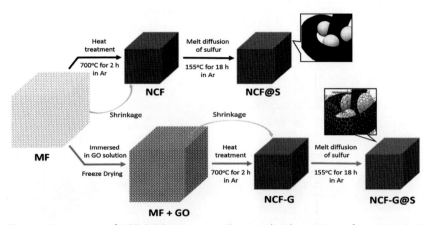

**Fig. 9.5** Preparation of NCF-G@S composite. *(Reprinted with permission from J. Lee, S.-K. Park, Y. Piao, N-doped carbon framework/reduced graphene oxide nanocomposite as a sulfur reservoir for lithium-sulfur batteries, Electrochim. Acta 222 (2016) 1345–1353, Copyright (2016) Elsevier.)*

### 9.3.3 Three-dimensional porous carbon composites

A new simple and scalable approach shown in Fig. 9.6 enables three-dimensional porous graphitic carbon composites (3D S@PGC) to be prepared in situ with a high sulfur content [13,14]. The approach uses Na$_2$S as a source of sulfur and Na$_2$S and NaCl as a basis for the porosity of the resulting composite. The nanoparticles of sulfur were distributed uniformly and merged covalently with 3D PGC, as verified by numerous microscopic and spectroscopic methods. Composite cathodes in modified Li—S batteries showed outstanding performance. In particular, use of more active sulfur loading (up to 90%), higher specific discharge capacities (1382, 1242, and 1115 mAh g$^{-1}$ at 0.5, 1, and 2C, respectively), and long cycling life (small capacity decay of 0.039% per cycle over 1000 cycles at 2C) were observed. We believe that the strategy can also motivate the researchers to use other 3D porous networks in other materials engineering fields, including separations, selective adsorption, sensing, and catalytic applications.

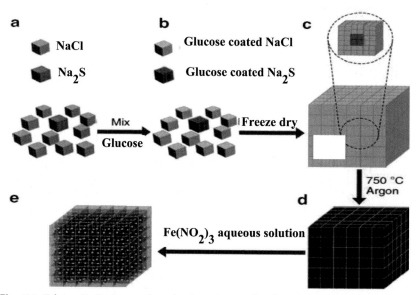

**Fig. 9.6** Schematic instance of an in situ strategy for the instruction of 3D S@PGC composites. *(Reprinted with permission from G. Li et al., Three-dimensional porous carbon composites containing high sulfur nanoparticle content for high-performance lithium-sulfur batteries, Nat. Commun. 7 (2016) 10601, doi:10.1038/ncomms10601, Copyright (2016) Springer Nature.)*

### 9.3.4 Micro-mesoporous graphitic carbon spheres

Zheng et al. have developed a multistage process using self-assembled $Fe_3O_4$ NP superlattices (Fig. 9.7) for manufacturing hybrid micro-mesoporous graphic spheres as a well-organized, bispatially curbed sulfur reservoir [14]. This research has shown that mesoporous carbon spheres with a suitable thickness of a microporous carbon shell can efficiently constrain polysulfide dissolution without sacrificing the well-organized use of active sulfur, providing high discharge capacity, exceptional rate performances, and outstanding cycling efficiency for Li—S cathodes even at high sulfur loadings. The S@M-MGCSs cathodes deliver superior electrochemical performance, including high reversible capacity (1122 mAh g$^{-1}$ at 0.1C), exceptional rate capability (340 and 300 mAh g$^{-1}$ at 10 and 12C, respectively), and good cycling stability (702 mAh g$^{-1}$ at 1C after 300 cycles and 440 mAh g$^{-1}$ at 4C after 500 cycles). Impressively, the S@M-MGCSs cathode still displays outstanding performance even at a high sulfur loading of 3.2 mg cm$^{-2}$.

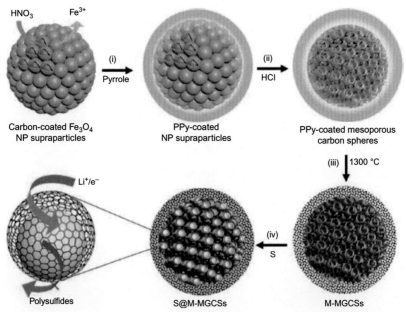

Fig. 9.7 Schematic example of the synthesis of M-MGCSs and S@M-MGCSs. *(Reprinted with permission from J. Zheng et al., Elaborately designed micro–mesoporous graphitic carbon spheres as efficient polysulfide reservoir for lithium–sulfur batteries, ACS Energy Lett. 2 (2017) 1105–1114, https://doi.org/10.1021/acsenergylett.7b00230, Copyright (2017) American Chemical Society.)*

### 9.3.5 Carbon nanotube cathodes

Carbon nanotubes (Fig. 9.8) impregnated with sulfur have been manufactured as active electrode material for the Li—S battery [15]. The attained cathodes (sulfur-carbon nanotubes) demonstrated significantly improved cyclic performance and high coulombic efficiency. Moreover, a new carbon-sulfur stabilization mechanism realized by heat treatment was indicated by the electrochemical characterization. Significant enhancements in the specific capacity, cycling stability, and long cycle life of lithium-sulfur batteries have undoubtedly been demonstrated by these materials. The sulfur-impregnated distorted carbon nanotube cathodes showed 96% coulombic efficiency and good cycle stability with 72.9% capacity retention after 100 cycles except the first one: after 30 cycles, the capacity stopped fading completely. However, the most significant drawback of these methods is that they are all quite complex, not cost-effective and therefore, not appropriate for practical applications.

Shao-Horn et al. synthesized multiwall carbon nanotube (MWNT)/manganese oxide ($MnO_2$) nanocomposites ultrathin film electrodes by

**Fig. 9.8** SEM and TEM snapshots of the disordered carbon nanotubes prepared via the template wetting method. *(Reprinted with permission from J. Guo, Y. Xu, C. Wang, Sulfur-impregnated disordered carbon nanotubes cathode for lithium-sulfur batteries, Nano Lett. 11 (2011) 4288–4294, Copyright (2011) American Chemical Society.)*

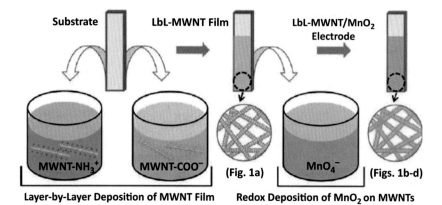

**Fig. 9.9** Preparation of multiwall nanotube as a host material for Li—S battery. *(Reprinted with permission from S.W. Lee, J. Kim, S. Chen, P.T. Hammond, Y. Shao-Horn, Carbon nanotube/manganese oxide ultrathin film electrodes for electrochemical capacitors, ACS Nano 4 (2010) 3889–3896, Copyright (2010) American Chemical Society.)*

means of continuous redox deposition of $MnO_2$ on layer-by-layer (LbL)-assembled MWNT films (Fig. 9.9) [16]. Such an approach generating the porous MWNT network via alternating LbL assembly allows the formation of fast electronic and ionic conducting channels in the presence of electrolyte. Also, the conformal coating of $MnO_2$ on MWNTs results in high capacitance thus, providing a basis for designing high-performance electrodes for electrochemical capacitor (EC) applications. The ability to generate high electrode capacitances and precise control of thickness and capacity from simple dipping processes at ambient conditions suggests a promising approach to the creation of ultrathin electrodes in a controlled manner. This approach to fabricate high-capacitance LbL-MWNT/$MnO_2$ nanostructured electrodes can be used for synthesis of novel electrode materials for EC and battery as well as various sensor applications. Also, LbL technique is highly adaptable to various types of substrates such as silicon, glass, and flexible substrates, which showcases binder-free MWNT/$MnO_2$ electrodes as promising electrodes for application of microelectromechanical flexible electronics.

The open-ended carbon nanotubes (N-O-*E*-CNTs) doped with nitrogen (Fig. 9.10) were prepared by L. Chen et al. by controllable synthetic cutting, followed by a facile hydrothermal process [17]. Compared to CNTs, the synthesized N-O-*E*-CNTs demonstrated increased surface area and porosity as well as an excellent conductivity owing to N-doping.

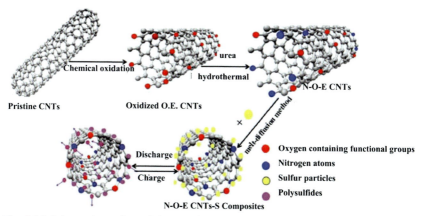

Fig. 9.10 Schematic outline of the amalgamation of N-O-E-CNTs/S composite and its electrochemical development in a lithium-sulfur energy storage device. *(Reprinted with permission from L. Chen et al., Chemical modification of pristine carbon nanotubes and their exploitation as the carbon hosts for lithium-sulfur batteries, Int. J. Hydrogen Energy 41 (2016) 21850–21860, Copyright (2016) Elsevier.)*

The N-O-*E*-CNTs/S cathode exhibited superior cyclability and higher rate capability over the O-E-CNTs/S and CNTs/S cathodes, profiting by every one of these benefits. The N-O-E-CNTs/S cathode exhibited promising cyclability with an initial discharge capacity of 1230 mAh/g and nearly 59% capacity retention (725 mAh/g) after 500 cycles, giving a cycling degradation rate of 0.08% per cycle. Furthermore, the coulombic efficiency of the N-O-*E*-CNTs/S cathode remained 99.5% after 500 cycles, demonstrating superior reversibility and efficiency. From one viewpoint, controllable chemical cutting for CNTs can increase the specific surface area and pore volume for N-O-E-CNTs in a viable manner while maintaining good conductivity that helps to support and accommodate sulfur particles. Moreover, the formed hollow structure allows impregnation of the S species within the inner nanotube cavity. Finally, N-doping of O-E-CNT helps not only enhance the conductivity of oxidized O-E-CNTs but also enables N-O-E-CNTs to provide better absorbability for S molecules, resulting in more efficient usage of mobile ion content and also in more effective polysulfide anchoring. Such developments not only provide a better carbon host but also offer another method of manufacturing S electrode with an innovative structure and advanced performance.

The integrally designed carbon nanotubes (CNT) and activated carbon nanofibers (ACNF) loaded with $MnO_2$ (CNT/ACNF@$MnO_2$) (Fig. 9.11) for active sulfur cathodes have been described by Xu et al. [18]. In this advanced design electrode, porous ACNFs serve as a reservoir for dissolved polysulfides and enable high sulfur loading, while CNTs act as a conductive network to improve the sulfur utilization and the mechanical stability of the highly flexible cathode. $MnO_2$ is easily grown onto the CNT/ACNF scaffold and binds with polysulfides to efficiently prevent them from diffusing out of the cathode. This integrally designed flexible host enables high sulfur loading, expands sulfur utilization, and subdues effectively the parasitic shuttle. The Li/dissolved polysulfide cells with high sulfur loading show a high-rate capacity of 780 mAh g$^{-1}$ at 2C rate and a high-capacity retention of 64% over 300 cycles. This innovative approach enables high sulfur loading from 2.4 to 7.2 mg cm$^{-2}$ with low polarization, higher rate capability, and better cycling performance at high current densities. The authors also discovered that trace amount of lithium polysulfides diffused out from the cathode acts

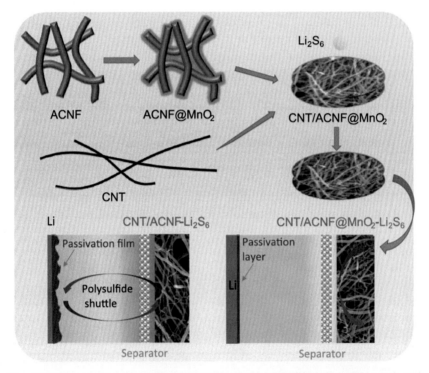

**Fig. 9.11** Preparation of carbon nanotubes with ACNF@MnO$_2$. *(Reprinted with permission from H. Xu, L. Qie, A. Manthiram, An integrally-designed, flexible polysulfide host for high-performance lithium-sulfur batteries with stabilized lithium-metal anode, Nano Energy 26 (2016) 224–232, Copyright (2016) Elsevier.)*

as an electrolyte additive to protect the lithium metal anode and thus prevent further Li-metal corrosion and electrolyte exhaustion. These outcomes show that this innovative electrode design is incredibly promising in the advancement of flexible and high-energy density Li—S batteries.

### 9.3.6 Hierarchical network macrostructure

Hu et al. successfully prepared (Fig. 9.12) 3D graphene progressive macrostructure via chemical vapor deposition (CVD) [19]. For Li—S battery applications, this approach combines the advantages of rGO, graphene, permeable materials, and 3D interconnected network framework. The permeable structure not only constitutes a high sulfur loaded cathode (14.36 mg cm$^{-2}$) with 89.4 wt% sulfur content ratio, it also functions as an electrolyte repository and adequately helps accommodate the volumetric expansion/contraction of S particles. The exceptionally conductive system can advance the rapid transport of ions and electrons, and the remaining oxygen atoms located at the rGO surface provide active sites for binding sulfur particles and holding the solvent polysulfide intermediates inside the cathode. Curiously enough, the GF–rGO/S cathodes with 9.8 mg cm$^{-2}$ of S load (83 wt% sulfur content ratio) exhibit a 10.3 mAh cm$^{-2}$ high areal capacity with long cycling performance at 0.2C. The areal limit is almost twice that of commercial Li-ion batteries and the cycling yield is far away superior to those recorded for lithium-sulfur energy storage devices with less

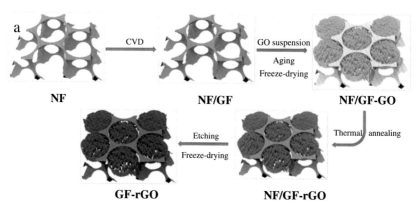

**Fig. 9.12** Preparation and fabricating scheme of 3D GF–rGO hierarchical macrostructure network. *(Reprinted with permission from G. Hu et al., 3D graphene-foam-reduced-graphene-oxide hybrid nested hierarchical networks for high-performance Li-S batteries, Adv. Mater. 28 (2016) 1603–1609, Copyright (2016) John Wiley & Sons.)*

sulfur loading and sulfur content ratio. They also demonstrate the utilization of GF–rGO/S cathodes for functional applications in large superior Li—S pouch cells.

Hussain et al. successfully synthesized nanoparticles of titanium nitride (TiN) and sulfur composite (TiN-NPs@S) via the dry freezing process as the cathode content is explored for high-performance Li—S batteries [20]. The ultrasmall nanoparticles (Fig. 9.13) have a large surface area with

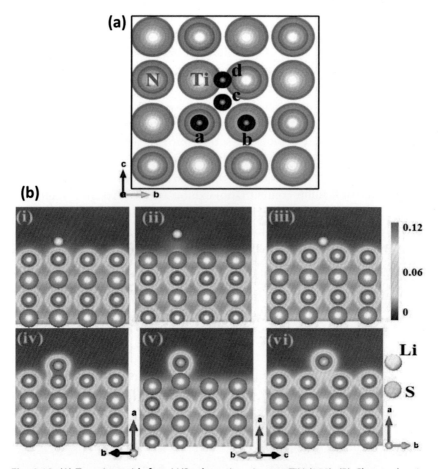

**Fig. 9.13** (A) Top view with four Li/S adsorption sites on TiN (100). (B) Charge density distribution maps for Li and S atoms on top of N, Ti, and hollow sites. *(Reprinted with permission from S. Hussain, et al., Robust TiN nanoparticles polysulfide anchor for Li–S storage and diffusion pathways using first principle calculations, Chem. Eng. J. 391 (2019) 123595, Copyright (2019) Elsevier.)*

narrow pores that function as conductive channels to exchange ions/electrons for storing electricity during electrochemical reactions. Density functional theory (DFT) results provide the distribution of density charges and the movement of charges as well as the study of structure geometries; these results demonstrate that the TiN surface exhibits a greater adsorption ability of S than the Li atoms and suggests TiN as a good polysulfide adsorbent [20].

After 100 cycles, the TiN cathode shows an excellent initial capacity of 1279 mAh g$^{-1}$ at 0.5C around 75.8% of the theoretical capacity (1562 mAh g$^{-1}$). The hot-to-cold synthesis process projects the development of ultrasmall nanoparticles illuminating the new horizons of Li—S rechargeable batteries in order to achieve excellent conductivity, high power, and energy density along with high efficiency, which is of great benefit for industrial applications.

Nazar et al. reported a nitrogen–sulfur-doped nanoporous carbon that was synthesized by self-templating sustainable cellulose-derived nanocrystals of biomaterials [21]; the nanomaterial exhibits an adjustable hierarchical porous structure with a high surface area (Fig. 9.14). Dual doping of porous carbon with S and N atoms greatly improves the chemisorption of lithium polysulfides (LiPSs). The XPS unraveled a synergistic effect of N—Li and S—S interactions. Li$^+$ and $S_n^{2-}$ are efficiently bonded to N, S dual-doped carbon, resulting in about twice much stronger LiPSs adsorption in comparison to N single-doped or nondoped carbon. The binding structure has also been confirmed by ab initio calculations.

This is the first time when a dual-doped carbon was used as a sulfur-host substrate with proven mechanisms for chemisorption. Furthermore, the electrical conductivity of nitrogen/sulfur-doped cellulose is also noticeably improved compared to undoped carbon, which additionally benefits the high-rate kinetics. Utilizing this double-doped carbon as a sulfur host, the

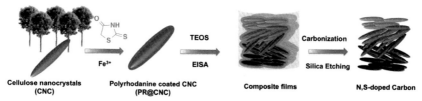

**Fig. 9.14** The schematic outline of the union of nitrogen/sulfur-doped cellulose (NSC). (Reprinted with permission from Q. Pang, et al., A nitrogen and sulfur dual-doped carbon derived from Polyrhodanine@Cellulose for advanced lithium-sulfur batteries, Adv. Mater. 27 (2015) 6021-6028, Copyright (2015) John Wiley & Sons.)

sulfur cathode could deliver a high capacity of $1370\,mAh\,g^{-1}$ at C/20 and charge/discharge at 2C current rate for 1100 cycles with an exceptionally low capacity fading of 0.052% per cycle. This is ascribed to the host's high pore volume and surface area, which provides sufficient number of active sites for effective LiPSs chemisorption.

Pan et al. have been successfully synthesized a micrometer-sized spherical carbon with a hierarchical structure of the macrohollow core and a microporous shell with $672\,m^2\,g^{-1}$ surface area from a amylose biomass material via a multistep pyrolysis route [22]. After sulfur is incorporated by a melting infiltration technique, sulfur is effectively confined in the shell's long and thin channels and the macrohollow core is primarily preserved. This S@C complex with a sulfur substance of 48 wt% demonstrates a high initial capacity of $1490\,mAh\,g^{-1}$ and a capacity of $798\,mAh\,g^{-1}$ at C/10 after 200 cycles as a cathode material for lithium-sulfur batteries. Also, the specific capacities of 681 and $487\,mAh\,g^{-1}$ are shown at 1C and 3C rates, respectively. During cycling, the S@C complex exhibits low polarization and fast redox kinetics. Sulfur confinement significantly mitigates the dissolution of polysulfides to electrolyte and greatly improves contact with the carbon base. The high surface area provides enough active sites to bind and store the insulating $Li_2S_2/Li_2S$. The microhollow core can absorb the sulfur volume change during lithiation. This special structure enhances the electron transfer along with lithium diffusion. One-step pyrolysis produces substandard permeable carbons; after the consolidation of sulfur results in subpar electrochemical properties. It should be noted that the one-step pyrolysis and the chemical reagent-assisted pyrolysis methods produce only poorly structured porous carbons, which offer inferior electrochemical properties after incorporating sulfur, while the multistep pyrolysis technique and synthesized porous spherical carbon are considered as having great potential for the development of exceptionally efficient carbonaceous S and PS storage materials derived from sustainable and inexhaustible biomass resources.

Li et al. demonstrated a multistep pyrolysis method to fabricate novel porous structured carbon spheres with a macrohollow core and a microporous shell. The nanomaterials were derived from a sustainable biomass (amylose) [23]. This hierarchically porous carbon shows a particle distribution of $2{-}10\,\mu m$ and a surface area of $672\,m^2\,g^{-1}$; see Fig. 9.15 [23].

It was found that a porous nanostructure decreases dissolution of polysulfides in the electrolyte, offering low electrical resistance during discharge/charge process and observed as operative host of sulfur for lithium–sulfur battery cathodes. As a sulfur cathode for Li—S battery, porous carbon

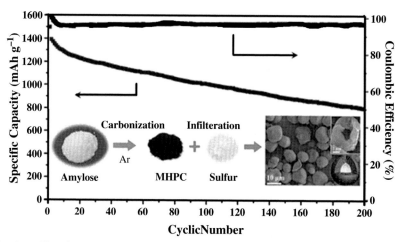

**Fig. 9.15** The description of the novel porous structured carbon spheres for Li—S cathode. *(Reprinted with permission from D. Guo et al., 3D hierarchical nitrogen-doped carbon nanoflower derived from chitosan for efficient electrocatalytic oxygen reduction and high performance lithium–sulfur batteries, J. Mater. Chem. A 5 (2017) 18193–18206, Copyright (2017) American Chemical Society.)*

nanosphere loaded 48 wt% sulfur content in its porous structure and demonstrated an exceptionally high capacity of 1490 mAh g$^{-1}$ @ 0.1C and retains a capacity of 798 mAh g$^{-1}$ after 200 cycles with 488 mA h g$^{-1}$ capacity retention at 3C. The porous channel and multistep pyrolysis provide an efficient and naturally benign approach to incorporate S-doping into the carbon framework for producing high-tap volumetric density for energy storage practical applications.

### 9.3.7 In situ wrapping process

Hu et al. proved an in situ wrapping technique for the sulfur cathode [24]. In conventional approaches, where the cathode material with a wrapping layer was first synthesized and assembled into batteries, there is a compromise between electrolyte infiltration and hampering polysulfide diffusion. The in situ wrapping approach eliminates this compromise by infusing the electrolyte into the open pores of the cathode material and then closing the pores by a surface reaction with special additives in the electrolyte. This idea was realized by means of the reaction between PANS and triphenylphosphine (Fig. 9.16).

The in situ wrapping mechanism has been verified using characterization tools and control experiments, and the polysulfide shuttle blocking effect has

**Fig. 9.16** Schematic representation of the in situ wrapping process. *(Reprinted with permission from C. Hu et al., In situ wrapping of the cathode material in lithium-sulfur batteries, Nat. Commun. 8 (2017) 1–9, Copyright (2017) Springer Nature.)*

been directly proven by shuttle current measurements. Batteries with such cathode materials demonstrated much better cycling stability with a decay rate of ∼0.030% per cycle within the 1000 cycles at a current of 1C, which is far slower than that of typical cathode materials coated using conventional preassembly wrapping approaches. Since there are many different types of the in situ wrapping reactions, such as photochemical and electrochemical methods, it opens virtually unlimited possibilities in the molecular design of in situ wrapping reagents and also improvement of the Li—S cathode stability.

## 9.4 Binders

Binders assume a significant role in lithium-sulfur batteries. Because most of the sulfur cathodes' inner surfaces are filled with binder, these binders' surface modification would have a significant impact on active cathodes. This section addresses various types of binders; each subsection discusses the types listed in the following order:
- Multifunctional polar binder
- Polyamidoamine dendrimer-based binders
- PAA/PEDOT: PSS as a functional binder

### 9.4.1 Multifunctional polar binder

Highly efficient Li—S batteries have been developed by Zhang et al. using polar polyethylenimine (PEI) polymer (Fig. 9.17) as a functional sulfur cathode binder [25]. With a high sulfur loading density of 8.6 mg cm$^{-2}$, an initial specific capacity of 1126.4 mAh g$^{-1}$ (9.7 mAh cm$^{-2}$) is achieved and the capacity is still kept at 744.2 mAh g$^{-1}$ (6.4 mAh cm$^{-2}$) after more than 50 cycles, which is superior to the performance of conventional PVDF-based cathode. In situ UV–visible spectroscopy and *operando* X-ray absorption spectra results show direct evidence that the amino groups in PEI

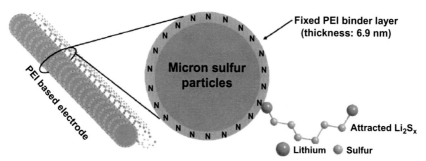

**Fig. 9.17** Trapping behavior of PEI polymer for polysulfide intermediates. *(Reprinted with permission from L. Zhang, M. Ling, J. Feng, G. Liu, J. Guo, Effective electrostatic confinement of polysulfides in lithium/sulfur batteries by a functional binder, Nano Energy 40 (2017) 559–565, Copyright (2017) Elsevier.)*

polymer can form strong electrostatic interaction with polysulfide intermediates, which effectively ameliorates the shuttle effect and enables the enhanced electrochemical performance. This study demonstrates an effective approach to enhance the cycling performance of Li/S batteries by engineering the functional binder, which promises to be beneficial for the commercial applications of high-energy Li/S batteries in the future.

A. Naseem et al. have ingeniously developed a water-soluble gelatin PEI composite (GPC) bifunctional binder that is able to match the typical characteristics of a commercial functional binder such as dispersion and adhesion capability, appropriate swelling capability, high conductivity, and polysulfide adsorption [11]. GPC binder (as shown in Fig. 9.18) keeps the cathode uniform and efficiently absorbs the volume change of the cathode during electrochemical cycling. GPC binder with nitrogen species exhibiting high affinity to polysulfide may effectively minimize polysulfide dissolution into the organic electrolyte. Modified cathode with GPC binder in lithium-sulfur batteries remarkably reduced the polysulfide shuttling effect and significantly enhanced the cycling performance with coulombic efficiency of 99% and capacity retention of ~100% after 100 cycles at a current rate of 1C.

A new type of amino functional group (AFG) binder by polymerization of hexamethylene diisocyanate (HDI) with PEI polymer (Fig. 9.19) was successfully developed and used by Chen et al. with a standard sulfur cathode [26]. Compared to a commercial binder (such as PVDF), the specific benefits of an AFG binder are notably hyperbranched network structures and numerous amine groups, which effectively absorb polysulfide species resulting in surprisingly improved cycling performance.

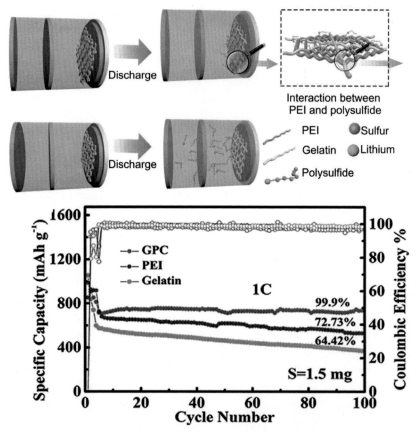

**Fig. 9.18** Schematic illustration diagram of Li—S batteries with GPC binder cathodes. *(Reprinted with permission from N. Akhtar et al., Gelatin-polyethylenimine composite as a functional binder for highly stable lithium-sulfur batteries, Electrochim. Acta 282 (2018) 758–766, Copyright (2018) Elsevier.)*

Pan et al. have successfully utilized LA-132, a polyacrylonitrile-based polymer, as a water-soluble binder for lithium-sulfur energy storage device [27]. The cathode with LA-132 binder has a high reversible capacity of 470 mAh g$^{-1}$ at C/2 after 100 cycles, showing surprisingly better cycling compared to the conventional PVDF binder containing cathode. Electrochemical studies show that the LA-132 binder containing sulfur cathodes has lower inner resistance and stronger kinetic properties than commercial PVDF binder containing cathodes. SEM images also show that the LA-132 binder has the ability to suppress the Li$_2$S$_2$/Li$_2$S cluster and preserve the porous structure of sulfur cathode. It can conclude that nonswellable

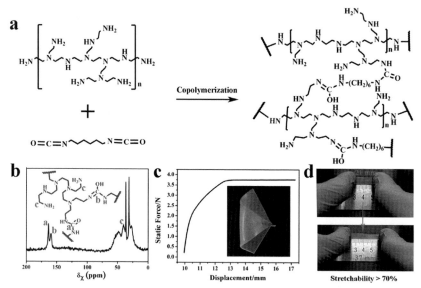

**Fig. 9.19** Copolymerization of polyethylenimine (PEI) polymer with hexamethylene diisocyanate (HDI). *(Reprinted with permission from W. Chen et al., A new type of multifunctional polar binder: toward practical application of high energy lithium sulfur batteries, Adv. Mater. 29 (2017) 1605160, Copyright (2017) John Wiley & Sons.)*

LA-132 binder not only acts as a strong dispersion and adhesion agent but also plays a key role in alleviating the framework of the cathodes. In addition, water-soluble LA-132 binder has low-cost, ecofriendly, and safety advantages. Based on this result, they made the conclusion that water-soluble binder such as LA-132 is a promising candidate for Li—S battery applications.

### 9.4.2 Polyamidoamine dendrimer-based binders

P. Bhattacharya et al. have explored the usage of polyamidoamine (PAMAM) dendrimers as functional binders in Li—S batteries [28]. High S loads (>4 mg cm$^{-2}$) can be conveniently obtained using the high surface functionality, interior porosity, and polarity of the PAMAM dendrimers (as seen in Fig. 9.20) by utilizing easy processing methods. The fabricated cell attained an extraordinary areal capacity of 4.32 mAh cm$^{-2}$ and a high specific capacity of 1045 mAh g$^{-1}$ at a C/20 discharge rate, with capacity retention of ~640 mAh g$^{-1}$ after 100 cycles (98% of the original capacity) at a C/5 discharge rate. Compared to conventional, cathodes with linear polymeric

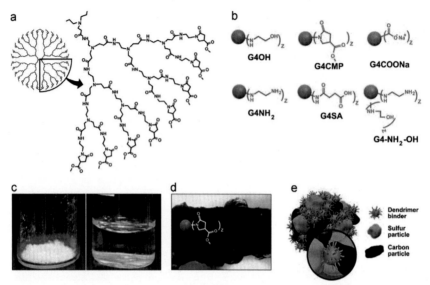

Fig. 9.20 (A) Generation schematic 3 PAMAM4-carbomethoxy pyrrolidone dendrimer, (B) surface group matrix analyzed, (C) dry G4CMP dendrimer image and dendrimer dissolved forming aqueous solution, (D) slurry-coated electrode image with G4CMP binder, and (E) potential dendrimer-C/S/MWCNT interactions cartoon. *(Reprinted with permission from P. Bhattacharya et al., Polyamidoamine dendrimer-based binders for high-loading lithium–sulfur battery cathodes, Nano Energy 19 (2016) 176–186, Copyright (2016) Elsevier.)*

binders, such as styrene-butadiene rubber (SBR) and carboxymethyl cellulose (CMC), improved interfacial interactions between C/S composite materials and dendrimers, as well as improved electrolyte wetting due to spherical molecular and porous dendrimer architectures, an extraordinary electrochemical cycling performance has been achieved.

High sulfur loading electrodes ($>6\,mg\,cm^{-2}$) have been successfully synthesized by Sun et al. [29] using traditional slurry casting by using a new binder and a new form of microporous carbon. Under low-rate coin cell discharge, the areal capacity of up to $12\,mAh\,cm^{-2}$ was achieved. The carbon particle size and its specific pore volume are considered as the key factors for the high sulfur loading cathode preparation and the high sulfur utilization during electrochemical testing of the Li—S battery. The cellulose family folio hydroxypropyl cellulose (HPC) has been identified to provide more adhesion between the cathode and the aluminum current collector in comparison to the broadly utilized polyvinylidene fluoride (PVDF). Since HPC is miscible with nonaqueous solvent NMP, it might decrease the danger of incorporation of residue water content when the aqueous process is used. While HPC slightly increases the resistivity of the anode system, it serves as a secondary binder to help improve the mechanical stability of thick and high sulfur loading electrodes.

In the interim, three different microporous carbons were explored as sulfur conductors for sulfur electrodes. The carbons were derived from natural sources, such as coconut shell (A5562), coal (A5583), and wood (A5597). It was noted that A5597 has the best ability to absorb sulfur during thermal fusion and also endows the cathode with the highest sulfur utilization during Li—S cell test. High specific pore volume instead of surface area was characterized as the key parameter that determines the better property of A5597 over the other two. In contrast to coconut shell (A5562) and coal (A5583), this high pore volume is attributed to the unique texture of wood (A5597). Thus, wood (A5597) is highly likely the best microporous carbon precursor for Li—S battery application among the three raw materials [29].

Considering that specific power and sulfur utilization are the most critical parameters, the small particle size sulfur-absorbing carbon is the best choice. However, preparation of thick sulfur electrodes with small carbon particle size demonstrating high mechanical stability is very difficult for achieving high energy density at the cell level. Larger microporous carbon particles such as A5597 provide an alternative option for sulfur incorporation in thick sulfur electrodes with micron-sized particles. At present, satisfactory performance of high sulfur loading electrodes prepared with the A5597 carbon and binary binder system can be achieved only at low current rate, and the kinetic barriers lie in both the transport of Li-ion/electron and the nucleation process of $Li_2S_2/Li_2S$ at the onset of the discharge plateau region [29]. Improvement in both aspects is necessary to make this concept applicable to practical applications. Future studies on microporous carbons with even higher porosity and particle size above 1 μm will also be extremely important for the production of efficient high sulfur loading electrodes.

### 9.4.3 PAA/PEDOT: PSS as a functional binder

J. Pan et al. developed a new binder based on a combination of polyacrylic acid (PAA) and poly (3,4-ethylenedioxythiophene): poly (styrene sulfonate) (PEDOT:PSS) (Fig. 9.21), which substantially improved the specific capacity and cycling stability of Li—S batteries through the synergistic effect of the

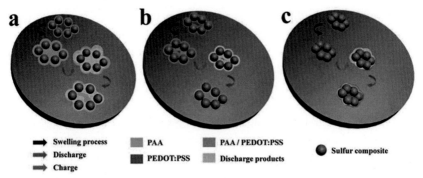

**Fig. 9.21** Sulfur cathodes scheme with various binders: (A) PAA, (B) PAA/PEDOT: PSS, and (C) PEDOT: PSS. *(Reproduced with permission from J. Pan et al., PAA/PEDOT:PSS as a multifunctional, water-soluble binder to improve the capacity and stability of lithium–sulfur batteries, RSC Adv. 6 (2016) 40650–40655, Copyright (2016) Royal Society of Chemistry.)*

different functional groups [30]. The conductive PEDOT:PSS effectively enhances electrons transfer and suppresses the polysulfide dissolution. Also, PAA enhances the solvent system for sulfur cathodes and improves the lithium-ion mobility. The sulfur cathode with PAA/PEDOT:PSS binder displayed an initial specific capacity of $1121\,mA\,h\,g^{-1}$ and $830\,mA\,h\,g^{-1}$ after 80 cycles at 0.5C. The electrochemical performance of this composite binder sulfur cathode is better than that of any other single-component binder-based electrode.

A room-temperature covalently cross-linked polyacrylamide (c-PAM) binder with high stretchability and multiple polar amide groups for high-performance Si anodes and sulfur cathodes has been effectively manufactured by Zhu et al. suitable for efficient lithium-ion energy storage devices [31]. The as-made c-PAM hydrogel is shown in Fig. 9.22, offering high flexibility and foldability for the Si electrode coatings.

In addition to providing significant enhancement of the strain resistance and maintaining the integrity of the working electrodes during cycling, the solid 3D binder network demonstrates also a strong propensity to bond with the nano-Si surface and capture the soluble $Li_2S_n$ intermediates, resulting in a dramatically improved cycling performance of both the Si anode and the sulfur cathode. In a wider context, such a groundbreaking route of high-performance Si and sulfur electrodes may be extended over other high-capacity electrode materials with substantial volume improvements across the long-term operating span.

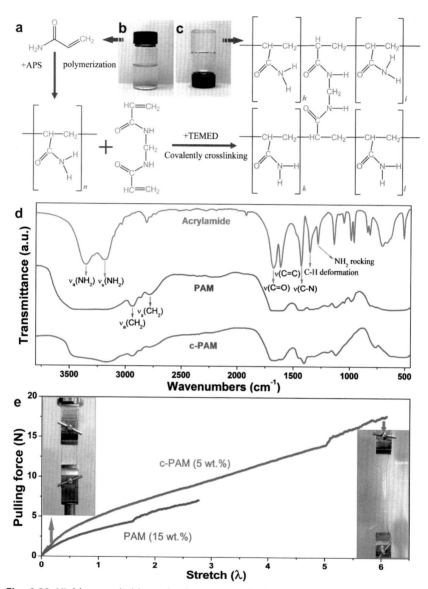

**Fig. 9.22** Highly stretchable and robust c-PAM hydrogel. (A) Schematic illustration showing the synthetic process of the c-PAM hydrogel. (B, C) Digital photographs of the initial acrylamide solution and as-prepared solid c-PAM hydrogel. *(Reproduced with permission from X. Zhu et al., A highly stretchable cross-linked polyacrylamide hydrogel as an effective binder for silicon and sulfur electrodes toward durable lithium-ion storage, Adv. Funct. Mater. 28 (2018) 1705015, Copyright (2018) John Wiley & Sons.)*

## 9.5 Conclusions and future directions

In the past decade, various multifunctional host materials and binders have been utilized for improved lithium-sulfur battery technologies to address the significant material and electrode challenges that exist for new battery chemistries. This chapter highlights the principles of materials used to design these host materials and functional binders and gives a broad picture of the developments of electrodes that facilitate the advancement of the battery and modern technology field. Future work on understanding the fundamental electrode and materials chemistry taking place in these electrode systems is needed. Detailed information about the electrochemical mechanisms involved in these battery systems is still absent due to their complexity. Meanwhile, investigation of the ion and electron kinetic transport at the electrode/electrolyte interface is also important; many electrochemical reactions are initiated at this interface. Further understanding of lithium-sulfur battery chemistry should enable an optimized modern structure design and lead to better electrochemical performance.

## References

[1] S. Zhang, K. Ueno, K. Dokko, M. Watanabe, Recent advances in electrolytes for lithium–sulfur batteries, Adv. Energy Mater. 5 (2015) 1500117.

[2] G. Benveniste, H. Rallo, L.C. Casals, A. Merino, B. Amante, Comparison of the state of Lithium-Sulphur and lithium-ion batteries applied to electromobility, J. Environ. Manage. 226 (2018) 1–12.

[3] X. Ji, K.T. Lee, L.F. Nazar, A highly ordered nanostructured carbon–sulphur cathode for lithium–sulphur batteries, Nat. Mater. 8 (2009) 500–506.

[4] X. Ji, A.F. Nazar, Advances in Li–S batteries, J. Mater. Chem. 20 (2010) 9821–9826.

[5] J. Hassoun, B. Scrosati, Moving to a solid-state configuration: a valid approach to making lithium-sulfur batteries viable for practical applications, Adv. Mater. 22 (2010) 5198–5201.

[6] Q. Zhao, X. Hu, K. Zhang, N. Zhang, Y. Hu, J. Chen, Sulfur nanodots electrodeposited on Ni foam as high-performance cathode for Li–S batteries, Nano Lett. 15 (2015) 721–726.

[7] S.-H. Chung, P. Han, R. Singhal, V. Kalra, A. Manthiram, Electrochemically stable rechargeable Lithium-Sulfur batteries with a microporous carbon nanofiber filter for polysulfide, Adv. Energy Mater. 5 (2015) 1500738.

[8] D.-W. Wang, Q. Zeng, G. Zhou, et al., Carbon–sulfur composites for Li–S batteries: status and prospects, J. Mater. Chem. A 1 (2013) 9382, https://doi.org/10.1039/c3ta11045a.

[9] N. Akhtar, X. Sun, M.Y. Akram, F. Zaman, W. Wang, et al., A gelatin-based artificial SEI for lithium deposition regulation and polysulfide shuttle suppression in lithium-sulfur batteries, J. Energy Chem. 52 (2020) 310–317.

[10] Z. Wu, W. Wang, Y. Wang, C. Chen, K. Li, et al., Three-dimensional graphene hollow spheres with high sulfur loading for high-performance lithium-sulfur batteries, Electrochim. Acta 224 (2017) 527–533, https://doi.org/10.1016/j.electacta.2016.12.072.

[11] N. Akhtar, H. Shao, F. Ai, Y. Guan, Q. Peng, et al., Gelatin-polyethylenimine composite as a functional binder for highly stable lithium-sulfur batteries, Electrochim. Acta 282 (2018) 758–766.

[12] J. Lee, S.-K. Park, Y. Piao, N-doped carbon framework/reduced graphene oxide nanocomposite as a sulfur reservoir for lithium-sulfur batteries, Electrochim. Acta 222 (2016) 1345–1353.

[13] G. Li, J. Sun, W. Hou, S. Jiang, Y. Huang, J. Gen, Three-dimensional porous carbon composites containing high sulfur nanoparticle content for high-performance lithium-sulfur batteries, Nat. Commun. 7 (2016) 10601, https://doi.org/10.1038/ncomms10601.

[14] J. Zheng, G. Guo, H. Li, L. Wang, B. Wang, et al., Elaborately designed micro–mesoporous graphitic carbon spheres as efficient polysulfide reservoir for lithium–sulfur batteries, ACS Energy Lett. 2 (2017) 1105–1114, https://doi.org/10.1021/acsenergylett.7b00230.

[15] J. Guo, Y. Xu, C. Wang, Sulfur-impregnated disordered carbon nanotubes cathode for lithium-sulfur batteries, Nano Lett. 11 (2011) 4288–4294.

[16] S.W. Lee, J. Kim, S. Chen, P.T. Hammond, Y. Shao-Horn, Carbon nanotube/manganese oxide ultrathin film electrodes for electrochemical capacitors, ACS Nano 4 (2010) 3889–3896.

[17] L. Chen, H. Zhou, C. Fu, Z. Chen, C. Xu, Y. Kuang, Chemical modification of pristine carbon nanotubes and their exploitation as the carbon hosts for lithium-sulfur batteries, Int. J. Hydrogen Energy 41 (2016) 21850–21860.

[18] H. Xu, L. Qie, A. Manthiram, An integrally-designed, flexible polysulfide host for high-performance lithium-sulfur batteries with stabilized lithium-metal anode, Nano Energy 26 (2016) 224–232.

[19] G. Hu, L.K. Ono, G. Tong, Y. Liu, Y. Qi, 3D graphene-foam–reduced-graphene-oxide hybrid nested hierarchical networks for high-performance Li–S batteries, Adv. Mater. 28 (2016) 1603–1609.

[20] S. Hussain, X. Yang, M.K. Aslam, A. Shaheen, M.S. Javed, et al., Robust TiN nanoparticles polysulfide anchor for Li–S storage and diffusion pathways using first principle calculations, Chem. Eng. J. 391 (2019) 123595.

[21] Q. Pang, J. Tang, H. Huang, X. Liang, C. Hart, K.C. Tam, L.F. Nazar, A nitrogen and sulfur dual-doped carbon derived from Polyrhodanine@Cellulose for advanced lithium–sulfur batteries, Adv. Mater. 27 (2015) 6021–6028.

[22] X. Li, X. Cheng, M. Gao, D. Ren, Y. Liu, et al., Amylose-derived macrohollow core and microporous shell carbon spheres as sulfur host for superior lithium–sulfur battery cathodes, ACS Appl. Mater. Interfaces 9 (2017) 10717–10729.

[23] D. Guo, H. Wei, X. Chen, M. Liu, Ding F., et al., 3D hierarchical nitrogen-doped carbon nanoflower derived from chitosan for efficient electrocatalytic oxygen reduction and high performance lithium–sulfur batteries, J. Mater. Chem. A 5 (2017) 18193–18206.

[24] C. Hu, H. Chen, Y. Shen, D. Lu, Y. Zhao, et al., In situ wrapping of the cathode material in lithium-sulfur batteries, Nat. Commun. 8 (2017) 1–9.

[25] L. Zhang, M. Ling, J. Feng, G. Liu, J. Guo, Effective electrostatic confinement of polysulfides in lithium/sulfur batteries by a functional binder, Nano Energy 40 (2017) 559–565.

[26] W. Chen, T. Qian, J. Xiong, N. Xu, X. Liu, et al., A new type of multifunctional polar binder: Toward practical application of high energy lithium sulfur batteries, Adv. Mater. 29 (2017) 1605160, https://doi.org/10.1002/adma.201605160.

[27] J. Pan, G. Xu, B. Ding, J. Han, H. Dou, X. Zhang, Enhanced electrochemical performance of sulfur cathodes with a water-soluble binder, RSC Adv. 5 (2015) 13709–13714.

[28] P. Bhattacharya, M.I. Nandasiri, D. Lv, A.M. Schwarz, J.T. Darsell, et al., Polyamidoamine dendrimer-based binders for high-loading lithium–sulfur battery cathodes, Nano Energy 19 (2016) 176–186.

[29] K. Sun, C.A. Cama, J. Huang, Q. Zhang, S. Hwang, et al., Effect of carbon and binder on high sulfur loading electrode for Li-S battery technology, Electrochim. Acta 235 (2017) 399–408.

[30] J. Pan, G. Xu, B. Ding, Z. Chang, A. Wang, H. Dou, X. Zhang, PAA/PEDOT:PSS as a multifunctional, water-soluble binder to improve the capacity and stability of lithium–sulfur batteries, RSC Adv. 6 (2016) 40650–40655.

[31] X. Zhu, F. Zhang, L. Zhang, L. Zhang, Y. Song, et al., A highly stretchable cross-linked polyacrylamide hydrogel as an effective binder for silicon and sulfur electrodes toward durable lithium-ion storage, Adv. Funct. Mater. 28 (2018) 1705015.

**PART IV**

# Future directions: Solid-state materials and novel battery architectures

# CHAPTER 10

# Future prospects for lithium-sulfur batteries: The criticality of solid electrolytes

**Patrick Bonnick and John Muldoon**
Toyota Research Institute of North America, Ann Arbor, MI, United States

## Contents

| | |
|---|---|
| 10.1 The advantages of lithium-sulfur batteries | 327 |
| 10.2 The challenges of conventional sulfur electrodes when used with liquid electrolytes | 331 |
|     10.2.1 Solid electrolytes in lithium-sulfur batteries | 334 |
| 10.3 Lithium metal electrodes in lithium-sulfur batteries | 339 |
| 10.4 Path forward | 345 |
| Dedication | 346 |
| References | 347 |

## 10.1 The advantages of lithium-sulfur batteries

The promise of an energy density greater than conventional lithium ion (Li-ion) batteries has long made lithium metal anodes a tantalizing prize [1–6]. Lithium metal is the lightest metallic element on the periodic table, yielding an impressively high specific capacity of 3861 mAh/g. It also has the lowest redox potential of any element ($-3.04$ V vs the standard hydrogen electrode), enabling cells employing lithium metal to provide higher discharge potentials than any other anode material. In fact, many "beyond Li-ion" battery chemistries rely on the realization of lithium metal as a stably rechargeable negative electrode to be used in place of the more ubiquitous graphite that is found in conventional Li-ion batteries [7]. Historically, graphite (i.e., $LiC_6$) has been a more stable negative electrode material, but the presence of carbon adds weight and volume, resulting in a theoretical maximum specific capacity of only $340 \, mAh/g_{LiC6}$ in the charged state. Stably rechargeable lithium metal electrodes could therefore increase the specific capacity of the negative electrode by up to a factor of 10. Similarly, the specific capacity of the positive

---

*Lithium-Sulfur Batteries*
https://doi.org/10.1016/B978-0-12-819676-2.00002-5

Copyright © 2022 Elsevier Inc.
All rights reserved.

327

328  Lithium-sulfur batteries

electrode can be dramatically increased using sulfur as a conversion-type active material. Sulfur prefers an oxidation state of $-2$, allowing it to "store" 2 electrons per sulfur atom, which translates into a theoretical specific capacity of $1672\,mAh/g_S$. This far surpasses the specific capacity of conventional Li-ion cathode materials, such as $LiCoO_2$ and $Li[Ni_{0.33}Mn_{0.33}Co_{0.33}]O_2$ (NMC), which have specific capacities ranging from 150 to 250 mAh/g, depending on the nickel content and maximum charge voltage [8,9]. When paired with a lithium metal negative electrode, the lithium-sulfur (Li-S) cell can deliver a specific capacity far higher than a conventional NMC-graphite cell. Furthermore, sulfur is quite plentiful in the Earth's crust; as an example, literal pyramids of unused sulfur, a byproduct of the petroleum industry, are being built in Alberta, Canada. This is in sharp contrast to cobalt, which is used in Li-ion batteries as $LiCoO_2$ or NMC. Unfortunately, approximately half of the world's supply of cobalt currently comes from the Democratic Republic of Congo, some of which is mined by child labor [10]. As such, replacing cobalt in batteries with a cheaper and more sustainable alternative is of high priority to the battery research community [9].

Theoretically, a block of pure sulfur combined with a sheet of pure lithium would have a gravimetric energy density of $2510\,Wh/kg_{cell}$ ($2741\,Wh/L_{cell}$), assuming an average discharge potential of 2.15 V. For comparison, a conventional Li-ion 18,650 cell delivers about $250\,Wh/kg_{cell}$ ($700\,Wh/L_{cell}$) with an average discharge potential of 3.6 V. The energy advantage of Li-S appears staggering at first, but unfortunately a pure block of sulfur is unrealistic. Realistic Li-S cells must contain current collectors, an electrically conductive network, porosity, electrolyte, and a separator. They must also contain a microchip (i.e., battery management system), pressure relief valve, and cell casing, but these will be ignored for the purposes of this discussion. To generate some perspective regarding the future of Li-S cells, we present Fig. 10.1, which models the volumetric and gravimetric energy densities of the following NMC and sulfur cell chemistries in the discharged state: conventional NMC∥carbon, NMC∥lithium metal, sulfur∥carbon, and sulfur∥lithium metal. Sulfur, NMC and carbon electrodes are modeled with 25% porosity, which is the expected electrode fabrication limit for NMC cells [11]. Sulfur electrodes were assumed to deliver 75% of their theoretical capacity (i.e., a utilization of 75%). The x-axes in both Fig. 10.1A and B depict the increase in utilized areal capacity density of the positive electrode as its thickness is increased. This is accompanied by an increase in the thickness of the negative electrode to maintain the capacity ratio between the negative and positive electrodes at $N/P = 1.25$ in this model.

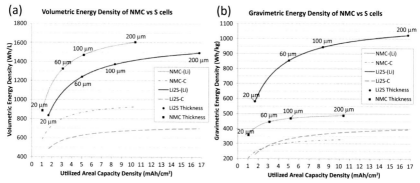

**Fig. 10.1** Cell modeling to compare the maximum feasible energy densities of lithium ion cells utilizing either an NMC or S positive electrode (in the discharged state). (A) Volumetric energy density, illustrating Li[Ni$_{0.33}$Mn$_{0.33}$Co$_{0.33}$]O$_2$ (NMC)||lithium cells (*dotted black*), NMC||carbon cells (*dash-dot dark gray*), sulfur (Li$_2$S)||lithium cells (*solid black*) and sulfur (Li$_2$S)||carbon cells (*dashed gray*). (B) Gravimetric energy density for the same cells. Note that the thickness of the cathode, which is determined by the active material and the areal capacity density, is indicated for the cells using lithium metal negative electrodes. The following parameters were used. Sulfur electrodes: 75 wt%, S; 20 wt%, carbon (2 g/mL); 5 wt%, PVDF (1.4 g/mL); 25%, porosity, 75%, utilization; and 2.15 V, average discharge potential. NMC electrodes: 96 wt%, NMC (4.77 g/mL); 2 wt%, carbon; 2 wt%, PVDF; 25%, porosity; 100%, utilization; and 3.6 V, average discharge potential. Carbon electrodes: 96 wt%, graphite; 2 wt%, carbon; 2 wt %, PVDF; 25%, porosity; and 100%, utilization. The separator was 16 μm thick and 0.561 g/mL with 40% porosity. All porosity was filled with liquid electrolyte, which had a density of 1.3 g/mL for NMC cells and 1.13 g/mL for Li$_2$S cells. Current collectors were 10 μm thick. 25% excess Li was used for lithium metal electrodes with no separate current collector. The pouch cell casing was ignored. X-axis values can be converted to sulfur loadings by multiplying by 0.797 mg$_S$/mAh. (*Reproduced with permission from P. Bonnick, J. Muldoon, Energy Environ. Sci. 13 (2020), 4808–4833, Copyright (2020) Royal Society of Chemistry.*)

Only positive electrode thicknesses between 20 μm and 200 μm are considered due to cost and likely fabrication limitations, respectively [11,12]. Firstly, the gray lines in both Fig. 10.1A and B demonstrate that Li-S cells with graphite negative electrodes will not surpass conventional Li-ion in terms of volumetric or gravimetric energy density. This is because the increase in areal capacity in the positive electrode, by changing from NMC to S, requires an identical increase in capacity of the negative electrode. Unfortunately, graphite is not a dense Li$^+$ storage material (i.e., 818 mAh/mL and 372 mAh/g$_C$) and, as a result, increasing the capacity of the graphite electrode increases the amount of non–energy-storing weight and volume in the cell at a rate that nearly cancels the gains granted using sulfur. As a result, Li–S batteries will only be able to compete with Li–ion in terms of energy density if lithium metal is used as the negative electrode; simply changing the positive electrode from NMC to S is not enough. Secondly, Fig. 10.1A demonstrates that if lithium metal electrodes become viable, then an NMC-Li cell will surpass a Li-S cell in terms of volumetric energy density. Fundamentally, this is because discharged S (i.e., $Li_2S$) is much less dense than NMC (i.e., 1.66 vs $4.77 g/cm^3$, respectively). On the other hand, Fig. 10.1B reveals that Li–S batteries could substantially surpass Li-ion in terms of gravimetric energy density if it achieves the modeled parameters. This is possible because $Li_2S$ has a specific capacity more than $7 \times$ greater than NMC: $1166 mAh/g$ vs $155 mAh/g$, respectively. Finally, Fig. 10.1B also clarifies that sulfur electrodes of equivalent thicknesses to conventional NMC electrodes (e.g., $60 \mu m$) will contain roughly 50% more capacity (i.e., $mAh/cm^2$). This means that charging or discharging the cell at a fast rate, such as fully charging the cell in 1 h, will necessitate using a current density roughly 50% higher than an NMC cell would require. Since the rate of (dis)charge has a major effect on conventional Li-ion capacity and cycle life, it should come as no surprise that requiring even faster (dis)charging rates in Li-S cells will also carry significant consequences that will become apparent later in this chapter. Decreasing the thickness of the electrode quickly reduces current densities but also quickly increases the amount of dead weight in the cell contributed by current collectors and separators; hence, why the solid black rises so sharply between 20 and $60 \mu m$ and why $60 \mu m$ is a relatively common electrode thickness to use in commercial Li-ion cells. In summary, Fig. 10.1 demonstrates that gravimetric energy density is where Li-S should strive to compete with Li-ion and that it will only be successful if lithium metal negative electrodes become viable and can support $>4 mAh/cm^2$, even at fast rates (e.g., $> 4 mA/cm^2$).

At this point in time, the challenges of Li-S batteries can be organized into three groups, which will be covered in the following section. (1) Challenges with sulfur electrodes when using liquid electrolytes. This section discusses the issue of low electrical conductivity of S as well as the infamous lithium polysulfide shuttle and the less acknowledged issue of low $Li_2S$ deposition kinetics at high lithium polysulfide concentrations. (2) The application of solid electrolytes in lithium-sulfur batteries. This section reiterates the importance of three-point contact regions, which is more difficult to achieve with solid electrolytes. It also explores the challenges and unexpected benefits of solid electrolytes, culminating in an explanation of the most pressing issue facing solid electrolyte systems: cracking.

(3) The application of lithium metal electrodes in lithium–sulfur batteries. This section confronts the realities of the 50-year-old challenge of lithium dendrite growth and the best methods to combat it.

## 10.2 The challenges of conventional sulfur electrodes when used with liquid electrolytes

Generally speaking, battery electrodes consist of active material (e.g., sulfur) particles mixed with electrically conducting carbon and some binder to both hold the electrode together and adhere it to the current collector. A thin (e.g., $20\,\mu m$) perforated plastic separator is placed between the positive and negative electrodes to prevent a short circuit, and then, the porosity in the electrode and separator is filled with a liquid electrolyte to conduct $Li^+$ ions. The main advantage of a liquid electrolyte over a solid electrolyte is an enhanced ionic conductivity. The ionic conductivity across a soaked separator is about $5–10\,mS/cm$, and liquids can maintain relatively high conductivities at low temperatures, generally until they start to freeze. Beyond this primary benefit, they can be made in large quantities relatively easily, and on the assembly line, liquid electrolytes are injected into nearly completed cells in a facile manner.

The first challenge of sulfur electrodes is the extremely low electrical conductivity of sulfur ($\sim 10^{18}\,S/cm$) and its discharge product, $Li_2S$ ($\sim 10^{-16}\,S/cm$). Their $Li^+$ conduction rates are also relatively low. In order for a sulfur atom to be reduced, it must simultaneously come into contact with an electron from the electrically conductive network (i.e., carbon) and $Li^+$ from the ionically conductive network (i.e., electrolyte). Areas where the active material, electron, and lithium can all reach can be referred to as three-point contact regions. Due to the poor electrical and ionic conductivity of S and $Li_2S$, sulfur electrodes must be meticulously engineered to expand three-point contact regions. This is in contrast to NMC, which conducts both electrons and $Li^+$ well enough that $e^-$ and $Li^+$ do not have to come directly from their respective networks. As a result, an NMC electrode, using about 10-$\mu m$–diameter NMC particles, can operate well with only about 1–2 wt% carbon to enhance the electrical conductivity, whereas sulfur electrodes require smaller particle sizes and significantly more carbon to deliver their full capacity (a.k.a. 100% utilization). A seminal publication on Li-S cells gained recognition because of the use of porous, macroscopic structures to ensure electrons reached the sulfur active material; Ji et al. infused a mesoporous carbon framework with melted sulfur through capillary action [13]. The enhanced electrical connectivity enabled the electrode to deliver about 79% of the available sulfur capacity (i.e., a utilization of 79%). Unfortunately, sulfur electrodes made with mesoporous carbons tend to have high carbon contents and therefore low specific capacity. For example, a sulfur electrode that is 50 wt% carbon only has a specific capacity of 837 mAh/g instead of 1675 mAh/g for pure sulfur. For researchers, this defines a delicate balance between non-active mass (e.g., carbon) and utilization, where the goal is to find the ideal amount of non-active mass to maximize the specific capacity (i.e., utilization/mass of electrode). In the models presented in Fig. 10.1, the sulfur electrode was assumed to be 75 wt% sulfur and capable of delivering 75% utilization, meaning its total specific capacity was 942 mAh/$g_{electrode}$. In our opinion, these values are generous but possible.

As research progressed, efforts shifted from sulfur-filled carbon to carbon-coated (or doped) sulfur in order to increase the weight fraction of sulfur. In effect, the uniform distribution of a minimal amount of carbon throughout the electrode is a fabrication challenge. Many groups sought to create individual sulfur particles (i.e., a powder) equipped with all the attributes they would need to enhance utilization and cycle life so that conventional NMC electrode fabrication processes could be used to create the final electrode. In one example, Bucur et al. formed truffle-shaped sulfur particles that were > 85 wt% sulfur, doped with carbon, and coated with an electrically conductive, polyelectrolyte membrane [14–17]. These membranes were about 30 nm thick, elastic to allow expansion of the sulfur active material, facilitated the passage of smaller $Li^+$ cations, and slowed the passage of larger lithium polysulfide anions. Cycling experiments demonstrated that truffle particles could effectively extend the cycle life of electrodes with high sulfur loadings (e.g., 3.5 mg/cm$^2$) out to 200 cycles while maintaining a utilization above 70%, but these electrodes were only 50 wt% sulfur [17]. Several more recent examples exist as well, with varying levels of success, but none with the combination of ~96 wt% active material and high utilization that NMC electrodes can achieve [18–20]. This does not doom the sulfur electrode; it simply means that 1675 mAh/$g_{electrode}$ is a fantasy; a goal of 1000 mAh/$g_{electrode}$ would be a more realistic, but still impressive, target.

The second challenge of sulfur electrodes used with liquid electrolytes is the lithium polysulfide shuttle, a process that has been thoroughly studied and described elsewhere for interested readers [2,3,21–24]. In short, during discharge, sulfur active material becomes increasingly lithiated, progressing from $S_8$ to $Li_2S_8$ and then progressively to $Li_2S$, where the number of

S atoms decreases stepwise from 8 to 1. These $Li_2S_x$ molecules are called lithium polysulfides, and the range $Li_2S_8$ to $Li_2S_4$ is soluble in ether-based liquid electrolytes and some polymer electrolytes. Once dissolved, they are free to diffuse across the separator to the negative electrode. In all batteries, it is detrimental to have active material from one electrode migrate to the other electrode, with the obvious exception of the ion that is intended to carry charge across the separator, such as $Li^+$. When dissolved $Li_2S_x$ molecules reach the negative electrode, they become reduced, which either deposits sulfur material onto the negative electrode or simply transfers electrons (i.e., a shuttle effect), bypassing the external circuitry. Losing sulfur to the negative electrode reduces the capacity of the sulfur electrode, while the deposited sulfur increases the resistance of the lithium electrode. Additionally, any electron transfer that bypasses the external circuit reduces the coulombic efficiency, an important metric that will be discussed later. Taken together, these effects dramatically reduce the cycle life of Li–S cells and have spurred a massive research effort to find a way to prevent the diffusion of lithium polysulfides across the separator. There have been two general approaches to reduce the shuttle effect in conjunction with a liquid electrolyte: (1) Shells around the sulfur particle to physically slow the outward diffusion of dissolved lithium polysulfides [2,13–17]. The polyelectrolyte layers in the truffle particle, discussed above, were an example of this strategy. (2) Non-active materials that bind to lithium polysulfides, preventing them from moving [25–27]. This approach has been excellently summarized by Pang et al., who explain how "the surface functionality, intrinsic polarity, electro-/nucleophilicity and/or redox potential play an important role in determining the strength of the interaction" between the lithium polysulfide and the non-active material [3]. The non-active material can double as another structure within the cell, including the separator, electrical conduction network, or sulfur-particle shell. Despite a plethora of creative ideas and enlightening research into the mechanisms of approaches (1) and (2), none of these proposed solutions has been able to prevent lithium polysulfide diffusion to the lithium electrode to the extent necessary to facilitate cycle lives similar to conventional Li-ion cells with high-loading sulfur electrodes. Indeed, many research papers only report $\sim$100 cycles in order to hide the abrupt failure of cells beyond that point.

Finally, the third challenge of sulfur electrodes when used with liquid electrolytes is the saturation limit of lithium polysulfides. During discharge, there is a period of time within the first 25% of discharge where all of the sulfur active material is in the form of $Li_2S_8$ to $Li_2S_4$ and thus attempts to dissolve into the ether-based liquid electrolyte simultaneously. If the solubility limit of lithium polysulfides is exceeded, Fan and Chiang have shown, in Fig. 10.2B and C, that the deposition kinetics of $Li_2S$ slows down dramatically due to either surface passivation or increased viscosity, which in turn increases the sulfur electrode resistance, prematurely terminating discharge [28]. Fig. 10.2B suggests that the electrolyte/sulfur (E/S) ratio at which the saturation limit is reached is approximately 4 mL/g, although the exact ratio will depend on the electrolyte formulation. This phenomenon is not commonly acknowledged in the literature due to its severity; most researchers use E/S ratios much larger than 4 mL/g in order to safely ignore this saturation complication because their focus is studying a different aspect of the electrode. However, some recent publications have probed the limits of how lean an electrolyte can be used, such as by Li *et al.* in Fig. 10.2D and E who used 2.8 mL/g at a C/40 rate [29]. Unfortunately, this C/40 discharge rate will not be useful in most practical applications for lightweight batteries, such as drones. The lowest E/S ratio that is capable of supporting high rates (e.g., 2C) is of crucial importance because energy density estimates are strongly dependent on the E/S ratio. For instance, Fig. 10.1 assumed an E/S ratio of 0.47 mL/g, in line with the volume of electrolyte used in NMC electrodes. This model is plotted again in Fig. 10.2A and F, along with identical models with increasing E/S ratios. Evidently, an E/S ratio of less than about 3 mL/g will be required to compete with NMC||Li cells, which seems unlikely to be achievable without also creating the detrimental lithium polysulfide saturation conditions discussed above.

## 10.2.1 Solid electrolytes in lithium-sulfur batteries

Since lithium polysulfide dissolution and saturation are such major challenges for reversible and high-rate lithium-sulfur cells, it is highly desirable to sidestep both issues by changing the discharge mechanism to progress directly from solid $S_8$ to solid $Li_2S$ without passing through a solvated state. One way to achieve this is by using a solid electrolyte that cannot dissolve lithium polysulfides but still conducts $Li^+$. Fig. 10.3 contains typical cycling curves for Li-S cells made with liquid (Fig. 10.3A) or solid (Fig. 10.3B) electrolytes, which demonstrates that the solid–solid process occurs at a slightly lower discharge potential and generally with a higher overpotential between discharge and charge [30]. Major advances in solid-state $Li^+$ conductors has occurred over the past 15 years, where some sulfide-based solid electrolytes

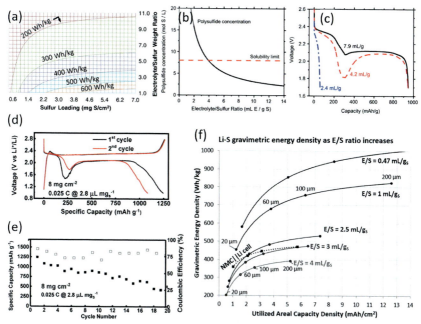

**Fig. 10.2** (A) Modeled gravimetric energy densities of a hypothetical Li-S cell with a 75 wt%, S; 20 wt%, carbon (2 g/mL); 5 wt%, PVDF (1.4 g/mL) positive electrode; a utilization of 75%; and an average discharge potential of 2.15 V. The following assumptions were also made: The liquid electrolyte had a density of 1 g/mL; the separator was 25 μm thick, had a dry density of 0.561 g/mL, and had a porosity of 40% (which was filled with electrolyte in the model); the positive electrode current collector was 10 μm thick; 100% excess Li was used with no separate current collector; and the pouch cell casing was ignored. (B) Lithium polysulfide concentration in the electrolyte if all sulfur is dissolved versus electrolyte/sulfur ratio. Note that below a ratio of about 4 mL/g, the lithium polysulfide concentration would exceed the solubility limit of typical ethers, producing a saturated solution and precipitate. The concentration is calculated by multiplying the molar mass of sulfur by the E/S ratio and inverting the product. (C) Galvanostatic (C/4) discharge curves for Li-S cells with sulfur-carbon composite cathodes at three different electrolyte/sulfur ratios. As the electrolyte/sulfur ratio decreases from 7.9 mL/g, the voltage drop at ~300 mAh/g, which corresponds to the onset of $Li_2S$ nucleation and growth, increases dramatically in size before disappearing altogether. Electrolyte: 1 DOL: 1 DME, 0.5 M LiTFSI, 0.15 M $LiNO_3$. (D) Charge/discharge voltage profile and (E) cycling performance at C/40 of a sulfur electrode with 8 $mg_S/cm^2$ and an E/S ratio of 2.8 mL/g. (F) Model of the gravimetric energy density of Li-S cells with increasing amounts of liquid electrolyte (density 1.13 g/mL) in comparison with an NMC∥Li cell (*dotted line*). The model parameters are identical to those within Fig. 10.1, except for the amount of electrolyte within the Li-S cell. Dots indicate the thickness of the positive electrode. (*A: Reproduced with permission from P. Bonnick, E. Nagai, J. Muldoon, J. Electrochem. Soc. 165 (2018), A6005, Copyright (2018) The Electrochemical Society; B and C: Reproduced with permission from F. Y. Fan, Y.-M. Chiang, J. Electrochem. Soc. 164 (2017), A917, Copyright (2017) The Electrochemical Society; D and E: Reproduced with permission from M. Li, Y. Zhang, Z. Bai, W.W. Liu, T. Liu, J. Gim, et al., Adv. Mater. 30 (2018), 1804271, Copyright (2018) John Wiley & Sons; F: Reproduced with permission from P. Bonnick, J. Muldoon, Energy Environ. Sci. 13 (2020), 4808–4833, Copyright (2020) Royal Society of Chemistry.*)

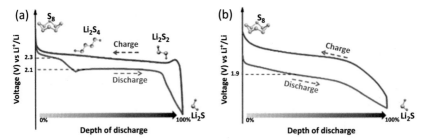

**Fig. 10.3** Typical charge/discharge potential curves of (A) solid–liquid dual-phase Li-S redox reactions and (B) solid-phase Li-S redox reactions. *(Reproduced with permission from X. Yang, J. Luo, X. Sun, Chem. Soc. Rev. 49 (2020), 2140–2195, Copyright (2020) Royal Society of Chemistry.)*

can support $Li^+$ conductivities equal to, or greater than, that of conventional liquid-electrolyte-soaked separators [31,32]. Unfortunately, sulfide-based solid electrolytes are only electrochemically stable up to between 2.3 and 2.6 V vs $Li/Li^+$, as calculated by Richards et al. [33]. When in contact with any positive electrode active material with a higher operating potential, the active material oxidizes the electrolyte, which usually creates a resistive layer of reaction products between them, an example of which is shown in Fig. 10.4A [33]. Interestingly, there is evidence that when sulfur active material is used, sulfide-based solid electrolytes will participate reversibly in the electrochemistry of the cell without detriment [34,35]. This is demonstrated by the first three half-cycles of the S-composite/nanocrystalline $Li_3PS_4 \cdot \frac{1}{2}LiI/Li$ cell shown in Fig. 10.4C. During the first discharge (black dashed line), there is only one potential plateau for the sulfur active material from about 2.1 to 1.5 V, as expected for a solid–solid process. When the cell is charged for the first time, the sulfur active material is delithiated first (black solid line) followed by some of the electrolyte above about 2.6 V vs $Li/Li^+$ [34]. Nagai et al. have shown evidence that lithium thiophosphate electrolytes can shed one or two sulfur atoms during charging above about 2.5 V vs $Li/Li^+$ via Reaction (10.1) and possibly Reaction (10.2) [35]:

$$2Li_3PS_4 \rightarrow Li_4P_2S_7 + 2Li^+ + S + 2e^- \qquad (10.1)$$

$$Li_4P_2S_7 \rightarrow Li_2P_2S_6 + 2Li^+ + S + 2e^- \qquad (10.2)$$

It was revealed during the second discharge in Fig. 10.4C that this charged $Li_3PS_4 \cdot \frac{1}{2}LiI$ "active material" was reversible by returning the 0.5 mAh/cm$^2$ via the higher potential plateau at about 2.2 V (gray dotted line). Furthermore, Fig. 10.4D shows that the reversible capacity provided

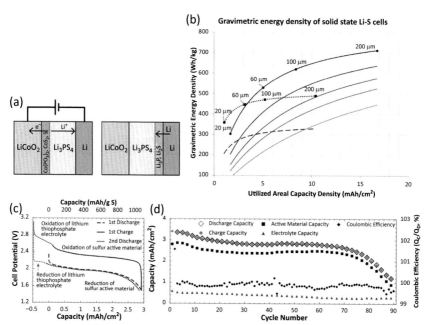

**Fig. 10.4** (A) Example of SEIs formed between Li$_3$PS$_4$ solid electrolyte and LiCoO$_2$ (left) or lithium metal (right). (B) Model of the gravimetric energy density of solid-state Li-S cells using Li$_3$PS$_4$·½LiI solid electrolyte in comparison with the NMC cells from Fig. 10.1. All model parameters are the same as in Fig. 10.1, except the Li-S separator is 100% Li$_3$PS$_4$·½LiI with a density of 2.07 g/mL and the separator thickness varies as indicated. The *dots* indicate the thickness of the positive electrode. (C) Discharge and charge curves as a function of capacity for the S composite/nanocrystalline Li$_3$PS$_4$·½LiI/Li cell at 60°C. The cell was discharged at 2.28 mA/cm$^2$ (C/2) and charged at 0.456 mA/cm$^2$ (C/10) with potential limits of 1.5 V and 3.1 V. (D) Evolution of the full-cell capacities (*gray circles* and *hollow diamonds*) and coulombic efficiency (*black diamonds*) as a function of cycle number for the cell shown in (C). The charge capacities of the sulfur active material (*black squares*) and solid electrolyte (*gray triangles*) are shown separately but add together to form the total charge capacity (*gray circles*). (A: Reproduced with permission from W.D. Richards, L.J. Miara, Y. Wang, J.C. Kim, G. Ceder, Chem. Mater. 28 (2016), 266–273, Copyright (2016) American Chemical Society; B: Reproduced with permission from P. Bonnick, J. Muldoon, Energy Environ. Sci. 13 (2020), 4808–4833, Copyright (2020) Royal Society of Chemistry; C and D: Reproduced with permission from P. Bonnick, K. Niitani, M. Nose, K. Suto, T.S. Arthur, J. Muldoon, J. Mater. Chem. A 7 (2019), 24173–24179. Copyright (2019) Royal Society of Chemistry.)

by the electrochemically active electrolyte (gray triangles) was stable for longer than the sulfur active material (black squares) [34]. Overall, these findings demonstrated that the reversible nature of sulfide solid electrolytes is likely a benefit to Li-S cells instead of a hindrance. Furthermore, the absence of overcharge signifies that no lithium polysulfide shuttle occurred; nearly all the capacity charged during the first charge (solid line) is returned during the second discharge (dotted line).

At the negative electrode side, the instability of sulfide solid electrolytes has been somewhat mitigated by the inclusion of halides in the solid electrolyte. When the solid electrolyte becomes reduced by lithium metal, a resistive interphase is formed between them, as demonstrated in Fig. 10.4A. Halides within the electrolyte form lithium halide products that act as electrical insulators and acceptable $Li^+$ conductors, thereby preventing further growth of the resistive interphase between the lithium and electrolyte. Halide-doped sulfide solid electrolytes, such as $Li_3PS_4 \cdot \frac{1}{2}LiI$, tend to be able to support $Li^+$ conductivities between about 1 and 3 mS/cm at room temperature, which could be sufficient for low-rate applications. Higher-rate applications, such as drone take-off, will require a further increase in the $Li^+$ conductivity.

Another important aspect of inorganic solid electrolytes is the thickness of the separator layer. Conventional battery separators are perforated, lightweight plastic layers about 15–25 μm thick. It is not trivial to make 25-μm-thick inorganic solid electrolyte layers, and as a result, the achievable energy density of an all-solid-state Li-S battery will strongly depend on the thickness of that layer. Fig. 10.4B uses the same model as in Fig. 10.1, except with $Li_3PS_4 \cdot \frac{1}{2}LiI$ ($\rho = 2.07$ g/mL) as the electrolyte and separator. It is evident that a separator thickness of about 25 μm and an areal capacity $> 4$ mAh/cm$^2$ would be required to compete with NMC||Li cells.

With solid electrolytes, the low electrical conductivity of $S_8$ and $Li_2S$ is still a challenge, with the same sorts of solutions discussed earlier in the case of liquid electrolytes. However, unlike liquid electrolytes, the solid electrolyte must be included in the formation of the electrode film. In a conventional electrode design, once the active material, electrically conductive additives, and solid electrolyte have been mixed together as uniformly as possible, this electrode material must be pressed under immense pressure in an attempt to reduce the porosity to zero. Any remaining porosity will act as a barrier to $Li^+$ conduction and increase the resistance of the electrode. Unfortunately, compact all-solid-state electrodes suffer from mechanical issues due to the expansion and contraction of the active material. For example, $S_8$ grows 80 vol% when it is lithiated into $Li_2S$. Since inorganic solid electrolytes tend to be somewhat brittle, all-solid-state electrodes generally cannot accommodate this expansion and crack [36]. As cycling continues, the spreading cracks act as barriers to both $e^-$ and $Li^+$ conduction, increasing the electrode resistance and reducing the capacity as active material becomes electrically or ionically marooned from the rest of the electrode. Several strategies have been investigated to overcome this issue of mechanical stress within the positive electrode, including (1) hybrid systems where a liquid electrolyte is used within the positive electrode and a solid electrolyte is used to separate the positive and negative electrodes [37,38]. This concept relies upon the ability of the liquid to "squeeze out" of the positive electrode into spare volume within the cell package when the active material expands. The challenge here becomes preventing the liquid electrolyte from contacting the lithium metal negative electrode. (2) Intentionally porous positive electrode structures to provide space for the active material to expand into [39]. The challenge here becomes microscopically engineering the structure to facilitate good electronic and ionic conductivity despite the porosity. (3) Elastic solid electrolytes that can be compressed to relieve mechanical stress within the electrode [40,41]. The challenge here becomes developing solid electrolytes, such as polymer electrolytes, that can be compressed and released repeatedly while maintaining a sufficiently high ionic conductivity to supply adequate power. Polymer electrolytes are attractive in general because of their inexpensiveness and potential to be easily incorporated into existing cell manufacturing processes; however, polymer electrolyte $Li^+$ conductivities have historically been <0.1 mS/cm at room temperature, which is too low for many applications. The risk of reintroducing the lithium polysulfide shuttle also exists when using polymer electrolyte within the separator layer; Song et al. demonstrated that the ubiquitous poly(ethylene oxide) (PEO) electrolyte dissolves and transmits lithium polysulfides [42]. This is not necessarily the case for all polymer electrolytes, but caution must be exercised to avoid negating the primary benefit of using a solid electrolyte in a lithium-sulfur cell. Overall, the choice of phase of the electrolyte in a Li-S cell dictates the challenges the cell must overcome. Liquid electrolytes do not exacerbate structural problems in the positive electrode but must contend with the lithium polysulfide shuttle. Inorganic solid electrolytes are the opposite; they prevent shuttling but lose electrical and ionic conductivity due to structural failure during cycling. Polymer-inorganic composite electrolytes may be able to provide elasticity while preventing the polysulfide shuttle, but more study will be required to determine whether they can withstand the physical stresses of cycling within the sulfur electrode.

## 10.3 Lithium metal electrodes in lithium-sulfur batteries

The majority of the Li-S literature is focused on overcoming lithium polysulfide dissolution within the sulfur positive electrode; however, the lucrative promise of high-energy-density Li-S batteries is also dependent on the reversible operation of a lithium metal negative electrode with a high areal capacity and coulombic efficiency [11,12,43]. Lithium metal has the highest gravimetric capacity density of any metallic element: 3861 mAh/g. As demonstrated in Fig. 10.1, this advantage must be leveraged for Li–S to have any hope of surpassing the energy density of conventional Li-ion technology. The primary challenge facing lithium metal electrodes is the growth of lithium dendrites during charging, which pierce through the electrolyte and separator to contact the positive electrode and cause a short circuit. In fact, graphite was adopted in 1991 as the negative electrode in conventional Li-ion cells instead of lithium metal precisely because lithium dendrites do not form when $Li^+$ intercalates into a host material like graphite.

A crucial metric when evaluating proposed lithium electrode technologies is the coulombic efficiency (CE), which is essentially a measurement of the stability of an electrode (or cell) as it is cycled and can be used to roughly predict the cycle life of the electrode (or cell). CE is usually defined as $Q_{Discharge}/Q_{Charge}$ for an electrode, where $Q$ is the capacity, but variations do exist [44,45]. Any deviation from a CE of 100% indicates that the cell is either experiencing a shuttle effect or degrading through the consumption of $Li^+$, active material, or electrolyte. In lithium batteries, it is common for $Li^+$ to be permanently consumed in whatever process is reducing the CE from 100%, which reduces the available $Li^+$ supply and thus the cell capacity ($Q$). If the CE is constant over many cycles ($n$), the longevity of the cell can be predicted using Eq. (10.3),

$$\frac{Q_{Remaining}}{Q_{Initial}} = CE^n \tag{10.3}$$

where $Q_{Remaining}/Q_{Initial}$ is the normalized remaining capacity and $n$ is the number of cycles. For example, this equation reveals that a lithium electrode with no excess capacity would be able to deliver 80% of its initial capacity after 1000 cycles only if the average CE were $> 99.978\%$. Similarly, a CE of 99.955% would be required for a cycle life of 500 cycles. For comparison, Bond et al. measured the CE of a conventional $LiCoO_2\|graphite$ cell to be 99.945% [46]. Unfortunately, many common potentiostats are incapable of measuring the CE accurately enough to see such small deviations from 100%, although such potentiostats are sufficient when the CE is below 99%. Three methods currently exist to enhance the precision of CE measurements: (1) (ultra-)high precision coulometry, which requires special equipment that is becoming easier to rent or buy commercially [47]. (2) Using symmetric cells assembled with identical electrodes except one is lithiated and the other is not. This tactic effectively doubles the "signal intensity"

Future prospects for lithium-sulfur batteries **341**

and therefore improve the chances that a conventional potentiostat can measure the coulombic inefficiency precisely [48]. (3) A precise average CE can be measured over many (n) cycles using either low depth-of-discharge cycling, as explained by Adams et al. [45], or full depth-of-discharge cycling using Eq. (10.3), mentioned previously. CE measurements are a key metric of lithium electrode research, and so it is imperative that some effort be made to make and report precise measurements.

Because lithium has such a low reduction potential ($-3.04$ V vs SHE), most electrolytes brought into contact with lithium will be reduced and a solid electrolyte interphase (SEI) will form between the lithium and electrolyte. This process consumes $Li^+$, but usually SEIs are stable and only grow thicker at a very slow rate. Consequently, further $Li^+$ loss (i.e., a CE $<$ 100%) occurs only when fresh lithium metal is exposed to the electrolyte, which occurs when new lithium metal is deposited and especially when high surface area lithium dendrites grow out of the deposit. Many approaches have been attempted to reduce lithium dendrite growth and improve the CE of lithium metal electrodes in liquid electrolytes. Examples include increasing the surface area of the lithium metal deposit [49], lithium nucleation aids [50], electrolyte additives [51], and artificial SEIs [52,53]; homogenizing the electric field at the deposition surface [54]; and increasing the $Li^+$ concentration at the deposition surface [55]. For interested readers, these approaches are reviewed in more details elsewhere [56–62]. Alas, no combination of these approaches has yet produced a CE $>$ 99.94% to match conventional Li-ion cells. For this reason, and because solid electrolytes might be required in Li-S cells to fully prevent the lithium polysulfide shuttle, the remaining discussion here will be focused on approaches involving solid electrolytes.

Solid electrolytes have two major theoretical advantages over liquid electrolytes when used with a lithium metal electrode: (1) Inorganic solid electrolytes have a transference number of 1 (i.e., $Li^+$ alone carries charge between the electrodes, instead of some ratio of an anion and $Li^+$). Theoretically, this means that concentration polarization, a driving factor of lithium dendrite growth, should not occur, since the concentration of $Li^+$ within the electrolyte at the deposition surface should remain constant. (2) Most inorganic solid electrolytes have a high Young's modulus, which was theorized by Monroe and Newman to physically prevent the penetration of lithium dendrites [63]. Despite these theoretical advantages, lithium dendrites have been demonstrated to grow through inorganic solid electrolytes, including a single crystal with a large Young's modulus [64,65].

Recently, the growth of lithium dendrites through sulfide glass solid electrolytes was correlated to the lithium plating current density, as shown in Fig. 10.5A. Electrochemomechanical models already include the current density as a crucial factor [66–68], but this demonstration of correlation is striking since the Li$^+$ conductivities on the x-axis reach notable levels (i.e., 6.7 mS/cm at 60°C) and yet the critical (i.e., maximum) current density (CCD) on the y-axis does not yet reach the required current densities for competitive Li-S cells (i.e., >4 mA/cm$^2$). Fig. 10.5A suggests that higher Li$^+$ conductivities will be required to facilitate high charging current densities, but these results may be limited to the electrolyte and pressure conditions under which this cell was tested. It is likely prudent to test other electrolytes and conditions to determine whether high current densities will be possible. A relatively easy way to stress-test the system is to perform critical current density (CCD) measurements, such as the one demonstrated in Fig. 10.5B.

Bonnick et al. explain that, in a typical CCD experiment, a constant current, such as 0.1 mA/cm$^2$, is applied across the cell for 1 h and, then, the current is reversed for the same amount of time to return the two electrodes to their initial state of charge. This cycle is then repeated, using a stepwise higher current density (e.g., 0.2, 0.3, 0.4... mA/cm$^2$), to push the separator/electrolyte/electrode system to its rate limit, where a sudden drop in potential (or area-specific resistance) indicates that a dendrite has caused a

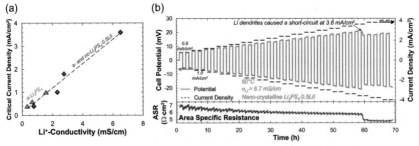

**Fig. 10.5** (A) Relationship between the onset of lithium dendrite growth (i.e., critical current density) and the Li$^+$ conductivity of Li$_3$PS$_4$ and Li$_3$PS$_4$·½LiI. The *dashed line* is a linear fit to all seven data points. (B) Critical current density test of a Li/nanocrystalline Li$_3$PS$_4$·½LiI/Li cell at 60°C. The current is held constant for two cycles at each current, starting at 0.8 mA/cm$^2$ and rising to 4.0 mA/cm$^2$ in stepwise increments of 0.2 mA/cm$^2$. The lower plot depicts the area-specific resistance (ASR = overpotential/current density). *(Reproduced with permission from P. Bonnick, K. Niitani, M. Nose, K. Suto, T.S. Arthur, J. Muldoon, J. Mater. Chem. A 7 (2019), 24173–24179, Copyright (2019) Royal Society of Chemistry.)*

soft short circuit. The current density at which this is observed is dubbed the critical current density, which can be thought of as an approximate maximum current density for the system tested [34]. This relatively easy test could be used to quickly determine whether a proposed solution to lithium dendrite growth is effective, no matter the form of the proposed solution.

Many of the same remedies to lithium dendrite growth that have been investigated for liquid electrolytes have also been attempted with solid electrolytes. Two approaches appear to hold some promise, including 3D current collectors and nucleation aids. Three-dimensional current collectors (a.k.a. lithium metal hosts or composite lithium metal anodes) provide two effective benefits: (1) They increase the surface area, thereby decreasing the local current density; (2) They allow the lithium to plate in voids, which reduces pressure on the lithium deposit and contact with the electrolyte. Reducing both the current density and pressure on the lithium deposit has been observed by Hitz et al. to allow current densities as high as $10 \, mA/cm^2$, with an electrolyte that supports a $Li^+$ conductivity of only about 1 mS/cm, as shown in Fig. 10.6 [39]. Care must be exercised when using 3D current collectors to ensure that they do not weigh too much, lest they reduce the energy density of the lithium electrode to something close to that of graphite. Yet they must also be strong enough to withstand repeated lithium expansion and contraction without breaking down over many cycles.

Nucleation aids, as explained by Yan et al., are metals that dissolve in lithium and reduce the nucleation overpotential during lithium deposition, thereby reducing lithium dendrite growth [50]. As an example, Lee et al. combined the concepts of a 3D current collector (carbon black) with a nucleation aid (silver nanoparticles) to achieve an unprecedented CE while cycling $6 \, mAh/cm^2$ at $3.4 \, mA/cm^2$ using a solid electrolyte ($Li_6PS_5Cl$ argyrodite) [69]. The capacity and CE along with a schematic of the lithium side of the cell are shown in Fig. 10.7A and B, respectively. Although Lee et al. only claimed to have achieved a $CE > 99.8\%$, it is likely that they were not using a high-precision coulometer to perform the measurement and so the true CE was actually higher. Since the cell only lost 11% capacity over 1000 cycles despite having no excess lithium, Eq. (10.3) suggests that the average CE is more likely 99.988%. Such a high value certainly surpasses the CE required to make a long cycle life cell. Understanding in detail how this lithium electrode works could provide direction for future research. For example, Lee et al. suggested that lithium intercalates and diffuses through the carbon black to electroplate between the stainless steel

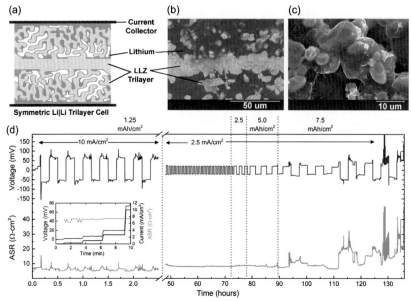

Fig. 10.6 (A) Diagram of trilayer lithium symmetric cell with 14-μm-thick separator layer. (B) Backscattered electron image of trilayer garnet (bright areas) with both porous layers partially filled with lithium metal (dark areas). (C) SEM image of lithium filling the pores before cycling. (D) Cell voltage response (*blue (black in the printed version)*) and area-specific resistance (ASR, *orange (gray in the printed version)*) during galvanostatic cycling at 10 mA/cm² and then 2.5 mA/cm². *(Reproduced with permission from G.T. Hitz, D.W. McOwen, L. Zhang, Z. Ma, Z. Fu, Y. Wen, et al., Mater. Today 22 (2019), 50–57, Copyright (2019) Elsevier.)*

current collector and the carbon black [69]. They also raised the issue of silver particle migration from within the carbon black toward the current collector and demonstrated that the electrode does not function nearly as well without either the carbon black or silver nanoparticles.

No matter what the proposed lithium electrode structure, one often overlooked consideration is that the lithium electrode must be lithiated prior to cell assembly. This is because $Li_2S$ is far more expensive than $S_8$ and so the lithium will not be able to originate in the positive active material the way it does in many conventional Li-ion materials like NMC. Several techniques of pre-lithiating electrodes have been investigated, including chemical and electrochemical methods, direct contact with lithium metal and sacrificial electrode additives. Techniques that use inexpensive sources of lithium

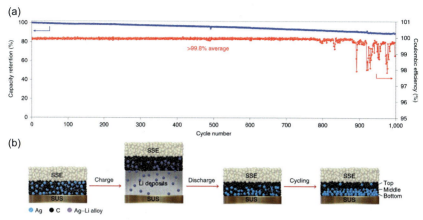

**Fig. 10.7** (A) Cycle life and coulombic efficiency of an NMC/Li$_6$PS$_5$Cl/Ag–C pouch cell (600 mAh). The charge/discharge rate was 0.5 C/0.5 C (0.5 C = 3.4 mA/cm$^2$) and the potential window was 2.5–4.25 V versus Li/Li$^+$ at 60°C. The areal capacity loading of the NMC cathode was 6.8 mAh/cm$^2$. (B) Schematic of Li plating/stripping on the current collector under a Ag–C nanocomposite layer during charging and discharging. *(Reproduced with permission from Y.-G. Lee, S. Fujiki, C. Jung, N. Suzuki, N. Yashiro, R. Omoda, et al., Nat. Energy 5 (2020), 299–308, Copyright (2020) Nature.)*

and can be performed in ambient or dry room conditions will likely be favored. For interested readers, Jin et al. and Holtstiege et al. have written informative reviews on this subject [70,71].

## 10.4 Path forward

Lithium-sulfur technology has been widely touted as the next generation of batteries beyond Li-ion due to the massive gravimetric capacity of both lithium metal and sulfur. Until recently, the progress of Li-S technology was hindered by the serious challenges associated with liquid electrolytes. For instance, the thoroughly studied dissolution of lithium polysulfides into ether-based liquid electrolytes has persisted at troublesome levels through countless attempts to prevent it, resulting in reduced cycle life. Furthermore, achieving 400 Wh/kg in a Li-S cell will require an electrolyte/sulfur weight ratio < 3; however, recent studies have demonstrated that Li$_2$S deposition rates (during discharge) at electrolyte/sulfur weight ratios < 4 cannot support reasonable current densities. Such revelations suggest that producing a Li-S cell with performance metrics surpassing conventional Li-ion technology will be extremely difficult—if not impossible—to achieve with the ether-based liquid electrolytes historically used in Li-S cells. Consequently, new approaches are needed.

Solid electrolytes are an alternative approach that eliminates both major challenges that plague ether-based liquid electrolytes. Inorganic solid electrolytes functionally do not dissolve lithium polysulfides, thereby removing the challenge of the polysulfide shuttle from the system and eliminating lithium sulfide deposition as part of the sulfur electrode discharge mechanism. Instead, sulfur and lithium react directly from solid S to solid $Li_2S$ without passing through a dissolved state. Solid electrolytes are not without their own challenges though. Within the sulfur electrode, the expansion and contraction of sulfur active material by 80 vol% during cycling applies immense pressure on the electrode causing cracking, which hinders both $Li^+$ and electron conduction. At the lithium metal side, the infamous hypothesis that solid electrolytes require only a high enough Young's modulus to physically prevent lithium dendrite growth has been proven false. Evidently, the electrochemomechanical properties of lithium in contact with solid electrolyte are crucial and continue to be investigated and clarified. In particular, the current density of lithium plating seems to strongly affect the growth of lithium dendrites through the solid electrolyte. Achieving a utilizable areal capacity $>4$ mAh/cm$^2$ at practically relevant current densities (i.e., $> 4$ mA/cm$^2$ in this case) continues to be a challenge, but 3D current collectors have been demonstrated to dramatically reduce the local current density, thereby facilitating faster charging without dendrite growth. An example of a feasible cell design could include a sulfur electrode crafted with an elastic solid polymer electrolyte to withstand active material expansion and contraction, an inorganic solid electrolyte separator layer with a high $Li^+$ conductivity, and a lithium electrode relying on both a 3D current collector and nucleation aids to facilitate non-dendritic lithium deposition. Such a system will have to ensure that the polymer and inorganic electrolytes are chemically compatible and that $Li^+$ transport across their interface is fast. Successfully realizing robust Li-S cell technology could unlock many real-world applications that require cheaper or lighter energy storage than the conventional Li-ion technology of today. With every new research effort, the challenges become clearer and less severe, revealing a bright future.

## Dedication

This chapter is dedicated to our loving wives, who supported us unconditionally throughout the research that contributed to this work.

# References

[1] M. Barghamadi, A.S. Best, A.I. Bhatt, A.F. Hollenkamp, M. Musameh, R.J. Rees, T. Rüther, Lithium–sulfur batteries—the solution is in the electrolyte, but is the electrolyte a solution? Energy Environ. Sci. 7 (2014) 3902–3920, https://doi.org/10.1039/C4EE02192D.

[2] Z.W. Seh, Y. Sun, Q. Zhang, Y. Cui, Designing high-energy lithium–sulfur batteries, Chem. Soc. Rev. 45 (2016) 5605–5634, https://doi.org/10.1039/C5CS00410A.

[3] Q. Pang, X. Liang, C.Y. Kwok, L.F. Nazar, Advances in lithium–sulfur batteries based on multifunctional cathodes and electrolytes, Nat. Energy 1 (2016) 16132, https://doi.org/10.1038/nenergy.2016.132.

[4] H.-J. Peng, J.-Q. Huang, X.-B. Cheng, Q. Zhang, Review on high-loading and high-energy lithium–sulfur batteries, Adv. Energy Mater. (2017) 1700260, https://doi.org/10.1002/aenm.201700260.

[5] A. Bhargav, J. He, A. Gupta, A. Manthiram, Lithium-sulfur batteries: attaining the critical metrics, Joule 4 (2020) 285–291, https://doi.org/10.1016/j.joule.2020.01.001.

[6] M. Hagen, D. Hanselmann, K. Ahlbrecht, R. Maça, D. Gerber, J. Tübke, Lithium–sulfur cells: the gap between the state-of-the-art and the requirements for high energy battery cells, Adv. Energy Mater. 5 (2015) 1401986, https://doi.org/10.1002/aenm.201401986.

[7] P. Bonnick, J. Muldoon, The Dr Jekyll and Mr Hyde of lithium sulfur batteries, Energy Environ. Sci. 13 (2020) 4808–4833, https://doi.org/10.1039/D0EE02797A.

[8] G.-L. Xu, X. Liu, A. Daali, R. Amine, Z. Chen, K. Amine, Challenges and strategies to advance high-energy nickel-rich layered lithium transition metal oxide cathodes for harsh operation, Adv. Funct. Mater. 30 (2020) 2004748, https://doi.org/10.1002/adfm.202004748.

[9] H. Li, M. Cormier, N. Zhang, J. Inglis, J. Li, J.R. Dahn, Is cobalt needed in Ni-rich positive electrode materials for lithium ion batteries? J. Electrochem. Soc. 166 (2019) A429, https://doi.org/10.1149/2.1381902jes.

[10] Cobalt Institute, Responsible Mining of Cobalt, 2019. https://www.cobaltinstitute.org/responsible-mining-of-cobalt.html. (Accessed 24 August 2020).

[11] J. Liu, Z. Bao, Y. Cui, E.J. Dufek, J.B. Goodenough, P. Khalifah, Q. Li, B.Y. Liaw, P. Liu, A. Manthiram, Y.S. Meng, V.R. Subramanian, M.F. Toney, V.V. Viswanathan, M.S. Whittingham, J. Xiao, W. Xu, J. Yang, X.-Q. Yang, J.-G. Zhang, Pathways for practical high-energy long-cycling lithium metal batteries, Nat. Energy 4 (2019) 180–186, https://doi.org/10.1038/s41560-019-0338-x.

[12] P. Albertus, S. Babinec, S. Litzelman, A. Newman, Status and challenges in enabling the lithium metal electrode for high-energy and low-cost rechargeable batteries, Nat. Energy 3 (2018) 16–21, https://doi.org/10.1038/s41560-017-0047-2.

[13] X. Ji, K.T. Lee, L.F. Nazar, A highly ordered nanostructured carbon–sulphur cathode for lithium–sulphur batteries, Nat. Mater. 8 (2009) 500–506, https://doi.org/10.1038/nmat2460.

[14] C.B. Bucur, J. Muldoon, A. Lita, J.B. Schlenoff, R.A. Ghostine, S. Dietz, G. Allred, Ultrathin tunable ion conducting nanomembranes for encapsulation of sulfur cathodes, Energy Environ. Sci. 6 (2013) 3286–3290, https://doi.org/10.1039/C3EE42739K.

[15] C.B. Bucur, J. Muldoon, A. Lita, A layer-by-layer supramolecular structure for a sulfur cathode, Energy Environ. Sci. 9 (2016) 992–998, https://doi.org/10.1039/C5EE02367J.

[16] N. Osada, C.B. Bucur, H. Aso, J. Muldoon, The design of nanostructured sulfur cathodes using layer by layer assembly, Energy Environ. Sci. 9 (2016) 1668–1673, https://doi.org/10.1039/C6EE00444J.

[17] C.B. Bucur, M. Jones, M. Kopylov, J. Spear, J. Muldoon, Inorganic–organic layer by layer hybrid membranes for lithium–sulfur batteries, Energy Environ. Sci. 10 (2017) 905–911, https://doi.org/10.1039/C7EE00398F.

[18] Q. Wang, Z.-B. Wang, C. Li, D.-M. Gu, High sulfur content microporous carbon coated sulfur composites synthesized via in situ oxidation of metal sulfide for high-performance Li/S batteries, J. Mater. Chem. A 5 (2017) 6052–6059, https://doi.org/10.1039/C6TA10163A.

[19] T.-G. Jeong, Y.-S. Lee, B.W. Cho, Y.-T. Kim, H.-G. Jung, K.Y. Chung, Improved performance of dual-conducting polymer-coated sulfur composite with high sulfur utilization for lithium-sulfur batteries, J. Alloys Compd. 742 (2018) 868–876, https://doi.org/10.1016/j.jallcom.2018.01.364.

[20] M. Sevilla, J. Carro-Rodríguez, N. Díez, A.B. Fuertes, Straightforward synthesis of Sulfur/N,S-codoped carbon cathodes for Lithium-Sulfur batteries, Sci. Rep. 10 (2020) 4866, https://doi.org/10.1038/s41598-020-61583-1.

[21] Z. Li, H.B. Wu, X.W. (David) Lou, Rational designs and engineering of hollow micro−/nanostructures as sulfur hosts for advanced lithium–sulfur batteries, Energy Environ. Sci. 9 (2016) 3061–3070, https://doi.org/10.1039/C6EE02364A.

[22] Y. Hu, W. Chen, T. Lei, Y. Jiao, J. Huang, A. Hu, C. Gong, C. Yan, X. Wang, J. Xiong, Strategies toward high-loading lithium–sulfur battery, Adv. Energy Mater. 10 (2020) 2000082, https://doi.org/10.1002/aenm.202000082.

[23] M. Jana, R. Xu, X.-B. Cheng, J.S. Yeon, J.M. Park, J.-Q. Huang, Q. Zhang, H.S. Park, Rational design of two-dimensional nanomaterials for lithium–sulfur batteries, Energy Environ. Sci. 13 (2020) 1049–1075, https://doi.org/10.1039/C9EE02049G.

[24] C. Fang, G. Zhang, J. Lau, G. Liu, Recent advances in polysulfide mediation of lithium-sulfur batteries via facile cathode and electrolyte modification, APL Mater. 7 (2019), https://doi.org/10.1063/1.5110525, 080902.

[25] L. Zhang, M. Ling, J. Feng, G. Liu, J. Guo, Effective electrostatic confinement of polysulfides in lithium/sulfur batteries by a functional binder, Nano Energy 40 (2017) 559–565, https://doi.org/10.1016/j.nanoen.2017.09.003.

[26] Y.-H. Lee, J.-H. Kim, J.-H. Kim, J.-T. Yoo, S.-Y. Lee, Spiderweb-Mimicking anion-exchanging separators for Li–S batteries, Adv. Funct. Mater. 28 (2018) 1801422, https://doi.org/10.1002/adfm.201801422.

[27] J. Yoo, S.-J. Cho, G.Y. Jung, S.H. Kim, K.-H. Choi, J.-H. Kim, C.K. Lee, S.K. Kwak, S.-Y. Lee, COF-Net on CNT-Net as a molecularly designed, hierarchical porous chemical trap for polysulfides in lithium–sulfur batteries, Nano Lett. 16 (2016) 3292–3300, https://doi.org/10.1021/acs.nanolett.6b00870.

[28] F.Y. Fan, Y.-M. Chiang, Electrodeposition kinetics in Li-S batteries: Effects of low electrolyte/sulfur ratios and deposition surface composition, J. Electrochem. Soc. 164 (2017) A917, https://doi.org/10.1149/2.0051706jes.

[29] M. Li, Y. Zhang, Z. Bai, W.W. Liu, T. Liu, J. Gim, G. Jiang, Y. Yuan, D. Luo, K. Feng, R.S. Yassar, X. Wang, Z. Chen, J. Lu, A lithium–sulfur battery using a 2D current collector architecture with a large-sized sulfur host operated under high areal loading and low E/S ratio, Adv. Mater. 30 (2018) 1804271, https://doi.org/10.1002/adma.201804271.

[30] X. Yang, J. Luo, X. Sun, Towards high-performance solid-state Li–S batteries: From fundamental understanding to engineering design, Chem. Soc. Rev. 49 (2020) 2140–2195, https://doi.org/10.1039/C9CS00635D.

[31] N. Kamaya, K. Homma, Y. Yamakawa, M. Hirayama, R. Kanno, M. Yonemura, T. Kamiyama, Y. Kato, S. Hama, K. Kawamoto, A. Mitsui, A lithium superionic conductor, Nat. Mater. 10 (2011) 682–686, https://doi.org/10.1038/nmat3066.

[32] Y. Kato, S. Hori, T. Saito, K. Suzuki, M. Hirayama, A. Mitsui, M. Yonemura, H. Iba, R. Kanno, High-power all-solid-state batteries using sulfide superionic conductors, Nat. Energy 1 (2016) 16030, https://doi.org/10.1038/nenergy.2016.30.

[33] W.D. Richards, L.J. Miara, Y. Wang, J.C. Kim, G. Ceder, Interface stability in solid-state batteries, Chem. Mater. 28 (2016) 266–273, https://doi.org/10.1021/acs.chemmater.5b04082.

[34] P. Bonnick, K. Niitani, M. Nose, K. Suto, T.S. Arthur, J. Muldoon, A high performance all solid state lithium sulfur battery with lithium thiophosphate solid electrolyte, J. Mater. Chem. A 7 (2019) 24173–24179, https://doi.org/10.1039/C9TA06971B.

[35] E. Nagai, T.S. Arthur, P. Bonnick, K. Suto, J. Muldoon, The discharge mechanism for solid-state lithium-sulfur batteries, MRS Adv. 4 (2019) 2627–2634, https://doi.org/10.1557/adv.2019.255.

[36] W. Zhang, D. Schröder, T. Arlt, I. Manke, R. Koerver, R. Pinedo, D.A. Weber, J. Sann, W.G. Zeier, J. Janek, (Electro)chemical expansion during cycling: Monitoring the pressure changes in operating solid-state lithium batteries, J. Mater. Chem. A 5 (2017) 9929–9936, https://doi.org/10.1039/C7TA02730C.

[37] S.-J. Cho, G.Y. Jung, S.H. Kim, M. Jang, D.-K. Yang, S.K. Kwak, S.-Y. Lee, Monolithic heterojunction quasi-solid-state battery electrolytes based on thermodynamically immiscible dual phases, Energy Environ. Sci. 12 (2019) 559–565, https://doi.org/10.1039/C8EE01503A.

[38] S.-H. Kim, J.-H. Kim, S.-J. Cho, S.-Y. Lee, All-solid-state printed bipolar Li–S batteries, Adv. Energy Mater. 9 (2019) 1901841, https://doi.org/10.1002/aenm.201901841.

[39] G.T. Hitz, D.W. McOwen, L. Zhang, Z. Ma, Z. Fu, Y. Wen, Y. Gong, J. Dai, T.R. Hamann, L. Hu, E.D. Wachsman, High-rate lithium cycling in a scalable trilayer Li–garnet-electrolyte architecture, Mater. Today 22 (2019) 50–57, https://doi.org/10.1016/j.mattod.2018.04.004.

[40] X. Chen, P.M. Vereecken, Solid and solid-like composite electrolyte for lithium ion batteries: engineering the ion conductivity at interfaces, Adv. Mater. Interfaces 6 (2019) 1800899, https://doi.org/10.1002/admi.201800899.

[41] Z. Zhang, Y. Zhao, S. Chen, D. Xie, X. Yao, P. Cui, X. Xu, An advanced construction strategy of all-solid-state lithium batteries with excellent interfacial compatibility and ultralong cycle life, J. Mater. Chem. A 5 (2017) 16984–16993, https://doi.org/10.1039/C7TA04320A.

[42] Y.-X. Song, Y. Shi, J. Wan, S.-Y. Lang, X.-C. Hu, H.-J. Yan, B. Liu, Y.-G. Guo, R. Wen, L.-J. Wan, Direct tracking of the polysulfide shuttling and interfacial evolution in all-solid-state lithium–sulfur batteries: a degradation mechanism study, Energy Environ. Sci. 12 (2019) 2496–2506, https://doi.org/10.1039/C9EE00578A.

[43] X.-B. Cheng, R. Zhang, C.-Z. Zhao, Q. Zhang, Toward safe lithium metal anode in rechargeable batteries: a review, Chem. Rev. 117 (2017) 10403–10473, https://doi.org/10.1021/acs.chemrev.7b00115.

[44] E. Talaie, P. Bonnick, X. Sun, Q. Pang, X. Liang, L.F. Nazar, Methods and protocols for electrochemical energy storage materials research, Chem. Mater. 29 (2017) 90–105, https://doi.org/10.1021/acs.chemmater.6b02726.

[45] B.D. Adams, J. Zheng, X. Ren, W. Xu, J.-G. Zhang, Accurate determination of Coulombic efficiency for lithium metal anodes and lithium metal batteries, Adv. Energy Mater. 8 (2018) 1702097, https://doi.org/10.1002/aenm.201702097.

[46] T.M. Bond, J.C. Burns, D.A. Stevens, H.M. Dahn, J.R. Dahn, Improving precision and accuracy in Coulombic efficiency measurements of Li–ion batteries, J. Electrochem. Soc. 160 (2013) A521, https://doi.org/10.1149/2.014304jes.

[47] J.R. Dahn, J.C. Burns, D.A. Stevens, Importance of Coulombic efficiency measurements in R&D efforts to obtain long-lived Li–ion batteries, Electrochem. Soc. Interface 25 (2016) 75, https://doi.org/10.1149/2.F07163if.

[48] J.C. Burns, L.J. Krause, D.-B. Le, L.D. Jensen, A.J. Smith, D. Xiong, J.R. Dahn, Introducing symmetric Li–ion cells as a tool to study cell degradation mechanisms, J. Electrochem. Soc. 158 (2011) A1417, https://doi.org/10.1149/2.084112jes.

[49] P. Shi, X.-Q. Zhang, X. Shen, R. Zhang, H. Liu, Q. Zhang, A review of composite lithium metal anode for practical applications, Adv. Mater. Technol. 5 (2020) 1900806, https://doi.org/10.1002/admt.201900806.

[50] K. Yan, Z. Lu, H.-W. Lee, F. Xiong, P.-C. Hsu, Y. Li, J. Zhao, S. Chu, Y. Cui, Selective deposition and stable encapsulation of lithium through heterogeneous seeded growth, Nat. Energy 1 (2016) 16010, https://doi.org/10.1038/nenergy.2016.10.

[51] B.D. Adams, E.V. Carino, J.G. Connell, K.S. Han, R. Cao, J. Chen, J. Zheng, Q. Li, K.T. Mueller, W.A. Henderson, J.-G. Zhang, Long term stability of Li-S batteries using high concentration lithium nitrate electrolytes, Nano Energy 40 (2017) 607–617, https://doi.org/10.1016/j.nanoen.2017.09.015.

[52] G. Ma, Z. Wen, M. Wu, C. Shen, Q. Wang, J. Jin, X. Wu, A lithium anode protection guided highly-stable lithium–sulfur battery, Chem. Commun. 50 (2014) 14209–14212, https://doi.org/10.1039/c4cc05535g.

[53] M. He, R. Guo, G.M. Hobold, H. Gao, B.M. Gallant, The intrinsic behavior of lithium fluoride in solid electrolyte interphases on lithium, PNAS 117 (2020) 73–79, https://doi.org/10.1073/pnas.1911017116.

[54] Z.L. Brown, S. Jurng, C.C. Nguyen, B.L. Lucht, Effect of fluoroethylene carbonate electrolytes on the nanostructure of the solid electrolyte interphase and performance of lithium metal anodes, ACS Appl. Energy Mater. 1 (2018) 3057–3062, https://doi.org/10.1021/acsaem.8b00705.

[55] G. Li, Z. Liu, Q. Huang, Y. Gao, M. Regula, D. Wang, L.-Q. Chen, D. Wang, Stable metal battery anodes enabled by polyethylenimine sponge hosts by way of electrokinetic effects, Nat. Energy 3 (2018) 1076–1083, https://doi.org/10.1038/s41560-018-0276-z.

[56] D. Lin, Y. Liu, Y. Cui, Reviving the lithium metal anode for high-energy batteries, Nat. Nanotechnol. 12 (2017) 194–206, https://doi.org/10.1038/nnano.2017.16.

[57] T. Tao, S. Lu, Y. Fan, W. Lei, S. Huang, Y. Chen, Anode improvement in rechargeable lithium–sulfur batteries, Adv. Mater. 29 (2017) 1700542, https://doi.org/10.1002/adma.201700542.

[58] X. Yu, A. Manthiram, Electrode–electrolyte interfaces in lithium-based batteries, Energy Environ. Sci. 11 (2018) 527–543, https://doi.org/10.1039/C7EE02555F.

[59] Y. Zhao, Y. Ye, F. Wu, Y. Li, L. Li, R. Chen, Anode interface engineering and architecture design for high-performance lithium–sulfur batteries, Adv. Mater. 31 (2019) 1806532, https://doi.org/10.1002/adma.201806532.

[60] F. Wu, J. Maier, Y. Yu, Guidelines and trends for next-generation rechargeable lithium and lithium-ion batteries, Chem. Soc. Rev. 49 (2020) 1569–1614, https://doi.org/10.1039/C7CS00863E.

[61] X. Zhang, Y. Yang, Z. Zhou, Towards practical lithium-metal anodes, Chem. Soc. Rev. 49 (2020) 3040–3071, https://doi.org/10.1039/C9CS00838A.

[62] J. Zheng, M.S. Kim, Z. Tu, S. Choudhury, T. Tang, L.A. Archer, Regulating electrodeposition morphology of lithium: towards commercially relevant secondary Li metal batteries, Chem. Soc. Rev. 49 (2020) 2701–2750, https://doi.org/10.1039/C9CS00883G.

[63] C. Monroe, J. Newman, The Impact of elastic deformation on deposition kinetics at lithium/polymer interfaces, J. Electrochem. Soc. 152 (2005) A396–A404, https://doi.org/10.1149/1.1850854.

[64] L. Porz, T. Swamy, B.W. Sheldon, D. Rettenwander, T. Frömling, H.L. Thaman, S. Berendts, R. Uecker, W.C. Carter, Y.-M. Chiang, Mechanism of lithium metal penetration through inorganic solid electrolytes, Adv. Energy Mater. (2017) 1701003, https://doi.org/10.1002/aenm.201701003.

[65] E. Kazyak, R. Garcia-Mendez, W.S. LePage, A. Sharafi, A.L. Davis, A.J. Sanchez, K.-H. Chen, C. Haslam, J. Sakamoto, N.P. Dasgupta, Li penetration in ceramic solid

electrolytes: Operando microscopy analysis of morphology, propagation, and reversibility, Matter 2 (2020) 1025–1048, https://doi.org/10.1016/j.matt.2020.02.008.

[66] A. Jana, S.I. Woo, K.S.N. Vikrant, R.E. García, Electrochemomechanics of lithium dendrite growth, Energy Environ. Sci. 12 (2019) 3595–3607, https://doi.org/10.1039/C9EE01864F.

[67] A.L. Davis, R. Garcia-Mendez, K.N. Wood, E. Kazyak, K.-H. Chen, G. Teeter, J. Sakamoto, N.P. Dasgupta, Electro-chemo-mechanical evolution of sulfide solid electrolyte/Li metal interfaces: Operando analysis and ALD interlayer effects, J. Mater. Chem. A 8 (2020) 6291–6302, https://doi.org/10.1039/C9TA11508K.

[68] R. Koerver, W. Zhang, L. de Biasi, S. Schweidler, A.O. Kondrakov, S. Kolling, T. Brezesinski, P. Hartmann, W.G. Zeier, J. Janek, Chemo-mechanical expansion of lithium electrode materials—on the route to mechanically optimized all-solid-state batteries, Energy Environ. Sci. 11 (2018) 2142–2158, https://doi.org/10.1039/C8EE00907D.

[69] Y.-G. Lee, S. Fujiki, C. Jung, N. Suzuki, N. Yashiro, R. Omoda, D.-S. Ko, T. Shiratsuchi, S. Sugimoto, S. Ryu, J.H. Ku, T. Watanabe, Y. Park, Y. Aihara, D. Im, I.T. Han, High-energy long-cycling all-solid-state lithium metal batteries enabled by silver–carbon composite anodes, Nat. Energy 5 (2020) 299–308, https://doi.org/10.1038/s41560-020-0575-z.

[70] L. Jin, C. Shen, A. Shellikeri, Q. Wu, J. Zheng, P. Andrei, J.-G. Zhang, J.P. Zheng, Progress and perspectives on pre-lithiation technologies for lithium ion capacitors, Energy Environ. Sci. 13 (2020) 2341–2362, https://doi.org/10.1039/D0EE00807A.

[71] F. Holtstiege, P. Bärmann, R. Nölle, M. Winter, T. Placke, Pre-lithiation strategies for rechargeable energy storage technologies: Concepts, promises and challenges, Batteries 4 (2018) 4, https://doi.org/10.3390/batteries4010004.

# CHAPTER 11

# New approaches to high-energy-density cathode and anode architectures for lithium-sulfur batteries

**Moni K. Datta[a,b], Ramalinga Kuruba[a,b], T. Prasada Rao[a,b], Oleg I. Velikokhatnyi[a,b], and Prashant N. Kumta[a,b,c,d]**

[a]Department of Bioengineering, Swanson School of Engineering, University of Pittsburgh, Pittsburgh, PA, United States
[b]Center for Complex Engineered Multifunctional Materials (CCEMM), Swanson School of Engineering, University of Pittsburgh, Pittsburgh, PA, United States
[c]Department of Chemical and Petroleum Engineering, Swanson School of Engineering, University of Pittsburgh, Pittsburgh, PA, United States
[d]Department of Mechanical Engineering and Materials Science, Swanson School of Engineering, University of Pittsburgh, Pittsburgh, PA, United States

## Contents

| | | |
|---|---|---:|
| 11.1 | Introduction | 354 |
| 11.2 | Novel confinement architectures for sulfur cathodes | 359 |
| | 11.2.1 Synthesis of Li-ion conductors on novel carbon framework-polymer-coated materials | 361 |
| | 11.2.2 Chemical and electrochemical characterizations of novel framework materials | 362 |
| | 11.2.3 Follow-on processing of complex framework materials | 363 |
| 11.3 | Assembly and testing of pouch cells | 366 |
| | 11.3.1 Overview of pouch cell fabrication process | 366 |
| | 11.3.2 Super P-containing pouch cell cycling capacity studies | 367 |
| | 11.3.3 YP-80F-containing pouch cell cycling capacity studies | 371 |
| 11.4 | Coin cells: Preparation of hybrid solid electrolyte-coated battery separators | 373 |
| | 11.4.1 Li plating and deplating studies | 374 |
| 11.5 | Directly deposited sulfur architectures | 376 |
| | 11.5.1 Advanced materials and processing approaches | 377 |
| | 11.5.2 Pouch cell fabrication | 380 |
| | 11.5.3 Applications for practical battery systems | 382 |
| | 11.5.4 Functional electrocatalysts for conversion of polysulfides | 383 |
| | 11.5.5 Directly doped sulfur architectures with higher loadings of sulfur | 385 |
| 11.6 | Computational studies to identify functional electrocatalysts | 393 |
| | 11.6.1 Theoretical methodology | 393 |
| | 11.6.2 Computational results | 395 |

*Lithium-Sulfur Batteries*
https://doi.org/10.1016/B978-0-12-819676-2.00014-1

Copyright © 2022 Elsevier Inc.
All rights reserved.

354 Lithium-sulfur batteries

| | | |
|---|---|---|
| 11.7 | Functional electrocatalysts and related materials for polysulfide decomposition | 399 |
| | 11.7.1 Functional electrocatalyst material preparation and characterization | 400 |
| | 11.7.2 Novel complex framework material processing and characterization | 407 |
| | 11.7.3 Synthetic polymer binder with carbon framework materials | 411 |
| | 11.7.4 Hybrid active material (HBA) synthesis and characterization | 413 |
| | 11.7.5 Inorganic framework materials | 416 |
| 11.8 | Engineering dendrite-free anodes for Li-S batteries | 421 |
| | 11.8.1 Theoretical strategies to overcome the diffusion barrier in structurally isomorphous alloys | 427 |
| | 11.8.2 Electrochemical cycling of Li-SIA alloys | 430 |
| | 11.8.3 Multicomponent alloys as dendrite-free anodes | 432 |
| 11.9 | Conclusions | 435 |
| Acknowledgments | | 436 |
| References | | 436 |

## 11.1 Introduction

Portable electronic devices and gadgets have become commonplace in today's modern electronic Internet era. In the late 1990's, camcorders were perceived as the major portable consumer device along with the perennial laptop and portable digital assistants. However, with the evolution of the Internet, the world also saw the emergence of the ubiquitous mobile or cell phone. Along with it, came the introduction of the "smart phone" concept, an ingenious combination of the mobile phone with a miniaturized microprocessor, and thanks to the vision of Steve Jobs, the Apple iPhone, a revolution in portable electronic devices (PEDs), was born. The continued dominance of the cell phone and the simultaneous impending push for reduction in carbon footprint and greenhouse gas emissions for climate change mitigation together resulted in the burgeoning emergence of electrification concepts for transportation [1]. This gave rise to the development of hybrid vehicles which then transitioned initially, into plug-in hybrids and now the all-pervasive electric vehicle (EV), the "Holy Grail" in transportation to completely replace the internal combustion engine (ICE) [1,2]. The drive to realizing the EV has also opened opportunities for electrification in other trasportation areas such as, space, aeronautics, and even in unmanned air vehicle or drone applications [3,4].

Development in PEDs including laptops, camcorders, mobile phones, and portable digital assistants as well as EVs is very much dependent on the deployment and moreover, the efficacy of one single component, namely the energy storage system [2–6]. Ever since the first Li–ion rechargeable battery (LIB) was commercialized by Sony in 1994, implementation of LIBs has grown in leaps and bounds, and Li ion is considered today as the uncontested flagship rechargeable energy storage system for all portable devices requiring long-lasting high energy and power in a single and continuous use. The field has also grown tremendously since 1994, with major research advances being witnessed in all five areas of the energy storage systems: anodes, electrolytes, cathodes, separators, and Li ion conducting solid state membranes [2–9].

The cathode area, largely dominated by the first generation of lithiated transition-metal oxides, namely $LiCoO_2$ (LCO), $LiNiO_2$ (LNO), and high voltage $LiMn_2O_4$ (LMO), experienced a rapid and intense growth of research into a plethora of new and mixed transition-metal oxide systems with the goal of identifying new systems with higher specific capacity [4–8]. Despite these efforts, the cathode area over the last three decades has, however, remained largely dominated by these three systems. At present, the major focus is to harness the theoretical potential of nickel-rich LNO, while minimizing the materials-, chemistry-, antisite defect-, and disorder-related problems to achieve the desired high energy density of 300 Wh/kg, coupled with the identification and deployment of suitable anodes [6–8].

While the cathode area witnessed major research emphasis, the anode has remained primarily dedicated to carbon and various carbonaceous phases. The primary reason for this steadfast adherence to carbon-based materials is driven by the safety of eliminating the use of lithium (Li) metal. The ubiquitous ability of carbon to intercalate Li ions, eliminating the hazards of using Li metal, is largely attributed to the unique layered ($sp^2$) crystal structure of graphite [5–7]. Correspondingly, lowering the electrochemical potential to the limit of maintaining the Li in the +1 ionic state in the intercalated form without compromising on the anode electrochemical potential enables the advantage of reversibly cycling Li ions between the two intercalated states of the lithiated transition-metal oxide cathode and the carbon anode [5–7]. Careful control of the overall electrochemical voltage prevented the electrochemical deposition of metallic Li and consequent formation of perilous dendrites puncturing the separator and short-circuiting the system, thereby leading to the flammability hazard issue, thus making the LIB systems the most preferred and largely, safe energy storage system. This unique ability of carbon and the elimination of Li metal deposition minimizing safety and flammability issues of metallic lithium have given carbon a significant advantage and preferred anode status. However, it has become abundantly clear recently that any major advances in energy density in the LIB system will not be possible without the identification of new anode systems exhibiting at least 10-fold higher specific capacity than afforded by carbon [9].

The late 1990s and early 2000s witnessed the first attempts at exploring alternative anodes to carbon; silicon (Si) as well as tin (Sn) were identified as potentially attractive alternates with theoretical capacity almost 10-fold and 3-fold that of graphite, respectively, without compromising the electrochemical potential, being similar to that of carbon [5–7,9–12]. However, the colossal volume expansion and solid electrolyte interphase (SEI) issues still plague these systems. The pursuit of high energy and power density energy storage systems has over the last decade focused on metallic lithium and other cost-effective cathode systems [13–20]. Oxygen and sulfur have therefore emerged due to their respective attributes of being a major constituent of the atmosphere in the case of the former and the latter, being an Earth-abundant element as well as a common by-product of factories and petrochemical processes including industrial waste with the latter. The Li-air system though promising, however, is plagued with the formation of the recalcitrant lithium oxide ($Li_2O$) and superoxide ($Li_2O_2$) requiring an efficient lightweight electrocatalyst to permit the reversible cycling of Li from the oxide; this, however, also risks the formation of and deleterious dendrite-related plating issues of the Li anode [19,20].

The lithium-sulfur (Li-S) system on the other hand, is therefore considered more promising for realizing the high energy density of 500 Wh/kg and even higher, widely considered the gold standard by the U.S. Department of Energy and other energy related agencies, required for meeting the growing energy storage demands of portable devices as well as EVs and other advanced systems. High-energy-density LIBs are warranted to match the power, energy, and the sustainability of ICE-driven vehicles. Additionally, a higher energy density system would also enable packing more capabilities and functionality into portable electronic devices including cell phones. Development of such systems is largely limited by the gravimetric and volumetric energy densities of LIB systems ($\sim$220 Wh/kg and $\sim$750 Wh/L). Improvements in energy density are therefore imperative to exploit the potential of these novel environmentally benign technologies. The Li-S battery systems could use elemental sulfur as the cathode, which has a theoretical specific capacity of 1675 mAh/g, and exhibit an attractive thermodynamic specific energy density of 2567 Wh/kg [14–16,18,21].

Lithium-sulfur battery (LSB) technology is widely investigated as an attractive alternative to currently used LIB chemistries for the PED/EV industries, due to the superior theoretical specific capacity and specific energy density of elemental sulfur. Furthermore, the natural abundance of sulfur in the earth's crust makes it a more economical and highly attractive alternative compared to the currently existing intercalation-based LIB cathode material systems. Lithium-sulfur batteries have the potential to meet the increased energy density requirements of EV technologies. However, the Li-S system is plagued by deleterious polysulfide shuttling amongst many others [14,18]. Lithium polysulfides (LPS) are formed during the discharge reaction of Li ions reacting with sulfur. These are compounds represented by the chemical formula of $Li_2S_x$ ($1 \leq x \leq 8$), wherein the stable form of $Li_2S$ is formed after a series of continued reactions of Li with sulfur during the discharge process resulting in the formation of higher polysulfide species, $Li_2S_8$, initially with sequential formation of lower order polysulfides until the stable form of $Li_2S$ is attained [18]. The higher S forms of polysulfides ($Li_2S_x$) with the exception of $Li_2S_2$ and $Li_2S$ tend to be liquid; these materials are unstable at room temperature, and are also highly soluble in polar organic solvents. As a result of this dissolution and continued loss of sulfur, there is a gradual decrease in capacity due to the loss of sulfur, the active material resulting in the eventual failure of the cell [22,23]. The solubility of the LPS species thus adds to the limitations associated with the LSB technology causing poor capacity retention. This is further exacerbated by the formed LPS species shuttling from the cathode to the anode undergoing further reduction to lower order short–chain polysulfides and back to higher order long–chain polysulfides at the cathode. To further add to the existing issues is the formation of unwanted and pernicious dendrites on the Li metal anode [13,24–26]. The aforementioned problems are also not alleviated by the inferior electronic conductivity of sulfur (a barrier to complete active material utilization) as well as the poor Li-ion conductivity of the elemental sulfur. Current generation, sulfur cathodes therefore exhibit low reversible specific storage capacity, poor charging rates, and low sulfur loading densities [15,16,22,27–29]. These limitations are further accentuated in the presence of lean electrolytes required to enhance the specific energy density to lower the overall mass of the fully assembled cell.

Various strategies have been reported in the literature to address these challenges [18,23,28–31]. Conductive carbon was introduced into the sulfur cathodes to increase the electrical conductivity and active material utilization of the electrodes, with an overall reduction in cathode resistance by high-conductivity carbon black incorporation in the active material mixture [30,32,33]. The active carbon possesses nanopores ($\sim$2–10 nm) with a high specific surface area ($\sim$500–2000 $m^2/g$) absorbing the LPS species, thereby preventing their dissolution into the electrolyte. Mesoporous carbon acts as an ordered encapsulation substrate for sulfur [30,32,34]. There has been systematic tuning and investigation of the pore sizes and pore volumes of several mesoporous carbon materials and the results have shown that the large pore size (~3–22 nm) of mesoporous carbon can accommodate higher sulfur loading (>80% S) and can also exhibit enhanced electrochemical cell performance under higher sulfur-loading situations. Transition–metal silicates, aluminum oxides, vanadium oxides, and transition–metal chalcogenides have also been utilized with sulfur cathodes to decrease the polysulfide diffusion and migration [27]. However, their electron transport property was limited by the large particle size that tends to decrease the electrochemical performance. This is further accentuated in the presence of lean electrolyte conditions of ~2-4 microliters of electrolyte (E) per milligram of sulfur (S) and high sulfur loadings of > 4 mg/cm$^2$. In addition to these inherent limitations in the sulfur cathodes, the lithium anode side, as mentioned earlier, is also plagued with limitations of dendrite formation posing a safety hazard as explained earlier [13,25,26].

Replacing the commonly used Li-S battery organic electrolyte (dioxolane (DOL)/dimethoxyethane (DME)) with a polyvinylidene fluoride-*co*-hexafluoropropene (PVDF-HFP)–based composite polymer electrolyte (CPE) has been demonstrated to trap the polysulfides due to the very low electrolyte content of the CPE (~1–2 µL electrolyte/mg sulfur) [35]. The safety and cyclic life of the anode were improved using polymer and solid-state electrolytes that protect lithium metal and minimize dendrite formation on the anode, hence leading to enhanced performance of Li-S batteries [36–38]. Nevertheless, polymer and solid-state inorganic, ceramic electrolytes generally suffer from low Li-ion conductivity due to the highly viscous nature of polymers hindering the lithium-ion transport due to the high energy barrier in the solid-state electrolytes as well as the refractory, ceramic nature of the solid-state electrolytes requiring high activation energy for Li-ion diffusion. The absence of liquid organic solvents in the solid-state electrolytes eliminates the problems of polysulfide dissolution seen in the current LSB systems making the all-solid-state Li-ion batteries extremely attractive, though presenting several other issues, the need for stack pressure and interfacial problems being the most prominent [38,39]. These different approaches mentioned above, though, lead to an increase in the utilization of active material in sulfur cathodes; however, they lack complete prevention of the dissolution of polysulfide species into the electrolyte. Furthermore, there is also the lingering problem of dendrite formation on the anodes combined with the corrosion of the Li metal anode.

Thus, there is a need for the design and development of new confinement architectures for the cathode and new concepts in designing stable and robust anode alloys resistant to dendrite formation. The generation of

Cathode and anode architectures for lithium-sulfur batteries **359**

new complex framework materials (CFM) based cathodes can, therefore, enable batteries with higher energy density and decreased capacity fade properties. Successful implementation of CFM-based sulfur cathodes in lithium-sulfur batteries would expedite the development of high-energy-density batteries. This chapter relates to the economical and scalable fabrication of CFMs developed by authors over the last few years for use as LPS confining architectures and structures in Li-S batteries. The chapter also discusses new alloy designs and approaches developed by the authors to produce dendrite-free alloy anodes for Li-S batteries combined with the results of coin cell and pouch cell fabrication and testing.

The following two sections address novel cathode and anode approaches, respectively. Section 11.2 discusses the new confinement architectures for polysulfide containment and encapsulation of sulfur including theoretical studies for identifying novel functional electrocatalysts (FC) and Li-ion conductors (LIC) for enabling improved transport and electrochemical conversion of polysulfides to $Li_2S$ and back to elemental Li and S in coin cell and pouch cell formats. Section 11.3 discusses the conceptual framework for creation of dendrite-free anodes, the engineering design and fabrication of new dendrite-free alloys for Li-S batteries. The remiander of the chapter discusses: coin cells (Section 11.4), sulfur architectures that are directly deposited (Section 11.5), computational studies (Section 11.6), electrocatalysts (Section 11.7), dendrite-free anodes (Section 11.8), and finally key conclusions to be drawn from our recent work (Section 11.9).

## 11.2 Novel confinement architectures for sulfur cathodes

In this section, we provide an introduction, and a summary of various CFM structures including S hosts, which includes the generation of a complex framework material combined with a coating applied to the CFM. The CFMs considered herein are largely made of a porous carbon matrix. The carbonaceous complex framework material essentially comprises Super P (Timcal, conductive carbon black), hereafter referred to simply as "Super P" or a highly porous high specific surface area, carbon (C) ranging from about 2000 to 3000 $m^2$/g, namely YP-80F (Kuraray). The coating includes one or more layers that comprise a component selected from the group consisting of an electronic conductor, a lithium-ion conductor, and a functional electrocatalyst. The CFM structure that was selected included an electronic-conducting polymer-coated complex framework material coated with an electronic conductor (EC-CFM), a Li-ion conductor-coated complex framework material (LIC-CFM), coated with a LIC known as LiOPAN, and finally, a functional electrocatalyst embedded complex framework material (FC–CFM), coated with a functional electro catalyst. All these complex framework materials coated with EC and LIC and containing the FC are then loaded with sulfur. This involves a secondary step, wherein the desired amount of sulfur is infiltrated into the complex framework material host to produce an EC–CFM–S, a LIC–CFM–S, and a FC–CFM–S electrode. The sulfur loading varies from 8 to 18 mg/cm$^2$ depending on the system.

The cathode architecture thus, includes a complex framework material-based composite cathode containing the functionalized carbon surface that is infiltrated with sulfur. Furthermore, the complex framework material-based cathode includes the foregoing complex framework material structure. The steps involved in generating the complex cathode architecture include a method of preparing a complex framework material structure. The method comprises forming a complex framework material host that includes providing a complex framework material; applying a coating to the complex framework material that includes one or more components selected from the group consisting of an electronic conductor, a lithium-ion conductor, and a functional electrocatalyst; and infiltrating the complex framework material host with sulfur. All these steps are aimed to be generated using a simple, single, one–pot economic synthesis step, at low to moderate temperatures of 240°C enabling vapor–liquid infiltration of sulfur into the carbon-based CFM while also ensuring the coating of the CFM with the EC and LIC, and at the same time, also incorporating the functional electrocatalyst, all achieved in a single scalable, and cost-effective process step.

We also describe testing of the composite complex framework architecture in a pouch cell that includes a CFM/S-composite cathode, which includes a CFM host, a coating applied to the CFM comprising one or more layers of a component selected as described above, from the group consisting of an electronic conductor, a lithium-ion conductor, and a functional electrocatalyst; sulfur infiltrated into the CFM host; a separator applied to the CFM/sulfur composite cathode; and a Li anode applied to the separator, wherein the CFM/sulfur cathode, the separator, and the lithium anode are in a stacked configuration. The complex framework material as described above is comprised of a porous carbon matrix. The carbon-based architectures that form the complex framework material comprise two forms of carbon: Super P and YP–80F . Both these carbon architectures when infiltrated with sulfur show the ability to bind polysulfide via the generated carbon-sulfur linkages. The authors have tested these architectures in coin cells as well as pouch cells which can include a single layer of each of the CFM/sulfur cathode and the lithium anode. Alternatively, the pouch cell

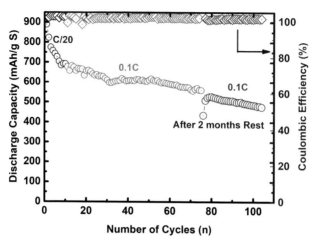

**Fig. 11.1** Cycling performance of LIC-CFM-S cycled at 0.1 C rate. *(Source: Authors.)*

can also include two layers as well as multilayers of each of the CFM/sulfur cathode and the lithium anode. Fig. 11.1 is a plot that shows the cycling performance of the LIC-CFM-S cycled at 0.1 C rate, in accordance with the description mentioned previously. The approach followed for creating these architectures and the characteristics of the synthesized systems are described later in the sections to follow.

### 11.2.1 Synthesis of Li-ion conductors on novel carbon framework-polymer-coated materials

A CFM (5.0 g) of Super P was dispersed in a mixed solution of deionized water (25 mL) and dimethyl sulfoxide (DMSO) (25 mL) under ultrasonication (Branson, 5800, United States) for 10 min. Acrylonitrile monomer (10 mL) and azobisisobutyronitrile (AIBN, 100 mg) were then added. The mixed solution was stirred under a $N_2$ atmosphere for 4 h at 65°C to initiate the polymerization reaction. The resulting solid was collected by filtration and washed with ethanol after polymerization. The product was vacuum-dried at 60°C for 24 h to yield polyacrylonitrile (PAN) coated on carbon framework (PAN/CFM). Any nitrogen-containing carbon backbone-based hydrocarbon polymer can also be selected.

The PAN/CFMs were dried under vacuum conditions at 60°C for 24 h to remove any residual solvent and water of crystallization from the synthesis process. The synthesized PAN/CFMs were then infiltrated with sulfur

## 362 Lithium-sulfur batteries

under vacuum, using the following procedure; PAN/CFM and S (mass ratio 90:10) and 10 wt% lithium salt such as lithium nitrate were ground together and then sealed and heated under argon at a rate of 5°C/min up to 240°C. The mixture was maintained at 240°C for 12 h to obtain LIC-CFM/S, also termed as LiOPAN-coated CFM-S, with the LiOPAN coating serving as a Li-ion conductor.

### 11.2.2 Chemical and electrochemical characterizations of novel framework materials

The crystal structure of the PAN/CFM and PAN/CFM-S before and after sulfur infiltration was analyzed by X-ray diffraction (XRD) in a Philips XPERT PRO system that uses Cu $K_\alpha$ ($\lambda = 0.15406$ nm) radiation. The samples were scanned in the $2\theta$ range of 10–90 degrees under a constant current and voltage of 40 mA and 45 kV, respectively. Scanning electron microscope (SEM) images of the CFMs were obtained using a Philips XL30 machine at 10 kV. An attenuated total reflectance-Fourier-transform IR (ATR-FTIR) Nicolet 6700 Spectrophotometer (Thermo Electron Corporation), which uses a diamond ATR smart orbit, was used to obtain the FTIR spectra of the samples. The FTIR spectra are collected at a resolution of 1 cm$^{-1}$, averaging 32 scans between the frequency of 400 and 4000 cm$^{-1}$. The X-ray photoelectron spectroscopy (XPS) analyses of the PAN/CFMs and PAN/CFM-S were performed using the ESCALAB 250 Xi system (Thermo Scientific). This XPS system consists of the monochromated Al $K_\alpha$ X-ray source and low-energy ($\leq$10 eV) argon ions and low-energy electron beams that provide the charge neutralization. The XPS measurements were carried out at room temperature, under an ultrahigh vacuum (UHV) chamber ($<5 \times 10^{-10}$ mBar) employing a spot size of $200 \times 200 \, \mu m^2$. The surface area and pore characteristics of all the CFM samples were analyzed using a Micromeritics ASAP 2020 Physisorption Analyzer, using the Brunauer, Emmett and Teller method (BET) isotherm generated.

The LIC-CFM-S or LiOPAN-coated CFM-S samples were cycled between 1.8 and 2.8 V (with respect to Li$^+$/Li) at a current rate of 0.1 C in a 2032 coin cell and pouch cells using the Arbin BT200 battery testing station to evaluate their electrochemical performance. The cathodes for electrochemical evaluation were prepared by manually coating a dispersion of 70 wt% LIC-CFM-S, 20 wt% acetylene black, and 10 wt% PVDF dispersed in $N$-methyl-2-pyrrolidone (NMP) on an aluminum foil, followed by vacuum-drying for 12 h at 60°C. All of the cathodes that were tested had a uniform sulfur loading of 4.0–6.0 mg/cm$^2$. Accordingly, 2032-coin cells were assembled with the LIC-CFM-S-coated cathodes as the working electrode, a Li foil as the counterelectrode, and Celgard 2400 polypropylene (PP) as the separator in an Innovative, Inc., glove box (UHP Argon, <0.1 ppm $O_2$, $H_2O$). 1.8 M LiTFSI (lithium bis (trifluoromethanesulfonyl)imide), and 0.4 M $LiNO_3$ dissolved in 50:50 vol% 1,3 dioxolane and 1,2 dimethoxyethane were used as the electrolytes. The lithium-ion conductor (LIC-CFM-S) coordination of interlayer structure and pore volume of LIC-CFM-S provide high activity for catalytic conversion of Li-polysulfides (LiPSs) and also sufficient entrapment of LiPSs. Fig. 11.2 shows the effective suppression of the shuttle effect, as a result, even under a C/20 rate under lean electrolyte condition of an electrolyte (E)-to-sulfur ratio (E:S) of (=) 4 µL/mg in a Li-S battery, the LIC-CFMS composite electrode still delivers a high initial capacity of 910.9 mAh/g that stabilizes to ∼750 mAh/g after 10 cycles.

### 11.2.3 Follow-on processing of complex framework materials

Lithium-sulfur cathodes in this chapter as mentioned above, consist of a uniquely designed CFM coated with a Li-ion conductor to form LIC-CFM. The LIC-CFM includes one or more porous carbon-based materials coated with a novel nitrogen-containing hydrocarbon or carbonaceous

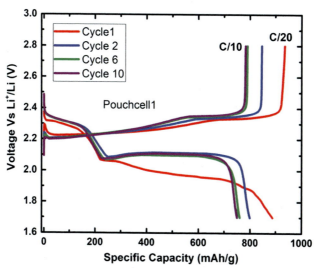

**Fig. 11.2** Single-layer pouch cell cycling results showing the voltage versus specific capacity response for 1st, 2nd, 6th, and 10th cycles corresponding to a sulfur loading of 6.26 mg/cm². *(Source: Authors.)*

polymer such as PAN modified by reaction with $LiNO_3$ resulting in a Li-ion-conducting coating combined with elemental sulfur. The composite LIC-CFM containing sulfur cathode is suitable for use as a stable, high cycle life and high-capacity cathode for Li-S batteries.

### 11.2.3.1 Polyacrylonitrile polymer processing

The PAN polymer on the surface of the porous carbon-based CFM is pyrolyzed in argon gas at 240°C in the presence of $LiNO_3$ and elemental sulfur (S), which leads to the formation of a cyclized PAN polymer network with structural changes including Li-ions attached to the carbon and nitrogen backbone on the surface of porous carbon matrix to form a LiOPAN coating. The entire synthesis of the cyclized PAN and $LiNO_3$ to form LiOPAN and all the chemical reactions are executed inside a sealed Swagelok container containing a dry mixture of the composite materials as described previously. Two major changes that occur simultaneously during the PAN pyrolysis process on the surface of porous carbon matrix are the following: (i) PAN converting into N-type PAN, nitrogen-doped PAN (N-type PAN) forming LiOPAN after decomposition of $LiNO_3$, (ii) N-type LiOPAN converting into sulfurized PAN with long-range N-type bonds, the N-type PAN undergoes sulfurization at 240°C and it is transformed into sulfurized PAN due to a cyclization reaction, which creates conjugated polymer structure with long-range N-type bonds. Hence, the LiOPAN-coated porous carbon matrix is converted to sulfurized LiOPAN coating on porous carbon matrix and the sulfurized LiOPAN coating also further develops cross-linked networks on the surface and within the bulk of the porous carbon matrix. In addition, the heat treatment at 240°C results in the creation of carbon-sulfur linkages leading to trapping of ensuing polysulfide species formed during the reaction with Li ions when cycled in the Li-S battery.

These C-S linkages are formed with the sulfur within the pores and on the surface of the porous carbon matrix filled with vaporized sulfur during vapor-phase sulfur infiltration. Hence, the final product of S cathode composite material consists of a uniform electrically conductive network of sulfurized PAN as well as Li ion reacted with PAN to form Li-ion conducting, LiOPAN on the surface and within the pores of the porous carbon matrix already filled with nanosized sulfur. The novel LiOPAN-coated CFM-S cathode provides the S cathode with required stability, high capacity, and cycle life improvements due to the synergetic effect of creating the sulfurized PAN network on porous carbon matrix. The sulfurized LiOPAN-coated porous carbon-based S cathode composite material can also be synthesized

Cathode and anode architectures for lithium-sulfur batteries **365**

in large scale (few kilograms per batch, depending on the size of the Swagelok container). The production is also possible at low cost using vapor-phase sulfur infiltration method. Commercial porous carbon materials used are porous carbon matrix for PAN coating and sulfur infiltration comprising Super P with surface area $\approx 62\,m^2/g$, activated carbon with surface area $\approx 2004\,m^2/g$ (YP-80F, Kuraray coal) and a mixture of Super P/YP-80F.

### 11.2.3.2 Polyacrylonitrile coating on super P and YP-80F

The commercial Super P/YP-80F (1.0 g) was mixed in a solution of deionized water (125 mL), dimethyl sulfoxide (DMSO, 125 mL), and stirred under argon (Ar) atmosphere at 80°C for 30 min. This is followed by the addition of 375.0 mg of 2,2′-azobisisobutyronitrile (AIBN) to the mixture, with the temperature of the mixture raised to 85°C and the entire mixture continuously stirred for 30 min. This is followed by the addition of the acrylonitrile (AN) monomer (50 mL) to the mixture and continuously stirred under Ar atmosphere at 85°C for 3 h to carry out the polymerization reaction. After polymerization, the milky white product was coated on Super P/YP-80F, and the resulting product was collected and washed with ethanol. The final product was then vacuum-dried at 60°C for 36 h to yield the PAN-coated Super P/YP-80F. A similar procedure can be applicable to any nitrogen-containing carbon backbone polymer.

### 11.2.3.3 Preparation of LiOPAN-coated super P/YP-80F-sulfur composites

The PAN-coated Super P/YP-80F (10 wt%), elemental sulfur (90 wt%), and lithium nitrate (LiNO$_3$, 10 wt%) were ground together, then sealed in Swagelok cell, and heated at 240°C (with a temperature ramp of 10°C/min) for 12 h to obtain LiPAN-coated Super P/YP-80F-S due to vapor-phase sulfur infiltration. The LiOPAN-S cathode was prepared by mixing LiOPAN-S, Super P/YP-80F, and PVDF with $N$-methyl-2-pyrrolidone (NMP) acting as a solvent to form a uniform slurry. The weight ratio of active material, Super P, and binder is 70:20:10 (wt%) and 72:18:10 (wt%).

The slurry was then coated onto a single side as well as on both sides to form double-sided coatings on a carbon-coated aluminum (Al) foil using the doctor-blade technique. The S cathode was then dried at 65°C in a heated vacuum oven for 18 h with optimized S areal loading of 4–6 mg/cm$^2$ per side. The cathode was then subjected to calendaring process to 80%–70% of its original thickness prior to use (i.e., 20%–30% calendaring).

The electrode was then punched into rectangular pieces (50 mm × 40 mm) for pouch cell use. The cathode porosity was calculated based on the density of element density of sulfur and carbon. The sulfur electrode thickness varied from 80 to 140 μm with respect to S loading given in mg/cm$^2$.

## 11.3 Assembly and testing of pouch cells

Single-sided 50-μm Li foil was calendared onto a 10-μm thick Cu foil and punched to rectangular pieces (50.2 mm × 40.2 mm) as the Li anode for the single-layer pouch cells. The separator for the pouch cell was Celgard 2400. Single-sided S cathode, separator, and single-sided were stacked as single-layer pouch cell. Double-sided 50-μm Li foil was calendared onto a 10-micron thick Cu foil and punched to rectangular pieces (50.2 mm × 40.2 mm) as the Li anode of pouch cell. Double-sided S cathode, separator, and double-sided Li anode were alternatively stacked together with two pieces of single-sided S cathode as the outer layer for the multilayer pouch cells.

The electrolyte used in this study was 1.8 M lithium bis(trifluoro-methanesulfonyl) imide (LiTFSI) in 1,3-dioxolane (DOL) and 1,2-dimethoxyethane (DME) with 0.4 M LiNO$_3$ as an additive and the E:S ratio used 4 μL/mg-S. The electrolyte injection and pouch cell sealing were carried out in a glove box. All the pouch cells were tested using the following constant current protocol: 2 formation cycles with C/20, 3 cycles with C/20, 100 cycles with C/10, and 100 cycles with C/5. More details of the individual electrodes and the performance of the fabricated cells are given in the different examples discussed below.

### 11.3.1 Overview of pouch cell fabrication process

Digital images of the steps used to fabricate the electrodes are shown in Fig. 11.3. Fabrication of crack-free electrode with a targeted porosity of ~50%–60% and a targeted sulfur loading of ~4–6 mg/cm$^2$ of the sulfur cathode is needed to facilitate and achieve efficient charge transport in the electrode. The single-layer Li-S pouch cell was fabricated with sulfurized LiOPAN-coated Super P:sulfur composite as the S cathode, with S loading of 6.26 mg/cm$^2$, a cathode porosity of 66.0% and 63.0 wt% of S content available in the cathode. The amounts of electrolyte and Li anode thickness are strictly controlled at electrolyte-to-sulfur ratios of 4.0 μL/mg-S and 50 μm, respectively.

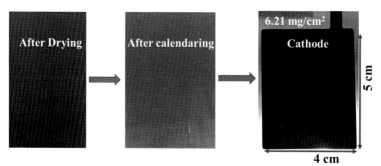

**Fig. 11.3** Digital image showing crack-/void-free electrode of sulfur cathode. *(Source: Authors.)*

### 11.3.2 Super P-containing pouch cell cycling capacity studies

The first cycle capacity was observed to be 900 mAh/g-S at C/20 rate, but the cathode was designed to have a nominal capacity of 1055.25 mAh/g-S, which is 63% of the theoretical capacity of 1675 mAh/g-S, and the observed capacity is indeed very close to the designed nominal capacity. The ideal discharge mechanism of a Li-S battery can be divided into four steps as follows: (i) reaction of elemental sulfur with lithium; (ii) reaction between the dissolved highest order polysulfide, $Li_2S_8$, and lithium; (iii) a transition of the dissolved polysulfide to $Li_2S_4$, eventually to insoluble $Li_2S_2$ and finally $Li_2S$; and (iv) an equilibrium reaction of the insoluble $Li_2S_2$ to $Li_2S$, see Fig. 11.2. Full-cell capacity, Coulombic efficiency, and energy density obtained are all shown in Fig. 11.4 with respect to cycle number. The full-cell capacity stabilized at ≈100 mAh at a current rate of C/10 rate with a reasonably long cycle life of 50 cycles with a capacity fade rate of 0.2% per cycle. Full-cell energy density observed ≈240 Wh/kg at a current rate of C/20 rate and specific capacity that stabilized at 655 mAh/g-S at the end of 50 cycles when cycled at a current rate of C/10 rate. Fig. 11.4A and B are plots that show cycling and energy density of single-layer Li-S pouch cells using sulfurized LiOPAN-coated Super P:sulfur composite as the S cathode with S loading of 6.26 mg/cm$^2$, in accordance with the earlier description.

Sulfurized LiOPAN-coated Super P:sulfur composite cathode with a total sulfur loading in the cathode of 4.59 mg/cm$^2$ was also used to fabricate the single-layer Li-S pouch cells and the data are shown in Figs. 11.5 and 11.6. The S cathode porosity was 70.0% with a thickness of 126 μm. The single-layer pouch cell with low loading shows similar results as in

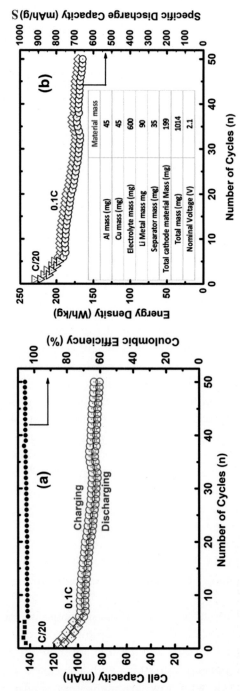

**Fig. 11.4** (A) Cycling response and (B) the energy density versus cycle number for a single-layer Li-S pouch cell using sulfurized LiOPAN-coated Super P:sulfur composite as the S cathode with S loading of 6.26 mg/cm$^2$. *(Source: Authors.)*

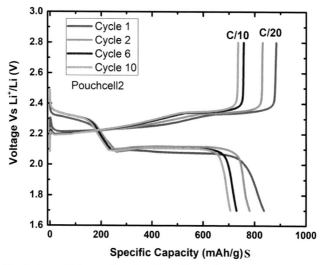

**Fig. 11.5** Single-layer Li-S pouch cell with sulfurized LiOPAN-coated Super P:sulfur composite as S cathode with sulfur loading of 4.59 mg/cm². *(Source: Authors.)*

Fig. 11.4. Hence, the data are reproducible within the S-loading range of 4.0–6.5 mg/cm². Fig. 11.5 is a plot that shows a single-layer Li-S pouch cell with sulfurized LiOPAN-coated Super P:sulfur composite as S cathode with sulfur loading of 4.59 mg/cm², in accordance with the aforementioned description. Fig. 11.6A and B are plots that show cycling and energy density of single-layer Li-S pouch cell with sulfurized LiOPAN-coated Super P:sulfur composite as the sulfur cathode with the sulfur loading of 4.59 mg/cm², in accordance with the aforementioned description.

Fig. 11.7A is a plot that shows two-layer pouch cell cycling data with sulfurized LiOPAN-coated Super P:sulfur composite as S cathode with a sulfur loading of 6 mg/cm², in accordance with the aforementioned description. Multilayer pouch cells were fabricated and tested to observe their charge-discharge cycling in order to further verify the cycling response of sulfurized LiOPAN-coated Super P:sulfur composite as cathode material. Two-layered pouch cell designed to exhibit a nominal capacity ≈200 mAh was fabricated and the cycling data obtained are shown in Fig. 11.7A. The sulfur loading was 6.0 mg/cm² on each sulfur cathode layer with a sulfur content of 63.0 wt%. The cycling data trend matches with that of Fig. 11.4. Fig. 11.7B is a plot that shows four-layer pouch cell cycling data with sulfurized LiOPAN-coated Super P:sulfur composite as S cathode with sulfur loading of 4.72 mg/cm², in accordance with the aforementioned description.

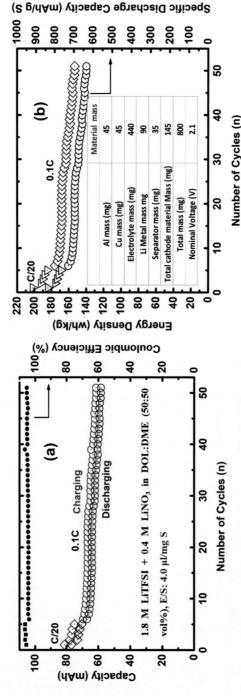

**Fig. 11.6** (A) Plot of capacity versus number of cycles. (B) Energy density versus cycle number for single-layer Li-S pouch cell with sulfurized LiOPAN-coated Super P:sulfur composite as the sulfur cathode with the sulfur loading of 4.59 mg/cm$^2$. (*Source: Authors.*)

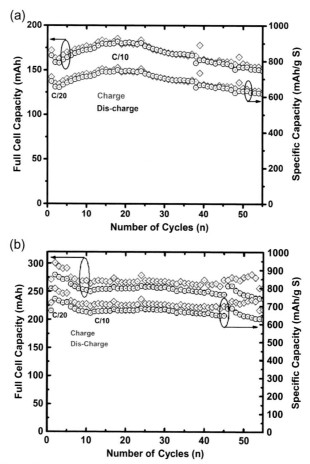

**Fig. 11.7** (A) Two-layer pouch cell cycling data with sulfurized LiOPAN-coated Super P: sulfur composite as S cathode with a sulfur loading of 6 mg/cm². (B) Four-layer pouch cell cycling data with sulfurized LiOPAN-coated Super P:sulfur composite S cathode with sulfur loading of 4.72 mg/cm². *(Source: Authors.)*

The four-layer pouch cell (stacking of multicells) designed to exhibit full-cell capacity of 300 mAh with sulfur loading of 4.72 mg/cm² on each S cathode was fabricated with sulfur content of 63.0 wt%.

### 11.3.3 YP-80F-containing pouch cell cycling capacity studies

The S-cathode LIC-CFM material was also prepared using sulfurized LiO-PAN coated on porous YP-80F as the active material with slurry composition of 72 (active material):18 (Super P):10 (PVDF). The single-layer pouch cell was fabricated using 50-μm-thick Li metal as anode. The sulfur cathode porosity is 62.0% with a total electrode thickness of 103.0 μm and a sulfur loading of 5.20 mg/cm$^2$ without calendaring and with a sulfur content of 64.80 wt%. The specific capacity stabilized at ≈700 mAh/g-S and the corresponding cycling data are shown in Fig. 11.8A. Fig. 11.8A is a plot that shows pouch cell cycling data with sulfurized PAN-coated YP-80F:sulfur composite as S cathode, in accordance with the aforementioned description. Single-layer pouch cell fabricated using calendared S cathode of LIC-CFM

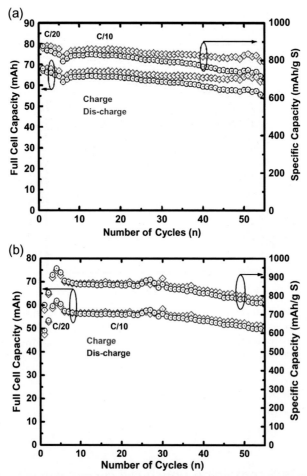

**Fig. 11.8** (A) Pouch cell cycling data with sulfurized polyacrylonitrile (LiOPAN)-coated YP-80F:sulfur composite as S cathode with sulfur loading of 5.2 mg/cm$^2$. (B) Pouch cell cycling data with sulfurized LiOPAN-coated YP-80F:sulfur composite as S cathode with calendaring incorporating a sulfur loading of 4.8 mg/cm$^2$. *(Source: Authors.)*

Cathode and anode architectures for lithium-sulfur batteries 373

comprising sulfurized LiOPAN coated on porous YP-80F as the active material with S loading of $4.8\,mg/cm^2$ and S content 64.8 wt%. The porosity of the S cathode decreased to 56.0% with thickness of $84.0\,\mu m$. The cycling data are shown in Fig. 11.8B. The cycling results indicate better performance after calendaring exhibiting a stable capacity of $\sim750-800\,mAh/g$-S. Fig. 11.8B is a plot that shows pouch cell cycling data with sulfurized LiOPAN-coated YP-80F:sulfur composite as S cathode with calendaring with sulfur loading of $4.8\,mg/cm^2$, in accordance with the description mentioned earlier.

## 11.4 Coin cells: Preparation of hybrid solid electrolyte-coated battery separators

In addition to the pouch cell systems studied as described above, using the CFM-based electrodes, efforts were also directed at utilizing hybrid solid electrolyte coated separators. The approaches and the specific systems considered are described in the following. Calculated amounts of polyethylene oxide (PEO) ($M_w$ $5 \times 10^6$) and $LiClO_4$ ([PEO]:[Li] = 15:1) were added to acetonitrile (ACN) and stirred for 24 h at ambient temperature. To this was added 50 wt% of Al-doped stabilized cubic $Li_7La_3Zr_2O_{12}$ (LLZO) ($Li_{6.25}Al_{0.25}La_3Zr_2O_{12}$) with submicron particle size ($0.1-0.5\,\mu m$) (MSE Supplies, Tucson, AZ) to form the PEO:$LiClO_4$ mixture. The polymer solution and the dispersed LLZO solution were mixed thoroughly for 24 h at room temperature in order to achieve a homogeneous and uniform viscous solution (PEO:$LiClO_4$-Al-doped LLZO). The uniform solution was then cast on one side of the Celgard separator as the substrate and then kept at room temperature in a closed chamber to evaporate the ACN solvent. The fully dried Celgard separator coated with hybrid composite solid polymer electrolyte (HSE) was then cut into circular disks and was used as the HSE-coated separator. The thickness of the HSE film on the Celgard separator varied from 5 to $20\,\mu m$.

The 2032 coin cells were assembled using the sulfurized LiOPAN-coated YP-80F:sulfur composite as the S cathode, and HSE was coated on the one side of the Celgard separator facing the Li anode and Li metal as the anode. The hybrid solid electrolyte was coated on the side of the separator facing the Li anode in order to suppress the polysulfide shuttling effect common in Li-S systems and to correspondingly help improve the cycling and the Coulombic efficiency of the Li-S battery. The liquid electrolyte was also used in the coin cell in addition to the HSE film (1.8 M lithium bis(trifluoromethanesulfonyl) imide (LiTFSI) in 1,3-dioxolane (DOL) and 1,2-dimethoxyethane (DME) with 0.4 M LiNO$_3$ as an additive to), and the electrolyte/sulfur ratio used was 4 μL/mg-S.

### 11.4.1 Li plating and deplating studies

Lithium plating and deplating studies shown in Fig. 11.9 conducted using cyclic voltammetry (CV) on a coin cell at the room-temperature configuration of Li metal || HSE coating + liquid electrolyte || Cu metal. Identical S cathodes used in this study included S loading 4.7 mg/cm$^2$, S content 64.8 wt%, and porosity 67%, respectively. These electrodes were used for assembling the coin cells. The voltage window stability of Li-S cell is shown in Fig. 11.10; the Li-S cell with HSE-coated separator shows a more stable voltage window than the Li-S cell cycled containing only liquid electrolyte. This can be clearly understood by comparing Fig. 11.10A and B.

The plots show 2032 coin cell data in Fig. 11.10A and B, which are: (A) HSE-coated separator and added liquid electrolyte, (B) with only liquid electrolyte, in accordance with the description mentioned previously. Fig. 11.11 is a plot that shows 2032 coin cell data comparing the initial cycling results of the cell containing the HSE-coated separator versus the cell cycled using only the liquid electrolyte of 4 μL/mg, in accordance with the aforementioned descriptions. Comparison of the Coulombic efficiency and

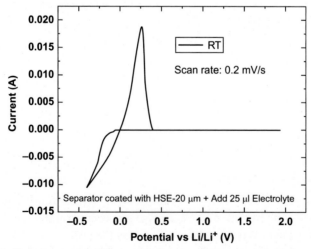

**Fig. 11.9** Cyclic voltammetry (CV) results showing plating and deplating of Li metal studied on the cell: Li metal || HSE coating + liquid electrolyte || Cu metal. *(Source: Authors.)*

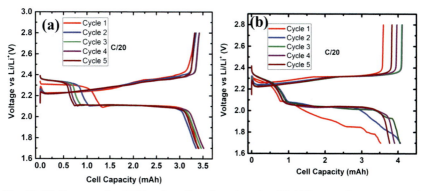

**Fig. 11.10** Comparative 2032 coin cell cycling data for: (A) HSE-coated separator and added liquid electrolyte and (B) identical cell with only liquid electrolyte. *(Source: Authors.)*

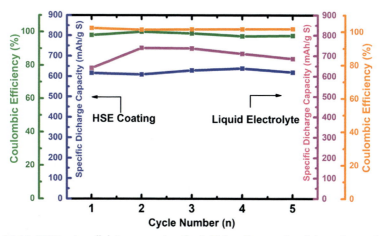

**Fig. 11.11** 2032 coin cell data comparing the initial cycling results of the cell containing the HSE-coated separator versus the cell cycled using only the liquid electrolyte of 4 microliters per mg-S representing the lean electrolyte conditions. *(Source: Authors.)*

the specific capacity of the coin cells cycled with the separator coated with the HSE and the coin cell cycled with the liquid electrolyte is shown in Fig. 11.11. From Fig. 11.11, it can be seen that the Coulombic efficiency is 99% for the HSE coating used in Li-S cells. In general, Li-S cells with liquid electrolyte shows a Coulombic efficiency greater than 100% due to the expected side reactions or polysulfide shuttling effect. The HSE coating on the separator facing Li anode demonstrates that the HSE network created by the Al-doped LLZO effectively traps the soluble polysulfide species and also helps to control any side reactions, thus improving the overall Coulombic efficiency.

The aforementioned description provided in the above sections relates to novel CFMs and CFM-based structures/architectures that include one or more of a CFM host for sulfur infiltration, a coating that includes an electronic conductor, a lithium-ion conductor, and/or a functional electrocatalyst. As discussed already, this includes the development of CFMs, which enable chemical binding of polysulfides and catalytic promoters for polysulfide conversion. The CFMs reduce or preclude, e.g., completely polysulfide dissolution into an electrolyte of a battery. The CFM-based architectures are therefore, used to form cathodes for use in lithium-sulfur batteries.

## 11.5 Directly deposited sulfur architectures

Another approach that was also pursued included generating sulfur directly onto the carbon architecture using a simple electrodeposition process. The approach provides the ability to generate high sulfur loadings onto the carbon structures creating directly deposited sulfur architectures (DDSAs) that result in binder-free cathodes and unique polysulfide trapping agent (PTA) configurations. Similarly, novel inorganic framework materials (IFMs) enabling high sulfur loading and polysulfide (PS) confinement, organic complex framework materials serving as improved sulfur hosts using wet-chemical methods have also been explored combined with high Li-ion-conducting and PS dissolution-resistant coatings on sulfur nanoparticles, and functional electrocatalysts determined by density functional theory (DFT) or first-principles theoretical calculations for rapid conversion of PS to $Li_2S$ and $Li_2S$ to Li and S [22].

All of the approaches discussed in this section provide high-throughput, high-yield, scalable, and commercially inexpensive processes for synthesizing electrochemically stable sulfur-based cathode materials/architectures. Suitable CFMs for use as described above include high specific surface area and high pore volume organic and inorganic materials, such as a porous carbon matrix, and any highly porous high specific surface area, e.g., from about 2000 to $3000\,m^2/g$, YP-80F. In certain forms as outlined above, the CFMs include but are not limited to carbon black (Super-P) to form carbon black (Super-P)-based CFM hosts. The approaches described above can also be extended to other carbon confining material structures.

## 11.5.1 Advanced materials and processing approaches

Suitable electrical conductors (ECs) for use in practical, improved systems include nitrogen- or sulfur- or phosphorus-containing carbon backbone-based aromatic or aliphatic ring polymers such as pyrolyzed form of PAN, polythiophene, and polyaniline. Suitable LICs include Li salt reacted with PAN to form LiOPAN as well as other S-, P-, and N-containing polymers could also serve as polysulfide trapping systems. Suitable functional electrocatalysts (FCs) include one or more of a transition-metal oxide, nitride, sulfide, boride, selenide, telluride, phosphide, bismuthide, antimonide, and arsenide, as well as any transition-metal-containing nonoxide with an ability to bind the negatively charged polysulfide species to form FC-CFM. The FC converts polysulfides to $Li_2S$ during the electrochemical discharge process, and $Li_2S$ to Li and S during the electrochemical charge process.

The approach can also be altered to include a single coating or a layer of EC or LIC or FC applied to the CFM, e.g., porous carbon matrix. In other modifications of the approaches, more than one coating or layer, e.g., a multilayer, of EC and/or LIC and/or FC or a combination thereof, is each separately and individually applied to the CFM, e.g., porous carbon matrix, not described in this chapter. The coated CFMs, for example, include EC-CFM, LIC-CFM, and FC-CFM nanocrystalline, porous architectures. Sulfur is then infiltrated into the CFM to form EC-CFM-S, LIC-CFM-S, and FC-CFM-S nanocrystalline, porous architectures. In other variations explored, the nanocrystalline porous architectures include porous carbon structures. The porosity of the architectures varies, and, in certain forms, the porosity varies from 10% to 90% or from 50% to 60%. The EC-CFM-S, LIC-CFM-S, and FC-CFM-S architectures include sulfur particles within at least a portion of the pores of the porous carbon structures. In certain alterations, the sulfur loading in the CFM architectures varies from 8 to $18\,mg/cm^2$.

Additionally, other variants that were explored and discussed in this chapter include coated EC-CFM, LIC-CFM, and FC-CFM nanocrystalline, porous architectures that are doped. A dopant, such as a nanostructured dopant, is added to the coating. In other variations, the coated EC-CFM, LIC-CFM, and FC-CFM nanocrystalline, porous architectures are undoped. Various conventional/traditional doping techniques are known in the literature [40–43]. In certain forms, using suitable doping techniques, such as a facile solid diffusion technique, a dopant is employed to interact with the LIC and/or EC coatings. Suitable dopants include many elements from the transition-metal series or main group IIIA, IVA, VA, VIA, and VIIA; examples include Ti, Au, Ag, Al, Mg, Ca, F, Nb, B, V, N, P, S, Se, Te, Mn, Sn, Bi, Ni, Co, and Fe [40,43–46].

The CFM host in the absence of the aforementioned coating(s) exhibits poor electrical conductivity, low specific surface area, and pore volume, as well as formation of low- and higher order polysulfides during lithiation of sulfur infiltrated CFM. Whereas the CFMs having the aforementioned EC coating(s) exhibit one or more of improved electrical conductivity, high specific surface area, and pore volume, as well as reduced polysulfide dissolution and high areal capacity following lithiation of the corresponding sulfur infiltrated CFM. The EC-CFM-S thus, demonstrates an ability to trap the polysulfide, thereby improving the areal capacity. The LIC-CFM-S having the aforementioned LIC coating(s) also exhibits one or more of improved ionic conductivity, high specific surface area, and pore volume, as well as high areal capacity. The FC-CFM-S having the aforementioned FC coating(s) exhibits one or more of high surface area, and pore volume, as well as a reduction or prevention of polysulfide dissolution and high areal capacity. All of these CFM systems have inherent PTA configurations, and hence, they demonstrate an improvement for confining polysulfide species and the ability to accommodate high sulfur loadings.

In particular structural variations, as previously explained, the CFM structure/architecture has an LIC coating deposited on a high specific surface area CFM (LIC-HCFM), wherein sulfur is infiltrated into the CFM to form a LIC-HCFM-S nanocrystalline, porous architecture. The EC and LIC are prepared using conventional wet chemical and low-temperature synthesis approaches. The ECs and LICs used to coat the CFMs exhibit high electronic conductivities ($>10^{-3}$ S/cm) and room-temperature lithium-ion conductivities ($>10^{-4}$ S/cm), respectively. The porous metal oxide and nonoxide FC ($>150 \mathrm{m}^2$/g) capable of catalyzing the polysulfide reduction and oxidation are generated using a simple room-temperature method. In certain variations, the oxide and nonoxide functional electrocatalysts (FC) are prepared using dry and wet chemical methods. The EC-CFM, LIC-CFM, and FC-CFM are then infiltrated with sulfur using low temperature under vacuum/inert conditions to form EC-CFM-S, LIC-CFM-S, and FC-CFM-S electrode architectures. These EC-CFM-S, LIC-CFM-S, and FC-CFM-S architectures exhibit a strong binding between the carbon atoms from the CFM hosts and the sulfur molecules forming —C—S— bonds. The creation of these —C—S— bonds inhibits polysulfide dissolution by binding them as is evident from XPS analysis of the sulfur-infiltrated CFMs.

Lithium-sulfur cathodes according to the approach developed and descibed here and earlier, consist of a uniquely designed CFM coated with a Li-ion conductor to form LIC-CFM. The LIC-CFM includes one or more

Cathode and anode architectures for lithium-sulfur batteries · **379**

porous carbon-based materials coated with a novel nitrogen-containing hydrocarbon or carbonaceous polymer such as PAN, modified by reaction with $LiNO_3$, resulting in a Li-ion-conducting coating (LiOPAN) as described earlier, combined with elemental S. The composite LIC-CFM-containing sulfur cathode is suitable for use as a stable, high cycle life, and high-capacity cathode for Li-S batteries.

The PAN polymer on the surface of the porous carbon-based CFM is pyrolyzed, e.g., in argon gas at a temperature of 240°C, in the presence of $LiNO_3$ and elemental sulfur leads to the formation of a cyclized PAN polymer network with structural changes including Li ions attached to the carbon and nitrogen backbone on the surface of porous carbon matrix to form a LiOPAN coating. In certain forms, the entire synthesis of the cyclized PAN and $LiNO_3$ to form LiOPAN and all the chemical reactions are executed in a single step inside a sealed Swagelok container containing a dry mixture of the composite materials. Two major changes that occur simultaneously during PAN pyrolysis process on the surface of porous carbon matrix are the following: (i) PAN converting into N-type PAN, namely nitrogen-doped PAN (N-type PAN) forming LiOPAN after decomposition of $LiNO_3$, and (ii) N-type LiOPAN converting into sulfurized PAN with long-range N-type bonds, the N-type PAN undergoes sulfurization, e.g., at a temperature of 240°C, and it is in turn, transformed into sulfurized PAN due to a cyclization reaction, which creates conjugated polymer structure with long-range N-type bonds. Hence, the LiOPAN-coated porous carbon matrix is converted to sulfurized LiOPAN coating on porous carbon matrix and the sulfurized LiOPAN coating also further develops cross-linked networks on the surface and within the bulk of the porous carbon matrix. In addition, the heat treatment, e.g., at a temperature of 240°C, results in the creation of carbon-sulfur linkages leading to binding or trapping of ensuing polysulfide species formed during the reaction with Li ions when cycled in the Li-S battery. Thus, the sulfurized LiOPAN coating serves to effectively trap polysulfides while also serving as Li-ion-conducting channels with the electron-conducting coatings and the FCs together providing electron transport and catalytic features enabling conversion of polysulfides to $Li_2S$ and back to Li and S.

These C-S linkages are formed with the sulfur within the pores and on the surface of the porous carbon matrix filled with vaporized sulfur during vapor-phase sulfur infiltration. Hence, the final product of S cathode composite material consists of a uniform electrically conductive network of sulfurized PAN as well as Li ion reacted with PAN to form LiOPAN on the surface and within the pores of the porous carbon matrix already filled with nanosized sulfur. The novel LiOPAN-coated sulfur infiltrated CFM-S cathode provides S cathode stability, high specific capacity, and cycle life improvements due to the synergetic effect of sulfurized PAN network present on porous carbon matrix. The sulfurized LiOPAN-coated porous carbon-based S cathode composite material can also be synthesized in large scale (few kilograms per batch, depending on the size of the Swagelok container). The structures can also be produced at a low-cost using vapor-phase sulfur infiltration method. Commercial porous carbon materials used as porous carbon matrix for PAN coating and sulfur infiltration comprise Super P with the specific surface area $\approx 62\,m^2/g$ (Timcal Super P), activated carbon with the specific surface area $\approx 2004\,m^2/g$ (YP-80F, Kuraray coal), and a mixture of Super P and YP-80F.

It is possible to fabricate crack-free electrodes as described earlier, with targeted porosity ($\sim$50%–60%) and a target sulfur loading of ($\sim$4–6 mg/cm$^2$); these improved sulfur cathodes facilitate and achieve efficient charge transport in the electrode as explained previously. Attainment of crack-free S cathode is critical and needed to improve the S utilization rate and confer cycling stability of the assembled Li-S cell. According to the approach developed here and described in this chapter, slurry coatings and 3D-printed (3DP) architectures are also used to form the cathodes. The slurry coatings are produced using conventional slurry coating methods. The 3D-printed architectures of the CFMs, e.g., EC-CFM-S, LIC-CFM-S, and FC-CFM-S, are produced by conventional 3D-printing techniques to form the CFM-based cathodes. A generated composite polymer electrolyte (CPE) membrane also replaces liquid electrolyte (e.g., CPE as electrolyte). In some modifications not described here, a lithium anode ($-$), CPE, and sulfur cathode ($+$) form a stacked configuration. The CPE is positioned (e.g., sandwiched) between the lithium anode and the sulfur cathode. In accordance with the description in this chapter, the authors have found for 3DP-LIC-CFM-S of nanocrystalline porous architecture with CPE, for which the CPE has a very low electrolyte/sulfur ratio and the 3DP-LIC-CFM-S separator has no polysulfide dissolution after cycling, the systems exhibit very high areal capacity. The CFM-based cathodes as described here within a lithium-ion battery provide a Li-S battery system with enhanced performance and properties.

## 11.5.2 Pouch cell fabrication

Pouch cells described earlier, containing such architectures are fabricated as described as follows. A single-sided Li foil is applied, e.g., calendared, onto a

Cu foil as the Li anode for the single-layer pouch cell. The separator used in for the pouch cell is a battery separator membrane or film of Celgard 2400 that is typically used. The single-sided CFM-S-based cathode, separator, and single-sided Li anode, together are arranged in a stacked configuration, such that the separator is "sandwiched" between the S cathode and the Li anode, as the single-layer pouch cell. In certain other forms, a multilayer pouch cell is fabricated. A double-sided Li foil is applied, e.g., calendared, onto a Cu foil as the Li anode of a pouch cell. A double-sided CFM-S-based cathode, separator, and double-sided Li anode are stacked together with two pieces or layers of the single-sided CFM-S-based cathode as the outer layer for the multilayer pouch cells. The electrolyte used was LiTFSI, commonly used in Li-S systems. The electrolyte injection and pouch cell sealing are carried out, such as, in a glove box. The CFM architectures are tested in coin cells and pouch cells under lean electrolyte conditions E:S $= 3-4\,\mu L/mg$-S; the results described here in this chapter have shown significant promise and feasibility.

In certain forms, as described earlier, the LIC-CFM-S is prepared by mixing the LIC-CFM and sulfur in a weight ratio from 10:90 to 20:80. Hybrid composite solid polymer electrolytes are synthesized as explained earlier by mixing sulfur, CFM, and functional monomer in a weight ratio from 70:20:10 to 80:10:10, to yield sulfur-rich CFM with sulfur polymer (S-CFM). In another approach, CFM-S is prepared using a synthetic polymer binder (SPB) to produce a cathode. The CFM-S material is synthesized using a vapor-phase infiltration method. This CFM-S has the ability to trap polysulfide ions; hence, SPB improves the cell performance. The CFM-S system enhances polysulfide trapping and increases sulfur utilization with optimal porosity. In another modification, an inorganic framework material (IFM) is mixed with sulfur in a ratio from 10:90 to 20:80. The IFM-S system is synthesized using a vapor-phase infiltration method. Bifunctional-catalyst (BC) cathode materials are also explored and are synthesized by mixing sulfur, polymer, IFM, and the functional electro catalyst in a ratio of 90:4:3:3 with ball milling as described later, to produce a BCCM-S system. The CFMs are tested in prismatic pouch cells against lithium metal anode using organic liquid electrolyte. It is contemplated and expected that materials and approaches described in the following here are not limited to the particular materials described herein and are applicable and extendable to include a wide variety of high specific surface area and high pore volume organic and inorganic CFM materials for confining sulfur and the generated polysulfides. The CFMs described in this work provide one or more unique features as compared to known materials, such as very high lithium-ion conductivity and electronic conductivity, high specific surface area and pore volume, polysulfide binding and trapping properties, very low fade rate cycling on account of prevention of polysulfide dissolution in Li-S batteries, and high sulfur loading and lean electrolyte (Hi-S/LE) testing properties.

## 11.5.3 Applications for practical battery systems

This section addresses disadvantages associated with known battery systems, including: (a) low overall electrode capacity (mAh/g-active material) occurring due to low electronic conductivity of sulfur, (b) poor cycling stability owing to LiPSs dissolution, (c) voltage drop due to PS transport across and the deposition at the lithium anode, and (d) poor Coulombic efficiency (CE). Thus, the study conducted here delivers either lithium metal/sulfur cells or lithium/[nickel (N)-manganese (M)-cobalt (Co) oxide (NMC)] cells exhibiting good energy density (e.g., in excess of 500 Wh/kg) and stable cycling (e.g., over 1000 cycles). The modification described already includes polysulfide confinement, sulfur electrocatalyst (SEC), DDSA electrodes of electronically conducting nanostructured doped sulfur. In certain descriptions, the approaches described here yield high sulfur loadings $>10 \text{ mg/cm}^2$ and sulfur cathodes displaying specific capacity $\geq 1200 \text{ mAh/g}$ electrode-level capacity, at $\geq 2.2 \text{ V}$ generating $\sim 500 \text{ Wh/kg}$ and like energy density systems. Full cells are constructed, and performance tested in pouch cells. The results include (a) targeted lithium–sulfur battery specific energy ($\geq 500 \text{ Wh/kg}$) and energy density ($\geq 750 \text{ Wh/L}$), (b) compact lightweight LSB, (c) economical and scalable precursors, and (d) excellent life cycle ($\sim 1000 \text{ cycle}$), calendar life ($\sim 15 \text{ years}$) and fast rechargeability. Further, in accordance with the description here, full cells with optimized sulfur cathodes and dendrite-free lithium metal anodes (LMA) are generated with the following performance metrics: specific energy $>500 \text{ Wh/kg}$, cyclability ($>1000 \text{ cycles}$), loss per cycle $<0.01\%$, Coulombic efficiency (CE): $>90\%$ capable of meeting MIL-STD-810G and IEC62133 industry safety standards.

The EC-CFM-S and LIC-CFM-S architectures exhibit polysulfide confinement. Due to the unique polysulfide confinement property of the EC-CFM-S and LIC-CFM-S, a significantly improved electrochemical cycling performance is observed in both coin cell and pouch cell configurations when tested using the Department of Energy (DOE)s Battery 500 protocol to test sulfur cathodes ($>4 \text{ mg/cm}^2$ S loading, $>4 \text{ mAh/cm}^2$ areal capacity, $>64\%$ electrode sulfur content, and $4 \mu\text{L/mg-S}$ electrolyte-to-sulfur (E:S) ratio). The EC-CFM-S systems, upon testing at $\sim5.6\,mg/cm^2$ sulfur loadings, show areal capacities of $3.4\,mAh/cm^2$ when cycled at C/10 rate under Battery 500 lean electrolyte testing conditions ($4\,\mu L/mg$-S). The LIC-CFM-S system shows $\sim4.3\,mAh/cm^2$ capacity for over 100 cycles ($\sim6\,mg/cm^2$) when tested under Battery 500 lean electrolyte conditions ($4\,\mu L/mg$-S).

Further, the FC-CFM-S architectures, upon testing at $\sim6.4\,mg/cm^2$ loadings, show areal capacities of $\sim4.3\,mAh/cm^2$ when cycled at C/10 rate under Battery 500 lean electrolyte conditions ($4\,\mu L/mg$). The 3DP-LIC-CFM electrodes with a sulfur loading of $\sim8\,mg/cm^2$ show high areal capacities of $\sim4.8\,mAh/cm^2$. Initial pouch cell testing of the LIC-CFM-S system shows specific capacities ($\sim450\,mAh/g$) at $\sim2.4\,mg/cm^2$ loadings and under E:S ratio ($10\,\mu L/mg$-S). The CFM architectures according to the work reported here confine polysulfide species with high S loadings ($\sim6\,mg/cm^2$) and areal capacity $\sim4\,mAh/cm^2$ under lean electrolyte ($4$–$10\,\mu L/mg$-S) conditions. These CMF-based nanocrystalline ($\sim10\,nm$) polysulfide confinement structures are cost-effective and easily scalable, sulfur host structures for Li-S batteries with high energy density. While specifically this chapter provides explanation and methodologies that have been described in detail, it will be also appreciated by electrochemistry experts, materials scientists as well as those skilled in the art that various modifications and alternatives to these details explained here could be developed in light of the overall approaches outlined here. Accordingly, the particular electrode systems and architectures described in this chapter are meant as an illustrative pathway to demonstrate the potential of the various systems for Li-S batteries, the fabrication techniques used, and the corresponding electrochemical performance responses. All of the methodologies, described in this chapter, provide a method to prevent polysulfide trapping and enable adequate Li transport.

### 11.5.4 Functional electrocatalysts for conversion of polysulfides

This approach includes: (i) the generation of novel high sulfur loading DDSAs binder-free cathodes and unique PTA configurations, (ii) the generation of novel inorganic framework materials (IFMs) enabling high sulfur loading and PS confinement; (iii) the development of organic CFMs serving as improved sulfur hosts using wet-chemical methods; (iv) the generation of high LIC and PS dissolution–resistant coatings on sulfur nanoparticles; and (v) the identification of FCs for rapid conversion of PS to $Li_2S$ and $Li_2S$ to Li and S, respectively. Identification and synthesis of FCs for rapid conversion of PSs to lithium disulfide was performed as well as optimization of the high loading DDSA electrodes.

To accelerate the PS conversion forward to $Li_2S$ and backward to form pure Li and S, FC materials were identified to facilitate both, the forward and backward reactions (Eq. 11.1) and correspondingly, significantly decrease or even completely eliminate the PS dissolution. Several compounds serving as the FCs were used.

$$2Li^+ + 2e^- + S_8 \leftrightarrow Li_2S_8 \leftrightarrow Li_2S_7 \leftrightarrow Li_2S_6 \leftrightarrow ... \leftrightarrow Li_2S, \qquad (11.1)$$

The decomposition energy $\Delta G$ of various PSs is different for different FCs. The FCs were evaluated experimentally. Electrochemical testing of PTA-coated DDSA electrodes was conducted using the Battery 500 protocol (lean electrolyte-3 $\mu L/mg$-S). Both forms of the DDSA electrodes, PTA-DDSA-1 and PTA-DDSA-2, showed an initial capacity of 1188 mAh/g and 1152 mAh/g, respectively, when cycled at C/20 rate. Upon prolonged cycling, both PTA-DDSA-1 and PTA-DDSA-2 showed the capacity of 870 mAh/g and 925 mAh/g, respectively, after 100 cycles at C/10 rate. An electrochemical performance study was performed of doped lithium-ion conductors (LICs) and their polysulfide shielding properties. The doped and undoped LIC-coated thick nanosulfur-pelletized electrodes ($\sim$7.5 mg-S/cm$^2$) were tested in a Li-S battery, and the changes in the charge transfer resistance ($R_{ct}$) before cycling at open circuit potential were analyzed by electrochemical impedance spectroscopy (EIS). EIS tests showed a steady $R_{ct}$ decrease for the LIC-coated electrodes with dopant additions due to higher lithium-ion conductivity of the doped LICs contrasted with undoped LIC predicted by DFT calculations. Doped and undoped LIC-coated S electrodes were then cycled at a charge and discharge rate of 0.2 C for 100 cycles to assess the influence of the dopants on the cycle life. Cycling results showed improved cycling stability for the doped LIC-coated electrodes compared to undoped LIC-coated electrodes. A doped LIC, in fact, yielded a capacity of 553 mAh/g, while undoped LIC displayed only 247 mAh/g after 100 cycles. These initial results reveal the ability of the doped LICs to improve the sulfur cathode cycling stability by decreasing the interface impedance and also anchoring the polysulfides from entering the electrolyte.

Cathode and anode architectures for lithium-sulfur batteries **385**

## 11.5.5 Directly doped sulfur architectures with higher loadings of sulfur

Directly doped sulfur architectures (DDSAs) with sulfur loadings of $\sim$8–18 mg/cm$^2$ were created using simple electrodeposition technique. The DDSA electrodes were then coated with a polysulfide trapping agent (PTA) to chemically prevent the dissolution of polysulfides. These free-standing cathodes were studied chemically and electrochemically to understand the mechanism of polysulfide dissolution in these structures. These PTA-coated DDSA showed a high initial capacity of $1170 \pm 18$ mAh/g and a stable capacity of $897 \pm 27$ mAh/cm$^2$ for over 100 cycles.

### 11.5.5.1 Synthesis

The synthesis of DDSA was accomplished by electrochemically depositing sulfur onto a conducting carbon nanofiber (CNF) matt. The CNF matt was prepared by electrospinning 1 M solution of polyacrylonitrile (PAN) into a nanofiber ($\sim$200 nm) matt at a high voltage of 25 kV and flow rate of 1 mL/h using an in-house-built electrospinning setup. The electro-spun PAN matt was subsequently carbonized at 700°C for 4 h in ultrahigh purity (UHP)-argon atmosphere (Matheson; 99.99%, flow rate of 100 cm$^3$/min) to form the CNF matt. Sulfur was electrodeposited onto the CNF matt under aqueous electrolyte conditions using a two-electrode setup. The electrolyte consisted of 4.8 g of sulfuric acid (98%, Sigma-Aldrich), 0.3 g of KOH (99.9%, Sigma-Aldrich), and 0.5 M thiourea (99.9%, Sigma-Aldrich) dissolved in 100 mL of deionized water. The CNF matt was used as the working electrode and a Pt foil was used as the counterelectrode and by applying a constant voltage of 5 V between the electrodes for 24 h using a current limiting AC to DC transformer (25A) from McMaster-Carr.

During the electrodeposition process, the sulfate ($SO_4^{2-}$) and hydroxy ($OH^-$) ions were intercalated into the CNF, while the thiourea molecules infiltrated into the CNF and were converted into elemental sulfur particles. The DDSA electrodes had an average sulfur loading of $\sim$8–18 mg/cm$^2$. The DDSA on CNF matt was then electrochemically coated with gold (Au) which was selected as the polysulfide trapping agent. The PTA was electro-deposited onto the DDSA using gold chloride (200 mg/dL deionized water) (Sigma-Aldrich) solution by applying a potential of 5 V between Pt foil working electrode and DDSA counterelectrode.

### 11.5.5.2 Characterization

The XRD patterns of the DDSA-PTA were collected using the Philips XPERT PRO system employing Cu K$_\alpha$ ($\lambda = 0.15406$ nm) between $2\theta$ (10–40 degrees) at 40 mA and 45 kV, respectively. The microstructure and elemental composition of the DDSA-PTA were analyzed using JEOL JSM 6610 LV low-vacuum scanning electron microscope (SEM) equipped with an energy dispersion spectrometer (EDS). The surface chemistry of the DDSA-PTA was probed by X-ray photoelectron spectroscopy (XPS) using an ESCALAB250Xi system (Thermo Scientific) equipped with a monochromated Al-K$_\alpha$ X-ray source. Uniform charge neutralization was provided by beams of low-energy ($\leq 10$ eV) Ar$^+$ ions and low-energy electrons guided by magnetic lens. The UV-Vis spectroscopic measurements were performed in UV-Vis Evol 600 using the organic electrolyte as the reference. Characterization results are summarized in Fig. 11.12.

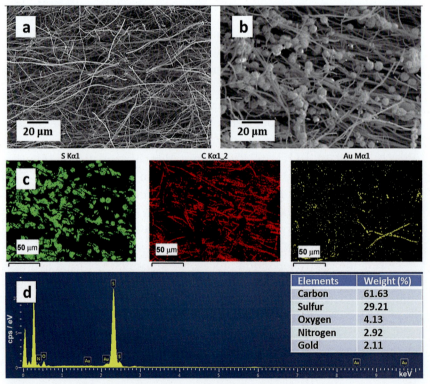

Fig. 11.12 (A) SEM image of the CNF matt; (B) SEM images of PTA-DDSA matt; (C) sulfur, carbon, and gold mapping of the DDSA-PTA matt; and (D) EDS and elemental composition results of the DDSA-PTA matt. *(Source: Authors.)*

## 11.5.5.3 Electrochemical performance

The electrochemical performance of the DDSA-PTA was evaluated in 2032 coin cell. The coin cells for the electrochemical cycling were assembled inside an argon-filled glove box using the DDSA-PTA electrode as the cathode, lithium foil as anode, and 1.8 M lithium trifluoromethanesulfonamide (LiTFSI) and 0.4 M LiNO$_3$ dissolved in dioxolane/dimethoxyethane (1:1 vol%) as electrolyte. The guidelines from the Department of Energy (DOE) for lean electrolyte testing conditions, electrolyte-to-sulfur (E:S) ratio of 4 µL/mg-S, were employed. The coin cells were tested in an Arbin BT200 battery cycler between 1.6 and 2.6 V (with respect to Li$^+$/Li) at 0.1 C current rate. EIS analysis was performed using a Gamry 600 potentiostat by varying the frequency between 100 kHz and 10 mHz at an amplitude of 10 mV with respect to the open circuit potential of ~2.2–2.4 V. The obtained EIS data were then fitted using the ZView software (Scribner and Associates). Electrochemical performance results are summarized in Fig. 11.13.

## 11.5.5.4 Discussion of characterization and electrochemical performance

The XRD analysis of the DDSA-PTA electrode showed peaks corresponding to crystalline sulfur confirming crystalline deposits of sulfur on the CNF matt. The microstructures of the CNF samples and DDSA samples before and after PTA electrodeposition were characterized by scanning electron microscopy. A SEM image (Fig. 11.12A) of the CNF matt showed that the CNF fibers were smooth and of uniform thickness (1–2 µm). The empty

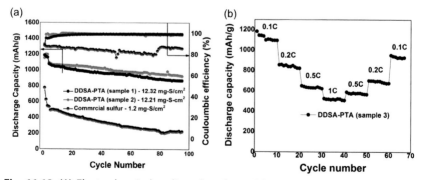

**Fig. 11.13** (A) Electrochemical cycling plot of two DDSA-PTA samples (sample 1 and sample 2) from two batches synthesized under identical deposition conditions. (B) Rate capability measurements on the DDSA-PTA sample 3. *(Source: Authors.)*

regions between the CNF fibers provided room for accommodating sulfur via sulfur electrodeposition. The STEM image (Fig. 11.12B) of the CNF matt after sulfur electrodeposition (DDSA-PTA) showed that the sulfur particles were of uniform size with an average diameter of 5–7 μm and the CNF matt was uniformly coated with the sulfur particles. The use of electrodeposition aided in preparing uniform sulfur deposits on the CNF matt. The sulfur, carbon, and gold mappings of the DDSA-PTA (Fig. 11.12C) showed that they match well with the STEM image (Fig. 11.12B), indicating that sulfur, carbon, and gold were distributed homogeneously throughout the CNF-S composites. The elemental composition analysis (Fig. 11.12D) showed the presence of ~30 wt% sulfur and ~2 wt% gold in the CNF matt comprising ~62 wt% carbon.

The electrochemical cycling plot of two DDSA-PTA samples (Fig. 11.13) from two batches synthesized under identical deposition conditions showed the cycling performance and rate capability of the DDSA electrodes cycled at 0.1 C rates. The discharge capacity was calculated based on the weight of sulfur in the electrode measured from EDS. The DDSA-PTA electrodes exhibited a relatively stable discharge capacity during electrochemical charge–discharge cycling. The 1st and 100th cycle discharge capacity of the DDSA-PTA (sample 1) is 1188 mAh/g and 870 mAh/g, respectively, corresponding to an areal capacity of ~10.71 mAh/cm$^2$. On the contrary, the 1st and 100th cycle discharge capacity of the DDSA-PTA (sample 2) is 1152 mAh/g and 925 mAh/g, respectively, giving an area capacity of ~11.29 mAh/cm$^2$. Correspondingly, the DDSA-PTA (sample 1) and DDSA-PTA (sample 2) electrodes exhibit a very low fade rate of 0.20%/cycle and 0.26%/cycle, respectively, while additionally exhibiting very high Coulombic efficiency of ~99.6%. The observation of slight fade in capacity is due to the formation of insulating Li$_2$S that is not completely oxidized upon charging and not due to polysulfide dissolution as can be expected.

It can be deduced that the sulfur electrodeposited onto CNF mattes by the electrochemical method is responsible for the stable electrochemical cycling performance due to its physical and chemical interactions with polysulfides. In addition, the sulfur electrodeposition at the solid/liquid (CNF/aqueous thiourea solution) interface can ensure intimate contact of sulfur particles with the CNF mattes, effectively confining lithium polysulfides from dissolving into the organic liquid electrolyte. The rate capability of the DDSA-PTA (sample 3) electrode at different current densities from 0.1 to 1 C rate was determined (Fig. 11.13B). A reversible capacity of

Cathode and anode architectures for lithium-sulfur batteries **389**

~825 mAh/g was obtained at a current density of 0.2 C rate, owing to the good electrical conductivity of the CNF ($1.81 \pm 0.17 \times 10^{-5}$ S/cm) and contact with the uniformly dispersed S. The value of the discharge capacity was ~589 and 492 mAh/g for 0.5 C and 1 C rate, respectively, and the discharge capacity returned to ~918 mAh/g at 0.1 C rate, the electrode almost recovering its original capacity. This value to the best of the authors' knowledge for discharge capacity at high sulfur loading is comparable to the best performance of sulfur cathode materials prepared by solution-based deposition technique and other methods.

### 11.5.5.5 Discussion of X-ray photoelectron and electrochemical impedance spectroscopy

X-ray photoelectron spectroscopy (XPS) was used to characterize the chemical state of sulfur in the DDSA-PTA separators and electrodes postcycling. The XPS spectra obtained from the DDSA-PTA separators and electrodes were compared with those obtained from corresponding separators and electrodes cycled with commercial sulfur cathode. The S2p XPS spectra (Fig. 11.14A) were obtained for the DDSA-PTA (sample 1 and sample 2) separators collected after 100 cycles. The commercial sulfur separator after 100 cycles exhibited S2p peaks at 168.53 eV [15,16,23,35], 167.14 eV [15,16,23], and 163.50 eV [23] corresponding to $CF_3SO_3^-$ groups from the LiTFSI salt, lower and higher order polysulfides, respectively. However, the DDSA-PTA separators after cycling showed a significant reduction in peak intensities at 167.14 eV and 163.50 eV confirming that there is almost negligible polysulfide dissolution in the DDSA-PTA system generated and described here.

Similarly, the commercial sulfur electrodes after cycling (Fig. 11.14B) exhibited S2p peaks at 168.53 eV, 167.14 eV, and 163.50 eV corresponding to $CF_3SO_3^-$ groups from the LiTFSI salt, lower and higher order polysulfides, respectively. However, the DDSA-PTA electrodes (sample 1 and sample 2) after cycling showed significant reduction in peak intensities at 163.50 eV, indicating the absence of higher order polysulfides on the DDSA-PTA cathode, which could be attributed to the polysulfide reduction property of electrodeposited Au. XPS results on the separators collected after 100 cycles of other DDSA-PTA electrodes were also compared to the commercial sulfur cycled separator after 100 cycles indicating similar results. The UV–Vis spectroscopic analysis conducted on the DOL-DME solvent-containing polysulfides added to the CNF matt, the DDSA and DDSA-PTA samples correspondingly also revealed the absence of higher order

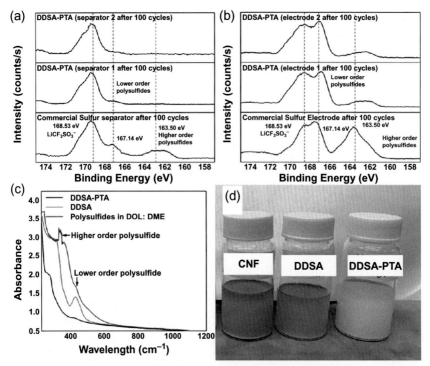

**Fig. 11.14** (A) XPS analysis of the separators, (B) electrodes from the two DDSA-PTA batteries, (C) UV-Vis spectroscopy analysis on the DDSA and DDSA-PTA samples, and (D) polysulfide solutions treated with CNF, DDSA, and DDSA-PTA. *(Source: Authors.)*

polysulfides in the polysulfide solution immersed in DDSA-PTA electrode (Fig. 11.14C and D).

The DDSA-PTA electrodes were further studied using EIS analysis before and after cycling to further understand their enhanced electrochemical performance. The Nyquist plots and the equivalent circuit used to fit the data showed that the experimental data fit well with the fitted data using the equivalent circuit (Fig. 11.15A and B). The Nyquist plots showed two semicircles, corresponding to the resistance of passivation film (interface resistance-$R_i$) of the discharge product in the high-frequency region and the charge transfer resistance $R_{ct}$ in the medium-frequency area. The $R_{ct}$ decreased considerably after cycling due to complete wetting of the electrode by the electrolyte and the rearrangement of the migrated active materials to the electrochemically favorable position.

The interfacial resistance $R_i$ of the pristine as well as the cells after 1st and 100th cycle remained almost constant at ~20 $\Omega$, before and after the 1st cycle

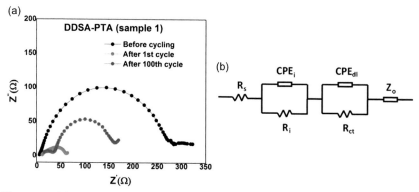

Fig. 11.15 (A) Nyquist plot of the DDSA-PTA battery before cycling, after 1st cycle and 100th charge-discharge cycles and (B) the equivalent circuit used to fit the Nyquist plots. (Source: Authors.)

while the charge transfer resistance ($R_{ct}$) undergoes significant change with lithiation (after the 1st cycle). This reduction in $R_{ct}$ is likely due to the immobilization of polysulfides by the Au nanoparticles, thereby restricting the formation of the solid electrolyte interface on the anode arising from the deposition of the low order polysulfides. All these electrochemical characterizations further suggest that the modest decoration of the cathode by the electrodeposition of Au nanoparticles has a profound influence on the improvement of the electrochemical performance of Li-S batteries.

The aforementioned results validate the use of a simple electrodeposition technique that was implemented to prepare PTA-coated DDSA electrodes that were used as free-standing cathodes in Li-S batteries. The PTA-DDSA cathodes exhibited significantly reduced polysulfide dissolution as was evident from the XPS analysis, in addition to displaying excellent cycling stability. The DDSA-PTA electrode 1 showed an initial capacity of 1188 mAh/g (14.6 mAh/cm$^2$ areal capacity) and a stable capacity of 870 mAh/g (10.71 mAh/cm$^2$ areal capacity) after 100 cycles at 0.1 C rate. The DDSA-PTA electrode 2 exhibited an initial capacity of 1152 mAh/g (14.06 mAh/cm$^2$ areal capacity) and a stable capacity of 925 mAh/g (11.29 mAh/cm$^2$ areal capacity) after 100 cycles at 0.1 C rates. The PTA-DDSA electrodes exhibited a very low fade rate of 0.23 ± 0.03%/cycle and significantly reduced polysulfides when examined by XPS and UV-Vis spectroscopy. The EIS impedance analysis of the DDSA-PTA before and after cycling also suggests polysulfide immobilization by Au nanoparticles. The development of PTA-DDSA electrodes enables the use of

392 Lithium-sulfur batteries

high-energy-density battery systems consisting of sulfur cathodes with superior capacity retention and stability. This approach serves to expedite developments in high-energy lithium-sulfur battery systems and help achieve the DOE's target of 500 Wh/kg. The key would be to extend this approach that was successfully demonstrated in coin cells to fabrication of pouch cells and further demonstration of similar success.

### 11.5.5.6 Sulfur-infiltrated sulfur-copper-bipyridine-derived complex framework materials

Similarly, sulfur-infiltrated sulfur-copper-bipyridine-derived complex framework material (S–Cu–bpy–CFM) was evaluated for their performance as cathodes. Correspondingly, the doped and undoped LICs coated on S–Cu–bpy–CFM pellet electrodes ($\sim3.8$ mg-S/cm$^2$) were tested in coin cells with Li metal in lean ($\sim3\,\mu L$/mg-S) electrolyte conditions. The effect of doping on cycling stability of the S–Cu–bpy–CFM cathodes cycled at 100 mA/g current and rate capability were evaluated by comparing the cycling of doped and undoped LIC-coated S–Cu–bpy–CFM cathodes. Undoped LIC-coated S–Cu–bpy–CFM cathodes displayed an initial capacity of 873 mAh/g and a stable capacity of 583 mAh/g after 100 cycles. The doped LIC-coated S–Cu–bpy–CFM cathodes on the contrary showed much better cycling and rate capability due to higher Li-ion conductivity of the doped LICs. Electrodes with D1 (Ca)-doped LIC showed an initial capacity of 871 mAh/g stabilizing to 702 mAh/g after 100 cycles. These results support the theory that doped LICs enhance the cycling stability and rate capability of S–Cu–bpy–CFM cathodes, by improving the sulfur utilization and anchoring polysulfides from entering the electrolyte.

EC and LIC-coated CFMs were constructed and tested. Novel doped LICs and ECs were chemically coated onto porous CFMs to improve their electronic conductivity ($\sim10^{-3}$ S/cm) along with room-temperature lithium-ion conductivity ($\sim10^{-4}$ S/cm). High-sulfur-loading ($\sim4.5$–7 mg/cm$^2$) cathodes with $\sim64$ wt% sulfur content was prepared from sulfur-infiltrated EC-CFM and LIC-CFM. These cathodes were cycled against lithium metal using 1.8 M LiTFSI and 0.2 M LiNO$_3$ solution under lean electrolyte conditions ($\sim4\,\mu L$/mg-S). The EC-CFM-S cathodes exhibited a stable specific capacity of 678 mAh/g and an areal capacity of 3.8 mAh/cm$^2$ after 70 cycles when cycled at C/10 rate. The LIC-CFM-S cathodes, on the contrary, exhibited a stable specific capacity of 662 mAh/g and areal capacity of 4.04 mAh/cm$^2$ after 70 cycles when cycled at C/10 rate.

## 11.6 Computational studies to identify functional electrocatalysts

An extensive ab initio study based on the density functional theory (DFT) was conducted for the identification of novel FCs for rapid conversion of the polysulfides, namely $Li_2S_m$ with $8 \geq m \geq 2$ to $Li_2S$ as the forward reaction during the discharge process and also to elemental Li and S as the backward reaction during the corresponding charge process. The computational methodology and the major results obtained from this theoretical study are summarized below.

### 11.6.1 Theoretical methodology

During the discharge process, the main sequence of higher to lower order PS conversion reactions forming $Li_2S$ (the end product) is as follows in Eq. (11.2):

$$2Li + (m/8)S_8 + xLi \leftrightarrow Li_2S_m + xLi$$
$$\leftrightarrow [m/(m-1)]Li_2S_{m-1} + \{x - [2/(m-1)]\}Li$$
$$\leftrightarrow PS\{m-2\}, ..., PS\{m-6\}$$
$$\leftrightarrow mLi_2S + [x - 2(m-1)]Li, \qquad (11.2)$$

where $m = 8$; $xLi$—source of lithium in the anode ($x$ is an arbitrary number), $PS\{n\}$—polysulfides of the $n$-th order.

The higher order PSs react with free Li ions forming new PSs with lower order (decreasing S/Li ratio). However, these are multistep reactions containing several relatively simple elementary steps. For example, the conversion of $Li_2S_4$ to $Li_2S_2$ could be written as $Li_2S_4 + 2Li \rightarrow 2Li_2S_2$, which may consist of the following sequence of intermediate specific simplified reactions:

$Li_2S_4 \rightarrow Li_2S_2 + 2S$ (step 1) (formation of the first $Li_2S_2$ molecule and free sulfur);

$2S + 2Li \rightarrow Li_2S + S$ (step 2) (formation of $Li_2S$ molecule and free sulfur);
$Li_2S + S \rightarrow Li_2S_2$ (step 3) (released S reacting with $Li_2S$ to give the second molecule of dilithium disulfide).

Since step 1 in the gaseous phase has sluggish kinetics [47], there is a need to consider the presence of a strong potential electrocatalyst material to improve the kinetics of the PS conversion. Similarly, for the reverse conversion of low-order PSs to high-order PSs and finally to pure S and Li, there will be the need to consider an equivalent elementary decomposition of the

PS with release of Li. For example, the backward conversion of $Li_2S_2$ to $Li_2S_4$ could be represented as $Li_2S_2 \rightarrow LiS_2 + Li$ followed by $LiS_2 + LiS_2 \rightarrow Li_2S_4$.

Such multistep reactions could be reduced to the following forward and backward elementary steps and may serve as the catalytic activity descriptors of the potential functional electrocatalysts:

$$\text{charge} \quad Li_2S_m{}^* \xrightarrow{\Delta G_{R1}} LiS_m{}^* + Li^* \quad \text{(Reaction 1)}$$

$$\text{discharge} \quad Li_2S_m{}^* \xrightarrow{\Delta G_{R2}} Li_2S_{(m-1)}{}^* + S^* \quad \text{(Reaction 2)}$$

with $m = 2\text{–}8$. All of the reactions considered in the present study are collected in Table 11.1. Oxidations of PSs (charging) are denoted as Reaction 1, while reductions (discharging) are denoted as Reaction 2.

The model of the electrocatalyst used for the calculations is shown in Fig. 11.16 and consists of a surface slab with the most stable crystallographic

**Table 11.1** Reactions that have been considered in the present study.

| Reaction 1 | Reaction 2 |
|---|---|
| $Li_2S_8 \rightarrow LiS_8 + Li^*$ | $Li_2S_8 \rightarrow Li_2S_6 + 2S^*$ |
| $Li_2S_6 \rightarrow LiS_6 + Li^*$ | $Li_2S_6 \rightarrow Li_2S_4 + 2S^*$ |
| $Li_2S_4 \rightarrow LiS_4 + Li^*$ | $Li_2S_4 \rightarrow Li_2S_2 + 2S^*$ |
| $Li_2S_2 \rightarrow LiS_2 + Li^*$ | $Li_2S_2 \rightarrow Li_2S + S^*$ |

Source: Authors.

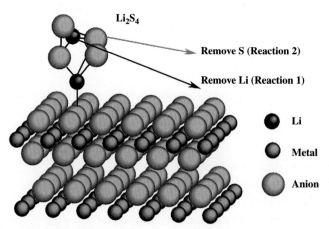

**Fig. 11.16** Polysulfide molecule at the surface of the potential functional electrocatalyst. *(Source: Authors.)*

Cathode and anode architectures for lithium-sulfur batteries    **395**

orientation containing the attached polysulfide molecule. The slab with a thickness of approximately 5–7 Å separated from its image perpendicular to the surface direction by ~20 Å to avoid their mutual interaction has been chosen for all the materials considered. The bottom two to three layers of the slab were fixed with lattice parameters corresponding to the bulk, while the remaining top layers along with the attached polysulfide molecule were allowed to completely relax. All the species are adsorbed on the electrocatalyst surface including Li and S atoms.

All the materials that have been computationally probed within this study are summarized in Table 11.2. DFT methodology implemented in the Vienna Ab initio Simulation Package (VASP) [48,49] has been utilized in all the calculations conducted in the present study.

## 11.6.2 Computational results

Free energies of both Reactions 1 and 2 for the various polysulfide molecules on the surface of all the functional electrocatalysts collected in Table 11.2 have been calculated and the values for several of the representative example materials are shown in Fig. 11.17. One can see that the $\Delta G$ trends for all the different types of materials considered in this study are similar in general but differ in the exact values for the specific polysulfide molecules considered. It is well known that transition metal (TM) sulfides are good functional electrocatalysts fostering rapid polysulfide conversion [50] and are considered as the control systems for comparison. Thus, those materials demonstrating a similar $\Delta G$ range of values to the TM-sulfides, for both Reactions 1 and 2 will be also considered as prospective functional electrocatalytic materials for the polysulfide oxidation/reduction. As for any electrocatalytic process, binding of the intermediate species to the electrocatalytic surface must be appropriate, that is "just right" (not too strong and not too weak) to propel the electrocatalytic reaction to occur in an optimal fashion.

Utilizing the aforementioned considerations combined with the results outlined in Fig. 11.17, one can make the following conclusions regarding the suitability of prospective candidate materials for promoting the polysulfide conversion:

- Since transition metal sulfides are known to be good catalysts for both Reactions 1 and 2, one can conclude that oxides adsorb Li and S atoms more strongly than sulfides. Thus, the decomposition of all PSs in both directions is not at the desired optimal rate.

**Table 11.2** Promising materials studied as potential functional electrocatalysts for PSs decomposition considered in this computational study.

| Sulfides | Oxides | Nitrides | Phosphides | Silicides | Selenides | Tellurides | Arsenides | Antimonides | Bismuthides | Borides | Carbides |
|---|---|---|---|---|---|---|---|---|---|---|---|
| $TiS_2$ | $TiO_2$ | TiN | TiP | $TiSi_2$ | $TiSe_2$ | $TiTe_2$ | TiAs | $Ti_2Sb$ | $Ti_2Bi$ | $Ti_2B$ | TiC |
| $VS_2$ | $V_2O_5$ | VN | VP | $VSi_2$ | $VSe_2$ | $VTe_2$ | Vas | $V_3Sb$ | $FeBi_2$ | VB | ZrC |
| FeS | $Fe_2O_3$ | $Fe_2N$ | $Fe_3P$ | FeSi | $FeSe_2$ | $FeTe_2$ | $FeAs_2$ | $FeSb_2$ | NiBi | BN | TaC |
| $CoS_2$ | CoO | $Co_2N$ | $Co_3P_2$ | $CoSi_2$ | $CoSe_2$ | $CoTe_2$ | $CoAs_2$ | $CoSb_2$ | | | SiC |
| $Ni_3S_2$ | NiO | $Ni_3N$ | $Ni_2P$ | $NiSi_2$ | NiSe | $NiTe_2$ | NiAs | NiSb | | | $Mo_2C$ |
| $SnS_2$ | $SnO_2$ | $Sn_3N_4$ | $Sn_4P_3$ | $Sn_2Si$ | SnSe | SnTe | SnAs | | | | $W_2C$ |
| ZnS | | | | | | | | | | | NbC |
| | | | | | | | | | | | $Fe_3C$ |

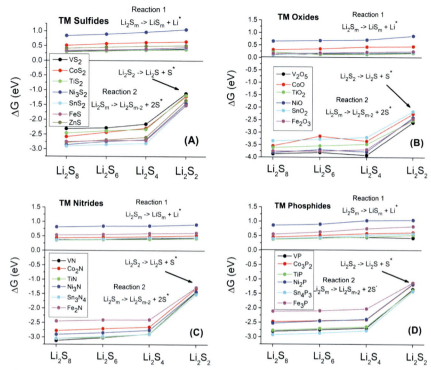

**Fig. 11.17** Free energies, ΔG of Reactions 1 and 2 for the different polysulfides on (A) transition-metal (TM) sulfides, (B) TM oxides, (C) TM nitrides, and (D) TM phosphides. *(Source: Authors.)*

- $TiO_2$, $V_2O_5$, and $Fe_2O_3$ are better than others. They are good anchoring materials but might not allow to desorb Li and S from the surface, thus resulting in catalyst poisoning and capacity loss.
- Phosphides and nitrides, on the contrary, are closer to sulfides and might serve as better catalysts than oxides. Phosphides indeed show weak binding of Li and S than oxides resulting in optimal free energy for both reactions similar to transition metal sulfides. The best materials are V-, Ti-, and Sn-phosphides enabling oxidation and reduction of $Li_2S$ in both directions.
- Nitrides bind Li and S similar to phosphides and sulfides making them good functional electrocatalyst candidates. Again, the best materials are V-, Ti-, and Sn-nitrides accelerating both Reactions 1 and 2.

The study conducted for all the materials considered here is collected in Table 11.2 and reveals the following trends for the Li-to-electrocatalyst and S-to-electrocatalyst binding strengths:

**Relative strength of binding of Li to the materials in decreasing order of strength**

Borides > Oxides > Arsenides > Sulfides ≥ Nitrides ≈ Phosphides > Selenides > Tellurides > Antimonides > Bismuthides >> Silicides ≈ Carbides

**Relative strength of binding S to the materials in decreasing order of strength**

Silicides ≈ Oxides ≥ Carbides > Borides > Nitrides ≈ Phosphides ≈ Sulfides ≈ Selenides ≥ Arsenides > Antimonides > Bismuthides ≈ Tellurides

The average $\Delta G$ calculated for the various binary compositions collected in Table 11.2 are shown in Fig. 11.18. The red dashed lines represent the average $\Delta G$ for both reactions in TM sulfides considered as the benchmarks for the comparison of the TM sulfides with other probed electrocatalytic materials. One can see that besides TM sulfides, other types of binary compounds, such as TM-nitrides and TM-phosphides, could also be potentially good functional electrocatalysts considered for *both* reactions. The transition-metal oxides and borides, however, could be good for Reaction 1 only, while transition-metal selenides and arsenides are expected to be good for Reaction 2 only. Other types of transition-metal-based materials, such as tellurides, antimonides,

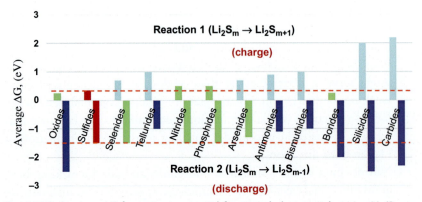

**Fig. 11.18** Average $\Delta G$ for various potential functional electrocatalysts in eV. *(Source: Authors.)*

Cathode and anode architectures for lithium-sulfur batteries    **399**

bismuthides, silicides, and carbides, are not expected to be optimal functional electrocatalysts for either reaction.

However, it should be noted that various solid solutions of individual binary compounds listed in Table 11.2 and multicomponent alloys and combinations of (Ni, Co, Mn, Fe, Ti, V, Sn, P, S, Se, Te, B, N, As, Sb, Si, and Bi) between the suboptimal materials may result in improved overall electrocatalytic properties for decomposition of polysulfides in both directions, simultaneously. Such ternary and multicomponent solid solution materials could be transition-metal oxotellurides, oxoarsenides, oxoantimonides, oxobismuthides as well as transition-metal borobismuthides, boroantimonides, boroarsenides, borotellurides, and other similar combinations thereof. All of the theoretical findings described earlier are under experimental validation and undergoing testing at present in this novel technologically relevant battery system serving as potential functional electrocatalysts for promoting the conversion of polysulfides to insoluble $Li_2S$ and back to elemental $Li + S$ during the discharge/charge processes, respectively of electrochemical cycling in rechargeable Li-S batteries.

## 11.7 Functional electrocatalysts and related materials for polysulfide decomposition

As discussed earlier, to improve the cycling characteristics and reduce the polysulfide dissolution in Li-S batteries, it is necessary to modify the electrode architecture. This was done by developing solid-state LICs identified using DFT calculations. Undoped LIC-coated S-Cu-bpy-CFM cathodes displayed an initial capacity of 873 mAh/g and a stable capacity of 583 mAh/g after 100 cycles. The doped LIC-coated S-Cu-bpy-CFM cathodes as mentioned earlier, showed much better cycling and rate capability due to higher Li-ion conductivity of doped LICs determined by theory and validated by experiments. Electrodes with doped LIC showed an initial capacity of 871 mAh/g stabilizing to 702 mAh/g after 100 cycles. In addition, sulfur architectures were developed by infiltrating sulfur into chemically coupled conducting CFMs. Various complex framework material systems (EC-CFM-S and LIC-CFM-S) were derived and evaluated as cathodes in lithium-sulfur batteries. The EC-CFM-S as mentioned earlier, showed a stable discharge capacity of 678 mAh/g after 70 cycles at C/10 rate, and LIC-CFM-S exhibited a high discharge capacity of 662 mAh/g after 70 cycles at C/10 rate.

400 Lithium-sulfur batteries

## 11.7.1 Functional electrocatalyst material preparation and characterization

### 11.7.1.1 Titanium oxide-based functional electrocatalyst material preparation

As discussed in the earlier sections, single–layer Li–S pouch cells were fabricated and tested with an optimized S electrode of thickness 150 μm (calculated porosity ~66%) with S loading of 6.26 mg/cm$^2$ under lean electrolyte (4 μL/mg-S) conditions. Initial discharge specific capacity of ≈900 mAh/g was observed at C/20 rate stabilizing to ≈650 mAh/g capacity at 0.1 C rate after 5 cycles. The energy density of the pouch cell was ≈200 Wh/kg at 0.1 C rate (neglecting the mass of pouch case and tabs). Other single-layer pouch cells fabricated with various mass loadings of S also displayed similar behavior. This section describes approaches used to generate functional electrocatalysts, incorporation of them into the base CFM architecture as well as other approaches used to generate various framework architectures containing bifunctional catalysts serving as PTA as well as catalyzing the PS conversion, synthetic polymer binders (SPB), functional monomer-based CFM architectures, 3-D printed architectures, and finally, inorganic framework materials (IFM).

The approaches followed for generating the titanium oxide-based functional electrocatalyst are described here. Two solutions were prepared; the first solution contained 60 mL of titanium butoxide dissolved in 250 mL methanol. Second solution contained 64 g of thiourea dissolved in 250 mL methanol. Solutions 1 and 2 were mixed together with magnetic stirrer at room temperature; after the evaporation of methanol, white powder was obtained. The white powder (S–TiO$_2$) was then calcined at 400°C (S–TiO$_2$–CFM) for 4–6 h (heating rate 5°C/min) to obtain sulfur-containing titanium oxide. The second step involved sulfur infiltration into the S-doped titanium oxide-CFM (S–TiO$_2$–CFM). The sulfur-doped titanium oxide carbon framework (S–TiO$_2$–CFM) was dried under vacuum conditions at 60°C for 12 h to remove residual solvent and the water of crystallization from the synthesis process. The synthesized S–TiO$_2$–CFMs were infiltrated with sulfur under vacuum using the following procedure. Precisely, S–TiO$_2$–CFM and S (mass ratio 90:10) were ground together and then sealed and heated under argon in a Swagelok cell at a heating rate of 5°C/min up to 240°C. The mixture was maintained at 240°C for 12 h to obtain S–TiO$_2$–CFM-S.

### 11.7.1.2 Characterization of titanium oxide-based functional electrocatalysts

The crystal structure of the S–TiO$_2$–CFM and S–TiO$_2$–CFM-S before and after sulfur infiltration was analyzed using X-ray diffraction in a Philips

XPERT PRO system that uses Cu $K_\alpha$ ($\lambda = 0.15406$ nm) radiation. The samples were scanned in the $2\theta$ range of 10–90 degrees under a constant current and voltage of 40 mA and 45 kV, respectively. The SEM images of the S-TiO$_2$-CFMs were obtained using a Philips XL30 machine at 10 kV. An attenuated total reflectance-Fourier-transform IR (ATR-FTIR, Nicolet 6700 Spectrophotometer, Thermo Electron Corporation) spectroscope, which uses a diamond ATR smart orbit, was used to obtain the FTIR spectra of the samples. The FTIR spectra are collected at a resolution of 1 cm$^{-1}$, averaging 32 scans between the frequency of 400 and 4000 cm$^{-1}$. The XPS analyses of the S-TiO$_2$-CFMs and S-TiO$_2$-CFM-S were performed using the ESCALAB 250 Xi system (Thermo Scientific). This XPS system consists of the monochromated Al $K_\alpha$ X-ray source and low-energy ($\leq 10$ eV) argon ions and low-energy electron beams that provide the charge neutralization. The XPS measurements were carried out at room temperature, under an ultrahigh vacuum (UHV) chamber ($<5 \times 10^{-10}$ mBar) employing a spot size of $200 \times 200$ $\mu$m$^2$. The surface area and pore characteristics of all the S-TiO$_2$-CFM samples were analyzed using a Micromeritics ASAP 2020 Physisorption analyzer, using the BET isotherm generated.

The S-TiO$_2$-CFM-S was cycled between 1.8 and 2.8 V (with respect to Li$^+$/Li) at a current rate of 0.05 C in a 2032-coin cell using the Arbin BT200 battery testing station to evaluate their electrochemical performance. The cathodes for electrochemical evaluation were prepared by manually coating a dispersion of 70 wt% S-TiO$_2$-CFM-S, 20 wt% acetylene black, and 10 wt% PVDF dispersed in $N$-methyl-2-pyrroli-done (NMP) on an aluminum foil, followed by vacuum-drying for 12 h at 60°C. All the cathodes that were tested had a uniform sulfur loading of 4.0–6.0 mg/cm$^2$. Accordingly, 2032-coin cells were assembled with the S-TiO$_2$-CFM-S-coated cathodes as the working electrode, a lithium foil as the counterelectrode, and Celgard 2400 polypropylene (PP) as the separator in an Innovative, Inc., glove box (UHP Argon, $<0.1$ ppm O$_2$, H$_2$O). An electrolyte solution of 1.8 M LiTFSI (lithium bis (trifluoromethanesulfonyl)imide) and 0.4 m LiNO$_3$ dissolved in 50:50 vol% 1,3 dioxolane and 1,2 dimethoxyethane was used as the electrolyte. The sulfur loading was relatively high at 4.3 mg/cm$^2$. The functional electrocatalyst (S-TiO$_2$-CFM-S) coordination configuration and interlayer area/pore volume of S-TiO$_2$-CFM-S provide the framework for catalytic conversion of LiPSs to Li$_2$S and sufficient entrapment of LiPSs. Fig. 11.19 shows the ability of the system to effectively suppress the shuttle effect, as a result, at C/20 rate under lean electrolyte condition of E:S = 4 $\mu$L/mg-S in a Li-S battery, the S-TiO$_2$-CFM-S composite electrode shows a stable capacity of $\sim$650 mAh/g. The system

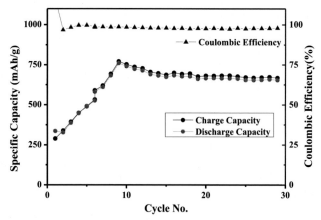

**Fig. 11.19** Cycling performance of S-TiO$_2$-CFM-S cycled at 0.05 C rate. *(Source: Authors.)*

however, showed a low initial capacity of 250 mAh/g steadily increasing to 750 mAh/g after 10 cycles stabilizing at ~650 mAh/g for over 20 cycles likely due to electrolyte wettability issues. The stable capacity of ~650 mAh/g shows the ability of the S-TiO$_2$ to catalyze and convert the higher order polysulfides to lower order polysulfides and still deliver a high reversible capacity of ~650 mAh/g. The low overall capacity would indicate the inability to successfully convert the polysulfides to Li$_2$S although, the stability would indicate that the Ti centers in the oxysulfide could serve to trap the polysulfides. More work is, however, necessary to show the true potential of the oxysulfide for effectively catalyzing as well as trapping the polysulfides.

### 11.7.1.3 Bifunctional electrocatalyst cathode material (BCCM) synthesis

Alternatively, bifunctional catalysts promoting PSs conversion and decomposition as well as suppressing the adverse polysulfide shuttling plaguing Li-S battery were added to the CFM to maximize S utilization. The CFM containing the bifunctional electrocatalyst derived by a simple solid-state scalable approach creating high surface area/pore volume likely provided a high activity for PS trapping, conversion, and decomposition. The 2032 coin cells were cycled with an optimized S electrode of 107 μm thickness, porosity 63%, and S loading 3.71 mg/cm$^2$. Initial capacity ≈650 mAh/g was observed, stabilizing at ~700 mAh/g at C/20 rate after 3 cycles. The approaches for generating the bifunctional electrocatalysts are outlined below.

Mixtures of PAN (Sigma-Aldrich, 99.5%), boron nitride (BN) (Sigma-Aldrich, 99.5%), titanium sulfide ($TiS_2$) (Sigma-Aldrich), and iron (Fe, 98%, Alfa Aesar) corresponding to the stoichiometric composition with urea ($\sim$10%) as a porogen were subjected to high-energy mechanical milling (HEMM) in a high-energy shaker mill for 1 h in a stainless steel (SS) vial using 20 SS balls of 2 mm diameter with a ball to powder weight ratio 10:1. The milled powder was then heat-treated at 700°C for 6 h (ramp rate of 5°C/min) in argon atmosphere. The synthesized BCCMs were infiltrated with sulfur under vacuum using the following procedure. The BCCM and S (mass ratio 90:10) were ground together and then sealed and heated under argon in a Swagelok cell at a rate of 10°C/min up to 240°C. The mixture was maintained at 240°C for 12 h to obtain BCCM-S.

### 11.7.1.4 Chemical and electrochemical characterization of bifunctional catalysts

Similar to the titanium oxide-based functional electrocatalysts, the bifunctional catalyst system was also characterized for structure and electrochemical response. The crystal structure of the BCCM and BCCM-S before and after sulfur infiltration was analyzed using the XRD in a Philips XPERT PRO system that uses Cu $K_\alpha$ ($\lambda = 0.15406$ nm) radiation. The samples were scanned in the $2\theta$ range of 10–90 degrees under a constant current and voltage of 40 mA and 45 kV, respectively. The SEM images of the CFMs were obtained using a Philips XL30 machine at 10 kV. An attenuated total reflectance-Fourier-transform IR (ATR-FTIR, Nicolet 6700 Spectrophotometer, Thermo Electron Corporation) spectroscope, which uses a diamond ATR smart orbit, was used to obtain the FTIR spectra of the samples. The FTIR spectra are collected at a resolution of 1 $cm^{-1}$, averaging 32 scans between the frequency of 400 and 4000 $cm^{-1}$. The XPS analyses of the BCCMs and BCCM-S were performed using the ES- CALAB 250 Xi system (Thermo Scientific). This XPS system consists of the monochromated Al $K_\alpha$ X-ray source and low-energy ($\leq$10 eV) argon ions and low-energy electron beams that provide the charge neutralization. The XPS measurements were carried out at room temperature, under an ultrahigh vacuum (UHV) chamber ($<5 \times 10^{-10}$ mBar) employing a spot size of $200 \times 200$ $\mu m^2$. The surface area and pore characteristics of all the BCCM-S samples were analyzed using a Micromeritics ASAP 2020 Physisorption analyzer, using the BET isotherm generated.

The BCCM-S samples were cycled between 1.8 and 2.8 V (with respect to $Li^+/Li$) at a current rate of 0.05C rate in a 2032-coin cell and pouch cells using the Arbin BT200 battery testing station to evaluate their electrochemical performance. The cathodes for electrochemical evaluation were prepared by manually coating a dispersion of 70 wt% BCCM-S, 20 wt% acetylene black, and 10 wt% PVDF dispersed in N-methyl-2-pyrrolidone (NMP) on an aluminum foil, followed by vacuum-drying for 12 h at 60°C. All the cathodes that were tested had a uniform sulfur loading of 3.6 mg/cm$^2$. Accordingly, 2032-coin cells were assembled with the BCCM-S-coated cathodes as the working electrode, a lithium foil as the counterelectrode, and Celgard 2400 polypropylene (PP) as the separator in an Innovative, Inc., glove box (UHP Argon, <0.1 ppm $O_2$, $H_2O$). A 1.8 M LiTFSI (lithium bis (trifluoromethanesulfonyl)imide), and 0.4 M $LiNO_3$ dissolved in 50:50 vol% 1,3 dioxolane and 1,2 dimethoxyethane were used as the electrolytes. The bifunctional electrocatalysts of ($TiS_2$-Fe) present in the complexed framework material of C with appropriate coordination configuration and high surface area/pore volume of $TiS_2$-Fe-C provide good activity for catalytic conversion of LiPS to $Li_2S$ and sufficient entrapment of

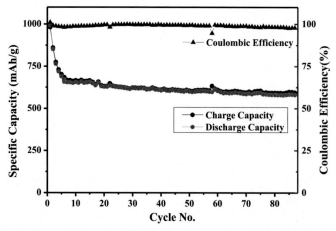

**Fig. 11.20** Cycling performance of BCCM-S cycled at 0.05 C rate. *(Source: Authors.)*

LiPS. Fig. 11.20 shows the ability of the system to effectively suppress the shuttle effect. As a result, even under a C/20 rate under lean electrolyte conditions of $E/S = 4\,\mu L/mg\text{-}S$ in a Li-S battery, the $S\text{-}TiS_2\text{-}Fe\text{-}C$ (BCCM-S) composite electrode still delivers a high initial capacity of $\sim 1000\,mAh/g$ stabilizing at $\sim 625\,mAh/g$ after 10 cycles up to 90 cycles.

### 11.7.1.5 Synthesis of lithium-ion conductor coated on bifunctional electrocatalyst cathode materials

The LIC coated systems were synthesized as follows. Mixtures of polyacrylonitrile (PAN) (Sigma–Aldrich, 99.5%), boron nitride (BN) (Sigma–Aldrich, 99.5%), titanium disulfide ($TiS_2$) (Sigma–Aldrich), and iron (Fe, 98%, Alfa Aesar) corresponding to the stoichiometric composition were subjected to high-energy mechanical milling (HEMM) in a high-energy shaker mill for 1 h in a stainless steel (SS) vial using 20 SS balls of 2 mm diameter with a ball-to-powder weight ratio 10:1. The milled powder was then heat-treated in the air at 700°C for 6 h (ramp rate = 5°C/min) in argon atmosphere. The LIC-coated BCCM-S electrode was generated following the two steps mentioned as follows.

In the first step, a BCCM (5.0 g) was dispersed in a mixed solution of deionized water (25 mL) and dimethyl sulfoxide (DMSO) (25 mL) under ultrasonication (Branson, 5800, United States) for 10 min. Acrylonitrile monomer (10 mL) and AIBN (100 mg) were then added. The mixed solution was stirred under a $N_2$ atmosphere for 4 h at 65°C to conduct the polymerization. The resulting solid was collected by filtration and washed with ethanol after polymerization. The product was vacuum-dried at 60°C for 24 h to yield PAN/BCCM. The second step involved sulfur infiltration into the PAN/BCCMs. The PAN/BCCMs were dried under vacuum conditions at 60°C for 12 h to remove residual solvent and the water of crystallization from the synthesis process. The synthesized PAN/BCCMs were infiltrated with sulfur under vacuum according to the following procedure. PAN/BCCM and S (mass ratio 90:10) and 10 wt% lithium salts of $LiNO_3$ were ground together and then sealed and heated under argon at a rate of 5°C/min up to 240°C. The mixture was maintained at 240°C for 12 h to obtain lithium-ion conductor coated on bifunctional catalyst cathode materials (LIC-BCCM-S).

## 11.7.1.6 Chemical and electrochemical characterization of Li-ion-coated bifunctional electrocatalysts

These systems were also characterized in a similar fashion. Thus, the crystal structure of the BCCM and LIC–BCCM-S before and after sulfur infiltration was analyzed using the XRD spectroscopy in a Philips XPERT PRO system that uses Cu $K_\alpha$ ($\lambda = 0.15406$ nm) radiation. The samples were scanned in the $2\theta$ range of 10–90 degree under a constant current and voltage of 40 mA and 45 kV, respectively. The SEM images of the CFMs were obtained using a Philips XL30 machine at 10 kV. An attenuated total reflectance Fourier transform IR spectroscope (ATR-FTIR, Nicolet 6700 Spectrophotometer, Thermo Electron Corporation), which uses a diamond ATR smart orbit, was used to obtain the FT-IR spectra of the samples. The FTIR spectra are collected at a resolution of 1 cm$^{-1}$, averaging 32 scans between the frequency of 400 and 4000 cm$^{-1}$. The XPS analyses of the BCCMs and LIC–BCCM-S were performed using the ESCALAB 250 Xi system (Thermo Scientific). This XPS system consists of the monochromated Al $K_\alpha$ X-ray source and low-energy ($\leq 10$ eV) argon ions and low-energy electron beams that provide the charge neutralization. The XPS measurements were carried out at room temperature, under an ultra-high vacuum (UHV) chamber ($< 5 \times 10^{-10}$ mBar) employing a spot size of $200 \times 200$ μm. The surface area and pore characteristics of all the BCCM samples were analyzed using a Micromeritics ASAP 2020 Physisorption analyzer, using the BET isotherm generated.

The LIC–BCCM-S samples were cycled between 1.8 and 2.8 V (with respect to Li$^+$/Li) at a current rate of 0.2C in a 2032-coin cell and pouch cells using the Arbin BT200 battery testing station to evaluate their electrochemical performance. The cathodes for electrochemical evaluation were prepared by manually coating a dispersion of 70 wt% LiC–BCCM-S, 20 wt% acetylene black, and 10 wt% PVDF dispersed in $N$-methyl pyrrolidone (NMP) on an aluminum foil, followed by vacuum-drying for 12 h at 60C. All the cathodes that were tested had a uniform sulfur loading of 4.0–6.0 mg/cm$^2$. Accordingly, 2032-coin cells were assembled with the LiC–BCCM-S-coated cathodes as the working electrode, a lithium foil as the counterelectrode, and Celgard 2400 polypropylene (PP) as the separator in an Innovative, Inc., glove box (UHP Argon, $<0.1$ ppm O$_2$, H$_2$O). A 1.8 m LiTFSI (lithium bis(trifluoromethanesulfonyl)imide), and

0.4 M LiNO$_3$ dissolved in 50:50 vol% 1,3 dioxolane and 1,2 dimethoxyethane were used as the electrolytes. The bifunctional electrocatalyst coated with LIC (LIC-BCCM-S) coordination configuration and high surface area/pore volume of LIC-TiS$_2$-Fe-C provide high activity for catalytic conversion of LiPS to Li$_2$S and sufficient entrapment of LiPS. As a result, even under a C/20 rate under lean electrolyte condition of E/S $= 4\,\mu$L/mg-S in a Li-S battery in a Li-S battery, the LIC-S-TiS$_2$-Fe-C composite electrode still delivers a high reversible capacity of 400–450 mAh/g.

## 11.7.2 Novel complex framework material processing and characterization

### 11.7.2.1 3D printing complex framework material-sulfure architecture

As mentioned above, 3D printing was also explored to study the electrochemical response as described herein. Sulfur infiltration into the S-CFM was accomplished via the following method. The CFM was dried under vacuum conditions at 60°C for 12h to remove residual solvent and the water of crystallization from the synthesis process. The synthesized CFMs were infiltrated with sulfur under vacuum, in the following the manner. The CFM)and S (mass ratio 30:70) were ground together and then sealed and heated under argon at a rate of 5°C/min up to 240°C. The mixture was maintained at 240°C for 12h to obtain CFM-S. The 3D-printed S-CFM cathode was fabricated using a custom-made 3D printer of Cellink equipped with a three-axis micropositioning stage printer. The printing process is as follows: S-CFM composite, polyvinylidene fluoride (PVDF)-hexafluoropropylene (HFP), and graphene with a weight ratio of 70:20:10 were first mixed with 1-methyl-2-pyrrolidone (NMP) to form an ink. The as-prepared ink was then loaded into a 10-mL syringe and extruded through a 150-μm-diameter nozzle. The 3D-printed S-CFM cathodes were printed onto a disk of diameter of 10mm at a print motion speed of 6mm/s. The printed electrode used two different treatments for comparison: firstly, printed electrodes were immediately immersed in a water (100mL) coagulation bath for 5min. It is noted that this step led to phase inversion during this process. The binder network, electron paths, and ion channels formed during the phase inversion can significantly improve the adhesive strength and facilitate electron/ion transport in the electrodes. Following this, freeze-drying was conducted at −50°C to

# 408  Lithium-sulfur batteries

maintain the structure formed during 3D printing and phase inversion. The obtained electrode named as 3D-PFDE.

### 11.7.2.2 Chemical and electrochemical characterization of 3D-printed materials

Similar to the other systems described earlier, the 3D printed structures were also characterized for structure and electrochemical response. Accordingly, the crystal structure of the CFMs and 3D-printed freeze-dried electrodes (3DPFDEs) was analyzed using the XRD spectroscopy in a Philips XPERT PRO system that uses Cu $K_\alpha$ ($\lambda = 0.15406$ nm) radiation. The samples were scanned in the $2\theta$ range of 10–90 degrees under a constant current and voltage of 40 mA and 45 kV, respectively. The SEM images of the 3DPFDE were obtained using a Philips XL30 machine at 10 kV. An attenuated total reflectance-Fourier-transform IR (ATR-FTIR, Nicolet 6700 Spectrophotometer, Thermo Electron Corporation) spectroscope, which uses a diamond ATR smart orbit, was used to obtain the FTIR spectra of the samples. The FTIR spectra are collected at a resolution of 1 cm$^{-1}$, averaging 32 scans between the frequency of 400 and 4000 cm$^{-1}$. The XPS analyses of the SCFMs and 3DPFDE were performed using the ESCALAB 250 Xi system (Thermo Scientific). This XPS system consists of the monochromated Al $K_\alpha$ X-ray source and low-energy ($\leq 10$ eV) argon ions and low-energy electron beams that provide the charge neutralization. The XPS measurements were carried out at room temperature, under an ultrahigh vacuum (UHV) chamber ($<5 \times 10^{-10}$ mBar) employing a spot size of $200 \times 200$ μm$^2$. The surface area and pore characteristics of all the SCFM and 3DPFDE samples were analyzed using a Micromeritics ASAP 2020 Physisorption analyzer, using the BET isotherm generated.

The 3DPFDEs were cycled between 1.8 and 2.8 V (with respect to Li$^+$/Li) at a current rate of 0.2 C in a 2032-coin cell using the Arbin BT200 battery testing station to evaluate their electrochemical performance. The cathodes for electrochemical evaluation were prepared by 3D-printing method using a dispersion of 70 wt% S-CFMs (S-infiltrated CFM), 20 wt% graphene, and 10 wt% PVDF dispersed in $N$-methyl-2-pyrrolidone (3DPink). All the cathodes that were tested had a uniform sulfur loading of 5.1 mg/cm$^2$. Accordingly, 2032-coin cells were assembled with the 3DPFDE-printed cathodes as the working electrode, a lithium foil as the counterelectrode, and Celgard 2400 polypropylene (PP) as the separator in an Innovative, Inc., glove box (UHP Argon, $<0.1$ ppm O$_2$, H$_2$O). An electrolyte solution

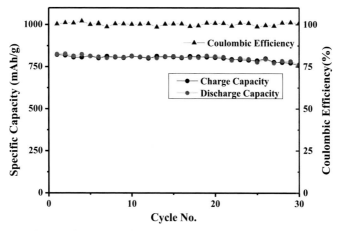

**Fig. 11.21** Cycling performance of 3DPFDE cycled at 0.05 C rate. *(Source: Authors.)*

containing 1.8 M LiTFSI (lithium bis(trifluoromethanesulfonyl)imide) and 0.4 M LiNO$_3$ dissolved in 50:50 vol% 1,3 dioxolane and 1,2 dimethoxyethane was used as the electrolyte. The 3DPFDE coordination interconnected materials and high surface area/pore volume of 3DPFDE and the C-S interactions provide high activity for the conversion of the polysulfides, LiPS to Li$_2$S, and sufficient entrapment of LiPS. Fig. 11.21 shows the ability of the system to effectively suppress the shuttle effect. As a result, even under a C/20 rate under lean electrolyte condition of E:S = 4 μL/mg-S in a Li-S battery, the 3D-printed architecture electrode still delivers a high reversible capacity of ~810 mAh/g for 25 cycles. The long-term stability of this system is currently under investigation. The result, nevertheless, shows the promise of using 3D printing to generate functional electrodes for Li-S batteries.

### 11.7.2.3 Synthesis of electrical conductor coated on complex framework materials

The electrical conductor coated CFM material is synthesized as follows. Complex framework material (CFM, 5.0 g) was dispersed in a mixed solution of deionized water (25 mL) and dimethyl sulfoxide (DMSO) (25 mL) under ultrasonication (BRANSON, 5800, United States) for 10 min. Acrylonitrile monomer (10 mL) and azobisisobutyronitrile (AIBN, 100 mg) were then added. The mixed solution was stirred under a N$_2$ atmosphere for 4 h at 65°C to conduct the polymerization. The resulting solid was collected by filtration and washed with ethanol after polymerization. The product was vacuum-dried at 60°C for 24 h to yield PAN/CFM. The composite material

## 410 Lithium-sulfur batteries

was then heat-treated in argon at 700°C for 6h (ramp rate $=5$°C/min). The synthesized EC-CFMs were then infiltrated with sulfur under vacuum, following a similar procedure as outlined previously as follows. The EC-CFM and S (mass ratio 90:10) were ground together and then sealed and heated under argon in a Swagelok cell at a rate of 5°C/min up to 240°C. The mixture was maintained at 240°C for 12h to obtain EC-CFM/S.

### 11.7.2.4 Chemical and electrochemical characterization of EC-CFMs

This system was also characterized for structure and electrochemical response similar to the other systems described earlier. The crystal structure of the EC-CFM and EC-CFM-S before and after sulfur infiltration was analyzed using the XRD spectroscopy in a Philips XPERT PRO system that uses Cu $K_\alpha$ ($\lambda = 0.15406$ nm) radiation. The samples were scanned in the $2\theta$ range of 10–90 degrees under a constant current and voltage of 40 mA and 45 kV, respectively. The SEM images of the EC-CFMs were obtained using a Philips XL30 machine at 10 kV. An attenuated total reflectance-Fourier-transform IR (ATR-FTIR, Nicolet 6700 Spectrophotometer, Thermo Electron Corporation) spectroscope, which uses a diamond ATR smart orbit, was used to obtain the FTIR spectra of the samples. The FTIR spectra are collected at a resolution of 1 cm$^{-1}$, averaging 32 scans between the frequency of 400 and 4000 cm$^{-1}$. The XPS analyses of the EC-CFMs and EC-CFMS were performed using the ESCALAB 250 Xi system (Thermo Scientific). This XPS system consists of the monochromated Al $K_\alpha$ X-ray source and low-energy ($\leq 10$ eV) argon ions and low-energy electron beams that provide the charge neutralization. The XPS measurements were carried out at room temperature, under an ultrahigh vacuum (UHV) chamber ($<5 \times 10^{-10}$ mBar) employing a spot size of $200 \times 200$ µm$^2$. The surface area and pore characteristics of all the EC-CFM and EC-CFMS samples were analyzed using a Micromeritics ASAP 2020 Physisorption analyzer, using the BET isotherm generated.

The EC-CFMS samples were cycled between 1.8 and 2.8 V (with respect to Li$^+$/Li) at a current rate of 0.2C in a 2032-coin cell using the Arbin BT200 battery testing station to evaluate their electrochemical performance. The cathodes for electrochemical evaluation were prepared by manually coating a dispersion of 70 wt% EC-CFMs, 20 wt% Super P, and 10 wt% PVDF dispersed in $N$-methyl pyrrolidone. All the cathodes that were tested had a uniform sulfur loading of 5.6 mg/cm$^2$. Accordingly, 2032-coin cells were assembled with the EC-CFMS-coated cathodes as the working electrode, a lithium foil as the counterelectrode, and Celgard 2400 polypropylene (PP) as the separator in an Innovative, Inc., glove box (UHP Argon,

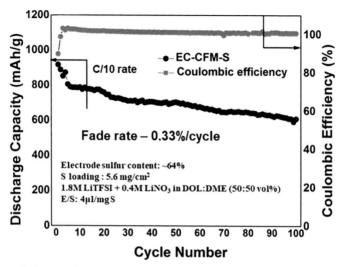

**Fig. 11.22** Cycling performance of EC-CFM-S cycled at 0.1 C rate. *(Source: Authors.)*

<0.1 ppm $O_2$, $H_2O$). A 1.8 M LiTFSI (lithium bis(trifluoromethanesulfonyl) imide), and 0.4 M $LiNO_3$ dissolved in 50:50 vol% 1,3 dioxolane and 1,2 dimethoxyethane were used as the electrolytes. The electronic conductor-coated electrode (EC-CFMS) coordination configuration and high surface area/pore volume of EC-CFMS provide high activity for the conversion of the polysulfides, LiPS to $Li_2S$, and sufficient entrapment of LiPS. Fig. 11.22 shows effective suppression of the shuttle effect. As a result, even under a C/10 rate under lean electrolyte condition of E:S = 4 μL/mg-S in a Li-S battery, the electronic conductor coated on EC-CFMS composite electrode still delivers an initial capacity of ~950 mAh/g that stabilizes after 20 cycles giving a high reversible capacity of ~700 mAh/g up to 50 cycles with a moderate fade to ~600 mAh/g at 100 cycles.

### 11.7.3 Synthetic polymer binder with carbon framework materials

This section describes the use of synthetic polymer binder to trap the polysulfides (PS). To study the effect of synthetic binder on the chemical affinity of PS as well as structural stability of the electrode, a blended polymer with improved mechanical properties, comprising chitosan (CS), and polyvinyl alcohol (PVA) (80:20) was used. Correspondingly, the CFMs were dried under vacuum conditions at 60°C for 12 h to remove residual solvent and the water of crystallization from the synthesis process. The synthesized CFMs were then infiltrated with sulfur under vacuum, following the similar procedure as mentioned previously. The CFM and S (mass ratio 10:90) were ground together and then sealed and heated under argon in a Swagelok cell at a rate of 5°C/min up to 240°C. The mixture was maintained at 240°C for 12h to obtain CFM-S. The synthetic polymer binder (SB) was prepared according to the procedure described as follows: chitosan (Sigma-Aldrich 99%) was dissolved in 1% acetic acid in water; poly-vinyl alcohol (PVA) (Sigma-Aldrich 99%) was dissolved in water; 80 wt% chitosan and 20 wt% PVA were then mixed together and stirred for 24h; this clear solution was used as the binder.

The crystal structure of the CFM and CFM-S before and after sulfur infiltration was analyzed using the XRD spectroscopy in a Philips XPERT PRO system that uses Cu $K_\alpha$ ($\lambda = 0.15406$ nm) radiation. The samples were scanned in the $2\theta$ range of 10–90 degrees under a constant current and voltage of 40 mA and 45 kV, respectively. The SEM images of the CFMs were obtained using a Philips XL30 machine at 10 kV. An attenuated total reflectance-Fourier-transform IR (ATR-FTIR, Nicolet 6700 Spectrophotometer, Thermo Electron Corporation) spectroscope, which uses a diamond ATR smart orbit, was used to obtain the FTIR spectra of the samples. The FTIR spectra are collected at a resolution of 1 cm$^{-1}$, averaging 32 scans between the frequency of 400 and 4000 cm$^{-1}$. The XPS analyses of the CFMs and CFM-S were performed using the ESCALAB 250 Xi system (Thermo Scientific). This XPS system consists of the monochromated Al $K_\alpha$ X-ray source and low-energy ($\leq 10$ eV) argon ions and low-energy electron beams that provide the charge neutralization. The XPS measurements were carried out at room temperature, under an ultrahigh vacuum (UHV) chamber ($<5 \times 10^{-10}$ mBar) employing a spot size of $200 \times 200$ µm$^2$. The surface area and pore characteristics of all the CFM and CFM-S samples were analyzed using a Micromeritics ASAP 2020 Physisorption analyzer, using the BET isotherm generated.

The CFM-S with the synthetic binder (SB) were cycled between 1.8 and 2.8 V (with respect to Li$^+$/Li) at a current rate of 0.05C in a 2032–coin cell using the Arbin BT200 battery testing station to evaluate their electrochemical performance. The cathodes for electrochemical evaluation were prepared by manually coating a dispersion of 70 wt% CFM-S, 20 wt% Super P, and 10 wt% synthetic polymer binder solution dispersed in water. All the cathodes that were tested had a uniform sulfur loading of 4.0 mg/cm$^2$. Accordingly, 2032–coin cells were assembled with the CFM-S coated cathodes as the working electrode, a lithium foil as the counter electrode, and Celgard 2400 polypropylene (PP) as the separator in an Innovative, Inc. glove box (UHP Argon, $<0.1$ ppm O$_2$, H$_2$O). A 1.8 M LiTFSI (lithium bis(trifluoromethanesulfonyl)imide) and 0.4 M LiNO$_3$ dissolved

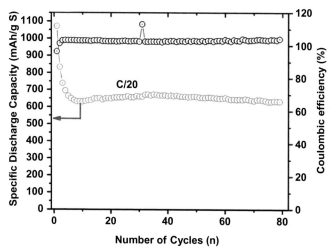

**Fig. 11.23** Cycling performance of CFM-S with synthetic binder cycled at 0.05 C rate. *(Source: Authors.)*

in 50:50 vol% 1,3 dioxolane and 1,2 dimethoxyethane were used as the electrolyte. The CFM-S electrodes with synthetic polymer binder and functional groups/pore volume of CFM-S provide high activity for conversion of the polysulfides, LiPS to Li$_2$S with sufficient entrapment of LiPS. Fig. 11.23 shows effective suppression of the shuttle effect. As a result, even under a C/20 rate under lean electrolyte condition of electrolyte (E) to sulfur (S) ratio of 4 microliter/mg-S in a Li-S battery, the CFM-S composite electrode in the presence of the synthetic binder still delivers a high initial capacity of ~1090 mAh/g that stabilizes to ~675 mAh/g after 20 cycles exhibiting ~625 mAh/g after 80 cycles indicating the beneficial influence of the binder although there is still the drop in capacity after the first cycle reflecting the inability of the CFM and the synthetic polymer binder to completely trap the soluble polysulfides.

### 11.7.4 Hybrid active material (HBA) synthesis and characterization

In this section, the use of a functional monomer to generate a hybrid active material is described. Correspondingly, surface functionalization and surface coating of CFM/S for PS encapsulation were also performed by polymerization of functional monomer on the surface of CFM/S. Surface functionalization/coating were performed by polymerization of tri-thio-cyanuric acid (Sigma-Aldrich 99%) monomer at 240°C during S infiltration within the CFM framework. Mixtures of complex framework materials (CFMs), sulfur and functional monomer (FM) of tri-thio cyanuric acid (Sigma Aldrich 99%), corresponding to the stoichiometric composition of 70:20:10 or 80:10:10 were used. Any N containing ring compound with at least three thiol groups can be used. The hybrid active materials were dried under vacuum conditions at 60°C for 12h to remove residual solvent and the water of crystallization from the synthesis process. Next, the hybrid active materials comprising sulfur, CFM, and FM (mass ratio 70:20:10) were ground together, then sealed and heated under argon in a Swagelok cell at a rate of 5°C/min up to 240°C. The mixture was maintained at 240°C for 12 h to obtain the hybrid active materials. To this mixture $LiNO_3$ can also be added to create a lithium ion conducting framework or essentially LiOPAN coated HBA as described in LiOPAN coated CFM-S above, if needed.

Similar to the other systems, the crystal structure of the sulfur, CFMs, FM, and hybrid active materials before and after sulfur infiltration was analyzed using the XRD spectroscopy in a Philips XPERT PRO system that uses Cu $K_\alpha$ ($\lambda = 0.15406$ nm) radiation. The samples were scanned in the 2q range of 10–90 degree under a constant current and voltage of 40 mA and 45 kV, respectively. The SEM images of the hybrid active materials were obtained using a Philips XL30 machine at 10 kV. An attenuated total reflectance Fourier transform IR spectroscope (ATR-FTIR, Nicolet 6700 Spectrophotometer, Thermo Electron Corporation), which uses a diamond ATR smart orbit, was used to obtain the FT-IR spectra of the samples. The FTIR spectra are collected at a resolution of 1 cm$^{-1}$, averaging 32 scans between the frequency of 400 and 4000 cm$^{-1}$. The XPS analyses of the starting materials and hybrid active materials were performed using the ESCA-LAB 250 Xi system (Thermo Scientific). This XPS system consists of the monochromated Al $K_\alpha$ X-ray source and low-energy ($\leq 10$ eV) argon ions and low-energy electron beams that provide the charge neutralization. The XPS measurements were carried out at room temperature, under an ultra-high vacuum (UHV) chamber ($<5 \times 10^{-10}$ mBar) employing a spot size of $200 \times 200 \mu m^2$. The surface area and pore characteristics of all the hybrid active materials were analyzed using a Micromeritics ASAP 2020 Physisorption analyzer, using the BET isotherm generated.

The hybrid active materials were cycled between 1.8 and 2.8 V (with respect to Li$^+$/Li) at a current rate of 0.05 C in a 2032-coin cell using the Arbin BT200 battery testing station to evaluate their electrochemical performance. The cathodes for electrochemical evaluation were prepared by manually coating a dispersion of 70 wt% HBA, 20 wt% acetylene black, and 10 wt

% PVDF dispersed in N-methyl 2-pyrrolidone (NMP) on an aluminum foil, followed by vacuum drying for 12h at 60°C. All the cathodes that were tested had a uniform sulfur loading of 4 mg/cm$^2$. Accordingly, 2032-coin cells were assembled with the hybrid active materials coated cathodes as the working electrode, a lithium foil as the counter electrode, and Celgard 2400 polypropylene (PP) as the separator in an Innovative, Inc. glove box (UHP Argon, <0.1ppm O$_2$, H$_2$O). A 1.8M LiTFSI (lithium bis(trifluoromethanesulfonyl)imide), and 0.4M LiNO$_3$ dissolved in 50:50vol% 1,3 dioxolane and 1,2 dimethoxyethane were used as the electrolyte. The hybrid active material coordination configuration and high surface area/pore volume of functional monomer provide high activity for the conversion of polysulfides, LiPS to Li$_2$S, and sufficient entrapment of LiPS. Fig. 11.24 shows the cycling response indicating effective suppression of the shuttle effect. As a result, even under a C/20 rate under lean electrolyte condition of E:S = 4 μL/mg-S in a Li-S battery, the hybrid active material architecture electrode still delivers a high initial capacity of 1300 mAh/g that stabilizes to ~900 mAh/g after 10 cycles showing the beneficial influence of the tri-thiocyanuric acid acting as the functional monomer. The presence of N-C functional groups within the functional monomer appears to help in reducing the polysulfide dissolution resulting in ~900 mAh/g capacity indicating the promise of this system although more work is needed to explore this system more thoroughly.

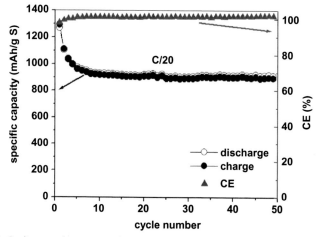

**Fig. 11.24** Cycling performance of HBA-S cycled at 0.05 C rate under lean electrolyte conditions of E:S = 4 μL/mg-S. *(Source: Authors.)*

# 11.7.5 Inorganic framework materials

## 11.7.5.1 Synthesis of boron nitride-S materials

In addition to porous carbon framework materials, porous ZSM-5 series zeolite powder (P-38, pore size: 5A, ACS Material) and hexagonal boron nitride (h-BN, Sigma-Aldrich 99%) was studied as a suitable inorganic framework material (IFM) for high degree of S infiltration and PS confinement. The synthesis of boron nitride containing IFM was conducted as follows.

The IFM (boron nitride, Sigma-Aldrich 99%) was dried under vacuum conditions at 60°C for 12h to remove the residual solvent and the water of crystallization from the synthesis process. The IFMs were infiltrated with sulfur under vacuum using the following procedure. The IFM and S (mass ratio 10:90) were ground together and then sealed and heated under argon in a Swagelok cell at a rate of 5°C/min up to 240°C. The mixture was maintained at 240°C for 12h to obtain IFM-S.

## 11.7.5.2 Characterization of boron nitride-S materials

The crystal structure of the IFM and IFM-S before and after sulfur infiltration was analyzed as with the other systems using the XRD spectroscopy in a Philips XPERT PRO system that uses Cu $K_\alpha$ ($\lambda = 0.15406$ nm) radiation. The samples were scanned in the $2\theta$ range of 10–90 degrees under a constant current and voltage of 40 mA and 45 kV, respectively. The SEM images of the IFMs were obtained using a Philips XL30 machine at 10 kV. An attenuated total reflectance-Fourier-transform IR (ATR-FTIR, Nicolet 6700 Spectrophotometer, Thermo Electron Corporation) spectroscope, which uses a diamond ATR smart orbit, was used to obtain the FTIR spectra of the samples. The FTIR spectra are collected at a resolution of 1 cm$^{-1}$, averaging 32 scans between the frequency of 400 and 4000 cm$^{-1}$. The XPS analyses of the IFMs and IFM-S were performed using the ESCALAB 250 Xi system (Thermo Scientific). This XPS system consists of the monochromated Al $K_\alpha$ X-ray source and low-energy ($\leq$10 eV) argon ions and low-energy electron beams that provide the charge neutralization. The XPS measurements were carried out at room temperature, under an ultrahigh vacuum (UHV) chamber ($<5 \times 10^{-10}$ mBar) employing a spot size of $200 \times 200\,\mu\text{m}^2$. The surface area and pore characteristics of all the IFM and IFM-S samples were analyzed using a Micromeritics ASAP 2020 Physisorption analyzer, using the BET isotherm generated.

The IFM-S samples were cycled between 1.8 and 2.8 V (with respect to $Li^+/Li$) at a current rate of 0.05C in 2032 and 2025 coin cells using the Arbin BT200 battery testing station to evaluate their electrochemical performance. The cathodes for electrochemical evaluation were prepared by manually coating a dispersion of 70 wt% IFM-S, 20 wt% Super P, and 10 wt% PVDF binder dispersed in N-methyl-2-pyrrolidone. All the cathodes that were tested had a uniform sulfur loading of 4.2 mg/cm$^2$. Accordingly, 2032 and 2025 coin cells were assembled with the IFM-S-coated cathodes as the working electrode, a lithium foil as the counterelectrode, and Celgard 2400 polypropylene (PP) as the separator in an Innovative, Inc., glove box (UHP Argon, <0.1 ppm $O_2$, $H_2O$). 1.8 M LiTFSI (lithium bis(trifluoromethanesulfonyl)imide), and 0.4 M $LiNO_3$ dissolved in 50:50 vol% 1,3 dioxolane and 1,2 dimethoxyethane were used as the electrolyte. The IFM-S-coated electrode (IFM-S) coordination and high surface area/pore volume of IFM-S provide high activity for the conversion of LiPS to $Li_2S$ with sufficient entrapment of LiPS. Fig. 11.25 shows the cycling response indicating effective suppression of the shuttle effect. As a result, even under a C/20 rate under lean electrolyte condition of E:S = 4 μL/mg-S in a Li-S battery, the CFM-S composite electrode still delivers a stable reversible capacity of 450–600 mAh/g in both cell configurations although there is an initial activation needed for the cell tested in 2032 coin cell possibly due to wettability of the lean electrolyte with the active material. This issue appears to be resolved as expected in the smaller 2025 coin cell.

The electrochemical cycling response of S infiltrated h-BN (S/h-BN) electrode with S content ~4 mg/cm$^2$ cycled @ C/20 rate with E:S = 4 μL/mg-S, shown in Fig. 11.25 shows the feasibility to achieve high initial specific capacity of ~900 mAh/g S. However, nonporous h-BN shows a significant fade in capacity within 20 cycles for the smaller 2025 configuration which suggest h-BN likely does not have strong chemical affinity for PS to minimize the PS dissolution. However, suitable LIC coating on h-BN/S could improve the cyclability due to encapsulation of PS within the LIC coated h-BN/S structure.

### 11.7.5.3 Synthesis of zeolite (ZSM-5)-S materials

The porous IFMs (ZSM-5, ACS Material) were dried under vacuum conditions at 60°C for 12 h to remove residual solvent and the water of crystallization from the synthesis process. The ZSMs were infiltrated with sulfur under vacuum, following the procedure as follows. The ZSM and S (mass ratio 5:95) were ground together and then sealed and heated under argon

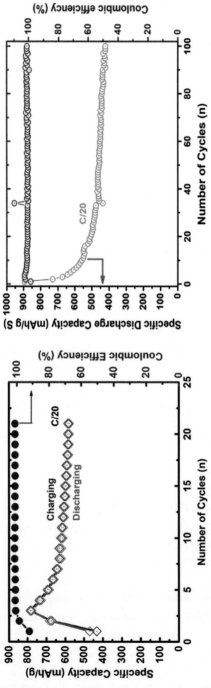

**Fig. 11.25** Cycling performance of IFM-S cycled at 0.05 C rate with 2032 (left) and 2025 (right) coin cells. *(Source: Authors.)*

Cathode and anode architectures for lithium-sulfur batteries  **419**

in a Swagelok cell at a rate of 5°C/min up to 240°C. The mixture was maintained at 240°C for 12h to obtain ZSM-S.

### 11.7.5.4 Characterization of zeolite-S materials

The crystal structure of the ZSM and ZSM-S before and after sulfur infiltration was analyzed using the XRD spectroscopy in a Philips XPERT PRO system that uses Cu $K_\alpha$ ($\lambda = 0.15406$ nm) radiation. The samples were scanned in the $2\theta$ range of 10–90 degrees under a constant current and voltage of 40 mA and 45 kV, respectively. The SEM images of the ZSMs were obtained using a Philips XL30 machine at 10 kV. An attenuated total reflectance-Fourier-transform IR (ATR-FTIR, Nicolet 6700 Spectrophotometer, Thermo Electron Corporation) spectroscope, which uses a diamond ATR smart orbit, was used to obtain the FTIR spectra of the samples. The FTIR spectra are collected at a resolution of 1 cm$^{-1}$, averaging 32 scans between the frequency of 400 and 4000 cm$^{-1}$. The XPS analyses of the ZSMs and ZSM-S were performed using the ESCALAB 250 Xi system (Thermo Scientific). This XPS system consists of the monochromated Al $K_\alpha$ X-ray source and low-energy ($\leq$10 eV) argon ions and low-energy electron beams that provide the charge neutralization. The XPS measurements were carried out at room temperature, under an ultrahigh vacuum (UHV) chamber ($<5 \times 10^{-10}$ mBar) employing a spot size of $200 \times 200\,\mu m^2$. The surface area and pore characteristics of all the ZSM and ZSM-S samples were analyzed using a Micromeritics ASAP 2020 Physisorption analyzer, using the BET isotherm generated.

The ZSM-S samples were cycled between 1.8 and 2.8 V (with respect to Li$^+$/Li) at a current rate of 0.05 C in a 2032-coin cell using the Arbin BT200 battery testing station to evaluate their electrochemical performance. The cathodes for electrochemical evaluation were prepared by manually coating a dispersion of 70 wt% ZSM-S, 20 wt% Super P, and 10 wt% PVDF binder dispersed in N-methyl pyrrolidone. All the cathodes that were tested had a uniform sulfur loading of 5.0–7.0 mg/cm$^2$. Accordingly, 2032-coin cells were assembled with the ZSM-S-coated cathodes as the working electrode, a lithium foil as the counterelectrode, and Celgard 2400 polypropylene (PP) as the separator in an Innovative, Inc., glove box (UHP Argon, <0.1 ppm O$_2$, H$_2$O). Electrolyte of 1.8 M LiTFSI (lithium bis(trifluoromethanesulfonyl)imide) and 0.4 M LiNO$_3$ dissolved in 50:50 vol% 1,3 dioxolane and 1,2 dimethoxyethane were used as the electrolytes. The sulfur infiltrated ZSM electrode (ZSM-S) coordination with metals (Al and Si) and high surface area/pore volume of ZSM-S provide high activity for the conversion of

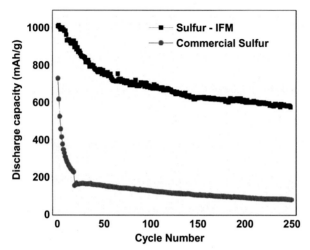

Fig. 11.26 Cycling performance of ZSM-5-S cycled at 0.05 C rate. *(Source: Authors.)*

LiPS to Li$_2$S with sufficient entrapment of LiPS. Fig. 11.26 shows the cycling response indicating effective suppression of the shuttle effect. As a result, even under a C/20 rate under lean electrolyte conditions of E:S = 4 μL/mg-S in a Li-S battery, the ZSM-S composite electrode still delivers a high reversible capacity of 600 mAh/g after 250 cycles. The ability of the system to exhibit a high initial capacity of 1000 mAh/g is a reflection of the aluminosilicate backbone of the zeolite that helps to catalyze the conversion of the polysulfides formed initially. The gradual drop in capacity, however, is a reflection of the insulating nature of the aluminosilicate framework compounded by the inability to completely utilize the sulfur due to the inferior electronic and ionic transport in the aluminosilicate insulator framework electrodes.

The inability to completely trap the polysulfides may also contribute to the initial fade in capacity. In comparison to S/h-BN, S infiltrated porous ZSM-5 (S/ZSM-5) shows high initial capacity of 1000 mAh/g S and excellent cyclability with low fade in capacity when cycled @ C/20 rate with 5–7 mg S/cm$^2$ loading. The system delivers a high reversible capacity of ~600 mAh/ g S upto 250 cycles. This results clearly suggest that the porous ZSM-5 or its derivatives are a better choice as IFM providing higher S utilization and high activity for entrapment of PS although there is still need for further improvement to eliminate the initial drop. Suitable LIC coating and enhancement of the electronic conductivity of S/ZSM-5 by suitable doping could improve the cyclability of S/ZSM-5 structure up to ~500 cycles.

## 11.8 Engineering dendrite-free anodes for Li-S batteries

In Section 11.2, we provided a collection of different concepts and strategies for confining the polysulfide species and preventing their dissolution in electrolytes combined with engineering new approaches to functionalize the surface of the carbon framework material to serve as Li–ion conductors, and electronic conductors helping to support high loadings of sulfur while also enabling to trap the polysulfides. Identification of functional electrocatalysts using DFT calculations and incorporation of them within the complex framework material dispersed uniformly within the active material helps further aid in enhancing the specific capacity and also retaining the capacity over longer cycles as shown by the initial studies described in Section 11.7.1.3 and Fig. 11.20.

As mentioned earlier, in as much as the polysulfide dissolution is a major impediment to the progress of Li-S batteries, the anode is also another equally important barrier preventing the successful implementation and deployment of the Li-S system. Dendrite formation on the Li metal anode particularly at high current rates is a serious problem. Furthermore, the reaction of the Li metal anode with the shuttling polysulfide species further leads to corrosion of the Li metal anode and irreversible consumption of the Li metal resulting in loss in capacity and eventual failure of the cell. Several researchers have attempted to address the issue by developing various coating strategies to protect the Li metal surface while also developing new electrolytes such as hydro–fluoro–ether (HFE)–based electrolytes that tend to be more resistant to polysulfide dissolution while also enabling the formation of a robust corrosion-resistant coating of LiF on the Li metal anode. Similarly, researchers from the pacific northwest national laboratory (PNNL) have developed localized high–concentration electrolytes (LHCEs) that also control the concentration gradients at the interface, thereby protecting the metal surface and preventing dendrites [51,52].

While all of these approaches do have merit, the efficacy and long-term stability of the Li metal interface during prolonged and extended cycles over a varied range of low, moderate, and high current densities are relevant and pertinent questions that will need to be addressed and overcome with ongoing research. The cathode issues thus continue to be researched intensely. At the same time, complementary to the cathode work described so far, the authors have developed a strategy focusing directly on the anode itself. In a recent publication [13], the authors have documented the various factors determining the interface morphology and stability, and the reasons for creating a stable and planar interface as opposed to a perturbed interface leading to formation of dendrites.

In this chapter, we have drawn the analogy to metal solidification for electrodeposition and studied the thermodynamic interplay between the concentration gradients and the electrochemical potential gradients similar to the temperature gradients in the case of metal solidification. In addition to this interplay that dictates the morphological instability at the interface, there is also the crystallographic symmetry criterion which is critical to preventing dendrites at the interface along with the interfacial energy of the substrate with the lithium metal that is being electrodeposited. This interfacial energy controlling the affinity of the plating metal to the substrate determined by the Gibbs–Thomson parameter (GTP) combined with the crystallographic symmetry and registry with the depositing Li plays a strong role in controlling the formation of dendrites. [13]. It is desirable to have the following criteria satisfied to prevent dendrites and create a dendrite-free metal anode alloy:

1. Substrate lattice structure should be similar or match that of the Li metal.
2. Energy for metal adsorption at the interface or interfacial enthalpy should be optimal, neither too positive nor too negative. A positive interfacial enthalpy for a metal such as iron will result in plating of metallic Li with a high nucleation underpotential resulting in detachment of plated Li causing dendrites while the latter with a negative interfacial enthalpy will indeed result in plating and deplating. However, there will be intermetallic phase formation due to alloying reaction causing colossal volume expansion and contraction issues ultimately leading to cell failure similar to the ubiquitous silicon and tin anodes.
3. Large Gibbs-Thomson parameter (GTP) would result in optimal interfacial enthalpy permitting an alloy to be resistant to dendrite formation.
4. The plating metal should exhibit good wetting with the underlying substrate.

Use of a current collector with positive $\Delta H_{mix}$ such as iron results in no solid solution formation leading to detachment of plated Li and nonplanar microstructure (Fig. 11.27A).

As shown in Fig. 11.27B, there is no intermetallic phase or solid solution formation. There is only plating phenomenon that occurs with a strongly negative nucleation potential of $-102\,mV$ and a growth potential of $-20\,mV$. As a result, there is detachment of the plated Li from the substrate after 100 cycles as shown in the SEM image in Fig. 11.28. This clearly shows the influence of the interfacial enthalpy or enthalpy of

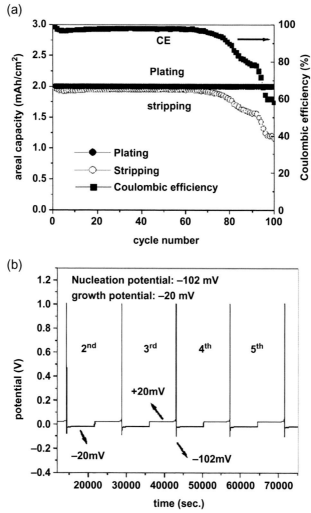

Fig. 11.27 (A) Cycling showing the plating and stripping of a metal substrate with low Gibbs-Thomson parameter and positive $\Delta H_{mix}$. (B) Potential-time plot showing the nucleation and growth potentials for a metal with a positive interfacial enthalpy. (Source: Authors.)

mixing which controls the interface characteristics of the plated metal. Alternatively, the use of a metal with substantially negative enthalpy of mixing or interfacial enthalpy such as zinc will lead to the formation of intermetallic phases resulting in volume expansion and contraction during cycling that could be sustained at low current densities as shown

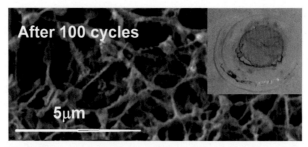

**Fig. 11.28** SEM image showing the nonplanar microstructure and detachment of plated Li on metal as described earlier (see Fig. 11.27A) after 100 cycles (inset: photograph). *(Source: Authors.)*

in Fig. 11.29 but would eventually cause failure and would be more pronounced at higher current densities. Such a current collector will have a high interfacial adhesion with Li metal, giving rise to good wetting and zero nucleation under potential (see Fig. 11.30) but with the formation of intermetallic phases that cause volume expansion and contraction and as a result, could lead to eventual formation of cracks and consequent delamination. This is because with prolonged cycling, there is continued formation of intermetallic or Zintl-type phases that will result in a significant change in crystal structure and volume change with continued cycling [9,53,54]. Depending on the system, the colossal volume expansion

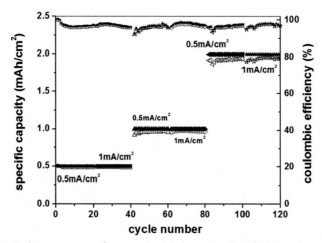

**Fig. 11.29** Cycling response for metal with negative interfacial enthalpy. *(Source: Authors.)*

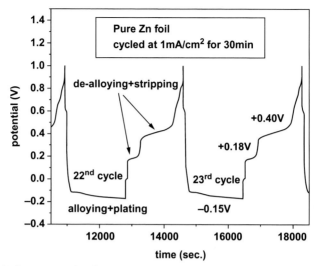

**Fig. 11.30** Voltage-time plot showing zero nucleation under potential. *(Source: Authors.)*

and contraction will result in pulverization after long-term cycling similar to Sn and Si. Both of these current collectors are therefore not desirable. Hence, there is a need for control of the interfacial energy and crystal structure match to result in solid solution formation inhibiting the formation of intermetallic phases.

An ideal solution is a material that has lattice match with Li and optimal interfacial energy for mixing and adsorption. There are very few metals that exhibit solid solubility with Li. Magnesium being a lightweight metal with a divalent charge has an ionic size very much similar to that of Li. Magnesium also exhibits good solid solubility with metallic Li. As a result, the Li-Mg system has been shown to be ideal for plating and deplating of Li. The solid solution behavior of Mg with Li results in a structurally isomorphous alloy (SIA) that exhibits very good wetting behavior as well [55]. In fact, the Li metal will instead of plating, rather alloy with the underlying metal, thereby showing very little volume expansion. Furthermore, the density of Mg-Li alloy is ~1.1 g/cm$^3$ almost ninefold lower than that of copper current collector. This would also lead to significant advantages with respect to overall weight of the cell being much lower than when using Cu. This weight loss will reflect a positive gain in the energy density of the overall fully assembled cell. We have therefore generated a Mg-Li alloy containing ~75 wt% Li and 25 wt% Mg. Such an SIA alloy has been cycled in a symmetric cell

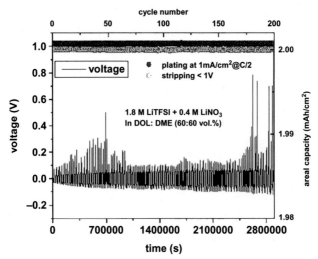

**Fig. 11.31** Symmetric cell cycling of SIA alloy showing ability to retain the aerial capacity of 2 mAh/cm² for 200 cycles without any dendrite formation. *(Source: Authors.)*

configuration and the result is shown in Fig. 11.31. As seen in the plot, the system shows excellent cyclability for 200 cycles showing an aerial capacity of 2 mAh/cm² without any dendrite formation.

The system can also cycle giving an aerial capacity of 5 mAh/cm² with alloying and dealloying at 2.5 mA/cm² for 2h, respectively (Fig. 11.32).

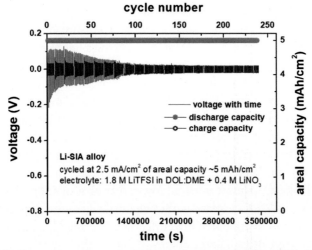

**Fig. 11.32** Cycling response in a symmetric cell showing the ability of SIA alloy to retain an aerial capacity of 5 mAh/cm² for more than 200 cycles with no dendrite formation. *(Source: Authors.)*

Cathode and anode architectures for lithium-sulfur batteries     427

This implies that at a given cycle, the system can withhold ~25-μm-thick Li deposit at each cycle without the formation of any dendrites. This is possible due to the solid solution formation with very little volume expansion or contraction. The solid solution formation results in easy access or diffusion of Li into the lattice with corresponding ease of removal as well. There is, albeit, an activation barrier to the initial insertion of the Li into the lattice as can be seen from the relatively large over potential observed in the initial few cycles. However, once it is overcome over a few cycles, and the lattice has been stabilized, the system is able to continuously cycle 5 mAh/cm$^2$ capacity for more than 200 cycles as shown in Fig. 11.32. This is indeed a remarkable result indicating the absence of dendrite formation and ability to withstand a large aerial capacity of 5 mAh/cm$^2$. This would imply that this system would have the potential to attain and realize the high energy densities for the Li-S system.

## 11.8.1 Theoretical strategies to overcome the diffusion barrier in structurally isomorphous alloys

The activation barrier to alloying and dealloying of Li in the initial few cycles seen in the Li–SIA earlier is a result of the poor diffusion of Li ions. One of the possible ways for improving Li diffusivity in the Li-based alloy anode is to alloy a small quantity of additional metallic elements that may exhibit weaker interatomic bonding between the individual atoms than Li-Li interatomic bonding in pure Li metal. This approach would decrease the potential energy required for overcoming the diffusion activation barrier during hopping of the lithium atoms between two neighbor unit cells of the crystal lattice via the vacancy hopping diffusion mechanism. Such alloying elements could be selected from the general consideration of their cohesive energies $E_{coh}$ which should be similar or lower than $E_{coh}$ of pure Li-bcc metal with $E_{coh} = 158$ kJ/mol. These metals could be Mg ($E_{coh} = 145$ kJ/mol), Na (107 kJ/mol), and Zn (130 kJ/mol). Also, it would be interesting to study the effect of introducing a metal with higher cohesive energy such as Al ($E_{coh} = 327$ kJ/mol) on the Li diffusion. All the $E_{coh}$ data are taken from Ref. [56].

The computational model for calculating the activation barriers consists of two body-centered cubic (bcc) elementary unit cells with alternate atomic layers of Li and doping metal layer, as shown in Fig. 11.33A. In this model, Li atom considered located in the center of the cubic bcc unit cell moves to

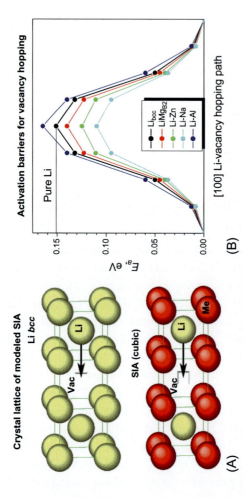

**Fig. 11.33** (A) Crystal lattice of modeled bcc-SIA. *Arrow* denotes Li-atom hopping path into the neighbor Li vacancy. (B) Potential energy profile of Li-vacancy hopping pathway for different alloying elements. *(Source: Authors.)*

the neighboring Li vacancy through the intermediate atomic layer. At the same time, the Li vacancy moves in the opposite direction. In the case of pure Li metal (upper portion of Fig. 11.33A), the intermediate layer consists of only Li atoms, but in the case of Li alloy, it would be the layer with foreign alloying element only, as shown in the bottom portion of Fig. 11.33A. All the computations have been executed using the climbing image nudged elastic band method [57] implemented in the VASP computational package within the projector-augmented wave (PAW) method [48,49] and the generalized gradient approximation (GGA) for the exchange–correlation energy functional in a form described by Perdew and Yue [58]. For all calculations, a $2 \times 2 \times 2$ supercell consisted of eight elementary *bcc* unit cells and containing one Li vacancy, seven Li atoms, and eight atoms of other metals has been constructed. For pure Li metal, the structure contained 1 vacancy and 15 Li atoms. The total path of the Li ion between centers of the adjacent unit cells was divided into eight equal intervals with calculations of the total energies of the system in each consecutive point from start to the end of this path. The standard PAW potentials were utilized for the elemental components, such that Li, Na, Mg, Zn, and Al potentials thus contained 1, 1, 2, 12, and 3 valence electrons, respectively. In the present theoretical analysis, to maintain the high precision for all the total energy and electronic structure calculations, the plane wave cutoff energy of 520 eV has been chosen. The relaxation procedure has been used to optimize the internal positions as well as the lattice parameters of atoms within the supercell. Also, the Monkhorst-Pack scheme has been used to sample the Brillouin zone and create the k-point grid for all the Li alloys used in the current study. The selection of appropriate numbers of k-points in the irreducible parts of the Brillouin zone was made on the grounds limiting the convergence of the total energy to 0.1 meV/atom.

The calculated potential energy profiles for the different alloying elements are shown in Fig. 11.33B. One can see that all the potential energy graphs have similar profiles with maximum energy values located at the middle of the Li hopping path that corresponds to the activation barriers for each specific alloying element. According to these results, the Li mobility in the modeled Li-Mg alloy is even slightly higher than in pure Li-bcc most likely due to the lower cohesive energy of pure Mg compared to pure Li, as mentioned earlier in this section, which makes Li atom easier to squeeze through the Mg-plane and, thus, improve the overall Li-ion conductivity of the Li-Mg alloy. Also, this graph shows

430    Lithium-sulfur batteries

that Na and Zn decrease the activation barriers for Li diffusion to a larger extent compared to the Li-Mg alloy. However, Na and Zn demonstrate much lower solubility with Li than Mg-Li system [59]. Hence, the use of these elements, although beneficial, could be possible only in much lower amounts in comparison with alloying of Li with Mg. The last plot in Fig. 11.33B of the present study belongs to Li-Al system. It features the highest activation barrier for Li diffusion due to a higher Al-Al interatomic bonding within the Al intermediate layer. This result is not surprising and is expected based on the comparison of the cohesive energies of pure Li and Al.

Thus, this computational study has demonstrated positive effects of introducing several alloying elements, such as Mg, Na, and Zn, on Li conductivity in Li-based SIA alloys. Realization of this alloying strategy would help substantially improve the electrochemical characteristics of the dendrite-free Li-SIA anodes as well as the overall performance of the whole Li-ion battery system. The present study only considered bulk lattice diffusion. There are, however, other modes of diffusion via grain boundary and within the grains as well as through interstitials that were not considered in the present study. Such modes could also serve to enhance the overall Li diffusion particularly, in alloys exhibiting reduced solid solubility of the alloying element.

## 11.8.2 Electrochemical cycling of Li-SIA alloys

Based on the positive initial results of the Li-SIA alloy discussed previously, we have therefore tested this system in a Li-S coin cell and pouch cell using the high surface area carbon framework material coated with LiOPAN as outlined in Section 11.2.3.2. The coin cell test results are shown in Fig. 11.34A and B. The result clearly shows the ability of this system to perform and display results matching that of metallic Li anode but without the formation of dendrites rendering the system extremely attractive for high-energy-density Li-S batteries. Both cells in the coin cell format were cycled in 1.8 M LiTFSI + 0.4 M LiNO$_3$ in 50:50 vol% DOL:DME electrolyte using the lean electrolyte condition of E:S = 4 µL/mg-S. The remarkably identical response shows the true potential of the Li-SIA alloy to display and achieve higher energy density desired for the Li-S batteries. The pouch cell response was also studied and compared. The results are shown in Fig. 11.34C and D. As seen from the plots, both systems show

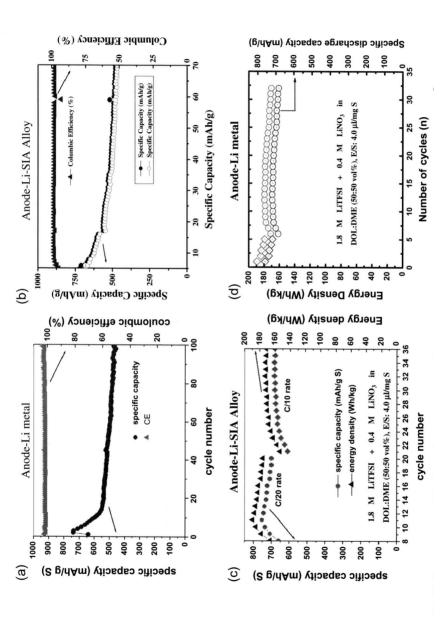

**Fig. 11.34** Cycling response of high surface area carbon framework material with S loading of: (A) 5.2 mg/cm² cycled with metallic Li, (B) 5.1 mg/cm² cycled with metallic Li-SIA, (C) 5.1 mg/cm² cycled with metallic Li-SIA alloy, and (D) 5.1 mg/cm² cycled with metallic Li. (*Source: Authors.*)

very similar responses in both coin cells and pouch cells. The response is, however, in single-layer pouch cells. The uncanny similarity in performance clearly shows the potential and promise of the Li-SIA system to be used as the next-generation anode system for achieving the high energy density needed for the practical implementation of the Li-S battery systems. There is, however, considerable research still needed to overcome the initial barrier to Li insertion or alloying and dealloying as discussed above, as well as the potential of the system to not react with any shuttling polysulfide species. These are certainly directions for future research.

## 11.8.3 Multicomponent alloys as dendrite-free anodes

Another strategy in line with the SIA alloys discussed above, satisfying the criteria previously mentioned for generating a morphological stable interface, preventing dendrite formation is an alloy design creating a BCC alloy lattice matched to metallic Li. Such a strategy for designing a multicomponent structurally ordered alloy (bcc solid solution) or ordered ternary compounds (multicomponent alloys, MCA) lattice-matched to Li is the $Fe_{50}Zn_{40}Mg_{10}$ system. The MCA of $Fe_{50}Zn_{40}Mg_{10}$ was identified to have the same BCC crystal structure matched to that of metallic Li. High-energy mechanical alloying (HEMA) was used to generate the $Fe_{50}Zn_{40}Mg_{10}$ alloy from individual elements in a mechanical mill using stainless steel vial and milling media of stainless steel balls with a ball/charge ratio of 10:1 for 6 h. The synthesized alloy has a lattice parameter of 2.97 Å very close to that of BCC, Li of 3.51 Å. The synthesized alloy was then pressed into a pellet in a hot press at 80°C for 10 min. A 13-mm-diameter pellet of 1 mm thickness was then used as the anode in a coin cell to test the ability of the MCA alloy to plate and deplate metallic lithium in a symmetric cell. Fig. 11.35A and B shows the plating and deplating response of the MCA pellet cycled again metallic Li using $1 mA/cm^2$ current density in 1.8 M LiTSI + 0.1 M $LiNO_3$ in DOL:DME (1:1 v/v) electrolyte.

Fig. 11.35A shows the ability of the MCA alloy to reversibly plate metallic Li giving an aerial capacity of $4 mAh/cm^2$ corresponding to $\sim 20$-μm-thick Li. Fig. 11.35B shows the plating and deplating profile for the first cycle corresponding to $1 mA/cm^2$ current for 0.5-h plating and deplating exhibiting excellent reversibility with almost zero nucleation overpotential for plating and almost zero growth underpotential of $\leq 0.01\%$ increases per cycle with cycle numbers. The results clearly show high Coulombic efficiency of $\sim 99.9\%$ at first cycle with the same efficiency being retained for up to 300 cycles showing the remarkable ability of the MCA alloy to reversibly cycle Li. Furthermore, there is no significant change in the plating and deplating growth underpotential up to 300 cycles. Fig. 11.35C shows the full symmetric cell plating and deplating response of the MCA electrode cycled in a coin cell at $1\,mA/cm^2$ current density for 4-h plating and deplating giving an areal capacity of $4\,mAh/cm^2$ for more than 300 cycles. As shown in the plot in Fig. 11.35D, the growth potential remains essentially invariant with time. Similarly, even at the 50th and 51st cycle of plating and deplating at $1\,mA/cm^2$ current density for 0.5 h plating and deplating intervals, the system exhibits excellent reversibility indicating no formation of any intermetallic phases. Finally, to confirm the ability of the MCA system to resist dendrite formation, the cycled electrode was observed under scanning electron microscopy after 100 cycles. As seen in Fig. 11.35E, the SEM image clearly shows the uniformly deposited granular morphology of the plated Li without the formation of any dendrites likely indicating that the designed MCA alloy of $Fe_{50}Zn_{40}Mg_{10}$ also exhibits high Gibbs-Thomson parameter with optimal interfacial energy for Li plating and deplating. The cycling response therefore, clearly demonstrates the excellent potential of the MCA alloys design to reversibly cycle $\sim 4\,mAh/cm^2$ capacity of Li corresponding to reversible deposition and stripping of $\sim 20\,\mu m$ of Li for over 300 cycles with no change in the nucleation and growth potential as well as exhibiting no dendritic morphology of the plated Li. The absence of Li in the MCA alloy provides the novelty of using these alloys as current collectors replacing the standard copper anode current collector permitting an anode-free design wherein the cathode would contain the required Li to be reversibly cycled between the MCA current collector and the lithiated cathode such as $LiCoO_2$, $LiNiO_2$, and $LiNi_xCo_yMn_zO_2$ as well as $Li_2S$ in Li-S batteries. These MCA alloys with a density of $\sim 6.9\,g/cm^3$ slightly lower than copper (density $\sim 8.9\,g/cm^3$) have the ability to cycle, and thus to replace Cu as the dendrite-free anode current collector. The MCA system, hence, also presents a much lighter option, thus enabling gain in weight for engineering high-energy-density full cells similar to the Li-SIA systems.

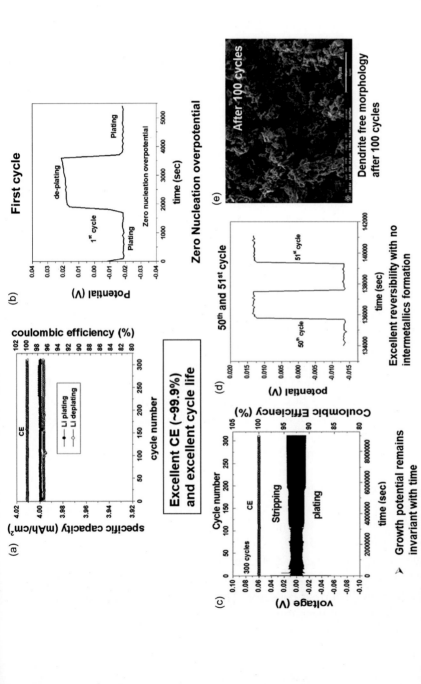

**Fig. 11.35** Cycling response of synthesized MCA alloy in symmetric cell configuration; (A) MCA electrode cycled at a current density of 1 mA/cm$^2$ for an aerial capacity of 4 mAh/cm$^2$ for over 300 cycles; (B) MCA electrode cycling response in the first cycle when cycled at a current density of 1 mA/cm$^2$ for 0.5-h plating and deplating showing the nucleation and growth potentials; (C) MCA electrode symmetric cell cycling response or over 300 cycles showing no change in growth potential; (D) MCA electrode cycling response in the 50th and 51st cycle when cycled at a current density of 1 mA/cm$^2$ for 0.5-h plating and deplating showing excellent reversibility; and (E) SEM image showing the dendrite-free morphology of MCA electrode after 100 cycles. All cells were tested in coin cell using 1.8 M LiTSI in 0.1 M LiNO$_3$ in DOL:DME (1:1 v/v) electrolyte. *(Source: Authors.)*

## 11.9 Conclusions

In this chapter, we describe the promising response of complex framework material structures (comprising a complex framework material host primarily derived from carbon), including a complex framework material that is carbon (as well as noncarbon) based, and a coating applied to the complex framework material comprising one or more layers (involving a component selected from the group consisting of an electronic conductor derived from polymer), a Li–ion conductor, and a functional electrocatalyst. The complex framework material structure can be made from a porous carbon matrix as well as other noncarbonaceous forms. Accordingly, the complex framework material structure is also made of Super P and YP-80F. Moreover, the complex framework material structure is then infiltrated with sulfur to demonstrate the ability of the complex framework material to serve as the host.

All of the confinement architectures, described in this chapter, have shown promising response in coin cell tests as well as in pouch cell format. Additionally, the promise of a separator applied to the complex framework material/sulfur composite cathode has been demonstrated. Furthermore, the performance of the lithium anode applied to the separator, wherein, the composite complex framework material/sulfur cathode, the separator, and the lithium anode are in a stacked configuration is also described. Both single and stacked pouch cells were fabricated and the potential of the CFM in these systems was shown. Finally, the promise of using structurally isomorphous alloys (SIA) containing Li as the next-generation anodes in generating high-energy-density Li-S batteries was demonstrated in both coin cells and pouch cells. Similarly, the promise of engineering alloy design utilizing the Gibbs-Thomson parameter (GTP) and lattice registry with Li was used to engineer new multicomponent alloys (MCAs) exhibiting the BCC structure lattice-matched to Li. The synthesized MCA of $Fe_{50}Zn_{40}Mg_{10}$ shows excellent ability to reversibly cycle $\sim20$-$\mu$m-thick Li giving an aerial capacity of $4 mAh/cm^2$ without any dendrite formation. The potential of these MCAs in a full cell is yet to be demonstrated. However, the prospects of using this system in anode-free cell design for reversibly cycling $4 mAh/cm^2$ areal capacity are quite encouraging, especially providing opportunities for generating high-energy-density Li-NMC cells as well as the next-generation Li-S cells using $Li_2S$ as the cathode.

## Acknowledgments

The authors would like to acknowledge the support of the US Department of Energy under DE-EE0008199 and DE-EE0006825 contracts awarded for support of this research. Prof. Prashant N. Kumta would also like to acknowledge the support of the Edward R. Weidlein Endowed Chair Professorship Funds and the Center for Complex Engineered Multifunctional Materials (CCEMM), Swanson School of Engineering, University of Pittsburgh, for partial support of several graduate students and use of equipment that made this research possible.

## References

[1] J. Lowry, J. Larminie, Electric Vehicle Technology Explained, Wiley, Hoboken, NJ, 2012.

[2] D. Andrea, Lithium-ion batteries and applications, in: A Practical and Comprehensive Guide to Lithium-Ion Batteries and Arrays, From Toys to Towns, vol. 2, Artech House, Norwood, MA, 2020.

[3] Y. Wu, Lithium-ion batteries: fundamentals and applications, in: Electrochemical Energy Storage and Conversion, first ed., CRC Press, Boca Raton, FL, 2015.

[4] J. Warner, The Handbook of Lithium-Ion Battery Pack Design: Chemistry, Components, Types and Terminology, Elsevier, Amsterdam, 2015.

[5] R.A. Huggins, Energy Storage: Fundamentals, Materials and Applications, second ed., Springer International, New York, NY, 2016.

[6] J.-M. Tarascon, P. Simon, Electrochemical energy storage, in: Energy Series: Energy Storage—Batteries and Supercapacitors Set, Wiley, Hoboken, NJ, 2012.

[7] C. Julien, A. Mauger, A. Vijh, K. Zaghib, Lithium Batteries; Science and Technology, Springer, 2016.

[8] A. Manthiram, A reflection on lithium-ion battery cathode chemistry, Nat. Commun. 11 (2020) 1550.

[9] P.N. Kumta, A.F. Hepp, O. Velikokhatnyi, M.K. Datta (Eds.), Silicon Anode Systems for Lithium-Ion Batteries, Elsevier, Cambridge, MA, 2021.

[10] B. Gattu, P.H. Jampani, M.K. Datta, R. Kuruba, P.N. Kumta, Water-soluble-template-derived nanoscale silicon nanoflake and nano-rod morphologies: stable architectures for lithium-ion battery anodes, Nano Res. 10 (2017) 4284–4297.

[11] R. Epur, M.K. Datta, P.N. Kumta, Nanoscale engineered electrochemically active silicon-CNT heterostructures-novel anodes for Li-ion application, Electrochim. Acta 85 (2012) 680–684.

[12] M.K. Datta, P.N. Kumta, Silicon and carbon based composite anodes for lithium ion batteries, J. Power Sources 158 (2006) 557–563.

[13] M.K. Datta, B. Gattu, R. Kuruba, P.M. Shanthi, P.N. Kumta, Constitutional underpotential plating (CUP)—new insights for predicting the morphological stability of deposited lithium anodes in lithium metal batteries, J. Power Sources 467 (2020) 228243.

[14] M. Wild, G.J. Offer, Lithium Sulfur Batteries, Wiley, Hoboken, NJ, 2019.

[15] P.M. Shanthi, P.J. Hanumantha, R. Kuruba, B. Gattu, M.K. Datta, P.N. Kumta, Effective bipyridine and pyrazine-based polysulfide dissolution resistant complex framework material systems for high capacity rechargeable lithium-sulfur batteries, Energy Technol. 7 (12) (2019) 1900141.

[16] P.M. Shanthi, P.J. Hanumantha, R. Kuruba, B. Gattu, M.K. Datta, P.N. Kumta, Sulfonic acid based complex framework materials (CFM): nanostructured polysulfide

immobilization systems for rechargeable lithium-sulfur battery, J. Electrochem. Soc. 166 (2019) A1827–A1835.

[17] L.T. Zhou, W. Zhang, Y. Wang, S. Liang, Y. Gan, H. Huang, J. Zhang, Y. Xia, C. Liang, Lithium sulfide as cathode materials for lithium-ion batteries: advances and challenges, J. Chem. 2020 (2020) 6904517.

[18] S.-H. Chung, A. Manthiram, Current status and future prospects of metal-sulfur batteries, Adv. Mater. 31 (2019) 1901125.

[19] Z. Ma, X. Yuan, L. Li, Z.-F. Ma, D.P. Wilkinson, L. Zhang, J. Zhang, A review of cathode materials and structures for rechargeable lithium-air batteries, Energy Environ. Sci. 8 (2015) 2144–2198.

[20] J. Lu, K.C. Lau, Y.-K. Sun, L.A Curtiss, K. Amine, Review–understanding and mitigating some of the key factors that limit non-aqueous lithium-air battery performance, J. Electrochem. Soc. 162 (2015) A2439–A2446.

[21] M. Rana, S.A. Ahad, M. Li, B. Luo, Review on areal capacities and long-term cycling performances of lithium sulfur battery at high sulfur loading, Energy Storage Mater. 18 (2019) 289–310.

[22] P.M. Shanthi, R. Kuruba, O.I. Velikokhatnyi, P.J. Hanumantha, B. Gattu, M.K. Datta, S.D. Ghadge, P.N. Kumta, Theoretical and experimental strategies for new heterostructures with improved stability for rechargeable lithium sulfur batteries, J. Electrochem. Soc. 167 (2020) 040513.

[23] P.M. Shanthi, P.J. Hanumantha, B. Gattu, M. Sweeney, M.K. Datta, P.N. Kumta, Understanding the origin of irreversible capacity loss in non-carbonized carbonate—based metal organic framework (MOF) sulfur hosts for lithium—sulfur battery, Electrochim. Acta 229 (2017) 208–218.

[24] Y.D. Liu, Q. Liu, L. Xin, Y. Liu, F. Yang, E.A. Stach, J. Xie, Making Li-metal electrodes rechargeable by controlling the dendrite growth direction, Nat. Energy 2 (2017) 17083.

[25] J. Zhang, Y. Su, Y. Zhang, Recent advances in research on anodes for safe and efficient lithium-metal batteries, Nanoscale 12 (2020) 15528–15559.

[26] D.H. Liu, Z. Bai, M. Li, A. Yu, D. Luo, W. Liu, L. Yang, J. Lu, K. Amine, Z. Chen, Developing high safety Li-metal anodes for future high-energy Li-metal batteries: strategies and perspectives, Chem. Soc. Rev. 49 (2020) 5407–5445.

[27] K. Mahankali, N.K. Thangavel, D. Gopchenko, L.M.R. Arava, Atomically engineered transition metal dichalcogenides for liquid polysulfide adsorption and their effective conversion in Li-S batteries, ACS Appl. Mater. Interfaces 12 (2020) 27112–27121.

[28] P.J. Hanumantha, B. Gattu, P.M. Shanthi, S.S. Damle, Z. Basson, R. Bandi, M.K. Datta, S. Park, P.N. Kumta, Flexible sulfur wires (Flex-SWs)—a new versatile platform for lithium-sulfur batteries, Electrochim. Acta 212 (2016) 286–293.

[29] P.J. Hanumantha, B. Gattu, O. Velikokhatnyi, M.K. Datta, S.S. Damle, P.N. Kumta, Heterostructures for improved stability of lithium sulfur batteries, J. Electrochem. Soc. 161 (2014) A1173–A1180.

[30] D. Gueon, J.T. Hwang, S.B. Yang, E. Cho, K. Sohn, D.-K. Yang, J.H. Moon, Spherical macroporous carbon nanotube particles with ultrahigh sulfur loading for lithium-sulfur battery cathodes, ACS Nano 12 (2018) 226–233.

[31] X.W. Yu, H. Wu, J.H. Koo, A. Manthiram, Tailoring the pore size of a polypropylene separator with a polymer having intrinsic nanoporosity for suppressing the polysulfide shuttle in lithium-sulfur batteries, Adv. Energy Mater. 10 (2020) 1902872.

[32] S. Choudhury, M. Zeiger, P. Massuti-Ballester, S. Fleischmann, P. Formánek, L. Borchardt, V. Presser, Carbon onion/sulfur hybrid cathodes via inverse vulcanization for lithium-sulfur batteries, Sustain. Energy Fuel 2 (2018) 133–146.

438 Lithium-sulfur batteries

[33] H.J. Peng, J.-Q. Huang, X.-B. Cheng, Q. Zhang, Review on high-loading and high-energy lithium-sulfur batteries, Adv. Energy Mater. 7 (2017) 1700260.

[34] Sion Power Demonstrates Key Electric Vehicle (EV) Battery Performance Requirements in Its Lithium-Metal Rechargeable Battery Cell Technology. https://sionpower.com/news/. (Accessed February 2021).

[35] P.M. Shanthi, P.J. Hanumantha, T. Albuquerque, B. Gattu, P.N. Kumta, Novel composite polymer electrolytes of PVdF-HFP derived by electrospinning with enhanced li-ion conductivities for rechargeable lithium-sulfur batteries, ACS Appl. Energy Mater. 1 (2018) 483–494.

[36] Y. Pang, J. Pan, J. Yang, S. Zheng, C. Wang, Electrolyte/electrode interfaces in all-solid-state lithium batteries: a review, Electrochem. Energy Rev. 4 (2021) 169–193.

[37] Y. Zheng, Y. Yao, J. Ou, M. Li, D. Luo, H. Dou, Z. Li, K. Amine, A. Yu, Z. Chen, A review of composite solid-state electrolytes for lithium batteries: fundamentals, key materials and advanced structures, Chem. Soc. Rev. 49 (2020) 8790–8839.

[38] H.C. Wang, L. Sheng, G. Yasin, L. Wang, H. Xu, X. He, Reviewing the current status and development of polymer electrolytes for solid-state lithium batteries, Energy Storage Mater. 33 (2020) 188–215.

[39] M.X. Wang, Y. Wu, M. Qiu, X. Li, C. Li, R. Li, J. He, et al., Research progress in electrospinning engineering for all-solid-state electrolytes of lithium metal batteries, J. Energy Chem. 61 (2021) 253–268.

[40] S.D. Ghadge, O.I. Velikokhatnyi, M.K. Datta, P.M. Shanthi, P.N. Kumta, Computational and experimental investigation of Co and S-doped $Ni_2P$ as an efficient electrocatalyst for acid mediated proton exchange membrane hydrogen evolution reaction, Cat. Sci. Technol. 11 (2021) 861–873.

[41] S.D. Ghadge, O.I. Velikokhatnyi, M.K. Datta, P.M. Shanthi, S. Tan, P.N. Kumta, Computational and experimental study of fluorine doped $(Mn_{1-x}Nb_x)O_2$ nanorod electrocatalysts for acid-mediated oxygen evolution reaction, ACS Appl. Energy Mater. 3 (2020) 541–557.

[42] S.D. Ghadge, O.I. Velikokhatnyi, M.K. Datta, P.M. Shanthi, S. Tan, K. Damodaran, P.N. Kumta, Experimental and theoretical validation of high efficiency and robust electrocatalytic response of one-dimensional (1D) (Mn,Ir)O-2:10F nanorods for the oxygen evolution reaction in PEM-based water electrolysis, ACS Catal. 9 (2019) 2134–2157.

[43] P.P. Patel, O.I. Velikokhatnyi, S.D. Ghadge, P.J. Hanumantha, M.K. Datta, R. Kuruba, B. Gattu, P.M. Shanthi, P.N. Kumta, Electrochemically active and robust cobalt doped copper phosphosulfide electro-catalysts for hydrogen evolution reaction in electrolytic and photoelectrochemical water splitting, Int. J. Hydrog. Energy 43 (2018) 7855–7871.

[44] P.P. Patel, P.H. Jampani, M.K. Datta, O.I. Velikokhatnyi, D. Hong, J.A. Poston, A. Manivannand, P.N. Kumta, $WO_3$ based solid solution oxide—promising proton exchange membrane fuel cell anode electro-catalyst, J. Mater. Chem. A 3 (2015) 18296–18309.

[45] K. Kadakia, M.K. Datta, O.I. Velikokhatnyi, P.H. Jampani, P.N. Kumta, Fluorine doped $(Ir,Sn,Nb)O_2$ anode electro-catalyst for oxygen evolution via PEM based water electrolysis, Int. J. Hydrog. Energy 39 (2014) 664–674.

[46] M.K. Datta, K. Kadakia, O.I. Velikokhatnyi, P.H. Jampani, S.J. Chung, J.A. Poston, A. Manivannan, P.N. Kumta, High performance robust F-doped tin oxide based oxygen evolution electro-catalysts for PEM based water electrolysis, J. Mater. Chem. A 1 (2013) 4026–4037.

[47] R.S. Assary, L.A. Curtiss, J.S. Moore, Toward a molecular understanding of energetics in Li-S batteries using nonaqueous electrolytes: a high-level quantum chemical study, J. Phys. Chem. C 118 (2014) 11545–11558.

[48] G. Kresse, J. Furthmuller, Efficient iterative schemes for ab initio total-energy calculations using a plane-wave basis set, Phys. Rev. B 54 (1996) 11169–11186.

[49] G. Kresse, J. Furthmuller, Efficiency of ab-initio total energy calculations for metals and semiconductors using a plane-wave basis set, Comput. Mater. Sci. 6 (1996) 15–50.

[50] G.M. Zhou, H. Tian, Y. Jin, X. Tao, B. Liu, R. Zhang, Z.W. Seh, et al., Catalytic oxidation of $Li_2S$ on the surface of metal sulfides for Li-S batteries, Proc. Natl. Acad. Sci. U. S. A. 114 (2017) 840–845.

[51] S. Chen, J. Zheng, D. Mei, K.S. Han, M.H. Engelhard, W. Zhao, W. Xu, J. Liu, J.-G. Zhang, High-voltage lithium-metal batteries enabled by localized high-concentration electrolytes, Adv. Mater. 30 (2018) e1706102.

[52] J. Zheng, P. Yan, D. Mei, M.H. Engelhard, S.S. Cartmell, B.J. Polzin, C. Wang, J.-G. Zhang, W. Xu, Highly stable operation of lithium metal batteries enabled by the formation of a transient high-concentration electrolyte layer, Adv. Energy Mater. 6 (2016) 201502151.

[53] M.K. Datta, R. Epur, P. Saha, K. Kadakia, S.K. Park, P.N. Kumta, Tin and graphite based nanocomposites: potential anode for sodium ion batteries, J. Power Sources 225 (2013) 316–322.

[54] R. Teki, M.K. Datta, R. Krishnan, T.C. Parker, T.-M. Lu, P.N. Kumta, N. Koratkar, Nanostructured silicon anodes for lithium ion rechargeable batteries, Small 5 (2009) 2236–2242.

[55] P.N. Kumta, B. Gattu, M.K. Datta, O. Velikokhatnyi, P.M. Shanthi, P.J. Hanumantha, High capacity, air-stable, structurally isomorphous lithium alloy multilayer porous foams, U.S. Patent Application US2020/0227736 A1, July 16, 2020.

[56] C. Kittel, Introduction to Solid State Physics, eighth ed., J. Wiley and Sons, Hoboken, NJ, 2005, p. 704.

[57] G. Henkelman, B.P. Uberuaga, H. Jonsson, A climbing image nudged elastic band method for finding saddle points and minimum energy paths, J. Chem. Phys. 113 (2000) 9901–9904.

[58] J.P. Perdew, W. Yue, Accurate and simple density functional for the electronic exchange energy—generalized gradient approximation, Phys. Rev. B 33 (1986) 8800–8802.

[59] T.B. Massalski, Binary Alloy Phase Diagrams, second ed., ASM International, Materials Park, OH, 1990.

# CHAPTER 12

# A solid-state approach to a lithium-sulfur battery

**Muhammad Khurram Tufail[a], Syed Shoaib Ahmad Shah[b], Shahid Hussain[c], Tayyaba Najam[d], and Muhammad Kashif Aslam[e]**

[a]Key Laboratory of Cluster Science of Ministry of Education Beijing Key Laboratory of Photoelectronic/Electrophotonic Conversion Materials, School of Chemistry and Chemical Engineering, Beijing Institute of Technology, Beijing, China
[b]Hefei National Laboratory for Physical Sciences at the Microscale, School of Chemistry and Material Science, University of Science and Technology of China, Hefei, People's Republic of China
[c]School of Materials Science and Engineering, Jiangsu University, Zhenjiang, China
[d]College of Physics and Optoelectronic Engineering, Shenzhen University, Shenzhen, People's Republic of China
[e]Faculty of Materials and Energy, Southwest University, Chongqing, China

## Contents

| | |
|---|---|
| 12.1 Introduction | 441 |
| 12.2 Solid electrolytes | 443 |
|     12.2.1 Solid polymer electrolytes | 443 |
|     12.2.2 Ceramic electrolytes | 446 |
| 12.3 Polymer/ceramic hybrid composite electrolytes | 456 |
| 12.4 Stable Li metal anodes for all-solid-state Li-S batteries | 458 |
|     12.4.1 Li anode/sulfide-based solid-state electrolyte | 460 |
|     12.4.2 Li anode/oxide-based solid-state electrolytes | 464 |
|     12.4.3 Li anode/solid polymer electrolytes | 466 |
| 12.5 Sulfur-based cathode composites for all-solid-state Li-S batteries | 467 |
|     12.5.1 Cathode/oxide-based electrolyte interface | 468 |
|     12.5.2 Cathode/sulfide-based electrolyte interface | 468 |
|     12.5.3 Cathode/solid polymer electrolytes interface | 474 |
| 12.6 All-solid-state thin-film batteries | 474 |
| 12.7 Conclusions | 477 |
| References | 478 |

## 12.1 Introduction

Engineering research suggests that high energy density and affordable energy storage devices have an opportunity to be utilized in new energy storage systems. Recent energy technology developments have been directed toward energy storage devices, particularly rechargeable batteries in applications

---

*Lithium-Sulfur Batteries*
https://doi.org/10.1016/B978-0-12-819676-2.00009-8

Copyright © 2022 Elsevier Inc.
All rights reserved.

**441**

such as portable electronics to electric vehicles (see discussion(s) in other chapters of this book). Over the last three decades, Li-ion battery development has made a significant contribution to the evolution of alternative energy storage devices. Although significant progress in development has been achieved, and dynamic research for more achievements continues, Li-ion batteries currently deliver a limited energy density (200–250 Wh kg$^{-1}$) [1]. For example, electric vehicles (EVs), which are powered by Li-ion batteries, resulting in an insufficient driving range of 100–160 km. Thus, mutual hybridization of low-cost Li-ion batteries with combustion engines of automobile vehicles resulting in ~500 km driving range (specific energy: 550 Wh kg$^{-1}$) could be an enticing solution to a green and pollution-free environment [2].

The Earth-abundant element sulfur (S) is a potential candidate to achieve energy storage goals. Lithium-sulfur batteries (LiSBs) provide an almost threefold gravimetric energy density related to ongoing lithium-ion batteries. The operation of the Li-S battery assembly at room temperature delivers valuable outputs of high energy density (100–500 Wh kg$^{-1}$), high capacity (1672 mAh g$^{-1}$) [3], and a theoretical energy density of 2600 Wh kg$^{-1}$ with low cost (about \$150 per ton). Sulfur is an excellent electrochemical candidate that can receive a pair of electrons per atom over Li/Li$^+$ at 2.1 V. In a typical LiSB system, sulfur is utilized as an active substance in the cathode. However, the development of LiSBs must address several technical obstacles that need to be overcome to realize their key advantages as energy storage devices. These challenges include, for example, limited cycle life, poor S utilization, polysulfide shuttling effect, and self-discharging [4]. A number of LiSB researchers have endeavored to overcome polysulfide shuttling and Li dendrite production of organic liquid electrolyte. The solid-state approach for electrolytes is preferable in LiSB applications because they are nonflammable and will offer the best solution to a safety concern for portable electronics and electric vehicles [5–8]. This device advancement is a potentially game-changing opportunity for LiSB technology [9–11]. The removal of the liquid portion from LiSBs results in a significantly improves the cyclic performance and safety for commercial applications. Unlike conventional organic liquid electrolytes, solid-state electrolytes are promising candidates for next-generation robust energy storage devices. However, volume changes in all-solid-state batteries cause a limited cycle life. To buffer the volume expansion differences between the electrodes and the solid electrolyte system, high-pressure, hot-press processing and several advanced binder materials (for example, polyvinylidene fluoride—PVDF) was used [12,13]. Indeed, a clear impetus in the progress of solid electrolytes is to unlock the control of limited energy density, and the capacity of electrodes, like sulfur-based cathode and the Li metal anode, for conventional energy storage systems using solid electrolytes [14,15]. In addition, the assembly of all-solid-state Li-S batteries (SS-LiSBs) by integrating bipolar type of electrodes enables reducing inactive lining compound, thereby increasing energy density at the cell level [16,17].

## 12.2 Solid electrolytes

The electrolytes are the main part of an assembled battery because they can potentially imbue batteries with higher energy density, longer lifetime, reliable safety concern, and small(er) packaging. An ideal solid electrolyte should have superior ionic conductivity at a wide range of operating temperatures, broad range of voltage window against $Li/Li^+$ at 5 V, and well compatible with the electrode as well. In other words, the application of the solid electrolytes would eliminate the issues of self-discharging and fading of capacity. The following requirements should be met to advance in the direction of solid electrolyte for all (SS-LiSBs):

**(1)** High Li-ion conductivity ($\sigma$) at room temperature
**(2)** Low activation energy ($E_a$) at room temperature
**(3)** Low electronic conductivity
**(4)** Sufficient mechanical strength to avoid the growth of Li dendrite
**(5)** Chemically stable with both cathodes and anodes
**(6)** Significant potential stability ranges with respect to the electrode materials
**(7)** Thermal stability at wide temperature ranges
**(8)** Most importantly, minimal interfacial resistance at the electrode/solid electrolyte interface
**(9)** Cost-effective and eco-friendly synthesis and assembling
**(10)** Minimal or preferably no safety issues

Recently, two prospective types of electrolytes have been inspired by strong scientific interest as well as commercial applications, one type is an organic solid polymer electrolytes, and the other is an inorganic glass-ceramics/crystal electrolytes.

### 12.2.1 Solid polymer electrolytes

Solid polymer electrolytes (SPEs) show greater versatility to mitigate defects and cost-effective economic synthesis relative to glass ceramic electrolytes

[18]. Furthermore, the considerable mechanical stability (106–108 Pa) [19–21] of SPEs has been observed to be efficient in suppressing the formation of lithium dendrites [22]. Such achievements will improve SPEs for SS-LiSBs in the future. Poly(ethylene oxide) (PEO) acts as a Li-ion conductor with lithium salts; thus, this combination attracts an interest from researchers studying SPEs due to strong coordination with metal ions [23]. In recent decades, the SPEs have been applied in SS-LiBs, where several key milestones have been accomplished from theoretical approaches to practical implementation.

Due to the multiplicity of polymer chains and their complicated structure-property relationship(s), investigators have not yet been able to attain detailed ion movements in SPEs. Several successful approaches such as Williams-Landel-Ferry (WLF) [24] and Vogel-Tammann-Fulcher (VTF) [25] equations have been employed to explore the ionic conductivity mechanism in SPEs. Nevertheless, there also have been models developed, which elaborate the particulate motions. One example is the dynamics bond percolation model (DBPM), and a second example is the free volume model [26] proposing the diffusion mechanism of the intramolecular interaction between ions. Among all, the Arrhenius equation is the best option to explain pathways of ionic conductivity in solid electrolytes, including SPEs, inorganic glass ceramics, as well as crystalline materials, utilizing $\log \sigma$ vs $1/T$ curves. The Arrhenius equation was primarily related to ion-jumping phenomena uncoupled from the wide scale movement of substrate, which can happen under the glass transition temperature $(T_c)$ in SPEs and their glass phase(s) [27].

### 12.2.1.1 PEO-based solid polymer electrolytes

Diversified candidates for SPEs such as polysiloxane, polycarbonate, and PEO have been discovered, among which PEO has garnered the most attention. The structure of $(-CH_2-CH_2-O-)_n$ in PEO enables the synthesis of their metal complexes with different lithium salts in which $LiCF_3SO_3$, LiSCN, and LiBr assist as appropriate candidates for the SPEs [28]. However, simple PEO-type SPEs cannot overcome many of challenging issues for solid batteries due to low ionic conductivity ($\sim 1.0 \times 10^{-5}\,S\,cm^{-1}$), chemical instability, and interface resistance. Elasticity-endorsed consolidation is not a key parameter to prevent the lithium dendrite formation, but ion migration is also crucial. Archer et al. [29] developed a new class of SPEs in which polyethylene (PE)/PEO reduces PEO crystallinity (see Fig. 12.1A). Solid-polymer electrolytes comprising 39% of polyethylene

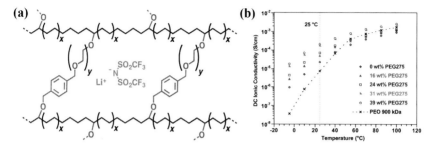

**Fig. 12.1** (A) Polyethylene/polymer (ethylene oxide) solid polymer electrolyte (SPE) synthesis. (B) Plot of DC lithium-ion conductor (S cm$^{-1}$) as a function of temperature for PEO electrolytes with different weight percent of PEG. *(Reproduced with permisison from R. Khurana, J.L. Schaefer, L.A. Archer, G.W. Coates, Suppression of lithium dendrite growth using cross-linked polyethylene/poly (ethylene oxide) electrolytes: a new approach for practical lithium-metal polymer batteries, J. Am. Chem. Soc. 136 (2014) 7395–7402, Copyright (2014) American Chemical Society.)*

glycol as the plasticizer (Fig. 12.1B) display excellent Li-ion conductivity ($1.0 \times 10^{-4}$ S cm$^{-1}$) at room temperature and suppress the formation of Li dendrites with considerable mechanical stability ($10^5$ Pa at 90°C, $G_{Li} > G_s$) [29]. Shin et al. reported (PEO)$_{10}$LiCF$_3$SO$_3$ SPEs with different wt.% of titanium oxide (Ti$_n$O$_{2n-1}$, $n=1, 2$) synthesized via ball milling as ceramic polymer electrolytes exhibited Li-ion conductivity (1.25–2.16 $\times 10^{-4}$ S cm$^{-1}$ at 90°C) for LiSBs [30].

### 12.2.1.2 Single-ion-conducting SPEs

There are several types of the dual-ion conductors, and a new type of SPE, named single-ion conductor, has recently garnered the most interest. Various anions can be bonded to the vertebral column by different types of bonding, such as covalent bond and electrostatic interaction, and immobilized by anion acceptors consistent with a Lewis acid-base concept [31]. The mechanism of conduction in single-ion conductor-type SPEs is a moveable cation bonded to the anions, losing attractive force(s); the anionic charge delocalization is responsible for ion movement between the electrodes through a hopping phenomenon, exhibiting a Li$^+$ transfer number reaching to unity ($t_{Li}^+ \approx 1$) [32].

Tikekar et al. [33] proposed a concept of an anion partially bonded during discharging and charging; the space charge can be reduced and electrodeposition of Li also occurred. Single-ion conductor-type SPEs are emerging as an opportunity to obtain a stabilized Li metal anode; this has

446 Lithium-sulfur batteries

an excellent potential for high–performance SS-LSBs. According to the current theory, a single-ion co-polymer with improved ionic conductivity has been proposed: poly(styrene trifluoromethanesulphonylimide) (poly [STFSILi]) co-block with PEO chain. Recently, Judez et al. [34] reported a novel-type SPE consisting of PEO and lithium bis(fluorosulfonyl) imide $(Li[N(SO_2F)_2]$ (LiFSI) that has been proposed for use in SS-LiSBs. This new type of SPE delivered the high capacity of $0.5\,mAh\,cm^{-2}$ and high specific capacity of $800\,mAh\,g^{-1}$ (Fig. 12.2A). The performance of the SS-LSBs with new SPEs $(Li[N(SO_2F)_2]$, LiFSI) was improved as compared to the lithium salt of LiTFSI $(Li[N(SO_2CF_3)_2])$ with polymer due to the better chemical stability between the Li metal anode and the as–prepared electrolyte (Fig. 12.2B) [34].

Another recent approach, the ultrahigh performance of the solid electrolyte, was synthesized via synergistic effect in between the polysulfide moieties formed during charge/discharge process and (fluorosulfonyl) (trifluoromethanesulfonyl) imide anions congenital from lithium salts in SPE. After the 60th cycle, the LiFTFSI/poly(ethylene oxide) (PEO) type of novel electrolyte in SS-LiSBs delivered a high specific capacity of $1394\,mAh\,g^{-1}$, high areal capacity of $1.2\,mAh\,cm^{-2}$, and also superior rate capability of $\sim 800\,mAh\,g^{-1}$. The XPS analysis represents the detailed understanding of electrochemical reduction of LiTFSI, LiFTFSI, and LiFSI. The most prominent components of LiF appeared at the inner part of the SEI layer on the Li metal anode surface. Besides, LiF is the primary building substance of the passivation layer especially for imide salts that contain an S—F bond. In summary, the SEI layer in LiFTFSI-contained electrolyte comprises an optimum number of inorganic materials (mainly LiF) and also meets with the requirement for an electrochemically stable SEI layer (Fig. 12.2C) [35].

## 12.2.2 Ceramic electrolytes

All-solid-state Li–S batteries attract a significant interest due to increased safety and reliability, and nonflammability of ceramics type of solid electrolyte. Also, inorganic ceramics electrolytes have a good cyclic performance for a wide temperature range, a lower electronic conductivity, and a broad range of electrochemical stability as compared to solid polymer electrolytes as well as liquid organic electrolytes. While these outstanding properties of ceramics electrolytes have been acknowledged, their low electrochemical stability and low ionic conductivity prevent them from being used in practical applications. Therefore, developing

A solid-state approach to a lithium-sulfur battery 447

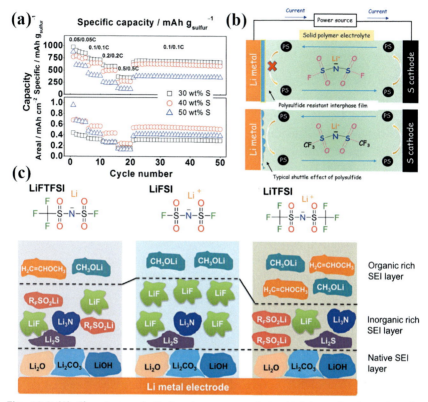

Fig. 12.2 (A) Charging and discharging curves of the Li-S batteries by using the PEO/LiFSI electrolyte at various rates and their Coulombic efficiency at 70°C. (B) Illustration of the shuttle effect of polysulfide of interface film with LiFSI and LiTFSI electrolyte systems. (C) Systematic representation on the SEI layer established on Li metal anode in three kind of salts. *(A and B) Reproduced with permission from X. Judez, H. Zhang, C. Li, J.A. González-Marcos, Z. Zhou, M. Armand, L.M. Rodriguez-Martinez, Lithium bis (fluorosulfonyl) imide/poly (ethylene oxide) polymer electrolyte for all solid-state Li–S cell, J. Phys. Chem. Lett. 8 (2017) 1956–1960, Copyright (2017) American Chemical Society. (C) Reproduced with permission from G.G. Eshetu, X. Judez, C. Li, M. Martinez-Ibañez, I. Gracia, O. Bondarchuk, J. Carrasco, L.M. Rodriguez-Martinez, H. Zhang, M. Armand, Ultrahigh performance all solid-state lithium sulfur batteries: salt anion's chemistry-induced anomalous synergistic effect, J. Am. Chem. Soc. 140 (2018) 9921–9933, Copyright (2018) American Chemical Society.*

new ceramic electrolyte composites has been a priority over the last few decades [36]. The Li-ion conductivity of ceramic electrolytes is known to be the major concern for their use as electrolytes in all-solid-state batteries with a wide range from $10^{-2}$ to $10^{-8}$ S cm$^{-1}$ at RT. The ion conduction phenomena and the calculation of Li ion conductivity for inorganic electrolytes can be explained via the Arrhenius equation (12.1).

$$\sigma = A \exp\left(-E_a/k_B T\right) \tag{12.1}$$

Here, the preexponential element A is proportional to the charge carrier quantity, $E_a$ represents the activation energy, $k_B$ expresses the Boltzmann constant, and $T$ is the absolute temperature in Kelvin. The activation energy is essential for ion conduction with a lower value of $E_a$, which usually leads to more ionic conductivity [37]. The mechanism of ionic conduction in ceramic electrolytes is completely different from liquid electrolytes, whereas ion movements in the ceramic electrolytes are controlled by a moving periodic bottleneck rather than moving solvent ions within liquid electrolytes [28]. Hence, the migration energy $(E_m)$ is an energy barrier. The trapping energy $(E_t)$ or defect formation energy $(E_f)$ also influences the activation energy $(E_a)$; here, $E_f$ specifies the quantity of interstitial positions in solid ion conductors. However, $E_t$ represents the trapping energy generated by the substitution of various aliovalent cations for more vacancies in conductive frameworks [37,38].

Inorganic solid electrolytes for Li-S battery mainly consist of oxide- and sulfide-based solid-state electrolytes. Sulfide-based solid-state electrolytes have attracted greater attention due to higher ionic conductivity than oxide-based electrolytes. Several different approaches have been taken for developing sulfide-based electrolytes, especially improving chemical stability for sulfide-based electrode materials and their electrochemical performance [39,40].

### 12.2.2.1 Oxide-based ceramics electrolytes

Research on oxide-based ceramics material as solid electrolytes has only been conducted recently because these oxide-based ceramics electrolytes are almost natural and easy to handle [41]. However, sulfide-based ceramics electrolytes are more chemically reactive and oxygen- as well as water-sensitive. Different types of oxysalts were discovered in the 1970s to demonstrate considerable $Li^+$ ion conductivity [42]—many with a $\gamma$-$Li_3PO_4$-type structure. A detailed study on the $\gamma$-$Li_3PO_4$ phase determined that there is a close relationship between the ionic conductivity and lattice volume; in other words, ionic conductivity improves with increasing lattice volume due to broader channel for ionic diffusion [43]. The peak value of the ionic conductivity measured for the oxysalts was $10^{-6} \, S \, cm^{-1}$.

Different oxide-based solid-state electrolytes such as 3D skeleton frameworks sodium (Na) superionic conductor (NASICON) were found to have the maximum ion conductivity in the order of $10^{-3}\,\mathrm{S\,cm^{-1}}$ [44]. Another class of oxide-based solid-state electrolytes perovskite structure was imitated to have the highest ion conductivity in the order of $10^{-3}\,\mathrm{S\,cm^{-1}}$ as well [45]. The garnet type of oxide-based electrolyte emerged as ceramics electrolyte that is a quite chemical stable with the lithium electrode [45]. However, the common disadvantage of all oxide-based solid-state electrolytes has not completely minimized, that is, high grain boundary resistance.

In 2003, the oxide-based garnet $Li_5La_3M_2O_{12}$ (M = Nb or Ta) was discovered. In 2007, a Li-rich garnet solid electrolyte $Li_7La_3Zr_2O_{12}$ (LLZO) was synthesized, it has been determined to be very chemically stable against Li metal and exhibited good ionic conductivity [45,46]. The structure of garnet $A_3B_3C_2O_{12}$ comprises the $B_3C_2O_{12}$ matrix framework of B cation at dodecahydrate positions and C cations at the octahedral positions. Several types of cation-based dopants with a range of ionic radii can be accommodated in the structure of garnet that induces more $Li^+$ site and results in higher ionic conductivity [47–50]. There are additional approaches to increase the ionic conductivity of garnet type of conductive framework; lanthanum and zirconium may be partially substituted by different transition metals to enhance the Li-ion mobility and lithium content. Li et al. disclosed that Li-rich garnet $Li_{56.4}La_3Zr_{1.4}Ta_{0.6}O_{12}$ exhibited a Li-ion conductivity of $1.0 \times 10^{-3}\,\mathrm{S\,cm^{-1}}$ with the lowest amount of activation energy of $0.35\,\mathrm{eV}$ [48]. Buschmann et al. reported that a cubic structure of LLZO could be chemically stabilized by doping of Al (1%wt) and showed the Li-ion conductivity of $4.0 \times 10^{-4}\,\mathrm{S\,cm^{-1}}$ (Al-doped LLZO) as contrasted to the pristine LLZO which exhibited the Li-ion conductivity of $2.0 \times 10^{-6}\,\mathrm{S\,cm^{-1}}$ with tetragonal geometry [49]. Bernuy et al. prepared Ga-substituted garnet under dry $O_2$ atmosphere, which leads to attain denser solid and prevent the chemical degradation, and they also showed the Li-ion conductivity as much as $1.3 \times 10^{-3}$ and $2.2 \times 10^{-3}\,\mathrm{S\,cm^{-1}}$ at 24°C and 42°C, respectively [51].

Xu et al. described a remarkable approach to the practical application of oxide-based garnet solid electrolytes for SS-LiSBs [52]. The all-solid-state cell architecture consists of a dense-porous-dense trilayer garnet solid-state electrolyte, and lithium metal anode and sulfur-based cathode were infiltrated. This all-in-one cell approach has many advantages, including continuous diffusion of $Li^+$ via the solid-state electrolyte, no formation of SEI at Li anode, limited volume changes, and prevention of dead sulfur and dead Li formation. With both sulfur-based cathode and lithium anode infused in the form of highly porous layer of the garnet solid electrolyte, the assembled SS-LiSBs presented a high discharge capacity of $\sim$1200 mAh g$^{-1}$ at a constant current of 50 mAg$^{-1}$ with the highest coulombic efficiency of 100% (Fig. 12.3A–I) [52].

A recent approach to the study of the durability of sodium superionic conductor (NASICON) type of oxide-based electrolyte $Li_{1+x}Ti_{2-x}Al_x(PO_4)_3$ offered good chemical stability over humid conditions. When $Li_{1+x}Ti_{2-x}Al_x(PO_4)_3$ becomes lithiated with $Li_2CO_3$, lithium polysulfide was formed on the surface, hindering the interfacial Li$^+$ conduction from the electrolyte. Lithium insertion into novel NASICON solid-state electrolytes weakens metal-oxygen bonds and increases crystalline volume; this produces framework cracking and encourages disintegration of the electrolyte. Further, these defected pores within ceramics produce a gateway for lithium polysulfide movements in between the electrodes, causing a sudden **B** degradation of the battery performance (Fig. 12.4A–D) [53]. Thus, for good cyclic performance of Li-S battery, discovery of ceramic-based solid-state electrolytes with chemical stability and highly stable grain boundary is an important milestone. Hence, several approaches need to be undertaken to improve the durability of (NASICON)type oxide electrolyte: (1) replace the Ti atom with other ions such as Ge and Zr in NASICON-conductive frameworks because Ti$^{4+}$ is reduced into Ti$^{3+}$ and (2) excretion of secondary phase and amorphous portion within grain boundary.

### 12.2.2.2 Sulfide-based ceramics electrolytes

A solid solution of $Li_4SiO_4$ and $\gamma$-$Li_3PO_4$ results in the formation of the so-called Li superionic conductor (LISICON). Replacing O atoms with the S atoms gives escalation to thio-LiSICON with three orders of higher Li$^+$ conductivity [37,54]. The argyrodite-type sulfide-based electrolyte is synthesized from the same ideology and undergoes identical conductivity mechanism [55,56]. It is stated that S$^{2-}$ owns higher polarizability and large atomic size than O$^{2-}$, which correspondingly increases the lattices volume for fast lithium-ion conduction [37]. Recently, the remarkable LISICON type of the electrolyte is $Li_{10}GeP_2S_{11}$ (LGPS) with very high Li$^+$ ion conductivity of $1.2 \times 10^{-2}$ S cm$^{-1}$ at room temperature (RT) [57]. The unique Li$^+$ ion-conductive frame consists of $(Ge_{0.5}P_{0.5})S_4$ tetrahedral bonded with LiS$_6$ octahedral by a simple edge-to-edge arrangement and forms a 1D chain that is more linked with PS$_4$ tetrahedral by the corner manner. The Li$^+$ can move from one position to another in a zigzag conduction pattern along

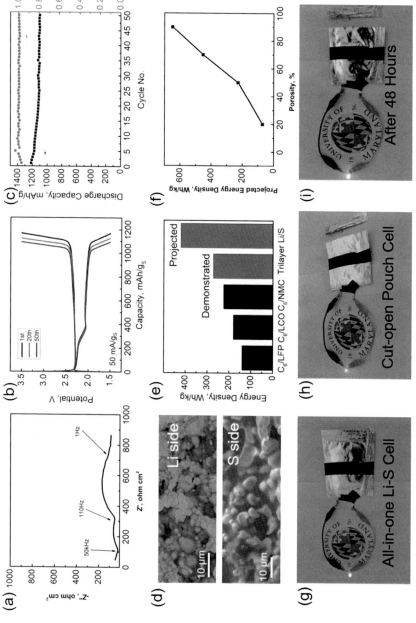

Fig. 12.3 See figure legend on next page

Fig. 12.3 (A) EIS of the all-solid-state Li-S cell based on garnet. (B) Voltage profiles of the 1st, 20th, and 50th cycle of the all-solid-state Li-S cell. (C) The cycling performance of the all-solid-state Li-S cell with 100% coulombic efficiency. (D) SEM image of anode side (top) and cathode side (bottom) of the trilayer garnet solid electrolyte after discharge. (E) Comparison of energy density on cell bases between trilayer all-solid-state Li-S battery and other Li-ion batteries. (F) Relation between porosity and energy density. Along 85% porosity, the energy density is 630 Wh kg$_{cell}^{-1}$. (G) Photo of the all-solid-state Li-S pouch cell. (H) Photo of cut-open all-solid-state Li-S pouch cell. The LED shows that the battery is working. (I) Picture of the all-solid-state Li-S pouch cell after cutting open for 48 h. The LED light displays the battery is still operational. *(Reproduced with permission from S. Xu, D.W. McOwen, L. Zhang, G.T. Hitz, C. Wang, Z. Ma, C. Chen, W. Luo, J. Dai, Y. Kuang, All-in-one lithium-sulfur battery enabled by a porous-dense-porous garnet architecture, Energy Storage Mater. 15 (2018) 458–464, Copyright (2018) Elsevier.)*

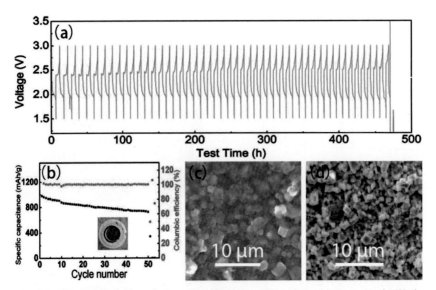

Fig. 12.4 Electrochemical analysis and performance: (A) voltage vs time graph; (B) the cyclic performance of the all-solid-state Li-S battery with the solid-state electrolyte. SEM graph of the all-solid-state electrolyte: (C) before and (D) after cycling performance. *(Reproduced with permission from S. Wang, Y. Ding, G. Zhou, G. Yu, A. Manthiram, Durability of the Li$_{1+x}$Ti$_{2-x}$Al$_x$(PO$_4$)$_3$ solid electrolyte in lithium–sulfur batteries, ACS Energy Lett. 1 (2016) 1080–1085, Copyright (2016) American Chemical Society.)*

c-axis. This high value to the Li$^+$ conductivity of LGPS is a nearly conventional liquid type of electrolytes. Nevertheless, the costly germanium (Ge) element replaces the opportunity for its commercial-scale development and its application in SS-LiSBs. In view of the current challenge, Ge has been substituted with Sn and Si to develop a cost-effective solid-state electrolyte [58]. In the recent report, $Li_{9.54}Si_{1.74}P_{1.44}S_{11.7}Cl_3$ was discovered to cumulatively exhibit extremely high $Li^+$ conductivity of $2.5 \times 10^{-2}\,S\,cm^{-1}$ and cost-effectiveness, which showcases it as an emerging candidate for a solid-state electrolyte for SS-LiSBs [17].

The binary system of $Li_2S$-$P_2S_5$ is an amazing ceramics solid-state electrolyte for solid-type high-energy batteries, particularly for SS-LiSBs, because it has good compatibility with high energy sulfur-based cathode. The $Li^+$ conductivity of $Li_2S$-$P_2S_5$ binary system can be tuned by varying the composition, annealing temperature as well as annealing time, and the stoichiometric ratio of $P_2S_5$ and $Li_2S$ [9]. For instance, the glass–ceramics of $Li_2S$-$P_2S_5$ electrolyte synthesized from cold-pressed and heat treatments exhibit high $Li^+$ conductivity of $1.7 \times 10^{-2}\,S\,cm^{-1}$ at room temperature due to low grain boundaries [59]. Another class of sulfide-based electrolyte, named argyrodites, derived from the $Ag_8GeS_6$ [60] may exhibit high $Cu^+$ and $Ag^+$ ions conductivity, as disclosed for $Cu_6PS_5Cl$ [61] and $Ag_9AlSe_6$ [62], respectively. First, Deiseroth et al. [63] published in 2008 a new class of $Li^+$-conductive argyrodites with the formula $Li_{7-x}PS_6X_x$ ($\leq x \leq 1$; $X=Cl$, Br, I), and $Li^+$ conductivity ranged from $10^{-2}$ to $10^{-3}\,S\,cm^{-1}$. This range of conductivity is similar to the conductivity of the commercial liquid type of electrolyte. Rao et al. disclosed a high $Li^+$ conductivity of $7 \times 10^{-3}\,S\,cm^{-1}$ and $3 \times 10^{-3}\,S\,cm^{-1}$ at RT for $Li_6PS_5Br$ and $Li_6PS_5Cl$, respectively [64].

$Li_3PS_4$ is another chemically stable sulfide-based crystalline phase obtained from a stoichiometric ratio of $Li_2S$-$P_2S_5$ mixture, which was reported by Tachez et al. in 1984, where this material was considered as an excellent lithium-ion conductor with a conductivity of $3 \times 10^{-7}\,S\,cm^{-1}$ at RT [65]. Between the two existing phases, the $\beta$-$Li_3PS_4$ phase is more $Li^+$ conductive than the $\gamma$-$Li_3PS_4$ phase. $\beta$-$Li_3PS_4$ exhibits high $Li^+$ conductive, but it can be changed to low $Li^+$-conductive $\gamma$-$Li_3PS_4$ phase when the temperature is less than 195°C. Nanoporous morphology of $\beta$-$Li_3PS_4$ has been synthesized, and the stability of this metastable phase was also achieved due to the nanocrystalline nature of $\beta$-$Li_3PS_4$, which shows a higher lithium-ion conductivity of $1.6 \times 10^{-4}\,S\,cm^{-1}$ at RT [66]. Xu et al. [67] explored that the $MoS_2$-doped $Li_2S$-$P_2S_5$ solid-state electrolyte ($Li_7P_{2.9}Mo_{0.01}S_{10.85}$) exhibited high $Li^+$ conductivity of $4.7\,mS\,cm^{-1}$ at RT and a broad-range electrochemical stability of 5 V (vs $Li/Li^+$). Assembled SS-LiSBs with solid-state electrolyte presented a highest discharge capacity of $1020\,mAh\,g^{-1}$ at the rate of 0.05 C from 1.0 to 3.0 V at RT [67]. A remarkable approach by Xu et al. [68] is that a newly designed Li-ion conductor $Li_7P_{2.9}Mn_{0.1}S_{10.7}I_{0.3}$ as sulfide-based solid-state electrolyte possesses a high Li–ion conductivity of $5.6\,mS\,cm^{-1}$ at RT and a broad electrochemical window for stability of $\sim 5\,V$ against $Li/Li^+$.

All–solid-state Li–S batteries assembled with as-prepared solid-state electrolyte and sulfur-based composite cathode material demonstrated a higher discharging capacity of $796\,mAh\,g^{-1}$ at the rate of $0.05\,C$ from the voltage window of 1.0 to $3.0\,V$ at RT, which are much better than those for the system with liquid-based electrolyte. The rate performance was examined at various current density values ranging from 0.05 to $0.5\,C$ at room temperature. The Columbic efficiency of the liquid Li–S battery decreased steadily to 90% due to the higher polysulfides solubility in the organic liquid-type of electrolyte, while the SS-LiSB retained up to 95% during cyclic performances (Fig. 12.5b) [68]. In terms of galvanometric cycling efficiency, the SS-LiSBs had advantages over organic liquid type of batteries in the Li–S devices (Fig. 12.5A–D) [68]. Due to the higher value of interfacial resistance between the electrolyte and their respective electrodes interface, the performance of the SS-LiSBs has been worse than that of the organic liquid electrolyte-based battery. In fact, the cycling performance and efficiency of the SS-LiSB were higher than those of the liquid type of battery due to the lack of dissolved polysulfide intermediates [69]. SS-LiSBs offer promising opportunities with impressive cyclic performance for future commercially safe energy storage devices.

To achieve chemical stability and high Li–ion conductivity of $5.28 \times 10^{-3}\,S\,cm^{-1}$ at 20°C, a new strategy of $SeS_2$-doped $Li_2S$-$P_2S_5$-based glass-ceramic electrolyte was developed by partially replacing $P_2S_5$ with $SeS_2$. A new strategy of $SeS_2$-doped $Li_2S$-$P_2S_5$-based glass-ceramic electrolyte has been developed by partially replacing $P_2S_5$ with $SeS_2$ to achieve chemical stability and high Li–ion conductivity of $5.28 \times 10^{-3}\,S\,cm^{-1}$ at 20°C [70]. The fabricated SS-LiSB based on $70Li_2S$-$29P_2S_5$-$1SeS_2$ solid-state electrolyte with rGO-S cathode composite shows a discharge capacity of $658.1\,mAh\,g^{-1}$ at $0.1\,mA\,cm^{-2}$ and an excellent rate performance at 30°C. Also, SEM analyses of SS-LiSBs showed that the doping of $SeS_2$ to the $Li_2S$-$P_2S_5$ glass-ceramics electrolyte can reduce the interfacial impedance between the solid-state electrolyte and the respective electrodes, and suppression of lithium dendrites growth. These findings showed that $1SeS_2$-$70Li_2S$-$29P_2S_5$ glass-ceramics electrolyte can be functioned as an emerging electrolyte candidate for the development of high-performance SS-LiSBs [70].

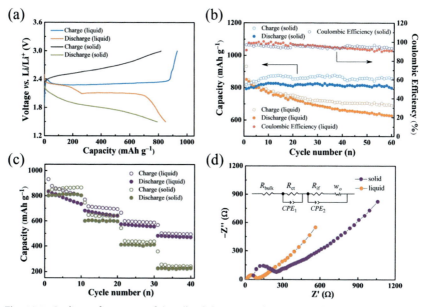

**Fig. 12.5** Cyclic performance of the all-solid-state Li-S batteries with Li$_7$P$_{2.9}$Mn$_{0.1}$S$_{10.7}$I$_{0.3}$ electrolyte and conventional liquid electrolyte: (A) charging and discharging at 0.05 C at RT; (B) cyclic performance at 0.05 C; (C) rate performance from 0.05 to 0.5 C; (D) impedance of all-solid-state Li-S battery and liquid battery with the equivalent circuit model. (Reproduced with permission from R.-C. Xu, X.H. Xia, S.H. Li, S.Z. Zhang, X.L. Wang, J.P. Tu, All-solid-state lithium-sulfur batteries based on a newly designed Li$_7$P$_{2.9}$Mn$_{0.1}$S$_{10.7}$I$_{0.3}$ superionic conductor, J. Mater. Chem. A 5 (2017) 6310–6317, Copyright (2017) Royal Society of Chemistry.)

The solid-state superionic argyrodite Li$_6$PS$_5$Cl was synthesized from Li$_2$S, P$_2$S$_5$, and LiCl precursors using a high-energy ball milling following the sintering method [69]. Li$_6$PS$_5$Cl with a Li-ion conductivity of $3.15 \times 10^{-3}$ S cm$^{-1}$ at RT was attained via the annealing method for 10 min at 550°C. The assembled SS-LISB based on the Li$_6$PS$_5$Cl electrolyte and multiwall carbon nanotube (CNT), nanosulfur, and Li$_6$PS$_5$Cl as cathode composite delivers a high discharge capacity of 1850 mAh g$^{-1}$ at 0.176 mA cm$^{-2}$ (0.1 C) and an excellent rate stability at 30°C; it retains 1393 mAh g$^{-1}$, and it retained 1393 mAh g$^{-1}$ after 50 charging/discharging cycles, as shown in Fig. 12.6A. The SS-LiSB maintained its columbic efficiency of 100%, which suggests high reversibility. During the first cycle of the assembled SS-LiSBs, the reversible capacity is larger than that of sulfur theoretical capacity. The extra capacity of the assembled cell could be attributed to sulfide-based Li$_6$PS$_5$Cl electrolyte in the cathode composite, which

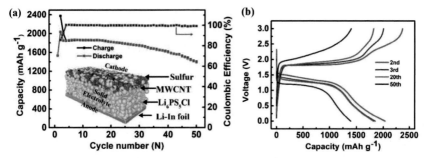

Fig. 12.6 (A) The charge and discharge performance of the all-solid-state Li-S battery with nanosulfur/MWCNT composite cathode and Li₆PS₅Cl solid-state electrolyte. (B) The discharge and charge voltage curves of the SS-LiSBs at 2nd, 3rd, 20th, and 50th cycles. (Reproduced with permission from S. Wang, Y. Zhang, X. Zhang, T. Liu, Y.-H. Lin, Y. Shen, L. Li, C.-W. Nan, High-conductivity argyrodite Li₆PS₅Cl solid electrolytes prepared via optimized sintering processes for all-solid-state lithium–sulfur batteries, ACS Appl. Mater. Interfaces 10 (2018) 42279–42285, Copyright (2018) American Chemical society.)

acts as an active substance, as shown in Fig. 12.6B. The redox electrochemistry of the sulfur is different in SS-LiSBs with the organic liquid electrolyte-based cell, because no discharge and charge curves associated with polysulfide production are found in the represented cyclic performance. These noteworthy results demonstrate that Li₆PS₅Cl is an excellent candidate for solid-state electrolytes for the construction of SS-LiSBs; these results are encouraging and promise further future improvement of solid-state energy storage devices [71].

## 12.3 Polymer/ceramic hybrid composite electrolytes

Despite significant research, the main challenges for the all-solid-state electrolytes made of bare polymer or ceramics have so far been associated with low Li ion conductivity, poor performance, and costly manufacturing process, which hinders further efforts to develop the ceramics polymers or hybrid-type electrolyte. The blend of both types of electrolytes combines the advantages of polymer and ceramics or even adds a little amount of liquid electrolyte as well because they have flammability issues and leakage but their conductivity are more than solid polymer electrolytes. Solid polymer electrolytes have wettability and excellent compatibility with the electrode;

their ionic conductivity at room temperature is currently too low for industrial applications. Polymers in ceramic hybrid electrolytes are engineered to address the key issues of a single electrolyte device. Several new types of polymer-in-ceramic hybrid composite electrolytes have recently been developed and offer significant promise of a new class of electrolyte for a solid approach toward Li-S batteries.

A novel design of polymer in ceramics type of electrolyte contains the three-layered sandwich structure of both ceramics electrolyte NASICON $Li_{1.3}Al_{0.3}Ti_{1.7}(PO_4)_3$ and polymer electrolyte methyl ether acrylate-cross-linked poly(ethylene glycol) [20]. This kind of structure like polymer-ceramics-polymer can improve the wettability in between the electrolyte and the respective electrodes. Another ceramic approach such as $Li_{1.3}Al_{0.3}Ti_{1.7}(PO_4)_3$ electrolyte layers can reduce ionic flow and minimizes the binary layers electric field adjacent to the electrodes. At this stage, the degradation of the polymer could be started. When a polymer/ceramics hybrid electrolyte is used in Li-S batteries, it could successfully depress (or diminish) the polysulfide shuttle effect by slowing the motion of polysulfide anions, achieving a high discharge capacity of 988 mAh g$^{-1}$ after 100 cycles [69]. Zhang et al. described a new electrolyte system that consists of lithium bis(trifluromethanesulfonyl) imide (LiTFSI), poly(ethylene oxide) (PEO), and nanoclay (montmorillonite) to improve flexibility and Li–ion conductivity; these have also been applied to SS-LiSBs (Fig. 12.7). The contents of the nanoclay and temperature are the main factors, i.e., 10 wt% nanoclay showed Li–ion conductivity in the order of $3.22 \times 10^{-3}$ S cm$^{-1}$ at 60°C. At the same temperature, SS-LiSBs using PEO/nanoclay ceramics in polymer represent better cyclic performance, delivering 988 mAh g$^{-1}$ after 100th cycles at the current rate of 0.1 C. At the current density of 0.5 C, the assembled SS-LiSBs still could transfer a discharge capacity of 643 mAh g$^{-1}$ at 60°C [72].

Tao et al. developed a new type of electrolyte via facile doping of $Al^{3+}$/ $Nb^{5+}$ in the cubic $Li_7La_3Zr_2O_{12}$ (LLZO), and these synthesized LLZO nanoparticles ornamented porous carbon by the sol-gel technique. The LLZO nanoparticle occupied poly(ethylene oxide) (PEO), and $LiClO_4$ exhibits a high conductivity of $1.1 \times 10^{-4}$ S cm$^{-1}$ at 70°C. The sulfur-containing cathode composite based on LLZO-PEO-LiClO$_4$ can show an excellent discharge capacity of 900 mAh g$^{-1}$ with the current density of 0.05 mA cm$^{-2}$ at 37°C and also presents 1210 mAh g$^{-1}$ and 1556 mAh g$^{-1}$ at 50°C and 70°C, respectively (Fig. 12.8A–C) [32].

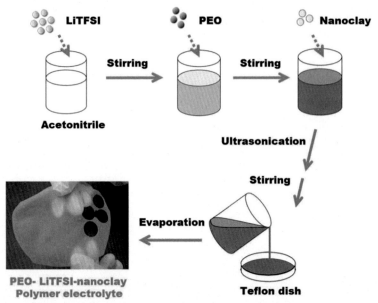

**Fig. 12.7** Schematic synthesis of PEO/MMT polymer electrolyte. *(Reproduced with permission from Y. Zhang, Y. Zhao, D. Gosselink, P. Chen, Synthesis of poly (ethyleneoxide)/nanoclay solid polymer electrolyte for all solid-state lithium/sulfur battery, Ionics 21 (2015) 381–385, Copyright (2015) Springer.)*

## 12.4 Stable Li metal anodes for all-solid-state Li-S batteries

Lithium metal anode is another big challenge for the exploitation of practical applications of all-solid-state Li-S batteries (SS-LiSBs) [73,74]. Although there is an extensive ride to completely resolve the technical issues of Li metal as an ideal anode, lithium is proposed as a remarkable anode candidate because of the highest theoretical specific capacity of $3860\,\text{mAh}\,\text{g}^{-1}$ and the lowest redox potential of $-3.04$ against standard hydrogen electrode (SHE) [75]. Nevertheless, there are several disadvantages to the Li metal as anode in the working battery. The inhomogeneous stripping/plating is a major reason for the dendrite formation. Lithium dendrites can puncture the separator between the electrode and the electrolyte, leading to the battery short circuit and safety issue. Moreover, the chemical reactions between lithium anode and the solid-state electrolyte, and chemical decomposition of liquid electrolyte at the Li metal surface reduce the performance of the lithium anode [76,77]. In the SS-LiSBs, the above-mentioned problems and challenges regarding Li metal as anode are still considerable [78,79].

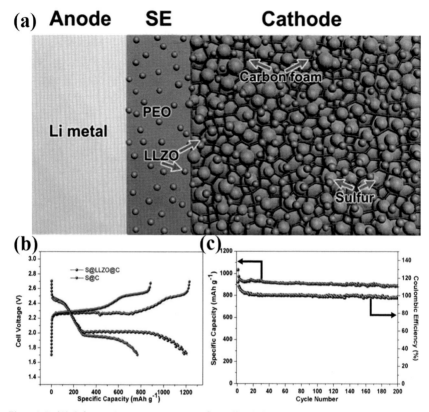

**Fig. 12.8** (A) Schematic representation of an all-solid-state Li-S battery based on LLZO nanostructure. The LLZO-PEO-LiClO$_4$ electrolyte is casting on the composite cathode (sulfur cathode using LLZO@C matrix with PEO binder). (B) The charge and discharge curves of S@LLZO@C and S@C with 0.1 mA cm$^{-2}$ at 50°C. (C) The columbic efficiency and cyclic performance of S@LLZO@C with 0.05 mA cm$^{-2}$ at 37°C. *(Reproduced with permission from X. Tao, Y. Liu, W. Liu, G. Zhou, J. Zhao, D. Lin, C. Zu, O. Sheng, W. Zhang, H.-W. Lee, Solid-state lithium–sulfur batteries operated at 37°C with composites of nanostructured Li$_7$La$_3$Zr$_2$O$_{12}$/carbon foam and polymer, Nano Lett. 17 (2017) 2967–2972, Copyright (2017) American Chemical Society.)*

The applications of Li alloys with different metals can decrease the activity and reducibility of Li anode. Li alloy as anode for SS-LiSBs is described in various publications, such as Li-Al, Li-Si, Li-In, and Li-Ge [80]. Molar ratio of lithium alloy with indium (Li/In = 0.79) is a potential candidate and its measured redox potential is +0.6 V (Li/Li$^+$)[81]. Nagata et al. reported Li-In as anode material and Li$_{10}$GeP$_2$S$_{12}$ sulfide type of the solid-state electrolyte that is used to fabricate the SS-LiSBs [9,82]. Zhang et al. [83] applied the

Li–B alloy as a negative electrode and also produced a passivation layer on the electrode surface, which impressively reduced the decay of performance due to the variation in volume of the alloy material. The battery has been assembled with the Li–B alloy as an anode and represented outstanding cyclic performance and transferred a specific discharge capacity of $1600 \text{ mAh g}^{-1}$. Kato et al. [84] fabricated Li/Li symmetric cell with compressed pelletize powder $Li_2S$-$P_2S_5$. Au makes the alloys with lithium anode, which was injected as a thin film to interface between the Li metal anode and the sulfide-based solid-state electrolyte. Au has an impressive compatibility with sulfide-based electrolytes [84]. Kato et al. studied XPS and SEM of injecting gold layer to the $Li_3PS_4$/Li-anode interface [85]. SEM analysis examined the effects of Au film on the alteration Li metal morphology and structure of solid electrolyte as well. XPS results suggested that $Li_3PS_4$ electrolyte showed a low stability due to Au film along with low resistance and good chemical stability as well. SEM results showed that the Au layer in between the solid-state electrolyte and Li anode stopped void formation after Li dissolution (Fig. 12.9) [85].

The major challenges of lithium metal anode/solid-state electrolyte interface include high interfacial resistance, the formation of the lithium dendrites, and unwanted chemical reactions of solid-state electrolyte with Li metal [86–88]. Different approaches, based on the ability and nature of the SSEs, such as passive layer on the Li anode, modification of the solid-state electrolyte, and cold-pressing, have been considered to solve the above-mentioned issues.

## 12.4.1 Li anode/sulfide-based solid-state electrolyte

The sulfide-based solid-state electrolyte (SSE) is not fully chemically stable over Li metal anode and therefore disintegrates to create an SEI layer with large interfacial resistance [39,89,90]. If the Li metal anode directly contacts with the surface of sulfide-based SSE, the nonuniform interface can stimulate Li deposition/dissolution and prevent the utility of Li metal during charging and discharging, leading to serious issues such as large interface impedance and limited durability of SS-LiSBs [40,91]. Recently, different advanced techniques such as X-ray absorption spectroscopy (XAS), X-ray photoelectron spectroscopy (XPS), Raman spectroscopy, scanning electron microscopy (SEM), and transmission electron microscopy (TEM) have been extensively used to investigate the mechanism of the formation of SEI layer and interfacial stability in between the Li metal anode and SSEs [90,92].

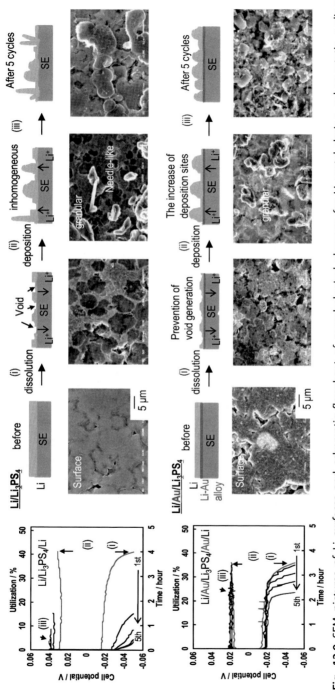

**Fig. 12.9** SEM pictures of Li surfaces and schematic flowsheet of morphological changes for Li metal during galvanostatic cycling examination. SEM observation was completed after three cell operations: (i) after the initial dissolution, (ii) the following deposition, and (iii) after 5th cycles. (Reproduced with permission from A. Kato, H. Kowada, M. Deguchi, C. Hotehama, A. Hayashi, M. Tatsumisago, XPS and SEM analysis between Li/Li$_3$PS$_4$ interface with Au thin film for all-solid-state lithium batteries, Solid State Ionics 322 (2018) 1–4, Copyright (2018) Elsevier.)

Nagao et al. [92] studied the interfacial aspects of the Li anode with the glass-ceramics type of $Li_2S-P_2S_5$ solid-state electrolyte by in situ SEM findings and explained the phenomena about Li metal anode, leading to create void spaces of the solid-state electrolyte during deposition/dissolution. Also, it explains that the lack of pores interconnection among the solid-state electrolytes plays a vital role in minimizing the Li dendrite growth. Wenzel et al. [90] observed the electrochemical reaction at the lithium metal anode and superionic conductor $Li_{10}GeP_2S_{11}$ solid-state electrolyte interface by in situ XPS-coupled time-dependent electrochemical impedance spectroscopy, disclosing that disintegration of the LGPS will mainly cause the development of an SEI layer. The disintegration of the LGPS leads to the production of interphase unwanted products such as $Li_2S$, $Li_3P$, and Li-Ge alloys, increasing the interfacial impedances, as shown in Fig. 12.10A and B. The growth of the interface between the lithium metal anode and $Li_7P_3S_{11}$ electrolyte was observed via in situ XPS-coupled time-resolved electrochemical impedance spectroscopy.

A qualitative method for the identification of the phase compositions at the solid-electrolyte interphase has been developed by utilizing a deep analysis of the XPS peaks [85]. Wenzel et al. [89] reported another approach to inspect the interphase formation and the chemical stability of $Li_7P_3S_{11}$ and argyrodite electrolyte $Li_6PS_5X$ (with $X = Cl$, Br, and I) with lithium metal anode. The XPS peaks represent $Li_2S$, $Li_3P$, and LiX, which is in good agreement with the theoretical prediction. The interfacial results of argyrodite were also compared to those of other superionic sulfide-based conductors such as $Li_7P_3S_{11}$ and $Li_{10}GeP_2S_{12}$. $Li_{10}GeP_2S_{12}$ and $Li_6PS_5I$ display higher SEI resistances as compared to $Li_6PS_5Cl$, $Li_6PS_5Br$, and $Li_7P_3S_{11}$. $Li_7P_3S_{11}$ appears to be the utmost promising material for the most chemical stable candidate for Li anode metal as a suitable solid-state electrolyte in SS-LiSBs.

Yamada et al. [93] reported a galvanostatic deposition/stripping measurement of Li/Li symmetric cell using $Li_3PS_4$ as a solid-state electrolyte with another symmetric cell using the mixture of 1M Li bis(trifluoromethane sulfonyl) imide (LiTFSI) and 1,3-dioxolane (DOL)-diethoxyethane (DEE). Using the liquid electrolyte, low charge transport resistance $(R_{ct})$ was observed after deposition/stripping of Li. But charge transport resistance $(R_{ct})$ for $Li_3PS_4$ solid-state electrolyte was also much decreased after the above-mentioned treatment. The activation energy and charge

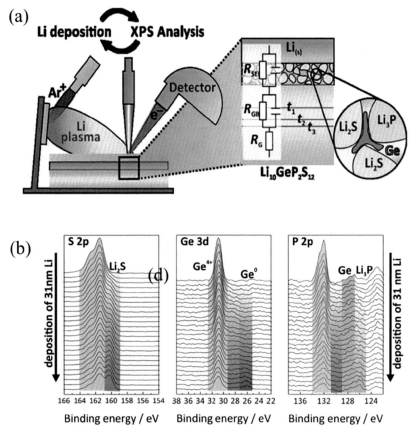

Fig. 12.10 (A) Schematic diagram of the in situ XPS to display the interactions between Li anode and $Li_{10}GeP_2S_{12}$ (LGPS) as well as the interphase formation. (B) XPS spectra scan during the deposition of 31-nm Li metal anode on LGPS. S 2p, Ge 3d, and P 2p/Ge 3p spectra are represented for various deposition states. With increasing Li deposition time, LGPS decomposes. The recognized species are marked and labeled in the spectra. *(Reproduced with permission from S. Wenzel, S. Randau, T. Leichtweiß, D.A. Weber, J. Sann, W.G. Zeier, J. Janek, Direct observation of the interfacial instability of the fast ionic conductor $Li_{10}GeP_2S_{12}$ at the lithium metal anode, Chem. Mater. 28 (2016) 2400–2407, Copyright (2020) American Chemical Society.)*

transport resistance of Li/Li symmetric battery using liquid electrolytes reported high values (68.40–67.60 kJ mol$^{-1}$). A subsequent deposition/stripping treatment of Li reduced the activation barrier and charge transport resistance of a symmetric cell using solid $Li_3PS_4$ electrolytes so that the reported values were much improved (52.1 – 44.5 kJ mol$^{-1}$) [93].

## 12.4.2 Li anode/oxide-based solid-state electrolytes

Oxide-based solid-state electrolytes (SSEs) are considered to be exceptionally chemically stable against lithium metal and also exhibit good Li-ion conductivity at RT. Nevertheless, the main challenge of an oxide-based solid-based electrolyte like garnet possesses the high interfacial impedance ranges ($10^2$–$10^3 \Omega \, cm^2$) due to poor contact between garnet type (LLZO) of SSEs and Li anode [94–96]. Due to the brittle nature of morphological impurities in garnets, such as lithium carbonate ($Li_2CO_3$), it is difficult to achieve an adequate wettability of Li metal at a garnet surface. The main cause of the chemical instability between garnet and Li metal is high interfacial resistance at the SSE/Li interface [97]. For addressing these issues, appropriate approaches are required.

High-pressure and heat treatments could be used to improve wettability between the SSEs and the Li anode. The sintering of Li metal/LLZO SSEs from RT to 175°C can decrease the interfacial impedance from 5822 to $514 \, \Omega \, cm^2$ [95]. Although this approach is not very useful because of the high brittleness of the LLZO SSEs, inferior contacted Li metal can create microscopic voids and poor interfacial current homogenous distribution. Another method for improving the contact between SSE and Li metal is to introduce an inorganic-type buffer layer containing various metals and metal oxides such as Au, Si, Ge, Al, ZnO, and $Al_2O_3$. These buffer layers are inserted between the oxide-type SSEs and Li anode, effectively densifying and closing the voids/spaces on the SSEs surface, which lead to the severs for internal resistance [96–100]. Han et al. [97] also reported a very effective and relatively cheap solution of interfacial problem in between the lithium anode and $Li_7La_{2.75}Ca_{0.25}Zr_{1.75}Nb_{0.25}O_{12}$ garnet via the deposition of a very thin film of aluminum oxide ($Al_2O_3$). A remarkable reduction of interfacial impedance from 1710 to $1 \, \Omega \, cm^2$ was detected at RT, which strongly reduces the oxide-based electrolyte garnet/Li metal interfacial resistance. Oxide coating permits extraordinary contact of garnet electrolyte and Li metal, and it also enables very effective $Li^+$ transportation (Fig. 12.11A–E) [97]. By the electrodeposition method, layer of Al on the surface of the doped garnet $Li_7La_{2.75}Ca_{0.25}Zr_{1.75}Nb_{0.25}O_{12}$ was designed and created a new interfacial layer composed of Li-Al alloy. Moreover, the contact of the doped garnet becomes lithiophilic, and there is a good agreement of wettability between Li anode and doped garnet. The EIS graph shows the reduction in the interfacial impedance from 950 to $75 \, \Omega \, cm^2$ in the assembled $Li/Al-Li_7La_{2.75}Ca_{0.25}Zr_{1.75}Nb_{0.25}O_{12}-Al/Li$ symmetric cell [100].

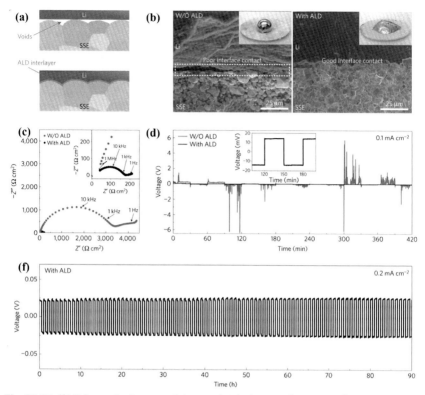

Fig. 12.11 (A) Schematic diagram of the wetting behavior of garnet surface with molten Li. (B) SEM images of the garnet solid-state electrolyte/Li metal interface. Without ALD-Al$_2$O$_3$ coating, garnet has a poor interfacial contact with Li metal even on heating. (C) Comparison of EIS curves of the symmetric Li nonblocking garnet cells. (D) Comparison of cycling for symmetric cells of Li/bare garnet/Li *(black curve)* and Li/ALD-treated garnet/Li *(red curve; gray* in print version) at a current density of 0.1 mA cm$^{-2}$. (e) Galvanostatic cycling of Li/ALD-treated garnet/Li with a current density of 0.2 A cm$^{-2}$. *(Reproduced with permission from X. Han, Y. Gong, K.K. Fu, X. He, G.T. Hitz, J. Dai, A. Pearse, B. Liu, H. Wang, G. Rubloff, Negating interfacial impedance in garnet-based solid-state Li metal batteries, Nat. Mater. 16 (2017) 572–579, Copyright (2017) Springer Nature)*

Shao et al. [101] reported a cost-effective and straightforward approach to minimize the interfacial resistance between lithium metal anode and garnet (LLZO)-based solid-state electrolyte by applying an ultrafine layer of graphite as a soft interface through a simple pencil. The graphite type, naturally soft interface, enables good wettability between LLZO electrolyte and lithium metal as a lithiated interface with improved electronic and Li$^+$

conductivities. EIS arc was used to measure the variations, and interfacial resistance decreased from 1350 to $105\,\Omega\,cm^2$ [101].

Liu et al. [98] described the insertion of a layer of $Li_3PO_4$ as thin SEI layer ($\sim$200 nm thickness) between the $Li_{1.3}Al_{0.3}Ti_{1.7}(PO_4)_3$ type of SSEs and lithium metal anode to shield the direct contact. The Li-ion conductivity of $Li_{1.3}Al_{0.3}Ti_{1.7}(PO_4)_3$ (LATP) type of SSEs exhibited $7.02 \times 10^{-4}\,S\,cm^{-1}$ at RT. The artificial SEI film of $Li_3PO_4$ is a powerful method to enhance the electrochemical performance of LATP as a promising candidate of electrolyte for SS-LiSBs and others solid-state energy storage devices [102]. Wang et al. [52] investigated the durability effects of LATP type of SSEs in contact with the polysulfide solution for Li-S battery. From the mechanism point of view, Li introduction into the conductive framework of LATP leads to the crystal volume expansion in an anisotropic way, which, in turn, results in weakening of the bonds within the crystal, cracking the ceramic matrix, erosion of grain boundaries, and formation of the unwanted pores. The pores develop an opening space for polysulfide movements from cathode to anode, resulting in the degradation of battery performance.

## 12.4.3 Li anode/solid polymer electrolytes

The sulfide-based SSEs and oxide-based SSEs exhibit high brittleness and poor interface contacts with the Li metal anode and sulfur-based cathode composite. The application of the ionic-conductive type of polymer interlayers between the lithium metal and the solid polymer minimizes the interfacial impedances. Though, the ionic-conductive polymer interlayer can prevent the dendrite growth due to homogenous $Li^+$ flux over Li/polymer interfaces. Moreover, the interlayer of the polymer between the Li anode and SSEs demonstrates better contact and improvement in wettability.

Fu et al. have priority in designing a state-of-the-art 3D Li-ion-conductive framework in a polyethylene oxide (PEO) composite based on a garnet type of oxide-based solid-state electrolyte $Li^+$ conductor with enhanced lithium-ion diffusion channels. [99]. These ceramic in polymer composites offer structural strengthening to increases the mechanical characteristics of the polymer-based templates. This flexible SSEs and polymer matrix membrane showed a lithium-ion conductivity of $2.4 \times 10^{-4}\,S\,cm^{-1}$ at RT. This composite membrane is effective in preventing lithium dendrites in the Li/Li symmetric battery at the time of Li plating/stripping over different current densities such as 0.3 and $0.5\,mA\,cm^{-2}$ over 500 h and 300 h, respectively [103]. The dense and thin layer of garnet is effectively limited by interfacial resistance with an excellent mechanical stability. The porous and thin layer role play of mechanical support to another thin layer providing a continuous channel for ion diffusion.

LLZO-based garnets chemically react with $CO_2$ and $H_2O$ in moist air to produce a $Li^+$-insulating $Li_2CO_3$ thin film on the surface; this results in enhancing interfacial impedances to $Li^+$ migration. In order to overcome this major issue, adding lithium fluoride (LiF) to LLZTO ($Li_{6.5}La_3Zr_{1.5}Ta_{0.5}O_{12}$) should improve the chemical stability of oxide-based SSEs. With 2%wt LiF in LLZTO, the coverage of $Li_2CO_3$ on garnet particles is reduced as is the interfacial impedance with Li metal anodes. For SS-LiSBs containing LLZTO with 2 wt.% LiF and also employing SSEs as a separator, polysulfide crossing between electrodes is dramatically reduced and a coulombic efficiency of 93% is observed after the hundredth cycle.

## 12.5 Sulfur-based cathode composites for all-solid-state Li-S batteries

Sulfur-based cathodes have three major disadvantages: (1) the intermediates of sulfur to lithium sulfide are followed by significant volume changes that can demolish the cathode framework; (2) the electronic and ionic conductivities of the active material are poor; (3) during the cyclic process in liquid electrolyte system, a high-order polysulfides diffuse toward the anode side and chemically react with Li metal and degrade to low-order lithium sulfide; showing the inadequate electrochemical performance and lowering coulomic efficiency [104,105]. In contrast to traditional Li-S batteries, SS-LiSBs exhibit single discharge and charge curves because of the direct conversion during the electrochemical process $S + 2Li^+ + 2e^- \leftrightarrow Li_2S$ in which unwanted polysulfides are not formed, thus improving the battery performances, including extended life, high capacity, and high energy density [106,107]. Nevertheless, the poor ionic and electronic conductivity of the $Li_2S$ and S as well as their poor contact between the active materials and additives (ionic conductors, in particular solid-state electrolytes and electronic conductor carbon-based materials) cause capacity fading and increase interfacial impedance in SS-LiSBs [82,108,109]. An effective approach to address these issues is a high-energy ball milling method to synthesize the sulfur-based cathode material; this technique can guarantee the complete mixing of all precursors to attaining the improvement in electronic as well as ionic conductivity [107]. Another approach is to resolve this issue by means of hybridizing the active material with SSEs and electronic-conductive additives, which can enhance ion as well as electron migration in the composite cathode for SS-LiSBs [108,110].

The interface in SS-LiSBs primarily comprises an electrode-electrolyte contact (solid/solid interface) showing huge resistance and restricted $Li^+$ migration. Moreover, strain and stress are other factors at the interface, which are significantly affected due to volume change of active materials like lithium sulfide and sulfur during cyclic process, leading to damage of the interface contact [8,107,111–114]. As a result, the performance of SS-LiSBs decreases, while improving the interaction between the sulfur-based electrode and SSE is a challenging task in the prospective development of SS-LiSBs.

## 12.5.1 Cathode/oxide-based electrolyte interface

In conventional Li-S battery, the production of polysulfide at the time of charging and discharging process is a main cause of poor performance of battery. But in SS-LiSBs, the produced polysulfides cannot cross the electrolyte, and the polysulfide shuttle behavior is successfully suppressed. However, the evolution of the novel type of SSEs with all the advantages involved continues to be resolved. The key issue of inorganic oxide-based ceramics type of electrolyte is the poor contact with electrodes. Hao et al. [115] assembled a SS-LiSB comprised of a sulfur-coated multiwalled carbon nanotube (S-CMWCNT)-based composite cathode with the NASICON type of solid-state electrolyte exhibiting $1.8 \times 10^{-4} \, S \, cm^{-1}$ at RT. Many SS-LiSB structures demonstrate a remarkable discharge capacity of $1510 \, mAh \, g^{-1}$ on the 1st cycle and retain $1400 \, mAh \, g^{-1}$ after 30 cycles [115]. In Table 12.1, the focus is primarily on the SS-LiSBs of different kinds of sulfur-based cathode materials with oxide-based SSEs and sulfide-based SSEs and their cyclic performances.

## 12.5.2 Cathode/sulfide-based electrolyte interface

The $S^{2-}$ is larger in size as compared to the $O^{2-}$ ion. The replacement of $O^{2-}$ in solid-state electrolyte by $S^{2-}$, leads to larger $Li^+$ diffusion channels and improves the ionic conductivity at RT. There are several approaches to unravel the interfacial problems by reducing the particle size and enhancing the intimate contact of the cathode composite made by homogeneous mixing of the SSEs, active material, and conductive carbon via high–energy ball milling. Choi et al, reported the cathode material in which conductive carbon AB, sulfur, and solid-electrolyte $70Li_2S$-$30P_2S_5$ glass ceramics were

**Table 12.1** The electrochemical performance of different sulfur-based composites in all-solid-state Li-S batteries with inorganic solid-state electrolytes.

| Active material | Conductive additives | Preparation method | Solid electrolytes | Conductivity S cm$^{-1}$ | Composition | Applied current density | Capacity and cycle | Refs. |
|---|---|---|---|---|---|---|---|---|
| _Inorganic solid-state electrolytes Li-S battery_ | | | | | | | | |
| Sulfur | NWCNT | Evaporation | NASICON-type LAGP | $1.8 \times 10^{-4}$ at RT | Sulfur 80wt% in NWCNT | $20 \, \text{mAg}^{-1}$ | 1400@30th | [116] |
| Sulfur | Graphite (MCMB) | Ball milling | 80Li$_2$S-20P$_2$S$_5$ glass-ceramics | $5 \times 10^{-3}$ at 80°C | S/Graphite = 1/1 | $84 \, \text{mAg}^{-1}$ | 1400@18th | [109] |
| Sulfur | Acetylene black | Ball milling | 80Li$_2$S-20P$_2$S$_5$ glass-ceramics | $\sim 10^{-3}$ at 25°C | Li$_2$S/C/SE = 25/25/5 | $0.06 \, \text{mA cm}^{-2}$ | 700@1st | [117] |
| Sulfur | Cu | Ball milling | 80Li$_2$S-20P$_2$S$_5$ | $\sim 10^{-3}$ at 25°C | Molar ratio of S/Cu = 3 | $64 \, \mu\text{A cm}^{-2}$ | 650@20th | [118] |
| Sulfur | Acetylene black | Ball milling | 80Li$_2$S-20P$_2$S$_5$ glass-ceramics | $\sim 10^{-3}$ at 25°C | S/C/SE = 50/21/20 | $0.06 \, \text{mA cm}^{-2}$ | 1050@50th | [119] |
| Sulfur | Acetylene black | Ball milling | 80Li$_2$S-20P$_2$S$_5$ glass-ceramics | $\sim 10^{-3}$ at 25°C | S/AB/SSE = 25:25:50 | $1.3 \, \text{mA cm}^{-2}$ | 850@200 | [106] |
| Sulfur | Ketjen black | Ball milling | 80Li$_2$S-20P$_2$S$_5$ | $5 \times 10^{-4}$ at 25°C | S/C/SE = 50/10/40 | $6.4 \, \text{mA cm}^{-2}$ | 565@1st | [9] |
| Sulfur | MCMB + super P | Ball milling | 80Li$_2$S-20P$_2$S$_5$ glass | $2.2 \times 10^{-4}$ at 25°C | S/C/SE = 5/15/2 | $0.074 \, \text{mA cm}^{-2}$ | 400@20th | [106] |
| Sulfur | Ketjen black | Ball milling | 60Li$_2$S-40P$_2$S$_5$ | $2 \times 10^{-5}$ at 25°C | S/C/SE = 50/10/40 | $0.64 \, \text{mA cm}^{-2}$ | 1568@1st | [9] |
| Sulfur | Ketjen black | Ball milling | 40Li$_2$S-60P$_2$S$_5$ | $2 \times 10^{-5}$ at 25°C | S/C/SE = 50/10/40 | $6.4 \, \text{mA cm}^{-2}$ | 1096@1st | [9] |
| Sulfur | Activated carbon | Ball milling | 70Li$_2$S-30P$_2$S$_5$ | $3.2 \times 10^{-3}$ at RT | S/C/SE = 25/25/50 | $0.13 \, \text{mA cm}^{-2}$ | 250@6th | [120] |
| Sulfur | Polyacrylonitrile | Thermal annealing | Li$_2$S-P$_2$S$_5$ | $9.3 \times 10^{-4}$ at RT | S/PAN = 1/2 | $26.5 \, \text{mAg}^{-1}$ | 650@50th | [121] |
| Sulfur | CNF | Ball milling | Li$_3$PS$_4$ | $\sim 10^{-4}$ at 25°C | S/C/SE = 30/10/60 | $0.025 \, \text{mA cm}^{-2}$ | 1500@10th | [93] |
| Sulfur | VGCF | Ball milling | Li$_3$PS$_4$ (amorphous) | N/A | S/VGCF/SE = 30/10/60 | $0.1 \, \text{mA cm}^{-2}$ | 1200@50th | [10] |
| Sulfur | VGCF | Ball milling | Li$_3$PS$_4$ | $2 \times 10^{-4}$ at RT | S/VGCF/SE = 30/10/60 | $0.1 \, \text{mA cm}^{-2}$ | 1230@50th | [122] |
| Sulfur | Ketjen black | Ball milling | Li$_3$PS$_4$ | N/A | S/KB/P$_2$S$_5$ = 50/10/40 | $0.64 \, \text{mA cm}^{-2}$ | 942 | [123] |
| Sulfur | FeS$_2$-AC | Ball milling | Li$_3$PS$_4$-LiI | $\sim 10^{-3}$ at RT | S/FeS$_2$/C = 30/30/40 | $83.5 \, \text{mAg}^{-1}$ | 1200@20th | [124] |
| Sulfur | MAXSORB-activated C | Ball milling | LiI–Li$_3$PS$_4$ | $1.2 \times 10^{-3}$ at 25°C | S/C/SE = 30/20/50 | $7 \, \text{mA cm}^{-2}$ | 1600 | [125] |
| Sulfur | MAXSORB-activated C | Ball milling | LiI-Li$_3$PS$_4$ | $1.2 \times 10^{-3}$ at 25°C | S/C/SE = 0.37/0.13/0.50 | $0.25 \, \text{mA cm}^{-2}$ (0.15C) | 1688@10th | [126] |

_Continued_

**Table 12.1** The electrochemical performance of different sulfur-based composites in all-solid-state Li-S batteries with inorganic solid-state electrolytes—cont'd

| Active material | Conductive additives | Preparation method | Solid electrolytes | Conductivity $S\,cm^{-1}$ | Composition | Applied current density | Capacity and cycle | Refs. |
|---|---|---|---|---|---|---|---|---|
| Sulfur | Acetylene black | Ball milling | $Li_{3.25}Ge_{0.25}P_{0.75}S_4$ | $2\sim4\times10^{-3}$ at RT | S/C/SE = 25/25/50 | $0.013\,mA\,cm^{-2}$ | 900@10th | [13] |
| Sulfur | CMK-3 | Ball milling | $Li_{3.25}Ge_{0.25}P_{0.75}S_4$ | $2\sim4\times10^{-3}$ at RT | S/C/SE = 15/35/50 | $0.023\,mA\,cm^{-2}$ | 1000@50th | [127] |
| Sulfur | Carbon replica CR12 | Ball milling | $Li_{3.25}Ge_{0.25}P_{0.75}S_4$ | $2\sim4\times10^{-3}$ at RT | S/C/SE = 15/35/50 | $0.065\,mA\,cm^{-2}$ | 366@20th | [128] |
| Sulfur | Acetylene black | Ball milling | $Li_{3.25}Ge_{0.25}P_{0.75}S_4$ | $2.2\times10^{-3}$ at RT | S/C/SE = 25/25/50 | $83.75\,mA\,g^{-1}$ | 1000@20th | [129] |
| Sulfur | Acetylene black | Ball milling | $Li_{3.25}Ge_{0.25}P_{0.75}S_4$ | $2.2\times10^{-3}$ at RT | S/C/SE = 05/25/50 | $83.75\,mA\,g^{-1}$ | 1100@1st | [130] |
| Sulfur | CMK-3 | Gas phase | $Li_{3.25}Ge_{0.25}P_{0.75}S_4$ | $1.1\times10^{-3}$ at RT | S/CR/SE = 8/22/70 | $0.13\,mA\,cm^{-2}$ | 700@30th | [127] |
| Sulfur | Acetylene black | Ball milling | $P_2S_50.82(Li_{1.5}\,PS_{3.3})$ $0.18LiI$ | $3.1\times10^{-3}$ at RT | S/PS/C/SE = 50/10/ 10/30 | $1.3\,mA\,cm^{-2}$ | 1400@100th | [82] |
| Sulfur | Super P | Ball milling | $Li_6PS_5Br$ | $\sim10^{-3}$ at RT | S/SP/SE = 20/10/70 | $0.38\,mA\,cm^{-2}$ | 1080@50th | [131] |
| Sulfur | Carbonized PVP + carbon black | Ball milling | $Li_6PS_5Cl$ $+80Li_2S\cdot20P2S_5$ glass-ceramics | $4\times10^{-3}$ at RT $1.3\times10^{-3}$ at RT | $Li_6PS_5Cl$-$Li_2SC$/CB/80 $Li_2S\cdot20P_2S_5$ = 60/10/ 30 | $0.18\,mA\,cm^{-2}$ | 1190@60th | [13] |
| Sulfur | Super P | Ball milling | $Li_6PS_5Cl$ | $1.1\times10^{-3}$ at RT | S–C/SP/SE = 4/2/4 | $64\,\mu A\,cm^{-2}$ | 400@20th | [132] |
| Sulfur | NMCNT | Ball milling | $Li_6PS_5Cl$ | $3.15\times10^{-3}$ at RT | S/NWCNT/SE = 2.5/1.5/6 | $0.176\,mA\,cm^{-2}$ | 1393@50th | [71] |
| Sulfur | Carbon black | Ball milling | $Li_{10}SnP_2S_{12}$ | $3.2\times10^{-3}$ RT | S/C/SE = $\gamma$/0.5$\gamma$/ $(100-1.5\gamma)$ | $40\,mA\,g^{-1}$ | 1607@1st | [133] |
| Sulfur | PEDOT/PSS | Solution method | $Li_{10}GeP_2S_{12}$ | N/A | S/PEDOT + PSS/SE = 60/10/30 | $0.64\,mA\,cm^{-2}$ | 300@10th | [134] |
| Sulfur | Graphene | Coating method | $Li_{10}GeP_2S_{12}$@75Li_2S- 24P_2S_5-20P_2O_5 | $8.27\times10^{-3}$@$8\times10^{-4}$ at RT | rGO@S/SE/AB = 30/50/20 | 0.05 C | 1500@30th | [107] |
| Sulfur | CNT | Ball milling | $Li_{10}GeP_2S_{12}$@75Li_2S- 24P_2S_5-20P_2O_5 | $8.27\times10^{-3}$@$8\times10^{-4}$ at RT | CNT@S/AB/SE = 30/50/20 | 0.1 C | 1193@1st | [135] |
| Sulfur | Cu and acetylene black | Ball milling | $60Li_2S-40Si_2S_5$ | N/A | S/Cu/AB = 64/42.8/3.2 | $64\,\mu A\,cm^{-2}$ | 980 | [136] |
| Sulfur | Acetylene black | Ball milling | $Li_7P_{2.9}S_{10.85}Mo_{0.01}$ | $4.8\times10^{-3}$ at RT | S/AB/SE = 30/10/60 | 0.05 C | 1020@1st | [67] |
| Sulfur | Reduced graphene oxide (rGO) | Ball milling | $Li_{9.54}Si_{1.74}P_{1.44}S_{11.7}Cl_{0.3}$ | $1.6\times10^{-2}$ at RT | S/rGO/SE = 15/45/40 | $80\,mA\,g^{-1}$ | 827@60th | [137] |

| | | | | | | | | |
|---|---|---|---|---|---|---|---|---|
| Sulfur | Carbon black | Ball milling | $Li_7P_{2.9}Mn_{0.1}S_{10.7}I_{0.3}$ | $5.6 \times 10^{-3}$ at RT | S/C/SE = 45/15/40 | 0.05 C | 796@1st | [68] |
| Sulfur | Ketjen black | Hot pressing | $Li_7P_3S_{11}$ | $1.7 \times 10^{-3}$ at RT | S/C/SE = 50/30/20 | 0.1C | 1370 | [138] |
| $Li_2S$ | Cu | Ball milling | $80Li_2S$-$20P_2S_5$ | $\sim10^{-3}$ at 25°C | $Li_2S$/Cu/SE + AB = 38/57/5 | $64\,\mu A\,cm^{-2}$ | 340@20 | [139] |
| $Li_2S$ | Acetylene black | Ball milling | $80Li_2S$-$20P_2S_5$ glass-ceramics | $\sim10^{-3}$ at 25°C | $Li_2S$/C/SE = 25/25/50 | $1.3\,mA\,cm^{-2}$ | 850@200th | [106] |
| $Li_2S$ | Acetylene black | Ball milling | $80Li_2S$-$20P_2S_5$ | $\sim10^{-3}$ at RT | $Li_2S$/AB/SE = 25/25/50 | $64\,\mu A\,cm^{-2}$ | 700@10th | [117] |
| $Li_2S$ | Acetylene black | Ball milling | $75Li_2S$-$25P_2S_5$ | $7.2 \times 10^{-4}$ | $Li_2S$/AB/SE = 25/25/50 | $64\,\mu A\,cm^{-2}$ | 600@10 | [140] |
| $Li_2S$ | LiI | Ball milling | $75Li_2S$-$25P_2S_5$ | $\sim10^{-4}$ at RT | $80Li_2S$-$20LiX$/VGCF/SE = 50/10/40 | $64\,mA\,cm^{-2}$ | 930@50 | [141] |
| $Li_2S$ | LiI-VGCF | Ball milling | $75Li_2S$-$25P_2S_5$ | $\sim10^{-4}$ at RT | $Li_2S$-LiX/VGCF/SE = 50/10/40 | 2C | 900@2000 | [142] |
| $Li_2S$ | VGCF | Ball milling | $78Li_2S$-$22P_2S_5$ | $6.3 \times 10^{-4}$ at RT | $Li_2S$/C/SE = 1/0.4/2 | $300\,mAh\,g^{-1}$ | 469@1st | [143] |
| $Li_2S$ | VGCF + Super P | Ball milling | $70Li_2S$-$30P_2S_5$ | $1.58 \times 10^{-3}$ at RT | $Li_2S$/C/VGCF/SE = 30/6.7/3.3/60 | $0.044\,mA\,cm^{-2}$ | 1055@1st | [144] |
| $Li_2S$ | Polyvinylpyrrolidone | Solution method | $Li_6PS_5Cl$ | $4 \times 10^{-4}$ at RT | $Li_2S$/PVP/SE = 2/2/1 | $50\,mA\,g^{-1}$ | 830@60th | [145] |
| $Li_2S$ | Acetylene black | Ball milling | $Li_7P_3S_{11}$ | N/A | $Li_2S$@C/AB/SE = 3/1/3 | $2\,mA\,cm^{-2}$ | 1067 | [146] |
| $FeS_2$ | VGCF + super P | Ball milling | $78Li_2S$-$22P_2S_5$ | $1.78 \times 10^{-3}$ at RT | $FeS_2$/VGCF/SP/SE = 30/3.3/6.7/60 | $0.044\,mA\,cm^{-2}$ | 535@50th | [147] |
| $TiS_2$ | N/A | Ball milling | $Li_{10}GeP_2S_{12}$ | $1.2 \times 10^{-2}$ at RT | $TiS_2$/SE = 30/70 | 1 C | 160 | [148] |
| $TiS_2$ | N/A | Physical mixing | $Li_3PS_4$ and $Li_{10}GeP_2S_{12}$ | $1.0 \times 10^{-3}$ and $6.3 \times 10^{-3}$ | $TiS_2$/SE = 50/50 | 20 C | 60 | [149] |
| $CuS_2$ | Carbon black | Ball milling | $Li_6PS_5Br$ | $7.0 \times 10^{-5}$ | $CuS_2$/C/SE = 40/4/56 | $11.2\,mA\,g^{-1}$ | 250@1st | [150] |
| $MoS_2$ | Super P | Ball milling | $Li_6PS_5Br$ | N/A | $MoS_2$/C/SE = 40/4/56 | $67\,mA\,g^{-1}$ | 650@1st | [151] |
| $Co_2S_3$ | Super P | Physical mixing | $Li_{10}GeP_2S_{12}$ and $70\%$ $Li_2S$-$29\%P_2S_5$-$1\%$ $P_2O_5$ | $8.27 \times 10^{-3}$ and $2.04 \times 10^{-3}$ | $Co_2S_3$@$Li_7P_3S_{11}$/C/SE = 4/1/5 | $0.5\,A\,g^{-1}$ | 685.5 | [152] |
| $Li_3PS_{4+5}$ | Carbon black | Solution method | $\beta$-$Li_3PS_4$ | $3.0 \times 10^{-5}$ at RT | Active material/C/PVC = 60/30/10 | $0.015\,mA\,cm^{-2}$ | 1200@300th | [153] |

*Continued*

**Table 12.1** The electrochemical performance of different sulfur-based composites in all-solid-state Li-S batteries with inorganic solid-state electrolytes—cont'd

| Active material | Conductive additives | Preparation method | Solid electrolytes | Conductivity $S\,cm^{-1}$ | Composition | Applied current density | Capacity and cycle | Refs. |
|---|---|---|---|---|---|---|---|---|
| $Li_2S@Li_3PS_4$ | WVA-1500 | Wet method | $\beta$-$Li_3PS_4$ | $\sim 10^{-7}$ at RT | Active material/C/PVC = 65/35/10 | $0.02\,mA\,cm^{-2}$ | 852@100th | [12] |
| $Li_3PS_{4+5}$ | WVA-1500 | Ball milling | $Li_3PS_{4+5}$ | $3 \times 10^{-5}$ at RT | $Li_3PS_{4+5}/C = 2/1$ | $0.036\,mA\,cm^{-2}$ | 700@300th | [153] |
| $80Li_2S$-$20P_2S_5$ | Acetylene black | Ball milling | $80Li_2S$-$20P_2S_5$ | N/A | SSE/AB = 2/1 | $0.064\,mA\,cm^{-2}$ | 170@1st | [154] |
| $80Li_2S$-$20P_2S_5$-Cu | Cu | Ball milling | $80Li_2S$-$20P_2S_5$ | $\sim 10^{-3}$ at 25°C | $Li_2S/Cu = 48/52$ | $641\,\mu A\,cm^{-2}$ | 109@50 | [155] |
| $78Li_2S$-$22P_2S_5$ | Super P carbon/VGCF | Ball milling | $78Li_2S$-$22P_2S_5$ | $1.78 \times 10^{-3}$ at RT | C/VGCF/SE = 20/10/70 | $0.176\,mA\,cm^{-2}$ | 480@60 | [156] |
| $Li_6PS_5Cl$ | NWCNTs | Ball milling | $Li_6PS_5Cl$ | $3 \times 10^{-3}$ at RT | 30 wt.% MWCNTs in the $Li_6PS_5Cl$ | 0.1 C ($56\,mA\,g^{-1}$) | 535@560th | [135] |

*RT*, room temperature.

milled via physical grinding and ball milling, and they compared the outcomes of assembled SS-LiSBs performance [120]. Other researchers reported that the composite cathode consisted of the SSEs such as Li$_3$PS$_4$ [157] as an active material with conductive carbon, which were milled through high-energy ball milling (Fig. 12.12A). Yao et al. [107] demonstrated another approach for solving the interfacial problem in which deposition of a sulfur nanoparticle layer approximately 2 nm on the rGO to

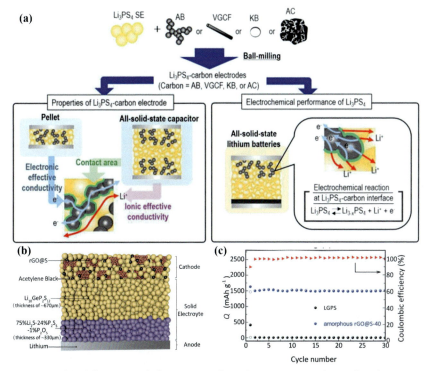

**Fig. 12.12** (A) Schematic of illustration of Li$_3$PS$_4$ composite electrodes that were synthesized via ball milling. (B) Schematic diagram of a trilayer all-solid-state Li-S battery. (C) Cycling performances of the amorphous rGO@S-40 composite by subtracting the Li$_{10}$GeP$_2$S$_{12}$ contribution. *((A) Reproduced with permission from T. Hakari, Y. Sato, S. Yoshimi, A. Hayashi, M. Tatsumisago, Favorable carbon conductive additives in Li$_3$PS$_4$ composite positive electrode prepared by ball-milling for all-solid-state lithium batteries, J. Electrochem. Soc. 164 (2017) A2804X, Copyright (2017) The Electrochemical Society; (B and C) Reproduced with permission from X. Yao, N. Huang, F. Han, Q. Zhang, H. Wan, J.P. Mwizerwa, C. Wang, X. Xu, High-performance all-solid-state lithium–sulfur batteries enabled by amorphous sulfur-coated reduced graphene oxide cathodes, Adv. Energy Mater. 7 (2017) 1602923, Copyright (2017) John Wiley & Sons.)*

achieve S/rGO hybrid to maintain good electronic conduction, and also then homogenously distribute this S/rGO nanocomposite into mixed conducting LGPS solid electrolyte-acetylene black to also attain high ionic conductivity of the cathode composite. The researcher also recommended that bilayer approach 1%$P_2O_5$-LGPS/75%$Li_2S$-24%$P_2S_5$ also can stop unfavorable reactions between the LGPS and lithium metal anode as shown in Fig. 12.12B and C. the SS–LiSBs of different cathode materials centered on sulfide-based electrolytes and their cyclic efficiency are emphasized in Table 12.1.

## 12.5.3 Cathode/solid polymer electrolytes interface

Solid polymer electrolytes are the leading electrolyte candidates for SS-LiSBs. The PEO with inorganic ceramics filler and different lithium salts has been used extensively as electrolyte with considerable conductivity and good mechanical properties. A composite of sulfur was used as cathode with PEO polymer and acetylene black carbon via mechanical milling at RT. The PEOs with inorganic filler $\gamma$-$LiAlO_2$ and lithium salts Li $(CF_3SO_2)_2N$ were also used as SPE for the solid battery showed average discharge capacity of 290 mAh g$^{-1}$ after 50th cycle [116]. A variety of sulfur-based cathode composites along with their compositions for SSLSBs, preparation methods and their cyclic performances, are summarized in Table 12.2.

## 12.6 All-solid-state thin-film batteries

All-solid-state thin-film batteries propose various benefits over traditional battery where a little and thin footprint is required, which is used for microchip power for portable medical devices, microsensor, and other wireless micro-electrochemical devices. All-solid-state thin-film battery constituents can be assembled by the different techniques, which include spattering and deposition. The solid-state materials used as cathodes and electrolytes require good flexibility for achieving high battery performance. The recent developments of the superionic thin-film solid electrolytes and high-energy thin-film cathodes demonstrate the outcoming prospects, which are fabulous for commercial application of thin-film solid-state batteries.

This battery architecture displays the intrinsic properties of solid-state batteries in which nonflammability and tolerance of temperature variations offer the significant benefits of a thin-film battery assembly, compared to a conventional cell, which may have a large volume but has a low-rate capability. Thin-film batteries with the standard footprint (100 µm × 100 µm)

**Table 12.2** The electrochemical performance of different sulfur-based composites in all-solid-state Li-S batteries with solid polymer electrolytes.

| Active material | Conductive additives | Preparation method | Solid polymer electrolytes | Conductivity S cm$^{-1}$ | Composition | Applied current density | Capacity and cycle | Refs. |
|---|---|---|---|---|---|---|---|---|
| **Solid polymer electrolytes-based Li-S batteries** | | | | | | | | |
| Sulfur | Ketjen black | Solvent casting | PEO + LiFSI | $\sim10^{-4}$ at 90°C | S/C/SPE = 40/15/45 | 0.1 C | 600@50th | [36] |
| Sulfur | Divinyl benzene | Polymerization | PEO + LiFSI | $6\times10^{-4}$ at 60–80°C | S/DVB = 80/20 | 0.1 C | 650@50th | [158] |
| Sulfur | Carbon black | Solution casting | PEO + LiTFSI | $1.5\times10^{-3}$ at RT | S/C/LiTFSI/PEO = 50/15/30/5 | 0.05 C | 270@10th | [159] |
| Sulfur | Ketjen black | Solution casting | PEO + LiFTFSI | $7\times10^{-4}$ at 70°C | S/C/LiX+PEO = 40/15/45 | 0.5 C | 800@60th | [35] |
| Sulfur | Black pearl | Milling device | 3% QA$^{+}$ Br$^{-}$ siloxane | $6.6\times10^{-4}$ at 30°C | S/C = 80/20 | 0.1 C / 167.5 mA g$^{-1}$ | 1082@1st | [160] |
| Sulfur | Acetylene black | Physical mixing | Composite gel electrolyte (u-CGE) | $2.1\times10^{-3}$ | S/AB/PVdF = 67.5/22.5/10 | 2 A g$^{-1}$ | 953@1st | [161] |
| Sulfur | OMC | Heating method | PEO + LiTFSI + 10wt.% SiO$_2$ | $5\times10^{-4}$ at 7°C | S-OMCs/AB/PEO = 60/20/20 | 0.1 C | 823@25th | [162] |
| Sulfur | Super P carbon | Ball milling (glass) | PEO + LTF + 10 mol% ZrO$_2$ | $10^{-3}$–$10^{-4}$ at 7°C | S-C/C/PVdF = 80/10/10 | 30 mA g$^{-1}$ | 500@30th | [163] |
| Sulfur | Acetylene black | Solution method | PVDF-based solid polymer | $\sim10^{-4}$ at 25°C | S/AB/PVdF = 80/10/10 | 0.15 C | 1160@1st | [164] |
| Sulfur | Super P | Solution method | Starch-based solid polymer | $3.39\times10^{-4}$ at 25°C | S/super P/binder = 70/20/10 | 0.1 C | 1442@1st | [165] |
| Sulfur | Polyacrylonitrile (PAN) | Solution method | Al$_2$O$_3$-coated PEO-LiTFSI | $3\times10^{-6}$ at RT | S/C/PAN = 50/21.42/28.5 | 0.2 C | 1090@1st | [166] |
| Sulfur | Acetylene black | Solution method | PEO-based IL@ZrO$_2$ | $2.32\times10^{-4}$ at 37°C | N-CNs-S/AB/PEO/ LiTFSI = 70/10/15/5 | N/A | $\sim800$@1st at 37°C | [167] |
| Sulfur | Carbon precursor | Solution method | PC-Li-Nafion | $2.1\times10^{-4}$ at 70°C | S/C = 4/6 | 0.05 C | 1072@1st at 70°C | [168] |
| Sulfur | Acetylene black | Solvent evaporation | P(EO)$_{20}$ Li (CF$_3$SO$_2$)2N-10wt.% γ-LiAlO$_2$ | $\sim10^{-4}$ at RT | S + PEO/AB/PVdF = 80/10/10 | 0.1 mA cm$^{-2}$ | 184@50th | [116] |
| Sulfur | Super carbon P | Solvent casting | PEO + LiTFSI + ZrO$_2$ | $1.11\times10^{-4}$ at 25°C | PANI@C+S/C/binder = 80/10/10 | 0.1 C | 745@100th | [169] |

*Continued*

**Table 12.2** The electrochemical performance of different sulfur-based composites in all-solid-state Li-S batteries with solid polymer electrolytes—cont'd

| Active material | Conductive additives | Preparation method | Solid polymer electrolytes | Conductivity $S\,cm^{-1}$ | Composition | Applied current density | Capacity and cycle | Refs. |
|---|---|---|---|---|---|---|---|---|
| Sulfur | Carbon black | Solvent evaporation | PEO + 10 wt.% montmorillonite | $3.22 \times 10^{-4}$ at 60°C | $S+PAN+Mg_{0.6}Ni_{0.4}O/$ AB/PEO/LiTFSI = 55/25/15/5 | 0.1 C | 634@100th at 60°C | [72] |
| Sulfur | Ketjen black | Solution casting | PEO + LiFSI | $9 \times 10^{-5}$ at 70°C | S/C/LiX + PEO = 50/15/35 | 0.1 C | 600@50 | [170] |
| Sulfur | Graphene | One-pot reaction | PEO + LiTFSI + MIL-53(Al) | N/A | GO-PEG@C/S/C/PAA = 70/20/10 | 0.2 C | 1060@100th | [171] |
| Sulfur | Super P | Solution casting | PEO + LiTFSI + MIL-53(Al) | N/A | S/PAA binder/C/SPE | 0.5 C | 793@50th | [172] |
| Sulfur | PANI@ super P | In situ polymerization | PEO + LiTFSI + Porous MOF (10 wt.% MIL-53(Al)) | N/A | PANI@C+S/C/binder = 90/10/10 | 0.5 C | 448@1000th | [173] |
| Sulfur | VGCF | Ball milling | $Li_2S-P_2S_5$ | $\sim 10^{-3}$ at RT | MWCNT-$HNO_3$/VGCF super P/SE = 20/10/10/60 | 0.176 mA $cm^{-2}$ 0.1 C | 720@100th | [174] |
| Sulfur | Ketjen black | Solution mixture | $Al_2O_3$-CPE + LICGC | $\sim 6 \times 10^{-3}$ at 90°C | S/C/PEO + LiFSI = 40/15/45 | 0.1 C | 512@50th | [175] |
| Sulfur | Activated carbon | Ball milling | PEO + $LiBF_4$ + 10 wt. % $Al_2O_3$ | $3 \times 10^{-4}$ at 80°C | S/C/PEO | 0.07 C | 40@10 | [176] |
| Sulfur | Super P | Solvent method | $Li_{6.5}La_{2.5}Ba_{0.5}ZrTaO_{12}$ | $1.6 \times 10^{-4}$ at 23°C | VCS/C/PVdF = 7.5/1.5/1 | 0.1 C | 936@1st | [177] |
| Sulfur | LLZO@carbon | Pechini sol-gel method | PEO + LiTFSI + LLZO | $1.1 \times 10^{-4}$ at 40°C | S-LLZO-C/C/ LLZO-PEO-$LiClO_4$ = 80/1/1 | 0.05 C | 800@200 | [32] |
| Sulfur | BP@2000 | Physical mixing | $Li_7P_3S_{11}$ | $2 \times 10^{-3}$ at RT | S/BP@2000/SE = 1.5/0.5/3 | 0.2 C | 1391@1st | [178] |
| Sulfur | Acetylene black | Solvent method | PEO-1%$Li_{10}SnP_2S_{11}$ | $1.69 \times 10^{-4}$ at 50°C | S/AB/LiTFSI + PEO = 40/15/45 | | 1000@1st | [179] |
| Sulfur | 3D flexible CNF | Solvent method | $Li_{0.33}La_{0.557}TiO_3$ (LLTO) nanofiber poly(ethylene oxide) (PEO) | $2.3 \times 10^{-4}$ at RT | | 0.05 C | 415@50th | [180] |
| Sulfur | Ketjen black | Solvent method | $Li_{1.5}Al_{0.5}Ge_{1.5}(PO_4)_3$-CPE- | $5 \times 10^{-4}$ at 20°C | S/KB = 60/40 | 0.1 C | 1080@150th | [181] |

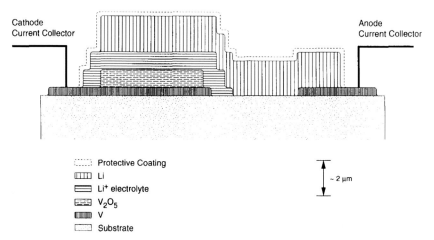

**Fig. 12.13** Schematic cross section of thin-film battery architecture. *(Reproduced with permission from J.B. Bates, G.R. Gruzalski, N.J. Dudney, C.F. Luck, X. Yu, Rechargeable thin-film lithium batteries, Solid State Ionics 70 (1994) 619–628, Copyright (1994) Elsevier.)*

are typically referred to as micro-batteries or thin-film microbatteries. Kanehori et al. explored the new solid type of battery with Li phosphosilicate electrolytes (Li$_{3.6}$Si$_{0.6}$P$_{0.4}$O$_4$) and TiS$_2$ as cathode, which exhibited an open-circuit voltage of 2.5 V [182]. Creus et al. [180] reported a smart thin-film battery, which consists of amorphous type of Li$_2$S-SiS$_2$-P$_2$S$_5$ solid electrolyte, V$_2$O$_2$-TeO$_2$ cathode, and lithium anode, which exhibited an open-circuit voltage of 3.1 V. Another research discovered that the poor cyclic performance of the assembled thin-film battery was caused by the chemical instability of lithium metal with electrolyte [183]. Bates and co-workers reported the rechargeable thin-film lithium battery consisting of inorganic amorphous solid electrolyte lithium phosphorous oxynitride (Li$_{3.3}$PO$_{3.8}$N$_{0.22}$) with 2 µS cm$^{-1}$ at 25 °C, cathode (TiS$_2$, V$_2$O$_5$ and Li$_x$Mn$_2$O$_4$) and lithium metal. The cross section of the assembled thin-film solid battery is shown in Fig. 12.13 [184]. The open-circuit voltage of fully charged thin-film batteries with cathode materials was 2.5 V, 3.7 V, and 4.2 V, respectively.

## 12.7 Conclusions

Although all-solid-state lithium-sulfur batteries based on solid electrolytes have shown significant potential for future energy storage devices, there is still a need for more research to explore the new superionic conductors and resolve the electrode-electrolyte interfacial resistance while ensuring safety issues. In this chapter, we have endeavored to take into account the latest developments in all-solid-state Li-S batteries. Due to their high energy content, they currently attract significant attention. The application of Li metal as an anode along with the development of novel superionic conductor solid electrolytes and a large stable electrochemical window, will be an additional impetus of the all-solid-state Li-S battery progress. In the meantime, potential solid electrolytes should demonstrate good chemical and electrochemical stabilities in the presence of Li anodes. Particle size regulation and homogeneity of both solid electrolyte and active material are also significant for attaining a satisfactory solid-solid interface. Enhancing the content of active mass in a composite cathode is a very successful approach for improving the power density and energy density of all-solid-state Li-S batteries.

## References

[1] C. Barchasz, F. Molton, C. Duboc, J.-C. Leprêtre, S. Patoux, F. Alloin, Lithium/sulfur cell discharge mechanism: an original approach for intermediate species identification, Anal. Chem. 84 (2012) 3973–3980.

[2] E. Climate, G., Project, Global Climate & Energy Project Technical Assessment Report, A Technical Assessment of High-Energy Batteries for Light-Duty Electric Vehicles, 2006.

[3] A. Manthiram, Y. Fu, Y.-S. Su, Challenges and prospects of lithium–sulfur batteries, Acc. Chem. Res. 46 (2013) 1125–1134.

[4] E. Peled, M. Goor, I. Schektman, T. Mukra, Y. Shoval, D. Golodnitsky, The effect of binders on the performance and degradation of the lithium/sulfur battery assembled in the discharged state, J. Electrochem. Soc. 164 (2016) A5001.

[5] R.C. Agrawal, G.P. Pandey, Solid polymer electrolytes: materials designing and all-solid-state battery applications: an overview, J. Phys. D Appl. Phys. 41 (2008), 223001.

[6] H. Danuta, U. Juliusz, Electric dry cells and storage batteries, United States Patent 3,043,896, June 10, 1962.

[7] D.A. Nole, V. Moss, Battery employing lithium-sulphur electrodes with non-aqueous electrolyte, United States Patent 3,532,543, October 6, 1970.

[8] E. Quartarone, P. Mustarelli, Electrolytes for solid-state lithium rechargeable batteries: recent advances and perspectives, Chem. Soc. Rev. 40 (2011) 2525–2540.

[9] H. Nagata, Y. Chikusa, Activation of sulfur active material in an all-solid-state lithium–sulfur battery, J. Power Sources 263 (2014) 141–144.

[10] S. Kinoshita, K. Okuda, N. Machida, M. Naito, T. Sigematsu, All-solid-state lithium battery with sulfur/carbon composites as positive electrode materials, Solid State Ionics 256 (2014) 97–102.

[11] S. Xiong, K. Xie, Y. Diao, X. Hong, Characterization of the solid electrolyte interphase on lithium anode for preventing the shuttle mechanism in lithium–sulfur batteries, J. Power Sources 246 (2014) 840–845.

[12] Z. Lin, Z. Liu, N.J. Dudney, C. Liang, Lithium superionic sulfide cathode for all-solid lithium–sulfur batteries, ACS Nano 7 (2013) 2829–2833.

[13] T. Kobayashi, Y. Imade, D. Shishihara, K. Homma, M. Nagao, R. Watanabe, T. Yokoi, A. Yamada, R. Kanno, T. Tatsumi, All solid-state battery with sulfur electrode and thio-LISICON electrolyte, J. Power Sources 182 (2008) 621–625.

[14] A. Manthiram, X. Yu, S. Wang, Lithium battery chemistries enabled by solid-state electrolytes, Nat. Rev. Mater. 2 (2017) 1–16.

[15] J. Janek, W.G. Zeier, A solid future for battery development, Energy 500 (2016) 300.

[16] K. Kerman, A. Luntz, V. Viswanathan, Y.-M. Chiang, Z. Chen, Practical challenges hindering the development of solid state Li ion batteries, J. Electrochem. Soc. 164 (2017) A1731.

[17] Y. Kato, S. Hori, T. Saito, K. Suzuki, M. Hirayama, A. Mitsui, M. Yonemura, H. Iba, R. Kanno, High-power all-solid-state batteries using sulfide superionic conductors, Nat. Energy 1 (2016) 16030.

[18] A. Mauger, M. Armand, C.M. Julien, K. Zaghib, Challenges and issues facing lithium metal for solid-state rechargeable batteries, J. Power Sources 353 (2017) 333–342.

[19] C. Wang, Y. Yang, X. Liu, H. Zhong, H. Xu, Z. Xu, H. Shao, F. Ding, Suppression of lithium dendrite formation by using LAGP-PEO (LiTFSI) composite solid electrolyte and lithium metal anode modified by PEO (LiTFSI) in all-solid-state lithium batteries, ACS Appl. Mater. Interfaces 9 (2017) 13694–13702.

[20] W. Zhou, S. Wang, Y. Li, S. Xin, A. Manthiram, J.B. Goodenough, Plating a dendrite-free lithium anode with a polymer/ceramic/polymer sandwich electrolyte, J. Am. Chem. Soc. 138 (2016) 9385–9388.

[21] X.-X. Zeng, Y.-X. Yin, N.-W. Li, W.-C. Du, Y.-G. Guo, L.-J. Wan, Reshaping lithium plating/stripping behavior via bifunctional polymer electrolyte for room-temperature solid Li metal batteries, J. Am. Chem. Soc. 138 (2016) 15825–15828.

[22] M.D. Tikekar, L.A. Archer, D.L. Koch, Stabilizing electrodeposition in elastic solid electrolytes containing immobilized anions, Sci. Adv. 2 (2016) e1600320.

[23] C.A. Vincent, Polymer electrolytes, Prog. Solid State Chem. 17 (1987) 145–261.

[24] Y. Aihara, S. Arai, K. Hayamizu, Ionic conductivity, DSC and self diffusion coefficients of lithium, anion, polymer, and solvent of polymer gel electrolytes: the structure of the gels and the diffusion mechanism of the ions, Electrochim. Acta 45 (2000) 1321–1326.

[25] X. Ollivrin, N. Farin, F. Alloin, J.F. Le Nest, J.Y. Sanchez, Physical properties of amorphous polyether networks, Electrochim. Acta 43 (1998) 1257–1262.

[26] H. Zhou, S. Dong, Determination of the standard heterogeneous rate constant in polyelectrolyte using steady state microdisk voltammograms, J. Electroanal. Chem. 425 (1997) 55–59.

[27] S.B. Aziz, T.J. Woo, M.F.Z. Kadir, H.M. Ahmed, A conceptual review on polymer electrolytes and ion transport models, J. Sci. Adv. Mater. Dev. 3 (2018) 1–17, https://doi.org/10.1016/j.jsamd.2018.01.002.

[28] Q. Li, J. Chen, L. Fan, X. Kong, Y. Lu, Progress in electrolytes for rechargeable Li-based batteries and beyond, Green Energy Environ. 1 (2016) 18–42.

[29] R. Khurana, J.L. Schaefer, L.A. Archer, G.W. Coates, Suppression of lithium dendrite growth using cross-linked polyethylene/poly (ethylene oxide) electrolytes: a new approach for practical lithium-metal polymer batteries, J. Am. Chem. Soc. 136 (2014) 7395–7402.

[30] J.H. Shin, K.W. Kim, H.J. Ahn, J.H. Ahn, Electrochemical properties and interfacial stability of $(PEO)_{10}LiCF_3SO_3-Ti_nO_{2n-1}$ composite polymer electrolytes for lithium/sulfur battery, Mater. Sci. Eng. B 95 (2002) 148–156.

[31] H. Zhang, C. Li, M. Piszcz, E. Coya, T. Rojo, L.M. Rodriguez-Martinez, M. Armand, Z. Zhou, Single lithium-ion conducting solid polymer electrolytes: advances and perspectives, Chem. Soc. Rev. 46 (2017) 797–815.

[32] X. Tao, Y. Liu, W. Liu, G. Zhou, J. Zhao, D. Lin, C. Zu, O. Sheng, W. Zhang, H.-W. Lee, Solid-state lithium–sulfur batteries operated at 37°C with composites

of nanostructured $Li_7La_3Zr_2O_{12}$/carbon foam and polymer, Nano Lett. 17 (2017) 2967–2972.

[33] M.D. Tikekar, L.A. Archer, D.L. Koch, Stability analysis of electrodeposition across a structured electrolyte with immobilized anions, J. Electrochem. Soc. 161 (2014) A847–A855.

[34] X. Judez, H. Zhang, C. Li, J.A. González-Marcos, Z. Zhou, M. Armand, L.M. Rodriguez-Martinez, Lithium bis (fluorosulfonyl) imide/poly (ethylene oxide) polymer electrolyte for all solid-state Li–S cell, J. Phys. Chem. Lett. 8 (2017) 1956–1960.

[35] G.G. Eshetu, X. Judez, C. Li, M. Martinez-Ibañez, I. Gracia, O. Bondarchuk, J. Carrasco, L.M. Rodriguez-Martinez, H. Zhang, M. Armand, Ultrahigh performance all solid-state lithium sulfur batteries: salt anion's chemistry-induced anomalous synergistic effect, J. Am. Chem. Soc. 140 (2018) 9921–9933.

[36] D. Liu, W. Zhu, Z. Feng, A. Guerfi, A. Vijh, K. Zaghib, Recent progress in sulfide-based solid electrolytes for Li–ion batteries, Mater. Sci. Eng. B 213 (2016) 169–176.

[37] J.C. Bachman, S. Muy, A. Grimaud, H.-H. Chang, N. Pour, S.F. Lux, O. Paschos, F. Maglia, S. Lupart, P. Lamp, Inorganic solid-state electrolytes for lithium batteries: mechanisms and properties governing ion conduction, Chem. Rev. 116 (2016) 140–162.

[38] C. Sun, J. Liu, Y. Gong, D.P. Wilkinson, J. Zhang, Recent advances in all-solid-state rechargeable lithium batteries, Nano Energy 33 (2017) 363–386.

[39] B. Zheng, J. Zhu, H. Wang, M. Feng, E. Umeshbabu, Y. Li, Q.-H. Wu, Y. Yang, Stabilizing $Li_{10}SnP_2S_{12}$/Li interface via an in situ formed solid electrolyte interphase layer, ACS Appl. Mater. Interfaces 10 (2018) 25473–25482.

[40] Z. Zhang, S. Chen, J. Yang, J. Wang, L. Yao, X. Yao, P. Cui, X. Xu, Interface re-engineering of $Li_{10}GeP_2S_{12}$ electrolyte and lithium anode for all-solid-state lithium batteries with ultralong cycle life, ACS Appl. Mater. Interfaces 10 (2018) 2556–2565.

[41] A.D. Robertson, A.R. West, A.G. Ritchie, Review of crystalline lithium-ion conductors suitable for high temperature battery applications, Solid State Ionics 104 (1997) 1–11.

[42] R.D. Shannon, B.E. Taylor, A.D. English, T. Berzins, New Li solid electrolytes, in: International Symposium on Solid Ionic and Ionic-Electronic Conductors, Elsevier, 1977, pp. 783–796.

[43] A.R. Rodger, J. Kuwano, A.R. West, Li + ion conducting $\gamma$ solid solutions in the systems $Li_4XO_4$-$Li_3YO_4$: X=Si, Ge, Ti; Y=P, As, V; $Li_4XO_4$-$LiZO_2$: Z=Al, Ga, Cr and $Li_4GeO_4$-$Li_2CaGeO_4$, Solid State Ionics 15 (1985) 185–198.

[44] N. Iminaka, Novel multivalent cation conducting ceramics and their application, J. Ceram. Soc. Jpn 113 (2005) 387–393.

[45] M. Itoh, Y. Inaguma, W.-H. Jung, L. Chen, T. Nakamura, High lithium ion conductivity in the perovskite-type compounds $Ln_{12}Li_{12}TiO_3$ (Ln= La, Pr, Nd, Sm), Solid State Ionics 70 (1994) 203–207.

[46] R. Murugan, V. Thangadurai, W. Weppner, Fast lithium ion conduction in garnet-type $Li_7La_3Zr_2O_{12}$, Angew. Chem. Int. Ed. 46 (2007) 7778–7781.

[47] H. Xie, J.A. Alonso, Y. Li, M.T. Fernández-Díaz, J.B. Goodenough, Lithium distribution in aluminum-free cubic $Li_7La_3Zr_2O_{12}$, Chem. Mater. 23 (2011) 3587–3589.

[48] Y. Li, J.-T. Han, C.-A. Wang, S.C. Vogel, H. Xie, M. Xu, J.B. Goodenough, Ionic distribution and conductivity in lithium garnet $Li_7La_3Zr_2O_{12}$, J. Power Sources 209 (2012) 278–281.

[49] H. Buschmann, J. Dölle, S. Berendts, A. Kuhn, P. Bottke, M. Wilkening, P. Heitjans, A. Senyshyn, H. Ehrenberg, A. Lotnyk, Structure and dynamics of the fast lithium ion conductor $Li_7La_3Zr_2O_{12}$, Phys. Chem. Chem. Phys. 13 (2011) 19378–19392.

[50] C.A. Geiger, E. Alekseev, B. Lazic, M. Fisch, T. Armbruster, R. Langner, M. Fechtelkord, N. Kim, T. Pettke, W. Weppner, Crystal chemistry and stability of $Li_7La_3Zr_2O_{12}$ garnet: a fast lithium-ion conductor, Inorg. Chem. 50 (2011) 1089–1097.

[51] A. Aguadero, F. Aguesse, C. Bernuy-López, W.W. Manalastas, J.M.L. del Amo, J.A. Kilner, Improvement of Transport Properties in Li-Conducting Ceramic Oxides, Meeting Abstracts, The Electrochemical Society, 2015, p. 503.

[52] S. Xu, D.W. McOwen, L. Zhang, G.T. Hitz, C. Wang, Z. Ma, C. Chen, W. Luo, J. Dai, Y. Kuang, All-in-one lithium-sulfur battery enabled by a porous-dense-porous garnet architecture, Energy Storage Mater. 15 (2018) 458–464.

[53] S. Wang, Y. Ding, G. Zhou, G. Yu, A. Manthiram, Durability of the $Li_{1+x}Ti_{2-x}Al_x(PO_4)_3$ solid electrolyte in lithium–sulfur batteries, ACS Energy Lett. 1 (2016) 1080–1085.

[54] M. Murayama, R. Kanno, M. Irie, S. Ito, T. Hata, N. Sonoyama, Y. Kawamoto, Synthesis of new lithium ionic conductor thio-LISICON—lithium silicon sulfides system, J. Solid State Chem. 168 (2002) 140–148.

[55] S. Kong, H. Deiseroth, J. Maier, V. Nickel, K. Weichert, C. Reiner, $Li_6PO_5Br$ and $Li_6PO_5Cl$: the first lithium-oxide-argyrodites, Z. Anorg. Allg. Chem. 636 (2010) 1920–1924.

[56] S. Boulineau, M. Courty, J.-M. Tarascon, V. Viallet, Mechanochemical synthesis of Li-argyrodite $Li_6PS_5X$ ($X = Cl$, Br, I) as sulfur-based solid electrolytes for all solid state batteries application, Solid State Ionics 221 (2012) 1–5.

[57] N. Kamaya, K. Homma, Y. Yamakawa, M. Hirayama, R. Kanno, M. Yonemura, T. Kamiyama, Y. Kato, S. Hama, K. Kawamoto, A lithium superionic conductor, Nat. Mater. 10 (2011) 682.

[58] P. Bron, S. Johansson, K. Zick, J. Schmedt auf der Günne, S. Dehnen, B. Roling, $Li_{10}SnP_2S_{12}$: an affordable lithium superionic conductor, J. Am. Chem. Soc. 135 (2013) 15694–15697.

[59] Y. Seino, T. Ota, K. Takada, A. Hayashi, M. Tatsumisago, A sulphide lithium super ion conductor is superior to liquid ion conductors for use in rechargeable batteries, Energy Environ. Sci. 7 (2014) 627–631.

[60] A. Weisbach, Argyrodit, ein neues Silbererz, 1886.

[61] A. Gagor, A. Pietraszko, D. Kaynts, Structural aspects of fast copper mobility in $Cu_6PS_5Cl$—the best solid electrolyte from $Cu_6PS_5X$ series, J. Solid State Chem. 181 (2008) 777–782.

[62] E. Gaudin, H.J. Deiseroth, T. Zaiß, The argyrodite $\gamma$-$Ag_9AlSe_6$: a non-metallic filled Laves phase, Z. Kristallogr. Cryst. Mater. 216 (2001) 39–44.

[63] H.J. Deiseroth, S.T. Kong, H. Eckert, J. Vannahme, C. Reiner, T. Zaiss, M. Schlosser, $Li_6PS_5X$: a class of crystalline Li-rich solids with an unusually high $Li^+$ mobility, Angew. Chem. Int. Ed. 120 (2008) 767.

[64] R.P. Rao, S. Adams, Studies of lithium argyrodite solid electrolytes for all-solid-state batteries, Phys. Status Solidi A 208 (2011) 1804–1807.

[65] M. Tachez, J.-P. Malugani, R. Mercier, G. Robert, Ionic conductivity of and phase transition in lithium thiophosphate $Li_3PS_4$, Solid State Ionics 14 (1984) 181–185.

[66] Z. Liu, W. Fu, E.A. Payzant, X. Yu, Z. Wu, N.J. Dudney, J. Kiggans, K. Hong, A.J. Rondinone, C. Liang, Anomalous high ionic conductivity of nanoporous $\beta$-$Li_3PS_4$, J. Am. Chem. Soc. 135 (2013) 975–978.

[67] R. Xu, X. Xia, X. Wang, Y. Xia, J. Tu, Tailored $Li_2S$–$P_2S_5$ glass-ceramic electrolyte by $MoS_2$ doping, possessing high ionic conductivity for all-solid-state lithium-sulfur batteries, J. Mater. Chem. A 5 (2017) 2829–2834.

[68] R.-C. Xu, X.H. Xia, S.H. Li, S.Z. Zhang, X.L. Wang, J.P. Tu, All-solid-state lithium-sulfur batteries based on a newly designed $Li_7P_{2.9}Mn_{0.1}S_{10.7}I_{0.3}$ superionic conductor, J. Mater. Chem. A 5 (2017) 6310–6317.

[69] Y. Li, B. Xu, H. Xu, H. Duan, X. Lü, S. Xin, W. Zhou, L. Xue, G. Fu, A. Manthiram, Hybrid polymer/garnet electrolyte with a small interfacial resistance for lithium-ion batteries, Angew. Chem. Int. Ed. 56 (2017) 753–756.

[70] Z. Wu, Z. Xie, A. Yoshida, X. An, Z. Wang, X. Hao, A. Abudula, G. Guan, Novel $SeS_2$ doped $Li_2S$-$P_2S_5$ solid electrolyte with high ionic conductivity for all-solid-state lithium sulfur batteries, Chem. Eng. J. 380 (2020), 122419.

[71] S. Wang, Y. Zhang, X. Zhang, T. Liu, Y.-H. Lin, Y. Shen, L. Li, C.-W. Nan, High-conductivity argyrodite $Li_6PS_5Cl$ solid electrolytes prepared via optimized sintering processes for all-solid-state lithium–sulfur batteries, ACS Appl. Mater. Interfaces 10 (2018) 42279–42285.

[72] Y. Zhang, Y. Zhao, D. Gosselink, P. Chen, Synthesis of poly (ethylene-oxide)/nanoclay solid polymer electrolyte for all solid-state lithium/sulfur battery, Ionics 21 (2015) 381–385.

[73] X. Cheng, R. Zhang, C. Zhao, F. Wei, J. Zhang, Q. Zhang, A review of solid electrolyte interphases on lithium metal anode, Adv. Sci. 3 (2016) 1500213.

[74] X.-B. Cheng, C. Yan, X. Chen, C. Guan, J.-Q. Huang, H.-J. Peng, R. Zhang, S.-T. Yang, Q. Zhang, Implantable solid electrolyte interphase in lithium-metal batteries, Chem 2 (2017) 258–270.

[75] Y. Guo, H. Li, T. Zhai, Reviving lithium–metal anodes for next-generation high-energy batteries, Adv. Mater. 29 (2017) 1700007.

[76] X. Cheng, T. Hou, R. Zhang, H. Peng, C. Zhao, J. Huang, Q. Zhang, Dendrite-free lithium deposition induced by uniformly distributed lithium ions for efficient lithium metal batteries, Adv. Mater. 28 (2016) 2888–2895.

[77] R. Zhang, X. Cheng, C. Zhao, H. Peng, J. Shi, J. Huang, J. Wang, F. Wei, Q. Zhang, Conductive nanostructured scaffolds render low local current density to inhibit lithium dendrite growth, Adv. Mater. 28 (2016) 2155–2162.

[78] C. Yang, K. Fu, Y. Zhang, E. Hitz, L. Hu, Protected lithium-metal anodes in batteries: from liquid to solid, Adv. Mater. 29 (2017) 1701169.

[79] B. Li, L. Kong, C. Zhao, Q. Jin, X. Chen, H. Peng, J. Qin, J. Chen, H. Yuan, Q. Zhang, Expediting redox kinetics of sulfur species by atomic-scale electrocatalysts in lithium–sulfur batteries, InfoMat 1 (2019) 533–541.

[80] M. Khurram Tufail, N. Ahmad, L. Zhou, M. Faheem, L. Yang, R. Chen, W. Yang, Insight on air-induced degradation mechanism of $Li_7P_3S_{11}$ to design a chemical-stable solid electrolyte with high $Li_2S$ utilization in all-solid-state Li/S batteries, Chem. Eng. J. 425 (2021), 130535.

[81] C.J. Wen, R.A. Huggins, Thermodynamic and mass transport properties of "LiIn", Mater. Res. Bull. 15 (1980) 1225–1234.

[82] H. Nagata, Y. Chikusa, An all-solid-state lithium–sulfur battery using two solid electrolytes having different functions, J. Power Sources 329 (2016) 268–272.

[83] X. Zhang, W. Wang, A. Wang, Y. Huang, K. Yuan, Z. Yu, J. Qiu, Y. Yang, Improved cycle stability and high security of Li-B alloy anode for lithium–sulfur battery, J. Mater. Chem. A 2 (2014) 11660–11665.

[84] A. Kato, A. Hayashi, M. Tatsumisago, Enhancing utilization of lithium metal electrodes in all-solid-state batteries by interface modification with gold thin films, J. Power Sources 309 (2016) 27–32.

[85] A. Kato, H. Kowada, M. Deguchi, C. Hotehama, A. Hayashi, M. Tatsumisago, XPS and SEM analysis between $Li/Li_3PS_4$ interface with Au thin film for all-solid-state lithium batteries, Solid State Ionics 322 (2018) 1–4.

[86] B. Wu, S. Wang, W.J. Evans IV, D.Z. Deng, J. Yang, J. Xiao, Interfacial behaviours between lithium ion conductors and electrode materials in various battery systems, J. Mater. Chem. A 4 (2016) 15266–15280.

[87] V. Thangadurai, W. Weppner, $Li_6ALa_2Ta_2O_{12}$ (A = Sr, Ba): novel garnet-like oxides for fast lithium ion conduction, Adv. Funct. Mater. 15 (2005) 107–112.

[88] M. Sakuma, K. Suzuki, M. Hirayama, R. Kanno, Reactions at the electrode/electrolyte interface of all-solid-state lithium batteries incorporating Li–M (M = Sn, Si) alloy electrodes and sulfide-based solid electrolytes, Solid State Ionics 285 (2016) 101–105.

[89] S. Wenzel, D.A. Weber, T. Leichtweiss, M.R. Busche, J. Sann, J. Janek, Interphase formation and degradation of charge transfer kinetics between a lithium metal anode and highly crystalline $Li_7P_3S_{11}$ solid electrolyte, Solid State Ionics 286 (2016) 24–33.

[90] S. Wenzel, S. Randau, T. Leichtweiß, D.A. Weber, J. Sann, W.G. Zeier, J. Janek, Direct observation of the interfacial instability of the fast ionic conductor $Li_{10}GeP_2S_{12}$ at the lithium metal anode, Chem. Mater. 28 (2016) 2400–2407.

[91] P. Hartmann, T. Leichtweiss, M.R. Busche, M. Schneider, M. Reich, J. Sann, P. Adelhelm, J. Janek, Degradation of NASICON-type materials in contact with lithium metal: formation of mixed conducting interphases (MCI) on solid electrolytes, J. Phys. Chem. C 117 (2013) 21064–21074.

[92] M. Nagao, A. Hayashi, M. Tatsumisago, T. Kanetsuku, T. Tsuda, S. Kuwabata, In situ SEM study of a lithium deposition and dissolution mechanism in a bulk-type solid-state cell with a $Li_2S–P_2S_5$ solid electrolyte, Phys. Chem. Chem. Phys. 15 (2013) 18600–18606.

[93] T. Yamada, S. Ito, R. Omoda, T. Watanabe, Y. Aihara, M. Agostini, U. Ulissi, J. Hassoun, B. Scrosati, All solid-state lithium–sulfur battery using a glass-type $P_2S_5$–$Li_2S$ electrolyte: benefits on anode kinetics, J. Electrochem. Soc. 162 (2015) A646.

[94] R. Sudo, Y. Nakata, K. Ishiguro, M. Matsui, A. Hirano, Y. Takeda, O. Yamamoto, N. Imanishi, Interface behavior between garnet-type lithium-conducting solid electrolyte and lithium metal, Solid State Ionics 262 (2014) 151–154.

[95] A. Sharafi, H.M. Meyer, J. Nanda, J. Wolfenstine, J. Sakamoto, Characterizing the $Li–Li_7La_3Zr_2O_{12}$ interface stability and kinetics as a function of temperature and current density, J. Power Sources 302 (2016) 135–139.

[96] W. Luo, Y. Gong, Y. Zhu, K.K. Fu, J. Dai, S.D. Lacey, C. Wang, B. Liu, X. Han, Y. Mo, Transition from superlithiophobicity to superlithiophilicity of garnet solid-state electrolyte, J. Am. Chem. Soc. 138 (2016) 12258–12262.

[97] X. Han, Y. Gong, K.K. Fu, X. He, G.T. Hitz, J. Dai, A. Pearse, B. Liu, H. Wang, G. Rubloff, Negating interfacial impedance in garnet-based solid-state Li metal batteries, Nat. Mater. 16 (2017) 572–579.

[98] K. Fu, Y. Gong, Z. Fu, H. Xie, Y. Yao, B. Liu, M. Carter, E. Wachsman, L. Hu, Transient behavior of the metal interface in lithium metal–garnet batteries, Angew. Chem. Int. Ed. 56 (2017) 14942–14947.

[99] W. Luo, Y. Gong, Y. Zhu, Y. Li, Y. Yao, Y. Zhang, K. Fu, G. Pastel, C. Lin, Y. Mo, Reducing interfacial resistance between garnet-structured solid-state electrolyte and Li-metal anode by a germanium layer, Adv. Mater. 29 (2017) 1606042.

[100] K.K. Fu, Y. Gong, B. Liu, Y. Zhu, S. Xu, Y. Yao, W. Luo, C. Wang, S.D. Lacey, J. Dai, Toward garnet electrolyte–based Li metal batteries: an ultrathin, highly effective, artificial solid-state electrolyte/metallic Li interface, Sci. Adv. 3 (2017) e1601659.

[101] Y. Shao, H. Wang, Z. Gong, D. Wang, B. Zheng, J. Zhu, Y. Lu, Y.-S. Hu, X. Guo, H. Li, Drawing a soft interface: an effective interfacial modification strategy for garnet-type solid-state Li batteries, ACS Energy Lett. 3 (2018) 1212–1218.

[102] J. Liu, T. Liu, Y. Pu, M. Guan, Z. Tang, F. Ding, Z. Xu, Y. Li, Facile synthesis of NASICON-type $Li_{1.3}Al_{0.3}Ti_{1.7}(PO_4)_3$ solid electrolyte and its application for

enhanced cyclic performance in lithium ion batteries through the introduction of an artificial $Li_3PO_4$ SEI layer, RSC Adv. 7 (2017) 46545–46552.

[103] K.K. Fu, Y. Gong, J. Dai, A. Gong, X. Han, Y. Yao, C. Wang, Y. Wang, Y. Chen, C. Yan, Flexible, solid-state, ion-conducting membrane with 3D garnet nanofiber networks for lithium batteries, Proc. Natl. Acad. Sci. 113 (2016) 7094–7099.

[104] T. Hou, X. Chen, H. Peng, J. Huang, B. Li, Q. Zhang, B. Li, Design principles for heteroatom-doped nanocarbon to achieve strong anchoring of polysulfides for lithium–sulfur batteries, Small 12 (2016) 3283–3291.

[105] L. Fan, H.L. Zhuang, K. Zhang, V.R. Cooper, Q. Li, Y. Lu, Chloride-reinforced carbon nanofiber host as effective polysulfide traps in lithium–sulfur batteries, Adv. Sci. 3 (2016) 1600175.

[106] M. Nagao, A. Hayashi, M. Tatsumisago, Sulfur–carbon composite electrode for all-solid-state Li/S battery with $Li_2S–P_2S_5$ solid electrolyte, Electrochim. Acta 56 (2011) 6055–6059.

[107] X. Yao, N. Huang, F. Han, Q. Zhang, H. Wan, J.P. Mwizerwa, C. Wang, X. Xu, High-performance all-solid-state lithium–sulfur batteries enabled by amorphous sulfur-coated reduced graphene oxide cathodes, Adv. Energy Mater. 7 (2017) 1602923.

[108] K. Fu, Y. Gong, S. Xu, Y. Zhu, Y. Li, J. Dai, C. Wang, B. Liu, G. Pastel, H. Xie, Stabilizing the garnet solid-electrolyte/polysulfide interface in Li–S batteries, Chem. Mater. 29 (2017) 8037–8041.

[109] M. Agostini, Y. Aihara, T. Yamada, B. Scrosati, J. Hassoun, A lithium–sulfur battery using a solid, glass-type $P_2S_5–Li_2S$ electrolyte, Solid State Ionics 244 (2013) 48–51.

[110] K. Suzuki, N. Mashimo, Y. Ikeda, T. Yokoi, M. Hirayama, R. Kanno, High cycle capability of all-solid-state lithium–sulfur batteries using composite electrodes by liquid-phase and mechanical mixing, ACS Appl. Energy Mater. 1 (2018) 2373–2377.

[111] C. Luo, X. Ji, J. Chen, K.J. Gaskell, X. He, Y. Liang, J. Jiang, C. Wang, Solid-state electrolyte anchored with a carboxylated azo compound for all-solid-state lithium batteries, Angew. Chem. Int. Ed. 57 (2018) 8567–8571.

[112] J. Yue, M. Yan, Y. Yin, Y. Guo, Progress of the interface design in all-solid-state Li–S batteries, Adv. Funct. Mater. 28 (2018) 1707533.

[113] K. Aso, A. Sakuda, A. Hayashi, M. Tatsumisago, All-solid-state lithium secondary batteries using NiS-carbon fiber composite electrodes coated with $Li_2S–P_2S_5$ solid electrolytes by pulsed laser deposition, ACS Appl. Mater. Interfaces 5 (2013) 686–690.

[114] X. Yu, A. Manthiram, Electrode–electrolyte interfaces in lithium–sulfur batteries with liquid or inorganic solid electrolytes, Acc. Chem. Res. 50 (2017) 2653–2660.

[115] Y. Hao, S. Wang, F. Xu, Y. Liu, N. Feng, P. He, H. Zhou, A design of solid-state Li–S cell with evaporated lithium anode to eliminate shuttle effects, ACS Appl. Mater. Interfaces 9 (2017) 33735–33739.

[116] X. Zhu, Z. Wen, Z. Gu, Z. Lin, Electrochemical characterization and performance improvement of lithium/sulfur polymer batteries, J. Power Sources 139 (2005) 269–273.

[117] M. Nagao, A. Hayashi, M. Tatsumisago, High-capacity $Li_2S$–nanocarbon composite electrode for all-solid-state rechargeable lithium batteries, J. Mater. Chem. 22 (2012) 10015–10020.

[118] A. Hayashi, T. Ohtomo, F. Mizuno, K. Tadanaga, M. Tatsumisago, All-solid-state Li/S batteries with highly conductive glass–ceramic electrolytes, Electrochem. Commun. 5 (2003) 701–705.

[119] M. Nagao, A. Hayashi, M. Tatsumisago, Electrochemical performance of all-solid-state Li/S batteries with sulfur-based composite electrodes prepared by mechanical milling at high temperature, Energy Technol. 1 (2013) 186–192.

[120] H.U. Choi, J.S. Jin, J.-Y. Park, H.-T. Lim, Performance improvement of all-solid-state Li-S batteries with optimizing morphology and structure of sulfur composite electrode, J. Alloys Compd. 723 (2017) 787–794.

[121] J.E. Trevey, J.R. Gilsdorf, C.R. Stoldt, S.-H. Lee, P. Liu, Electrochemical investigation of all-solid-state lithium batteries with a high capacity sulfur-based electrode, J. Electrochem. Soc. 159 (2012) A1019–A1022.

[122] S. Kinoshita, K. Okuda, N. Machida, T. Shigematsu, Additive effect of ionic liquids on the electrochemical property of a sulfur composite electrode for all-solid-state lithium–sulfur battery, J. Power Sources 269 (2014) 727–734.

[123] N. Tanibata, H. Tsukasaki, M. Deguchi, S. Mori, A. Hayashi, M. Tatsumisago, A novel discharge–charge mechanism of a S–$P_2S_5$ composite electrode without electrolytes in all-solid-state Li/S batteries, J. Mater. Chem. A 5 (2017) 11224–11228.

[124] U. Ulissi, S. Ito, S.M. Hosseini, A. Varzi, Y. Aihara, S. Passerini, High capacity all-solid-state lithium batteries enabled by pyrite-sulfur composites, Adv. Energy Mater. 8 (2018) 1801462.

[125] S.M. Hosseini, A. Varzi, S. Ito, Y. Aihara, S. Passerini, High loading CuS-based cathodes for all-solid-state lithium sulfur batteries with enhanced volumetric capacity, Energy Storage Mater. 27 (2020) 61–68.

[126] Y. Aihara, S. Ito, R. Omoda, T. Yamada, S. Fujiki, T. Watanabe, Y. Park, S. Doo, The electrochemical characteristics and applicability of an amorphous sulfide-based solid ion conductor for the next-generation solid-state lithium secondary batteries, Front. Energy Res. 4 (2016) 18.

[127] M. Nagao, Y. Imade, H. Narisawa, T. Kobayashi, R. Watanabe, T. Yokoi, T. Tatsumi, R. Kanno, All-solid-state Li–sulfur batteries with mesoporous electrode and thio-LISICON solid electrolyte, J. Power Sources 222 (2013) 237–242.

[128] M. Nagao, K. Suzuki, Y. Imade, M. Tateishi, R. Watanabe, T. Yokoi, M. Hirayama, T. Tatsumi, R. Kanno, All-solid-state lithium–sulfur batteries with three-dimensional mesoporous electrode structures, J. Power Sources 330 (2016) 120–126.

[129] K. Suzuki, D. Kato, K. Hara, T. Yano, M. Hirayama, M. Hara, R. Kanno, Composite sulfur electrode prepared by high-temperature mechanical milling for use in an all-solid-state lithium–sulfur battery with a $Li_{3.25}Ge_{0.25}P_{0.75}S_4$ electrolyte, Electrochim. Acta 258 (2017) 110–115.

[130] K. Suzuki, D. Kato, K. Hara, T. Yano, M. Hirayama, M. Hara, R. Kanno, Composite sulfur electrode for all-solid-state lithium–sulfur battery with $Li_2S–GeS_2–P_2S_5$-based thio-LISICON solid electrolyte, Electrochemistry (2017) 17–55.

[131] M. Chen, S. Adams, High performance all-solid-state lithium/sulfur batteries using lithium argyrodite electrolyte, J. Solid State Electrochem. 19 (2015) 697–702.

[132] C. Yu, L. van Eijck, S. Ganapathy, M. Wagemaker, Synthesis, structure and electrochemical performance of the argyrodite $Li_6PS_5Cl$ solid electrolyte for Li-ion solid state batteries, Electrochim. Acta 215 (2016) 93–99.

[133] J. Yi, L. Chen, Y. Liu, H. Geng, L.-Z. Fan, High capacity and superior cyclic performances of all-solid-state lithium–sulfur batteries enabled by a high-conductivity $Li_{10}SnP_2S_{12}$ solid electrolyte, ACS Appl. Mater. Interfaces 11 (2019) 36774–36781.

[134] H. Nagata, Y. Chikusa, All-solid-state lithium–sulfur batteries using a conductive composite containing activated carbon and electroconductive polymers, Chem. Lett. 43 (2014) 1335–1336.

[135] S. Wang, X. Xu, X. Zhang, C. Xin, B. Xu, L. Li, Y.-H. Lin, Y. Shen, B. Li, C.-W. Nan, High-performance $Li_6PS_5Cl$-based all-solid-state lithium-ion batteries, J. Mater. Chem. A 7 (2019) 18612–18618.

[136] N. Machida, K. Kobayashi, Y. Nishikawa, T. Shigematsu, Electrochemical properties of sulfur as cathode materials in a solid-state lithium battery with inorganic solid electrolytes, Solid State Ionics 175 (2004) 247–250.

[137] R. Xu, Z. Wu, S. Zhang, X. Wang, Y. Xia, X. Xia, X. Huang, J. Tu, Construction of all-solid-state batteries based on a sulfur-graphene composite and $Li_{9.54}Si_{1.74}P_{1.44}S_{11.7}Cl_{0.3}$ solid electrolyte, Chem. Eur. J. 23 (2017) 13950–13956.

[138] M.R. Busche, D.A. Weber, Y. Schneider, C. Dietrich, S. Wenzel, T. Leichtweiss, D. Schröder, W. Zhang, H. Weigand, D. Walter, In situ monitoring of fast Li-ion conductor $Li_7P_3S_{11}$ crystallization inside a hot-press setup, Chem. Mater. 28 (2016) 6152–6165.

[139] A. Hayashi, R. Ohtsubo, T. Ohtomo, F. Mizuno, M. Tatsumisago, All-solid-state rechargeable lithium batteries with $Li_2S$ as a positive electrode material, J. Power Sources 183 (2008) 422–426.

[140] M. Nagao, A. Hayashi, M. Tatsumisago, T. Ichinose, T. Ozaki, Y. Togawa, S. Mori, $Li_2S$ nanocomposites underlying high-capacity and cycling stability in all-solid-state lithium–sulfur batteries, J. Power Sources 274 (2015) 471–476.

[141] T. Hakari, A. Hayashi, M. Tatsumisago, Highly utilized lithium sulfide active material by enhancing conductivity in all-solid-state batteries, Chem. Lett. 44 (2015) 1664–1666.

[142] T. Hakari, A. Hayashi, M. Tatsumisago, $Li_2S$-based solid solutions as positive electrodes with full utilization and superlong cycle life in all-solid-state Li/S batteries, Adv. Sustain. Syst. 1 (2017) 1700017.

[143] M. Eom, S. Son, C. Park, S. Noh, W.T. Nichols, D. Shin, High performance all-solid-state lithium–sulfur battery using a $Li_2S$-VGCF nanocomposite, Electrochim. Acta 230 (2017) 279–284.

[144] Y. Zhang, K. Chen, Y. Shen, Y. Lin, C.-W. Nan, Synergistic effect of processing and composition x on conductivity of $xLi_2S$-$(100-x)P_2S_5$ electrolytes, Solid State Ionics 305 (2017) 1–6.

[145] F. Han, J. Yue, X. Fan, T. Gao, C. Luo, Z. Ma, L. Suo, C. Wang, High-performance all-solid-state lithium–sulfur battery enabled by a mixed-conductive $Li_2S$ nanocomposite, Nano Lett. 16 (2016) 4521–4527.

[146] H. Yan, H. Wang, D. Wang, X. Li, Z. Gong, Y. Yang, In situ generated $Li_2S$–C nanocomposite for high-capacity and long-life all-solid-state lithium sulfur batteries with ultrahigh areal mass loading, Nano Lett. 19 (2019) 3280–3287.

[147] Y. Zhang, R. Chen, T. Liu, Y. Shen, Y. Lin, C.-W. Nan, High capacity, superior cyclic performances in all-solid-state lithium-ion batteries based on $78Li_2S$-$22P_2S_5$ - glass–ceramic electrolytes prepared via simple heat treatment, ACS Appl. Mater. Interfaces 9 (2017) 28542–28548.

[148] W.J. Li, M. Hirayama, K. Suzuki, R. Kanno, Fabrication and all solid-state battery performance of $TiS_2/Li_{10}GeP_2S_{12}$ composite electrodes, Mater. Trans. 57 (2016) 549–552.

[149] B.R. Shin, Y.J. Nam, D.Y. Oh, D.H. Kim, J.W. Kim, Y.S. Jung, Comparative study of $TiS_2/Li$-In all-solid-state lithium batteries using glass–ceramic $Li_3PS_4$ and $Li_{10}GeP_2S_{12}$ solid electrolytes, Electrochim. Acta 146 (2014) 395–402.

[150] M. Chen, R.P. Rao, S. Adams, The unusual role of $Li_6PS_5Br$ in all-solid-state CuS/$Li_6PS_5Br$/In–Li batteries, Solid State Ionics 268 (2014) 300–304.

[151] M. Chen, X. Yin, M.V. Reddy, S. Adams, All-solid-state MoS 2/Li 6 PS 5 Br/In–Li batteries as a novel type of Li/S battery, J. Mater. Chem. A 3 (2015) 10698–10702.

[152] J. Shi, G. Liu, W. Weng, L. Cai, Q. Zhang, J. Wu, X. Xu, X. Yao, $Co_3S_4@Li_7P_3S_{11}$ hexagonal platelets as cathodes with superior interfacial contact for all-solid-state lithium batteries, ACS Appl. Mater. Interfaces 12 (2020) 14079–14086.

[153] Z. Lin, Z. Liu, W. Fu, N.J. Dudney, C. Liang, Lithium polysulfidophosphates: A family of lithium-conducting sulfur-rich compounds for lithium–sulfur batteries, Angew. Chem. Int. Ed. 52 (2013) 7460–7463.

[154] T. Hakari, M. Nagao, A. Hayashi, M. Tatsumisago, Preparation of composite electrode with $Li_2S–P_2S_5$ glasses as active materials for all-solid-state lithium secondary batteries, Solid State Ionics 262 (2014) 147–150.

[155] A. Hayashi, R. Ohtsubo, M. Nagao, M. Tatsumisago, Characterization of Li2S–P2S5–Cu composite electrode for all-solid-state lithium secondary batteries, J. Mater. Sci. 45 (2010) 377–381.

[156] Y. Zhang, R. Chen, T. Liu, B. Xu, X. Zhang, L. Li, Y. Lin, C.-W. Nan, Y. Shen, High capacity and superior cyclic performances of all-solid-state lithium batteries enabled by a glass–ceramics solo, ACS Appl. Mater. Interfaces 10 (2018) 10029–10035.

[157] T. Hakari, Y. Sato, S. Yoshimi, A. Hayashi, M. Tatsumisago, Favorable carbon conductive additives in $Li_3PS_4$ composite positive electrode prepared by ball-milling for all-solid-state lithium batteries, J. Electrochem. Soc. 164 (2017) A2804.

[158] I. Gracia, H. Ben Youcef, X. Judez, U. Oteo, H. Zhang, C. Li, L.M. Rodriguez-Martinez, M. Armand, S-containing copolymer as cathode material in poly (ethylene oxide)-based all-solid-state Li-S batteries, J. Power Sources 390 (2018) 148–152.

[159] D. Marmorstein, T.H. Yu, K.A. Striebel, F.R. McLarnon, J. Hou, E.J. Cairns, Electrochemical performance of lithium/sulfur cells with three different polymer electrolytes, J. Power Sources 89 (2000) 219–226.

[160] J.-H. Hong, J.W. Kim, S. Kumar, B. Kim, J. Jang, H.-J. Kim, J. Lee, J.-S. Lee, Solid polymer electrolytes from double-comb poly (methylhydrosiloxane) based on quaternary ammonium moiety-containing crosslinking system for Li/S battery, J. Power Sources 450 (2020), 227690.

[161] C. Ding, L. Huang, Y. Guo, Y. Yu, X. Fu, W.-H. Zhong, X. Yang, An ultra-durable gel electrolyte stabilizing ion deposition and trapping polysulfides for lithium-sulfur batteries, Energy Storage Mater. 27 (2020) 25–34.

[162] Y.-Z. Sun, J.-Q. Huang, C.-Z. Zhao, Q. Zhang, A review of solid electrolytes for safe lithium-sulfur batteries, Sci. China Chem. 60 (2017) 1508–1526.

[163] J. Hassoun, B. Scrosati, Moving to a solid-state configuration: a valid approach to making lithium-sulfur batteries viable for practical applications, Adv. Mater. 22 (2010) 5198–5201.

[164] C. Long, L. Li, M. Zhai, Y. Shan, Facile preparation and electrochemistry performance of quasi solid-state polymer lithium–sulfur battery with high-safety and weak shuttle effect, J. Phys. Chem. Solids 134 (2019) 255–261.

[165] Y. Lin, J. Li, K. Liu, Y. Liu, J. Liu, X. Wang, Unique starch polymer electrolyte for high capacity all-solid-state lithium sulfur battery, Green Chem. 18 (2016) 3796–3803.

[166] Z. Fan, B. Ding, T. Zhang, Q. Lin, V. Malgras, J. Wang, H. Dou, X. Zhang, Y. Yamauchi, Solid/solid interfacial architecturing of solid polymer electrolyte–based all-solid-state lithium–sulfur batteries by atomic layer deposition, Small 15 (2019) 1903952.

[167] O. Sheng, C. Jin, J. Luo, H. Yuan, C. Fang, H. Huang, Y. Gan, J. Zhang, Y. Xia, C. Liang, Ionic conductivity promotion of polymer electrolyte with ionic liquid grafted oxides for all-solid-state lithium–sulfur batteries, J. Mater. Chem. A 5 (2017) 12934–12942.

[168] J. Gao, C. Sun, L. Xu, J. Chen, C. Wang, D. Guo, H. Chen, Lithiated Nafion as polymer electrolyte for solid-state lithium sulfur batteries using carbon-sulfur composite cathode, J. Power Sources 382 (2018) 179–189.

[169] Y. Lin, X. Wang, J. Liu, J.D. Miller, Natural halloysite nano-clay electrolyte for advanced all-solid-state lithium-sulfur batteries, Nano Energy 31 (2017) 478–485.

[170] H. Zhang, C. Liu, L. Zheng, F. Xu, W. Feng, H. Li, X. Huang, M. Armand, J. Nie, Z. Zhou, Lithium bis (fluorosulfonyl) imide/poly (ethylene oxide) polymer electrolyte, Electrochim. Acta 133 (2014) 529–538.

[171] C. Zhang, Y. Lin, Y. Zhu, Z. Zhang, J. Liu, Improved lithium-ion and electrically conductive sulfur cathode for all-solid-state lithium–sulfur batteries, RSC Adv. 7 (2017) 19231–19236.

[172] Y. Zhu, J. Li, J. Liu, A bifunctional ion–electron conducting interlayer for high energy density all-solid-state lithium-sulfur battery, J. Power Sources 351 (2017) 17–25.

[173] C. Zhang, Y. Lin, J. Liu, Sulfur double locked by a macro-structural cathode and a solid polymer electrolyte for lithium–sulfur batteries, J. Mater. Chem. A 3 (2015) 10760–10766.

[174] C. Zhou, S. Bag, T. He, B. Lv, V. Thangadurai, A 20°C operating high capacity solid-state Li-S battery with an engineered carbon support cathode structure, Appl. Mater. Today 19 (2020), 100585.

[175] X. Judez, H. Zhang, C. Li, G.G. Eshetu, Y. Zhang, J.A. González-Marcos, M. Armand, L.M. Rodriguez-Martinez, Polymer-rich composite electrolytes for all-solid-state Li–S cells, J. Phys. Chem. Lett. 8 (2017) 3473–3477.

[176] S.S. Jeong, Y.T. Lim, Y.J. Choi, G.B. Cho, K.W. Kim, H.J. Ahn, K.K. Cho, Electrochemical properties of lithium sulfur cells using PEO polymer electrolytes prepared under three different mixing conditions, J. Power Sources 174 (2007) 745–750.

[177] S. Bag, C. Zhou, P.J. Kim, V.G. Pol, V. Thangadurai, LiF modified stable flexible PVDF-garnet hybrid electrolyte for high performance all-solid-state Li–S batteries, Energy Storage Mater. 24 (2020) 198–207.

[178] Q. Han, X. Li, X. Shi, H. Zhang, D. Song, F. Ding, L. Zhang, Outstanding cycle stability and rate capabilities of the all-solid-state Li–S battery with a $Li_7P_3S_{11}$ glass-ceramic electrolyte and a core–shell S@BP2000 nanocomposite, J. Mater. Chem. A 7 (2019) 3895–3902.

[179] X. Li, D. Wang, H. Wang, H. Yan, Z. Gong, Y. Yang, Poly (ethylene oxide)–$Li_{10}SnP_2S_{12}$ composite polymer electrolyte enables high-performance all-solid-state lithium sulfur battery, ACS Appl. Mater. Interfaces 11 (2019) 22745–22753.

[180] P. Zhu, C. Yan, J. Zhu, J. Zang, Y. Li, H. Jia, X. Dong, Z. Du, C. Zhang, N. Wu, Flexible electrolyte-cathode bilayer framework with stabilized interface for room-temperature all-solid-state lithium-sulfur batteries, Energy Storage Mater. 17 (2019) 220–225.

[181] W. Li, Q. Wang, J. Jin, Y. Li, M. Wu, Z. Wen, Constructing dual interfacial modification by synergetic electronic and ionic conductors: toward high-performance LAGP-based Li-S batteries, Energy Storage Mater. 23 (2019) 299–305.

[182] K. Kanehori, Y. Ito, F. Kirino, K. Miyauchi, T. Kudo, Titanium disulfide films fabricated by plasma CVD, Solid State Ionics 18 (1986) 818–822.

[183] R. Creus, J. Sarradin, R. Astier, A. Pradel, M. Ribes, The use of ionic and mixed conductive glasses in microbatteries, Mater. Sci. Eng. B 3 (1989) 109–112.

[184] J.B. Bates, G.R. Gruzalski, N.J. Dudney, C.F. Luck, X. Yu, Rechargeable thin-film lithium batteries, Solid State Ionics 70 (1994) 619–628.

PART V

# Applications: System-level issues and challenging environments

# CHAPTER 13

# State estimation methodologies for lithium-sulfur battery management systems

**Faten Ayadi, Daniel J. Auger, Abbas Fotouhi, and Neda Shateri**
School of Aerospace, Transport and Manufacturing, Cranfield University, Bedford, United Kingdom

## Contents

| | |
|---|---|
| **13.1** Introduction | 491 |
| **13.2** Lithium-sulfur battery models | 493 |
|     **13.2.1** Li-S battery electrochemical models | 493 |
|     **13.2.2** Li-S battery equivalent circuit network models | 496 |
| **13.3** Li-S BMS: State estimation methods | 497 |
|     **13.3.1** Weakness of direct methods for Li-S SoC estimation | 497 |
|     **13.3.2** Indirect methods based on control theory and computer science for Li-S SoC estimation | 500 |
| **13.4** Performance of state estimation methods | 507 |
|     **13.4.1** Li-S battery testing | 507 |
|     **13.4.2** Estimation results analysis for recursive Bayesian filters | 508 |
|     **13.4.3** Estimation results analysis for computer science techniques | 511 |
| **13.5** Conclusions and outlook | 525 |
| Acknowledgments | 527 |
| References | 527 |

## 13.1 Introduction

Much recent and current research on the vehicle electrification process has focused on the development of electrical energy storage systems [1–3]. For electric vehicles, the present consensus seems to be that lithium batteries are among the most promising devices in terms of energy storage [4]; indeed, the lithium-ion battery (LIB) is almost ubiquitous in electric vehicles today [5]. At the present time, the driving range or journey distance of gasoline or diesel-powered cars is still much better than that of a LIB. Thus, there is a strong reason to explore new technologies with the potential to achieve much higher energy densities such as lithium–sulfur batteries, particularly

*Lithium-Sulfur Batteries*
https://doi.org/10.1016/B978-0-12-819676-2.00006-2

Copyright © 2022 Elsevier Inc.
All rights reserved.   **491**

in scenarios where weight (rather than volume) is important [6]. In applications where weight is critical, a mature the lithium-sulfur battery is potentially attractive as a replacement for today's LIBs thanks to its high theoretical specific capacity 1675 mAh g$^{-1}$ and energy density (2500 Wh kg$^{-1}$) [6]. While it may not be possible to achieve the theoretical specific capacity for practical reasons, this figure is noticeably higher than that of lithium-ion which is 387 Wh kg$^{-1}$ [7,8].

Lithium-sulfur (Li-S) batteries need suitable care in EV applications, as they are potentially vulnerable to damage due to overcharging or discharging [9]. Depending on the application context, these may well cause damage, degraded function, and possibly even compromise safety [10]. Therefore, the battery management system (BMS) is an essential control element for lithium-sulfur cell operation in EVs [11]. The implementation of BMS in EVs has attracted significant interest from researchers in EV applications. For lithium-ion batteries, there are many proposed methods for state estimation which can be easily implemented in a BMS to ensure a good operation of the cell in EV application, and they are well described in the literature [4,12,13]. However, lithium-sulfur has its distinct characteristics that mean not all "general" lithium BMS techniques are applicable, and special techniques may be needed [6]. This chapter describes the techniques that are particularly relevant to lithium-sulfur batteries.

Fig. 13.1 defines the functional modules in the BMS for a lithium-sulfur EV containing battery state estimation and battery charging control. As shown in Fig. 13.1, Li-S cell charging can be ensured by a simple charge operation control [14]. As cited in [14], this technique can significantly increase the cycle life of Li-S batteries under long-term and repeated cycling conditions. Another key function of a BMS is ensuring the safety of battery operation. If unusual situations or battery states occur in operation, the

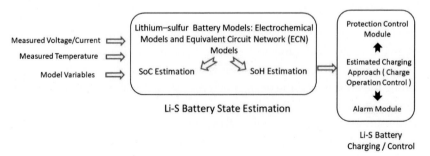

**Fig. 13.1** Functional modules in lithium-sulfur BMS for EVs.

protection control and alarm modules operate to record or remove these dangerous situations.

A critical task for lithium-sulfur BMS, which is state estimation. In lithium-sulfur, state estimation is a challenging issue for several reasons [15]:

**(a)** The absence of direct measurement methods for battery states.

**(b)** The presence of complex electrochemical reactions inside the battery.

**(c)** The change of battery characteristics over its lifetime.

This chapter surveys the different methods of estimating state of charge (SoC) in Li-S batteries from the literature. First, the key aspects of Li-S batteries that are a particular challenge and the functional modules in Li-S BMS are identified. Then, this chapter presents several approaches to this problem and discusses their experimental evaluations.

## 13.2 Lithium-sulfur battery models

Many studies have been conducted to model Li–ion batteries. Although different modeling techniques have been developed, some have been found particularly suitable for state estimation [16,17]. In comparison with Li–ion batteries, models for Li-S battery suffer from greater complexity due to the variety of chemical reactions which take place inside the battery [18]. In fact, the chemical mechanisms of the discharge/charge for Li-S battery are not fully known due to its characteristic features like the polysulfide shuttle phenomenon [18,19]. In the literature for Li-S to date, electrochemical models and equivalent circuit models have been described. Ongoing work seeks to develop these and other Li-S models further, particularly those which can be easily implemented within a BMS.

### 13.2.1 Li-S battery electrochemical models

Electrochemical models attempt to describe the inner battery reactions and use this to predict behavior and performance. They can be considered the most "precise" models thanks to their chemical basis [20]. However, these models are complex, because they need a wide number of physical and chemical variables, which are almost always difficult to acquire. Moreover, they require a large amount of computational power to run. For Li-S batteries, the complex reduction of sulfur presents a challenge, and there are many considerations on making a detailed mechanistic model that is being explored by current research [21]. For BMS implementations, to simplify the complex reactions of Li-S cells, researchers have used a simplified two-step Li-S model of two-step reaction consisting of a "high plateau"

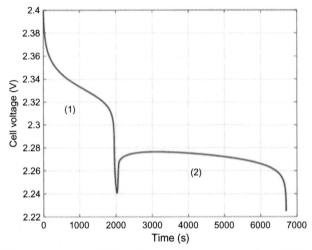

**Fig. 13.2** Voltage response of Li-S battery and two-step reaction: (1) instantaneous ohmic part and (2) dynamic part.

at high states of charge with a significant concentration of high-order polysulfides, and then a "low plateau" at lower states of charge with different reactions dominating in each region. Such a model can be related to the measurable characteristics of the battery, as shown in Fig. 13.2 [20], which gives an example of a Li-S battery voltage's response to a step current discharge of 1.7 A magnitude.

In Ref. [22], an example of electrochemical model for lithium-sulfur battery during charge and discharge as a "zero-dimensional model" is given. This model is "zero dimensional" in that there is no spatial dimension. All spatial variations are neglected. Such models predict the electrochemical dynamics of the Li-S batteries through Nernst formulations. These models can represent a battery's equilibrium voltage, its dynamic behavior with current, the polysulfide shuttle, and precipitation (Fig. 13.3 [20,23]). As shown in Fig. 13.3, in the voltage curve, there are two voltage plateaus with different properties:
- The "high" voltage plateau: The solid state sulfur $S_8(s)$ melts the liquid electrolyte and forms the liquid state sulfur $S_8(l)$. Then, $S_8(l)$ is reduced to soluble species $S_4^{2-}$.
- The "low" voltage plateau: The soluble species $S_4^{2-}$ are reduced to insoluble lithium sulfide $Li_2S$.

Applications of such models are of potential interest for Li-S state estimation, though the authors are not aware of any successful attempt to deploy such techniques with real Li-S cells.

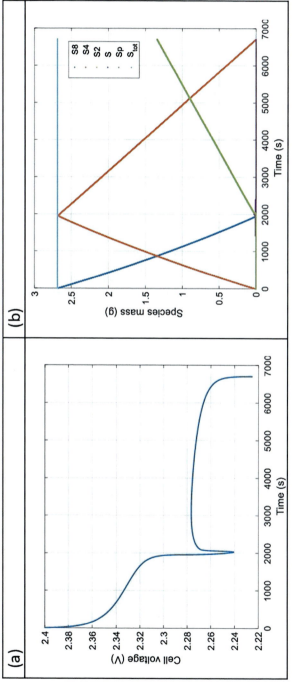

**Fig. 13.3** (A) Li-S battery voltage's response to a step current discharge of 1.7 A magnitude predicted by mechanistic model. (B) The following species are portrayed: S8 (teal blue), S4 (brown), S2 (green), and Stot(sub) (sky blue), S (purple), and Sp(sub) lie on the x-axis and are difficult to discern.

## 13.2.2 Li-S battery equivalent circuit network models

For BMS designers, battery modeling is a challenge because of the chemistry's complex dynamic behavior. Modeling for BMS applications aims to identify suitable models which can be implemented in software and used for accurate real-time computations. Thus, researchers have identified the equivalent circuit network (ECN) modeling as a useful approach that gives a good balance between complexity and reliability especially in automotive application. Primarily, the ECN modeling approach has been applied for SoC and state-of-health (SoH) estimation. In such applications, the relationship between the variable of interest (e.g., SoC) and battery model's local equivalent circuit network parameters is identified and used [17,20,24].

Following a pattern tried and tested for Li-ion cells, a good balance between and reliability has been found in the Thevenin model shown in Fig. 13.4 [20,23]. These models represent dynamic behavior over time with a simple set of electrical components. From Fig. 13.4, the region A represents the open-circuit voltage $V_A$. It designates the voltage between the electrodes of the battery when no current is drawn from or applied to the battery. Furthermore, the region B represents the effect of "ohmic" internal resistance in opposition to the movement of current in the battery. The region C represents dynamic effects (traditionally identified as "polarization"), modeled by a parallel $RC$ circuit [25]. In the literature of state identification, equivalent circuit network models are the most commonly used topology for modeling Li-S batteries, and techniques for identification of parameters are available [26,27].

A comparison between a Li-S cell's terminal voltage predicted from an ECN model and experimental measurement during a pulse test is described in Refs. [20, 28]. From Ref. [28], there is a good match between the model

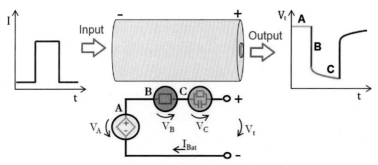

**Fig. 13.4** Input-output relationship of the ECN model.

and the experimental test data, which demonstrate that the terminal voltage can be well described by state-dependent model variables which account for variation through the discharge process. The literature contains published ECN models for Li-S batteries identified during discharge at multiple temperature points. These models need adaptation for use on a BMS. The suitability of a battery model for SoC estimation can be related to the dynamic system's observability [24]. It can be shown that in such models, state of charge is not fully observable from open-circuit voltage measurements alone due to the flat shape of battery's open-circuit voltage curve, a feature that differentiates Li-S batteries from many other types of battery. Thus, the estimation of a Li-S cell's SoC can be challenging. This topic is discussed in more detail in the following sections.

## 13.3 Li-S BMS: State estimation methods

State estimation is a key function of a BMS and especially for state-of-charge (SoC) and state-of-health (SoH). Formal definitions are available in the literature [20]. For lithium-ion batteries, SoC estimation of lithium-ion batteries through BMS is possible for any of several techniques. For lithium-sulfur batteries, their SoC estimation presents a difficult task for BMS due to the complexity of chemical mechanisms [29]. However, with appropriate techniques, SoC estimation is possible, and Fig. 13.5 shows the techniques that have been considered.

### 13.3.1 Weakness of direct methods for Li-S SoC estimation

For many Li-ion batteries, it is easy to manage a battery using direct methods such as measurement of open-circuit voltage (OCV) and coulomb counting

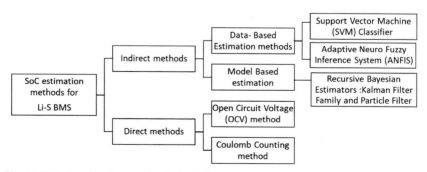

**Fig. 13.5** SoC estimation methods for Li-S BMS.

[30]. As well as being simple, these techniques are widely understood and have been extensively applied in industry [20]. In the next subsections, the application(s) of these techniques are explained; unfortunately, there are significant limitations of these methods for Li–S SoC estimation due to the complexity of electrochemical reactions which happen inside the Li–S batteries and the variation of remaining usable capacity. This leads to the necessity of more powerful techniques, which will be presented later in the paper.

### 13.3.1.1 Coulomb counting method

This method is the most widespread conventional method applied as a benchmark for interpretation of other methods. The principle of this technique is based on the integral of the current over time of battery discharge to estimate the SoC. It is determined by Eq. (13.1), if the initial value of the state of charge $x_0$ is known at time zero [20]:

$$x(t) = x_0 + \frac{1}{Q} \int_0^t \left( I_{\text{Load}}(t) - I_{\text{Loss}}(t) \right) d\tau \tag{13.1}$$

where $x(t)$ is the present state of charge, $Q$ is the capacity (ampere–second), $I_{\text{Load}}(t)$ is the instantaneous external load current (amperes), and $I_{\text{Loss}}(t)$ is the instantaneous loss current.

This technique has limitations in practical applications due to the vulnerability to errors caused by initial SoC estimate errors and accumulated measurement errors [20]. For Li–S SoC estimation, the coulomb counting method cannot be relied on to provide accurate results because of uncertainties in short- and long-term capacity arising from the variation in reaction pathways in Li–S chemistry and high rates of self-discharge.

### 13.3.1.2 Open-circuit voltage method

The OCV method is another popular method for SoC estimation of the battery. Its principle consists of evaluating the relationship between OCV and SoC by using lookup tables or polynomials when the cell is in equilibrium with a zero current. The relationship between OCV and SoC can be defined as follows (Eq. 13.2), for all states of charge $SoC \in [0, 1]$ [20]:

$$V_{oc} = h_{oc}(SoC) \tag{13.2}$$

where $h_{oc}(SoC)$ is a strictly monotonic function of SoC—in other words, from Eq. (13.3):

$$\frac{dh_{oc}}{dSoC} \succ 0 \text{ for all } SoC \in [0, 1] \tag{13.3}$$

If this is true, then $h_{oc}$ is invertible on [0, 1], and given a measured OCV value $\hat{V}_{oc}$, the estimated $\hat{SoC}$ is determined by Eq. (13.4):

$$\hat{SoC} = h_{oc}^{-1}(\hat{V}_{oc}) \tag{13.4}$$

A significant benefit of this method is the easiest determination of the SoC without the need for initial conditions or long measurements [20]. This technique is often successful for several battery types, but it is not facile when used for Li-S cells. In fact, this technique fails to predict the SoC of Li-S battery due to the presence of large flat region in the OCV-SoC curve, depicted in Fig. 13.6 [24].

Fig. 13.6 shows that the curve of OCV-SoC contains two different parts, referred to as the "high plateau" and the "low plateau." In the high plateau, there is significant presence of high-order polysulfide species like $Li_2S_8$. In this range, it is possible to get good OCV measurements as the strict monotonicity condition of Eq. (13.3) holds in this part. In the low plateau at lower SoC values, the flat shape makes the state of charge in the system unobservable; the strict monotonicity condition is not observed. Thus, the control of Li-S cell presents a very challenging task and the OCV method fails to be effective for the SoC of Li-S batteries.

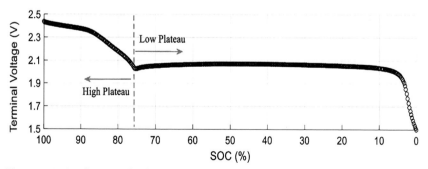

**Fig. 13.6** Li-S cell terminal voltage during discharge at C/30—two separate regions are observable: high plateau and low plateau.

## 13.3.2 Indirect methods based on control theory and computer science for Li-S SoC estimation

Research into efficient methods for Li-S battery SoC and SoH estimation is ongoing. According to the literature, so far two groups of methods are developed for Li-S state estimation: those based on control theory and those based on computer science [20]. In the following subsections, both methods are explained in more detail.

### 13.3.2.1 Li-S state estimation methods based on control theory

For SoC estimation in extremely dynamic areas, model-based estimation is applied. It consists of a predetermined off-line model, which is incorporated in an observer which is then used for optimization to estimate the states through an algorithm from the errors between the measured observations and their model-based predictions. This makes use of well-known formulations from control theory and estimates states in real time from measurements of current and voltage. Fig. 13.7 describes the principle of the model-based estimation for the battery [20,23].

From Fig. 13.7, the predicted voltage of the observer $\hat{V}(k)$ is higher than the measured terminal voltage of the battery $V(k)$. To redress this, the SoC needs to be modified to arrange the estimation. Such techniques are common for automotive SoC estimation in general and are widespread even outside the Li-S domain [24,31,32]. The most common algorithms are recursive Bayesian filters [33,34]. In this category, a suitable and accurate battery model is required to ensure good estimation of the SoC. Such a model will incorporate the relationship between model parameters and SoC.

When used in such algorithms, the ECN models can provide clear information of the batteries' internal states with low computational effort and simple measurements of current and terminal voltage. Examples of such algorithms are the extended and unscented Kalman filters and Particle Filter. They have been applied to Li-S batteries for the first time by Propp et al. in [28]. The principles of these estimation algorithms are presented in Figs. 13.8 and 13.9 [28].

Such methods are well-suited to the particular problem at hand. Due to the complexity of chemical reactions for Li-S cell, there is an uncertainty about internal dynamics and models are imperfect, but this is explicitly accounted for in the filter algorithms. Thus, recursive Bayesian filters are considered a robust technique even in the presence of mildly uncertain

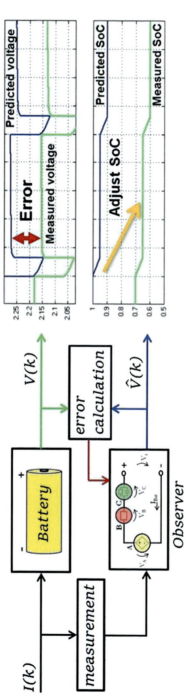

**Fig. 13.7** Principle of model-based estimation.

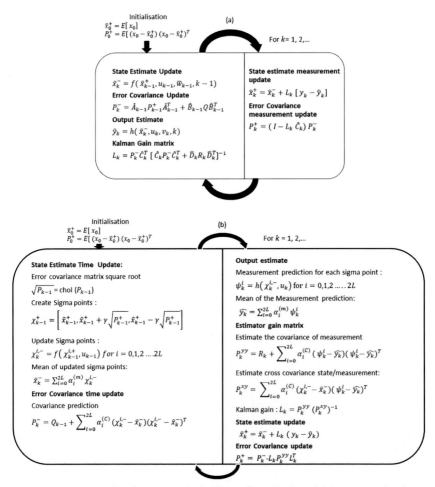

**Fig. 13.8** The principle of: (A) extended Kalman filter (EKF) and (B) unscented Kalman filter (UKF).

dynamics and noise [35]. These methods deal with the dynamic model states $x$ and the observations $y$ as stochastic parameters with affiliated probability density functions [36], with details as per Figs. 13.8 and 13.9.

From Ref. [28], for SoC estimation, these filters are combined with the following equivalent circuit model network, which is defined by Eqs. (13.5), (13.6) as follows:

$$\dot{x}(t) = A(t)x(t) + B(t)u(t)$$
$$y(t) = C(t)x(t) + D(t)u(t) \tag{13.5}$$

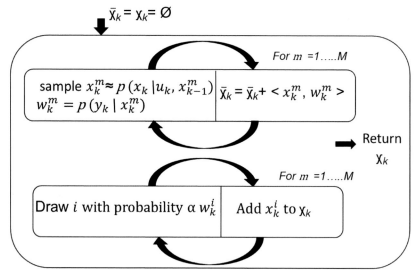

**Fig. 13.9** The principle of particle filter.

where $x = [x_1 \; x_2]^T$ is the dynamic states of the system, $x_1$ the voltage of the RC circuit, and $x_2$: SoC (X), determined by coulomb counting method.

$$A = \begin{pmatrix} \dfrac{-1}{f_{R_p}(X) f_{C_p}(X)} & 0 \\ 0 & 0 \end{pmatrix} \quad B = \begin{bmatrix} \dfrac{1}{f_{C_p}(X)} \\ \dfrac{-1}{3600 Q_{cap}} \end{bmatrix} \quad (13.6)$$

$$C = [-1 \; f_{OCV}(X)] \qquad D = [f_{R_0}(X)]$$

The functions $f_{OCV}$ and $f_{R_0}$ are the polynomials of open-circuit voltage (OCV) and internal resistance over SoC accordingly. The following Eqs. (13.7), (13.8) are expressed in terms of partial sinusoidal function $\gamma_{m,c}$ [28]:

$$f_{OCV}(X) = (1 - \gamma_{m,c}(X)) f_{OCV-\text{low}}(X) + \gamma_{m,c}(X) f_{OCV-\text{high}}(X) \quad (13.7)$$

$$f_{R_0}(X) = (1 - \gamma_{m,c}(X)) f_{R_0-\text{low}}(X) + \gamma_{m,c}(X) f_{R_0-\text{high}}(X) \quad (13.8)$$

The applicability of the recursive filters to Li-S batteries aims to estimate their state of charge.

## 504 Lithium-sulfur batteries

### Extended Kalman filter for SoC estimation

In combination with the Li–S battery model, there is a necessity to determine the Jacobian of the matrices $A$ and $C$ that are defined for convenience by Eqs. (13.9)–(13.11) [28]:

$$\hat{A}(1,1) = \frac{-1}{f_{C_p}(x_2)f_{R_p}(x_2)} \quad \hat{A}(2,1) = 0 \quad \hat{A}(2,2) = 0 \tag{13.9}$$

$$\hat{A}(1,2) = \left[ \frac{\dot{f}_{C_p}(x_2)}{f_{C_p}(x_2)^2 f_{R_p}(x_2)} + \frac{\dot{f}_{R_p}(x_2)}{f_{C_p}(x_2)f_{R_p}(x_2)^2} \right] x_1 - \left[ \frac{\dot{f}_{C_p}(x_2)}{f_{C_p}(x_2)^2} \right] I_{\text{Bat}} \tag{13.10}$$

$$\hat{C}(1,1) = -1 \quad \hat{C}(1,2) = \dot{f}_{OCV}(x_2) - \dot{f}_{R_0}(x_2) I_{\text{Bat}} \tag{13.11}$$

where SoC is defined as the second state of the model $x_2$. The derivatives of $f_{OCV}, f_{R0}, f_{Rp},$ and $f_{Cp}$ are defined in [28]; for extended Kalman filter (EKF), the values of the initial error covariance $P_0$ and system noise $Q$ are expressed by Eq. (13.12) as follows [28]:

$$P_{0_{EKF}} = \begin{pmatrix} 10 & 0 \\ 0 & 10 \end{pmatrix} \quad Q_{EKF} = \begin{pmatrix} 0.1 & 0 \\ 0 & 0.0000003 \end{pmatrix} \tag{13.12}$$

The system noise values $Q$ were results of consecutive iterations, which had been performed at EKF level to provide good results, and the values applied in the measurement noise matrix $R$ are equal to 0.15 [28].

### Unscented Kalman filter for SoC estimation

SoC estimation is performed under considered conditions during UKF operation [28]:

(a) The weights $\alpha_i^{(m)}$ and $\alpha_i^{(c)}$ are vectors including real constant scalars with the following condition that $\sum_{i=0}^{p} \alpha_i^{(m)}$ and $\sum_{i=0}^{p} \alpha_i^{(c)}$ are equal to 1 [37] and they are expressed through the following expression (Eq. 13.13):

$$\alpha_i^{(m)} = \alpha_i^{(c)} = \frac{1}{2(L+\lambda)}, \lambda \text{ the scaling value } \left(\lambda = \alpha^2(L+k) - L\right) \tag{13.13}$$

The initial weights are defined by Eq. (13.14) as follows:

$$\alpha_0^{(m)} = \frac{\lambda}{L+\lambda}, \alpha_0^{(c)} = \frac{\lambda}{L+\lambda} + \left(1 - \alpha^2 + \beta\right) \tag{13.14}$$

**(b)** The values of initial error covariance $P_0$ are determined via Eq. (13.15) as follows:

$$P_{0_{UKF}} = \begin{pmatrix} 1 & 0 \\ 0 & 0.014 \end{pmatrix} \tag{13.15}$$

**(c)** The values of system noise $Q$ for UKF are calculated by Eq. (13.16):

$$Q_{UKF} = \begin{pmatrix} 0.0005 & 0 \\ 0 & 0.0000007 \end{pmatrix} \tag{13.16}$$

**(d)** The values considered in the measurement noise matrix $R$ are equal to 0.3.

### Particle filter for SoC estimation

According to Ref. [28], using the Particle Filter, the SoC estimation for Li-S battery is achieved under following conditions:

**(a)** The Gaussian probability density function for SoC estimation is considered during the Particle Filter operation. It is defined by Eq. (13.17) as follows:

$$f(x) = \frac{1}{\sigma\sqrt{2\pi}} e^{-\frac{(x-\mu)^2}{2\sigma^2}} \tag{13.17}$$

**(b)** The standard deviations to sample the model states in the estimation step are selected in the same pattern as EKF and UKF to account for the uncertainties of the model and coulomb counting. The standard deviation values of each state of the model are defined by Eq. (13.18) as follows:

$$Std_{x_1} = 0.004 \; Std_{x_2} = 0.0003 \tag{13.18}$$

The performance of these recursive filters at Li-S cell SoC estimation level has been calibrated against pulse–discharge tests and validated against tests based on the New European Driving Cycle (NEDC) [28]; this is discussed further in Section 13.4.

### 13.3.2.2 Nonmodel Li-S state computer science-based estimation techniques

Due to the limitations of Li-S cell's SoC observability [37], a Li-S cell's SoC estimation is difficult compared to many other types of the battery. One approach that has been explored is based on quick online battery parameter identification coupled with a system to relate the identified parameters to state of charge. The literature explores the use of tools such as an adaptive neuro-fuzzy inference system (ANFIS) [24] or a support vector machine (SVM) classifier [38].

An ANFIS is a kind of artificial neural network based on a Takagi-Sugeno fuzzy inference system. The principle, the structure, and the used algorithms for training ANFIS are detailed in Ref. [20]. Fig. 13.10 shows an ANFIS Li-S battery SoC estimator [38]. In that method, battery terminal voltage "$V_t$" and battery current "$I$" are used for battery model parameter identification in real time. For calibration, it is possible to use an off-line process such as an online method forgetting factor recursive least squares (FFRLS) [38]. After model parameterization, real-time SoC estimation is performed by ANFIS through a mapping function "$g$" (Eq. 13.19) which is a function of battery parameters [24]:

$$SoC = g(P_1, P_2, \ldots P_n) \qquad (13.19)$$

where $P_n = f(SoC)$ $i = 1\ldots n$; $\{P_1, P_2, \ldots P_n\}$: identified cell parameters.

The efficiency of the ANFIS battery SoC estimator has been validated under two types of experimental tests, which are pulse discharge test and Urban Dynamometer Driving Schedule (UDDS) drive cycle test as discussed in Ref. [24]. The estimation results are detailed in the next section. More details about battery state estimation using ANFIS can be found in Refs. [24, 38].

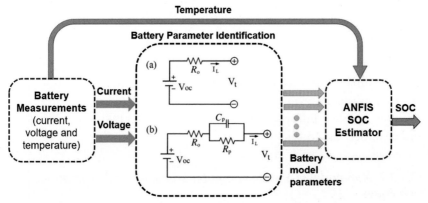

**Fig. 13.10** Equivalent circuit battery models: (a) internal resistance model, and (b) Thevenin model (1RC Model) and online battery measurement, identification, and SoC estimation.

An alternative approach to ANFIS is the support vector machine (SVM) Classifier. The SVM is a set of supervised learning techniques used for classification. Its approach is described in Ref. [39]. This technique has some similarities to the ANFIS one in that it uses online parameter estimates, but an SVM method is used instead of ANFIS for state estimation. More information about SVM battery SoC estimator is defined in Ref. [38]. The performance of SVM classifier in SoC estimation has been demonstrated at pulse test and Millbrook London Transport Bus (MLTB) drive cycle test [38]. Its state judgment results are explained in the following section.

## 13.4 Performance of state estimation methods
### 13.4.1 Li-S battery testing

This section presents the performance of Li-S BMS algorithms for SoC estimation in experimental environment. The considered prototype Li-S cells, which have been investigated here, had been manufactured by OXIS Energy Ltd. [40]. The tested prototype cells have had 3.4 Ah and 19 Ah nominal capacities. More details of the small cell are described in Refs. [20, 38].

Fig. 13.11 shows the test facilities used in the literature for Li-S battery testing [20,24,28]. As illustrated in the figure, a PC is used to program

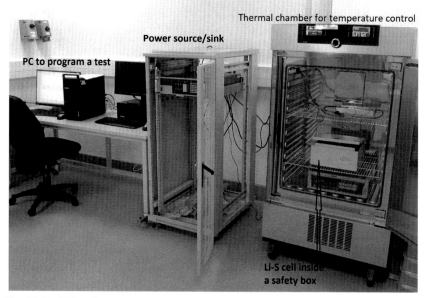

**Fig. 13.11** Li-S cell test equipment: PC to program a test, power source/sink, and Li-S cell inside the thermal chamber.

508 Lithium-sulfur batteries

charge/discharge test profiles, a power source/sink device is used to physically apply the current profiles to the cell/battery, and a thermal chamber is to control the temperature during the tests. For battery modeling purposes, the current signal is considered as the "input" whereas the terminal voltage is assumed as the "output." Both of those signals are recorded with a minimum frequency of 1 Hz during all tests. Two types of tests were conducted: (i) pulse test and (ii) drive cycling test. In the following sections, experimental evaluation of Li-S BMS algorithms is presented for both groups of SoC estimation techniques, which were discussed before.

## 13.4.2 Estimation results analysis for recursive Bayesian filters

This section presents the estimation results of the recursive Bayesian filters with mixed pulse and the New European Driving Cycle (NEDC) current profile during the discharge phase of the tested battery [28]. Here, a 3.4 Ah long-life chemistry pouch cell from OXIS was tested with current pulses of 290, 1450, and 2900 mA at 20°C (Fig. 13.12A). The test hardware included a Maccor 4000 battery tester with cells constantly held at temperature in sealed aluminum boxes inside a Binder KB53 thermal chamber. In addition, the tested cell was discharged with a current profile based on the New European Driving Cycle (NEDC) (Fig. 13.12B). The test hardware used in this case was a Kepco BOP100-10MG programmable power source/sink discharging a battery at room temperature (23°C). These discharge experiments are explained in Ref. [28].

For this work, a Thevenin model was used. The identification procedure was detailed in Ref. [28]. Recursive Bayesian filters for SoC estimation were employed, as described in Section 13.3.2.1. The performance of these filters was interpreted qualitatively at the level of their convergence time with imprecise initial states and quantitatively by their prediction accuracy through the root-mean-squared error (RMSE). Its Eq. (13.20) is defined as follows [28]:

$$\text{RMSE} = \frac{1}{\sqrt{n}} \left( \sum_{i=1}^{n} \left( SoC_{t,i} - So\hat{C}_{t,i} \right)^2 \right)^{0.5} \tag{13.20}$$

where $n$ is the number of data points, $SoC_{t,i}$ is the reference $SoC$ from the measurement, and $So\hat{C}_{t,i}$ is the estimated SoC by recursive Bayesian filters.

State estimation methodologies for Li-S BMSs  509

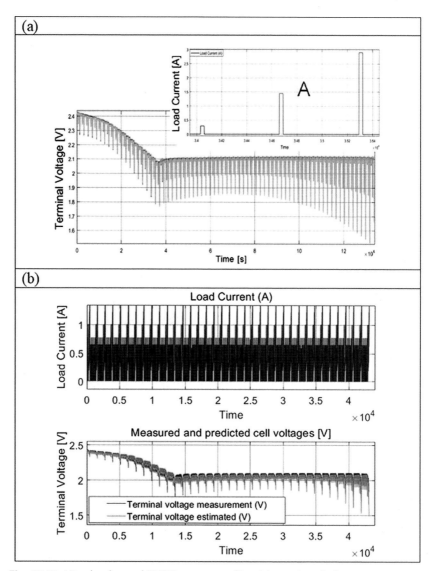

**Fig. 13.12** Mixed pulse and NEDC current profile with test installation.

The operation of the extended Kalman filter (EKF), unscented Kalman filter (UKF), and particle filter (PF) for Li-S batteries' SoC estimation was tested by considering a number of potential challenges:

**(a)** Due to the self-discharge phenomenon that exists in Li-S cell, it can be difficult to specify precise initial condition for the SoC estimation.

Lithium-sulfur batteries

Therefore, each test is accomplished with three different initial SoC values.

**(b)** The two initial SoC values are chosen at the level of high plateau ($SoC_0 = 1$, 0.7), and the other value is located after the breakpoint (0.68) at ($SoC_0 = 0.6$) to check the filters capacity to converge in between both plateaus.

**(c)** All filters assume the predetermined capacity found in the mixed pulse discharge profile tests during the model identification (9778 As), which is not adapted, even if it is known that the effective capacity with a different duty cycle will differ. This tests the robustness against current profile-induced changes to realizable capacity.

The estimation results for EKF, UFK, and PF with a mixed-pulse and a NEDC drive cycle current profile are presented in Ref. [28] under two cases of initial reference $SoC$: (1) fully charged battery-initial reference $SoC = 1$ and (2) initially partly discharged battery-initial reference $SoC = 0.6$. These results demonstrate the performance of these filters qualitatively at the level of their convergence time with imprecise initial states. Here, in case of starting with a fully charged battery, the UKF has the best efficiency in terms of the SoC convergence and estimation accuracy, which is shown in Table 13.1 [28].

From Table 13.1, EKF and PF perform differently according to the initial condition. When the initial condition is located in the high plateau, the EKF and PF provide a good estimation accuracy, as shown by the low values of the RMSE in operating modes with mixed pulse or NEDC drive cycle current profile. Otherwise, the EKF and PF converge slowly and the estimation

**Table 13.1** RMSE SoC estimation with EKF, UKF, and PF with initially fully charged battery.

| Algorithm | $SoC_0$ | Pulse RMSE | NEDC RMSE |
|---|---|---|---|
| EKF | 1 | 0.0114 | 0.0217 |
| | 0.7 | 0.0160 | 0.0267 |
| | 0.6 | 0.2986 | 0.2732 |
| UKF | 1 | 0.0347 | 0.0280 |
| | 0.7 | 0.0444 | 0.0537 |
| | 0.6 | 0.0705 | 0.1199 |
| PF | uni | 0.0576 | 0.0195 |
| | 0.7 | 0.0532 | 0.0694 |
| | 0.6 | 0.3997 | 0.3354 |

**Table 13.2** RMSE SoC estimation of EKF, UKF, and PF with partly discharged battery.

| Algorithm | $SoC_0$ | Pulse RMSE | NEDC RMSE |
|-----------|---------|------------|-----------|
| EKF       | 1       | 0.1593     | 0.1696    |
|           | 0.7     | 0.0860     | 0.0535    |
|           | 0.6     | 0.1203     | 0.0745    |
| UKF       | 1       | 0.0887     | 0.1743    |
|           | 0.7     | 0.0240     | 0.0687    |
|           | 0.6     | 0.0189     | 0.0332    |
| PF        | uni     | 0.0281     | 0.0561    |
|           | 0.7     | 0.1661     | 0.1176    |
|           | 0.6     | 0.0383     | 0.0320    |

accuracy of both is unsatisfactory. Furthermore, in case of partially discharged initial states, the performance of EKF and PF was evaluated both qualitatively and quantitatively and it is explained in Ref. [28].

Table 13.2 presents the estimation accuracy for EKF, UKF, and PF with a mixed-pulse and a NEDC drive cycle current profile with partly discharged cells [28]. From Table 13.2, the UKF provides the best estimation accuracy compared to the other filters. The estimator that is based on equivalent circuit network can provide valuable results. In conclusion, in case of applying the recursive Bayesian filters, the unscented Kalman filter provides a good performance, with an acceptable computational effort [28].

## 13.4.3 Estimation results analysis for computer science techniques

### 13.4.3.1 Estimation results for ANFIS with discharge current pulses

According to Ref. [24], the initial experiments take the form of a discharge pulse test. This consists of applying consecutive discharge current pulses to the battery and measuring its terminal voltage at 25°C as shown in Fig. 13.13. The aim of using the ANFIS is defining the relationship between the identification results and SoC. For Li-S SoC estimation, the experimental test is performed on three similar cells [24]. Data from the first cell are applied for ANFIS training. During this training, the weight numbers and membership functions (MFs) are adjusted so that the least error between true value of SoC obtained by coulomb counting and ANFIS estimation is to be optimized.

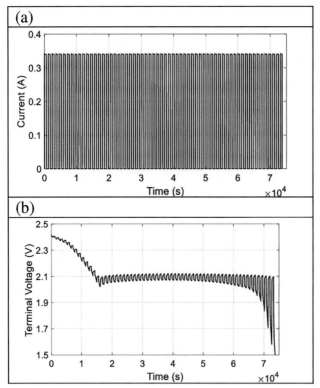

**Fig. 13.13** Load current (input) and Li-S cell's terminal voltage (output) during an experiment.

The other two cells are used for training and testing. Fig. 13.14 depicts the optimization results including the actual SoC and the estimated SoC by ANFIS, which are compared for both training and test data [24]. In addition, these results consider a set of ANFIS specifications, which is defined in Table 13.3 [24]. The ANFIS performance is evaluated through changing the number of inputs of ANFIS structure (Fig. 13.14).

The three tests are obtained from three Li-S cells, so at this level, the situation becomes much more difficult due to the variety of the total capacity values in the experiments. From Fig. 13.14, in this case, the results applying both $V_{oc}$ and $R_0$ seem more satisfactory especially for training data (Fig. 13.14 (3) (a)). Here, the mapping function can be defined between the model variables and the SoC. Otherwise, the obtained results do not present the desired performance because of the lack of information related to the dynamic behavior of the Li-S cell in the low plateau. In addition, there is other issue,

State estimation methodologies for Li-S BMSs   513

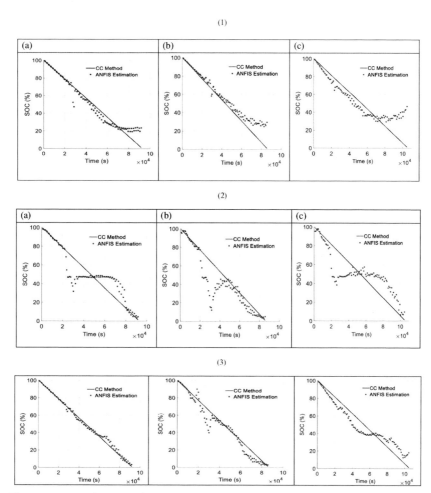

**Fig. 13.14** Actual SOC and ANFIS estimations using (1) $V_{OC}$, (2) $R_0$, (3) $V_{OC}$ and $R_0$. (a) Training data. (b) Test data 1. (c) Test data 2.

**Table 13.3** ANFIS specifications.

| Parameter | Description |
|---|---|
| Inputs | $R_0$, $\frac{dR_0}{dSoC}$, $V_{oc}$ |
| Output | $SoC$ |
| Input MF type | Generalized bell-shaped |
| Output MF type | Linear |
| Number of MFs | 3, 2, 5 |
| Number of rules | 30 |
| Training epoch number | 500 |

which is the complexity of the structure. Indeed, more complexity indicates less speed or more computational effort in real-time applications.

Thus, to save the same complexity while having more information for SoC estimation, the derivative of resistance with respect to SoC, $\frac{dR_0}{dSoC}$, is applied as another input. In fact, it includes additional information that ameliorates the estimation accuracy, which is acceptable in the high plateau and inadequate in the low plateau due to the flat shape of $V_{oc}$ in this region. So, the mapping function is expressed as follows (Eq. 13.21):

$$SoC = g\left(R_0, \frac{dR_0}{dSoC}, V_{OC}\right) \quad (13.21)$$

where $R_0$, $\frac{dR_0}{dSoC}$, and $V_{oc}$ are the three inputs of ANFIS structure as described in Fig. 13.15 [24].

The following process is utilized:
(a) The inputs ($R_0$, $\frac{dR_0}{dSoC}$, $V_{oc}$) are firstly determined (See Fig. 13.15), and then they estimate the SoC, which is the output of ANFIS.
(b) A generalized bell-shaped MF is chosen for the inputs. A hybrid learning approach is applied, combining the gradient-descent method with the least squares estimate to regulate ANFIS variables.
(c) From Table 13.3, the numbers of MFs used for $R_0$, $\frac{dR_0}{dSoC}$, and $V_{oc}$ inputs are 3, 2, and 5 accordingly.

The MFs of the three inputs after training are described in Fig. 13.16 [24].

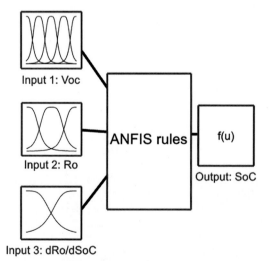

**Fig. 13.15** ANFIS structure using 3 inputs and 30 rules.

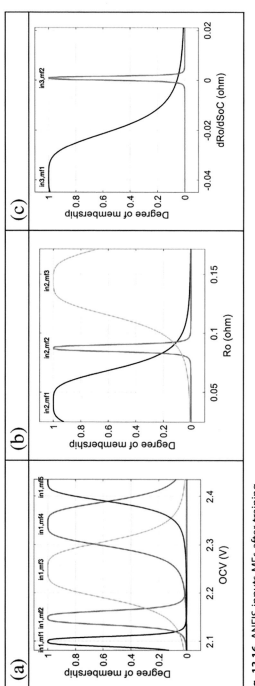

**Fig. 13.16** ANFIS inputs MFs after training.

Fig. 13.17 and Table 13.4 present the suitable results of ANFIS using the three inputs $R_0$, $\frac{dR_0}{dSoC}$, and $V_{oc}$ for SoC estimation at the level of training and test data [24]. From Table 13.4, the structure of ANFIS that is based on three inputs provides a good estimation accuracy. Besides, ANFIS can estimate the SoC value with average and maximum errors of 5% and 14%, respectively. The operation of ANFIS for Li-S battery's SoC estimation will be achieved at different temperature values and aging ranges to explicit more about the variation of the breakpoint's location and total capacity and their impact on state estimation [24]. The efficiency of the proposed ANFIS estimator for SoC is evaluated in a more realistic scenario for EV application that is defined in the next subsection.

### 13.4.3.2 Estimation results for ANFIS with UDDS (urban dynamometer driving schedule) cycle current profile

To evaluate the performance of the proposed ANFIS estimator, the Li-S cell was tested using scaled-down current profiles derived from the EV simulation [24]. Fig. 13.18 shows a Li-S battery discharge test based on Urban Dynamometer Driving Schedule (UDDS) drive cycle [24]. The current and terminal voltages were measured at sampling rate of 1 Hz.

UDDS speed profile is indicated at top, which is used to calculate the EV battery pack power demand. A scaled-down current profile is described in Fig. 13.18B, in which the maximum current is around 1.2 A. Fig. 13.18E depicts the terminal voltage of the Li-S cell during one UDDS cycle in response to the current demand. In Fig. 13.18C and D, cell's terminal voltage and current profile are illustrated during the whole test. Each test is established by repeating the UDDS cycle from 100% SoC to depleted state. Test time, which depends on the current profile, was more than 22 h in this case. A number of tests were conducted at different temperature and current levels [24]. The pattern of identified Li-S cell's parameters at different SoC during UDDS tests is described in Fig. 13.19 [24].

This identification is performed at 30°C during the three UDDS tests with different maximum current levels, which are illustrated in Table 13.5 [24]. Then, the identified Li-S cell's parameters are used by ANFIS for SoC estimation. An error in the identification part will cause an error in SOC estimation as well. Hence, the total SOC estimation error can be defined in Eq. (13.22) [24] as follows:

$$SoC_{\text{estimation error}} = SoC_{\text{estimation error due to identification uncertainty}} + \text{estimator}(ANFIS)_{\text{error}} \tag{13.22}$$

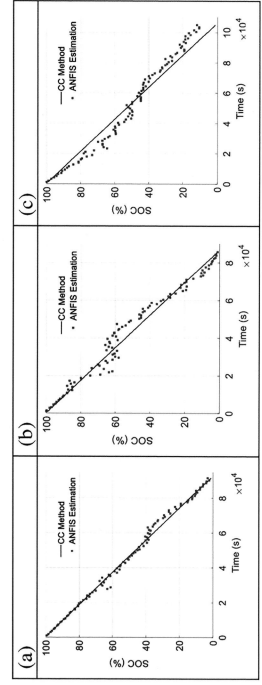

**Fig. 13.17** Actual SOC and ANFIS estimations using $R_0$, $\frac{dR_0}{dSoC}$, and $V_{oc}$: (A) Training data. (B) Test data 1. (C) Test data 2.

**518** Lithium-sulfur batteries

**Table 13.4** Li-S cell SoC estimation error using ANFIS.

| Inputs | Number of rules | Estimation error (%) in training data | Estimation error (%) in test data 1 | Estimation error (%) in test data 2 |
|---|---|---|---|---|
| $V_{oc}$ | 9 | Mean: 4.61 | Mean: 6.38 | Mean: 9.91 |
| | | Max.: 22.21 | Max.: 28.28 | Max.: 45.25 |
| $R_0$ | 9 | Mean: 8.88 | Mean: 9.54 | Mean: 13.82 |
| | | Max.: 35.55 | Max.: 51.39 | Max.: 36.98 |
| $R_0, V_{oc}$ | 35 | Mean: 1.74 | Mean: 4.84 | Mean: 7.81 |
| | | Max.: 8.08 | Max.: 25.61 | Max.: 16.87 |
| $R_0, V_{oc}, \frac{dR_0}{dSoC}$ | 30 | Mean: 1.55 | Mean: 3.93 | Mean:5.11 |
| | | Max.: 8.23 | Max.: 13.76 | Max.: 13.77 |

Fig. 13.20 describes the Li–S cell SoC optimization of the three UDDS tests using coulomb counting and ANFIS estimation, and its estimation accuracy is presented in Table 13.5 [24]. Here, the coulomb counting is applied as a benchmark for validation of the estimation results. It is assumed that the total capacity, needed for coulomb counting, is available; however, it is not in a real application. Ideally, the total capacity is calculated after finishing each test. This is a good theoretical method to evaluate the estimation results. Otherwise, the ANFIS SoC estimator does not need knowledge of the total capacity in advance, and it is useful in real application.

From Table 13.5, there is a big difference between the Li–S cell's SoC estimation results of three UDDS tests and the Li–S cell's SoC optimization results of the previous tests (See Table 13.4). The reason of this is the simplicity of the pulse discharge data comparing to a real discharge profile for automotive application. However, the SOC estimation results are not suitable, and they are not bad as an outcome of a first study in this area [24]. Therefore, an ANFIS estimator suffers from variations that cause a failure at accuracy level in a short time interval while the mean accuracy is acceptable. Moreover, in a real driving scenario, the results prove that ANFIS estimation error can increase to 20% in few points, which needs to be ameliorated.

However, there is a solution that consists of a hybrid between ANFIS and another method which helps to overcome the limitations and improve accuracy [24]. Its principle is exploiting the advantages of ANFIS and coulomb counting methods at the same time. This technique is accurate and useful in real conditions. The combination of ANFIS and coulomb counting methods ensures a suitable operation of the proposed method. Indeed, thanks to this combination, there is an ability of determining the initial

State estimation methodologies for Li-S BMSs 519

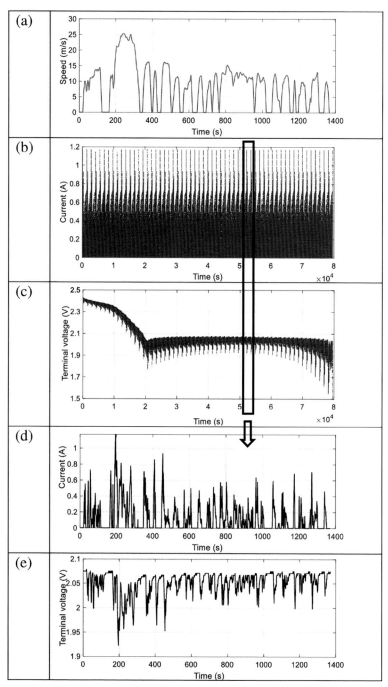

**Fig. 13.18** Li-S cell discharge test based on UDDS drive cycle: (A) UDDS speed profile, (B) repeating UDDS current profile, (C) Li-S cell's terminal voltage during the whole UDDS test, (D) Li-S cell current profile during one UDDS cycle, and (E) terminal voltage of a Li-S cell during one UDDS cycle.

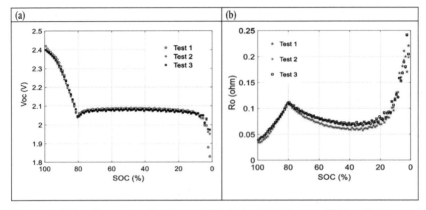

**Fig. 13.19** Li-S cell parameterization at 30°C during UDDS tests: (A) $V_{OC}$, (B) $R_0$.

SoC and compensation of the effect of measurement noise, although there is less fluctuation due to the use of coulomb counting as a limiting bound for ANFIS [24]. The operation of the hybrid estimation method is expressed through the following Eqs. (13.23), (13.24) [24]:

$$SoC_H = \frac{W_1 \, SoC_{ANFIS} + W_2 \, SoC_{coulomb-counting}}{W_1 + W_2} \quad (13.23)$$

$$WR = \frac{W_2}{W_1} \quad (13.24)$$

where $SoC_H$ is the value of SoC given by the hybrid technique, $SoC_{ANFIS}$ is the ANFIS estimation, and $SoC_{coulomb-counting}$ is the value of SoC based on coulomb counting. The weight ratio (WR) defines the role of each technique in this formulation.

The results of this hybrid estimation method over UDDS test at 30°C from different initial conditions are presented in Table 13.6 [24]. The hybrid technique can converge to the right value of SOC from any initial condition, and it can keep going in a limited bound around it. Therefore, applying

**Table 13.5** Li-S cell SoC estimation results of UDDS tests.

|  | Test 1 | Test 2 | Test 3 |
|---|---|---|---|
| Temperature (°C) | 30 | 30 | 30 |
| Maximum current (A) | 1.17 | 1.75 | 1.75 |
| Average SOC estimation error (%) | 7.1 | 6.2 | 6.0 |
| Maximum SOC estimation error (%) | 20.9 | 18.3 | 21.1 |

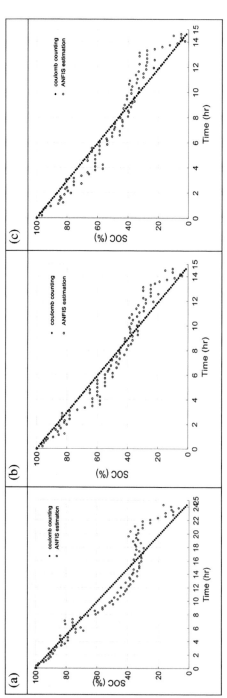

**Fig. 13.20** Li-S cell SOC calculation using coulomb counting and ANFIS estimation at 30°C during UDDS tests. (A) Test 1. (B) Test 2. (C) Test 3.

## 522 Lithium-sulfur batteries

**Table 13.6** Li-S cell SOC estimation accuracy using the hybrid technique over UDDS test at 30°C with various initial SoC.

| Initial SOC (%) | Average SOC estimation Error | Maximum SOC estimation Error |
|---|---|---|
| 100 | 4.18 | 6.64 |
| 85 | 4.03 | 6.66 |
| 60 | 4.13 | 6.93 |
| 40 | 2.91 | 6.01 |
| 15 | 3.08 | 4.51 |

this method in a realistic driving scenario, a Li-S cell's SOC can be estimated with a mean error of 4% and a worst case error of 7% [24].

### 13.4.3.3 Estimation results for SVM classifier with the Millbrook London transport bus (MLTB) test

The efficiency of the support vector machine (SVM) had been validated under drive cycle test: the Millbrook London Transport Bus (MLTB) Test [38] (See Fig. 13.21). This test had been applied to the Li–S battery which has had 19 Ah nominal capacity. More details are defined in Ref. [38].

In this study, the identified variables of Li-S cell model are used in SVM method to estimate the SoC. The parameterization of the battery model is achieved using the identification algorithm forgetting factor recursive least squares (FFRLS). The parameterization procedure is explained in Ref. [38]. The SVM method is operated as the classifier for Li-S battery SoC estimation. The SoC classification concept is described in Ref. [38]. Table 13.7 presents the values of the two types of accuracy measures "hard clustering" and "soft clustering" for the training and testing data sets using SVM classifier for a MLTB test at 25°C [38].

From Ref. [38], the accuracy for training and testing data sets is expressed by the ratio of the number of correctly classified data segments (Eq. 13.25) over the whole number of segments:

$$\text{Accuracy}(\%) = \frac{N_{\text{correct}}}{N_{\text{total}}} \times 100 \tag{13.25}$$

Besides, there is a consideration of different combination of inputs to obtain the best configuration that provides the highest accuracy. Thus, from Table 13.7, using the input Thevenin variables ($V_{OC}$, $R_0$, $R_p$, and $C_p$) for SoC classification gives moderately higher level of accuracy.

Furthermore, the performance of the classifier is indicated through the confusion matrices which are defined in the form of colored cells showing

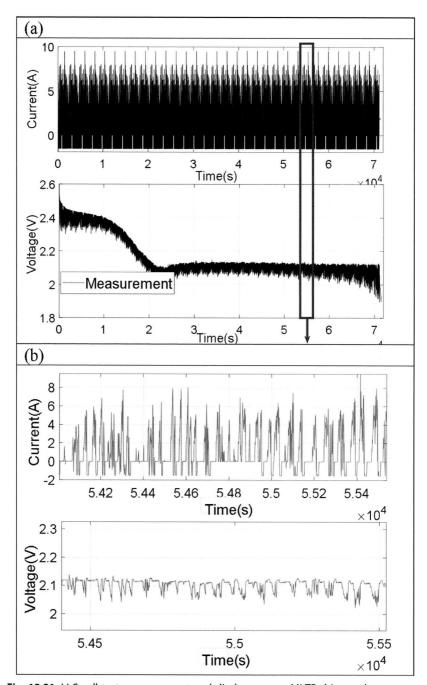

**Fig. 13.21** Li-S cell test measurement and discharge over MLTB drive cycle.

**Table 13.7** SoC classification results using SVM and different inputs.

| Input parameters used for SoC classification | $V_{oc}/$ $R_0$ | $V_{oc}/$ $R_P$ | $V_{oc}/$ $C_P$ | $V_{oc}/$ $R_0/R_P$ | $V_{oc}/$ $R_0/C_P$ | $V_{oc}/R_0/$ $R_P/C_P$ |
|---|---|---|---|---|---|---|
| Training accuracy (%) | 76 | 78 | 68 | 87 | 79 | 90 |
| Testing accuracy (hard clustering) (%) | 58 | 70 | 66 | 67 | 61 | 73 |
| Testing accuracy (soft clustering) (%) | 79 | 91 | 90 | 89 | 80 | 93 |

the levels of uncertainty at different SoC levels. More information can be found in Ref. [38]. Here, the confusion matrices of SVM battery SoC classifier using all the Thevenin variables demonstrate the best performance of the classifier. In this case, the results are inadequate, but they are encouraging. Moreover, there is an evaluation of SVM classifier performance using the input Thevenin parameters for SoC optimization at different values of temperature as shown in Table 13.8 [38].

From Table 13.8, the classification accuracy is slightly affected by temperature, and the suggested algorithm operates well in the whole range of temperature. In addition, the confusion matrices of the best classifier in case of using the input Thevenin variables at different temperature values show that the results are not precisely similar in the whole range of the temperature and generally they are quite promising [38]. Thus, there are two main issues which should be addressed to improve the performance of SVM classifier for Li-S cell SoC estimation, and they are as follows:

**(a)** The output of the classifier is a label that varies between 1 and 10, and the SoC value can only be defined in the form of fractions of 10%.

**(b)** Lack of consideration for the history of the estimation due to the fact that the classifier only applies the identification results at each time step. It leads to the fluctuations in the identification outcomes and then in SoC optimization.

**Table 13.8** SoC classification results using SVM and different temperatures for $V_{oc}/R_0/R_P/C_P$ inputs.

| Temperature values (°C) | 10 | 15 | 20 | 25 | 30 |
|---|---|---|---|---|---|
| Training accuracy (%) | 86 | 88 | 91 | 90 | 87 |
| Testing accuracy (hard clustering) (%) | 65 | 82 | 78 | 73 | 70 |
| Testing accuracy (soft clustering) (%) | 97 | 93 | 96 | 93 | 96 |

Therefore, a hybrid method which consists of combination between coulomb counting and SVM classifier is used to ameliorate the efficiency of the classifier for Li–S battery SoC estimation [38]. The mathematical formulation (Eq. 13.26) of this hybrid technique is a function of fusion gains $W_1$ which is defined as follows [38]:

$$SoC_H = \frac{W_1 \, SoC_{SVM} + W_2 \, SoC_{coulomb-counting}}{W_1 + W_2} \qquad (13.26)$$

Hence, the performance of the hybrid method depends on the fusion gains which is demonstrated in Table 13.9.

The results prove that Case 5 presents a suitable operational condition for hybrid technique. Indeed, the best performance of the hybrid method is reached in case of $W_1 = 0.9$ and $W_2 = 0.1$ at the beginning, and they vary to $W_1 = 0.1$ and $W_2 = 0.9$ after convergence. Hence, the hybrid method operates well in case of the variation of the fusion gains in real times. Therefore, despite the difficulties in Li–S cell SoC estimation, the performance of hybrid technique has been demonstrated through promising results that are validated by using experimental test data [38]. This method is able to optimize the SoC of Li–S cell from any initial condition with maximum error of 2.63% and average error less than 1% [38].

## 13.5 Conclusions and outlook

This chapter has presented a literature review on state-of-the-art state estimation methodologies for lithium-sulfur battery management systems in automotive applications. This is very different from the state of the art in lithium-ion batteries since two popular techniques for battery SoC estimation (open-circuit voltage and coulomb counting) do not work well for Li–S technologies due to the flat shape of the OCV–SoC feature and the variation of the battery capacity derived from lithium-sulfur's chemistry under various conditions like temperature variation. This has created a need for specific techniques specially adapted to lithium-sulfur.

In the literature, there are two groups of estimation methods that—with appropriate adaptations and specific modeling techniques—are showing promise. The first group includes SOC estimation methods based on recursive Bayesian filters (extended Kalman filter (EKF), unscented Kalman filter (UKF), and particle filter (PF)); and the second group includes approaches based on computer science such as adaptive neuro-fuzzy inference system (ANFIS) and a support vector machine (SVM) classifier. In the literature,

**Table 13.9** SoC classification results using hybrid method with different fusion gains.

| | Initial fusion gains | Fusion gain after convergence | Average error (%) after convergence | Maximum error (%) after convergence | Convergence rate (number of iterations) |
|---|---|---|---|---|---|
| Case 1 | $W_1=0.3,\ W_2=0.7$ (constant) | | 1.04 | 4.03 | 8 |
| Case 2 | $W_1=0.7,\ W_2=0.3$ (constant) | | 2.03 | 4.45 | 4 |
| Case 3 | $W_1=0.7,\ W_2=0.3$ | $W_1=0.3,\ W_2=0.7$ | 1.07 | 3.68 | 4 |
| Case 4 | $W_1=0.8,\ W_2=0.2$ | $W_1=0.2,\ W_2=0.8$ | 0.75 | 2.63 | 4 |
| Case 5 | $W_1=0.9,\ W_2=0.1$ | $W_1=0.1,\ W_2=0.9$ | 0.59 | 3.76 | 3 |
| Case 6 | $W_1=0.95,\ W_2=0.05$ | $W_1=0.05,\ W_2=0.95$ | 0.99 | 3.36 | 3 |
| Case 7 | $W_1=1,\ W_2=0$ | $W_1=0,\ W_2=1$ | 3.72 | 6.38 | 2 |

the performance of these techniques for SoC estimation has been demonstrated and validated by using experimental test data.

There is much scope for ongoing research in lithium-sulfur battery, both through developing improved techniques in the light of new scientific understanding and industrial experience, and in responding to new developments in lithium-sulfur cell chemistry. The authors' current work is exploring the industrial deployment of the algorithms, techniques for measuring lithium-sulfur battery state of health, algorithmic improvements and enhancements, and specific adaptations for particular operating conditions.

## Acknowledgments

Part of this work is being undertaken as part of the project "Lithium Sulfur for Safe Road Electrification" (LISA) funded by the European Commission under grant number 814471, to whom the authors express their thanks. We also thank the various industry and academic partners whom we have collaborated with on our research projects for their help and support.

## References

[1] M.A. Hannan, M.M. Hoque, A. Mohamed, A. Ayob, Review of energy storage systems for electric vehicle applications: issues and challenges, Renew. Sustain. Energy Rev. 69 (2017) 771–789.

[2] S. Snigdha, K.P. Amrish, M.M. Tripathi, Storage technologies for electric vehicles, J. Traffic Transp. Eng. 3 (2020) 340–361.

[3] E. Mancini, M. Longo, W. Yaici, D. Zaninelli, Assessment of the impact of electric vehicles on the design and effectiveness of electric distribution grid with distributed generation, Appl. Sci. 10 (15) (2020) 5125, https://doi.org/10.3390/app10155125.

[4] K. Liu, K. Li, Q. Peng, C. Zhang, A brief review on key technologies in the battery management system of electric vehicles, Front. Mech. Eng. 14 (2019) 47–64.

[5] C. Weidong, L. Jun, Y. Zhaohua, L. Gen, A review of lithium-ion battery for electric vehicle applications and beyond, Energy Procedia 158 (2019) 4363–4368.

[6] A. Fotouhi, D.J. Auger, L. O'Neill, T. Cleaver, S. Walus, Lithium-sulfur battery technology readiness and applications—a review, Energies 10 (12) (2017) 1937, https://doi.org/10.3390/en10121937.

[7] V. Knap, D.J. Auger, K. Propp, A. Fotouhi, D.-I. Stroe, Concurrent real-time estimation of state of health and maximum available power in lithium-sulfur batteries, Energies 11 (8) (2018) 2133, https://doi.org/10.3390/en11082133.

[8] P. Bruce, S. Freunberger, L. Hardwick, J. Tarascon, Li-$O_2$ and Li-S batteries with high energy storage, Nat. Mater. 11 (2012) 19–29.

[9] C. Lyness, Sources of risk: Chapter 7—lithium-secondary cell: sources of risks and their effects, in: Electrochemical Power Sources: Fundamentals, Systems, and Applications; Li–Battery Safety, Elsevier, 2018, pp. 143–166.

[10] R. Kumar, J. Liu, J.-Y. Hwang, Y.-K. Sun, Recent research trends in Li–S batteries, J. Mater. Chem.A 6 (25) (2018) 11582–11605.

[11] M. Brandl, H. Gall, M. Wenger, V. Lorentz, M. Giegerich, F. Baronti, G. Fantechi, L. Fanucci, R. Roncella, R. Saletti, S. Saponara, A. Thaler, M. Cifrain, W. Prochazka,

528 Lithium-sulfur batteries

Batteries and battery management systems for electric vehicles, design, automation & test, in: Europe Conference & Exhibition (DATE), IEEE, 2012, pp. 971–976.

[12] Z. Gao, C.S. Chin, J.H.K. Chiew, J. Jia, C. Zhang, Design and implementation of a smart lithium-ion battery system with real-time fault diagnosis capability for electric vehicles, Energies 10 (10) (2017) 1503, https://doi.org/10.3390/en10101503.

[13] Y. Miao, P. Hynan, A. von Jouanne, A. Yokochi, Current Li-ion battery technologies in electric vehicles and opportunities for advancements, Energies 12 (6) (2019) 1074, https://doi.org/10.3390/en12061074.

[14] Y.S. Su, Y. Fu, T. Cochell, M. Arumugam, A strategic approach to recharging lithium-sulphur batteries for long cycle life, Nat. Commun. 4 (2013) 456–459, https://doi.org/10.1038/ncomms3985.

[15] T. Li, X. Bai, U. Gulzar, Y.-J. Bai, C. Capiglia, W. Deng, X. Zhou, Z. Liu, Z. Feng, Z. Remo Proietti, A comprehensive understanding of lithium–sulfur battery technology, Adv. Funct. Mater. 29 (2019) 1901730, https://doi.org/10.1002/adfm.201901730.

[16] H. He, R. Xiong, X. Zhang, F. Sun, J. Fan, State-of-charge estimation of the lithium-ion battery using an adaptive extended Kalman filter based on an improved Thevenin model, IEEE Trans. Veh. Technol. 60 (4) (2011) 1461–1469.

[17] Y. Tian, B. Xia, W. Sun, Z. Xu, W. Zheng, A modified model-based state of charge estimation of power lithium-ion batteries using unscented Kalman filter, J. Power Sources 270 (2014) 619–626.

[18] A. Fotouhi, D.J. Auger, K. Propp, S. Longo, M. Wild, A review on electric vehicle battery modelling: from lithium-ion toward lithium–sulphur, Renew. Sustain. Energy Rev. 56 (2016) 1008–1021.

[19] D. Moy, A. Manivannan, S.R. Narayanan, Direct measurement of polysulfide shuttle current : a window into understanding the performance of lithium-sulfur cells, J. Electrochem. Soc. 162 (2015) A1–A7, https://doi.org/10.1149/2.0181501jes.

[20] D.J. Auger, A. Fotouhi, K. Propp, S. Longo, Battery management systems—state estimation for lithium–sulfur batteries, in: Lithium–Sulfur Batteries, 2019, pp. 249–272.

[21] M. Wild, L. O'Neill, T. Zhang, R. Purkayastha, G. Minton, M. Marinescu, G. Offer, Lithium sulfur batteries, a mechanistic review, Energ. Environ. Sci. 8 (2015) 3477–3494.

[22] M. Marinescu, T. Zhang, G. Offer, A zero dimensional model of lithium–sulfur batteries during charge and discharge, Phys. Chem. Chem. Phys. 18 (2015) 584–593, https://doi.org/10.1039/C5CP05755H.

[23] K. Propp, Advanced State Estimation for lithium-Sulfur Batteries (Ph.D. thesis), Cranfield University, 2017.

[24] A. Fotouhi, D.J. Auger, K. Propp, S. Longo, Lithium–sulfur battery state-of-charge observability analysis and estimation, IEEE Trans. Power Electron. 33 (8) (2018) 5847–5859.

[25] A. Jossen, Fundamentals of battery dynamics, J. Power Sources 154 (2) (2006) 530–538.

[26] A. Fotouhi, D.J. Auger, K. Propp, S. Longo, R. Purkayastha, L. O'Neill, S. Waluś, Lithium–sulfur cell equivalent circuit network model parameterization and sensitivity analysis, IEEE Trans. Veh. Technol. 66 (9) (2017) 7711–7721.

[27] K. Propp, M. Marinescu, D.J. Auger, L. O'Neill, A. Fotouhi, K. Somasundaram, G. Offer, S. Longo, M. Wild, V. Knap, Multi-temperature state-dependent equivalent circuit discharge model for lithium–sulfur batteries, J. Power Sources 328 (2016) 289–299.

[28] K. Propp, D.J. Auger, A. Fotouhi, S. Longo, V. Knap, Kalman-variant estimators for state of charge in lithium-sulfur batteries, J. Power Sources 343 (2017) 254–267.

[29] V. Knap, D-I. Stroe, R. Purkayastha, S. Walus, D.J. Auger, A. Fotouhi, K. Propp, Reference performance test methodology for degradation assessment of lithium-sulfur batteries, J. Electrochem. Soc. 165 (2018) A1601–A1609.

[30] J.P. Rivera-Barrera, N. Muñoz-Galeano, H.O. Sarmiento-Maldonado, SoC estimation for lithium-ion batteries: review and future challenges, Electronics 6 (4) (2017) 102, https://doi.org/10.3390/electronics6040102.

[31] W. He, N. Williard, C. Chen, M. Pecht, State of charge estimation for electric vehicle batteries using unscented Kalman filtering, Microelectron. Reliab. 53 (2013) 840–847.

[32] F. Sun, X. Hu, Y. Zou, S. Li, Adaptive unscented Kalman filtering for state of charge estimation of a lithium-ion battery for electric vehicles, Energy 36 (2011) 3531–3540.

[33] C. Zheng, Y. Fu, C. Mi, State of charge estimation of lithium-ion batteries in electric drive vehicles using extended Kalman filtering, IEEE Trans. Veh. Technol. 62 (2013) 1020–1030.

[34] H. Rahimi-Eichi, F. Baronti, M.Y. Chow, Online adaptive parameter identification and state-of-charge coestimation for lithium-polymer battery cells, IEEE Trans. Ind. Electron. 61 (2014) 2053–2061.

[35] S. Thrun, W. Burgard, D. Fox, Probabilistic Robotics, MIT Press, 2005.

[36] R. Karlsson, Particle Filtering for Positioning and Tracking Applications, Linköping University Electronic Press, 2005.

[37] A. Fotouhi, D.J. Auger, K. Propp, S. Longo, Electric vehicle battery parameter identification and SOC observability analysis: NiMH and Li-S case studies, in: 8th IET International Conference on Power Electronics, Machines and Drives (PEMD 2016), Glasgow, UK, 2016, pp. 1–6.

[38] N. Shateri, Z. Shi, D.J. Auger, A. Fotouhi, Lithium-sulfur cell state of charge estimation using a classification technique, IEEE Trans. Veh. Technol. 70 (2021) 212–224.

[39] I. Steinwart, A. Christmann, Support Vector Machines, Springer, New York, NY, 2008.

[40] OXIS Energy Ltd, Our Cell and Battery Technology Advantages, https://oxisenergy.com/technology/ (Accessed March 30, 2022).

# CHAPTER 14

# Batteries for aeronautics and space exploration: Recent developments and future prospects

**Aloysius F. Hepp[a], Prashant N. Kumta[b,c,d], Oleg I. Velikokhatnyi[c], and Moni K. Datta[c]**

[a]Nanotech Innovations, LLC, Oberlin, OH, United States
[b]Department of Chemical and Petroleum Engineering, Swanson School of Engineering, University of Pittsburgh, Pittsburgh, PA, United States
[c]Department of Bioengineering, Swanson School of Engineering, University of Pittsburgh, Pittsburgh, PA, United States
[d]Department of Mechanical Engineering and Materials Science, Swanson School of Engineering, University of Pittsburgh, Pittsburgh, PA, United States

## Contents

| | | |
|---|---|---|
| **14.1** | Introduction | 532 |
| **14.2** | Energy storage for (solar-) electric aircraft and high-altitude airships | 533 |
| | 14.2.1 Batteries for solar-electric aircraft | 533 |
| | 14.2.2 All-electric battery-powered aircraft | 533 |
| | 14.2.3 Batteries for high-altitude airships | 539 |
| | 14.2.4 High-altitude platforms: Power considerations and alternative technologies | 543 |
| **14.3** | Overview of energy storage for space exploration | 552 |
| **14.4** | Recent NASA missions to Mercury, Mars, and small bodies | 553 |
| | 14.4.1 Mercury Surface, Space Environment, Geochemistry and Ranging (MESSENGER) mission | 557 |
| | 14.4.2 Battery technologies for Mars surface rovers | 559 |
| | 14.4.3 Notable NASA exploration missions to comets and asteroids | 563 |
| **14.5** | Radiation issues and exploration missions to the Jupiter region | 565 |
| | 14.5.1 Radiation in space and impact on rechargeable batteries | 567 |
| | 14.5.2 Upcoming missions to Jupiter and several of its icy moons | 570 |
| **14.6** | Next generation(s) of battery technologies for space exploration | 572 |
| | 14.6.1 Future space exploration: Battery technology options and considerations | 572 |
| | 14.6.2 Upcoming missions to three major classes of asteroids | 575 |
| | 14.6.3 Aerial exploration of other planetary bodies: Mars, Titan, and Venus | 578 |
| | 14.6.4 Off-world utilization of local resources for inhabited settlements | 581 |
| **14.7** | Conclusions | 581 |
| References | | 584 |

*Lithium-Sulfur Batteries*
https://doi.org/10.1016/B978-0-12-819676-2.00011-6

Copyright © 2022 Elsevier Inc.
All rights reserved.

## 14.1 Introduction

Over the past half-century, rechargeable batteries have been employed for a wide range of applications involving electronics, electric vehicles, and aerospace systems [1–6]. Lithium-ion batteries (LIBs) are prevalent among rechargeable options [7–12]; they offer significant advantages for future space missions as a result of their high energy density, improved reliability, environmental compatibility, and lower power system life cycle costs [3–5,12]. However, LIBs are known to have potential safety and performance issues [10,13–15]; recent reviews highlight attempts to address these issues [16–18].

While this monograph focuses on Li-S battery technologies [19–28], we address aerospace applications of a number of advanced battery systems, highlighting Li-S batteries whenever possible. Over the past several decades, our research groups have examined aspects of advanced (nano)materials and aerospace power applications [3,4,19,21,29–35]. These applications present unique challenges such as temperature fluctuations, rapid gravitational fluctuations, high-energy particles and radiation environments, atomic oxygen, hard-ultraviolet light, thermal management, and the necessity of weight- and space savings [31,34–44].

In Section 14.2, we review recent and anticipated future uses of battery technologies for (solar-)electric aircraft and high-altitude airships [45–50]. Section 14.3 provides an overview of energy storage issues for space exploration [36,37,39,40]. The two following sections on space exploration focus on recent NASA missions to Mercury, Mars, and several small bodies (Section 14.4) as well as radiation issues and exploration missions to the Jupiter Region (Section 14.5) [36,51,52].

Appropriate battery technologies for specific destinations in the solar system [36] is surveyed in Section 14.6. A consideration of unique aspects of future missions to asteroids, and ocean worlds, including off-world atmospheric flight exploration, and Venus mission concepts, and concludes with some comments on the long-term prospects for humanity to explore and possibly settle in habitable regions of our solar system [53]. An important take-home lesson is the criticality of energy storage for advanced aerospace power applications including small off-world exploration vehicles such as rovers, helicopters, drones, and even airships. It is important to note that while a variety of battery technologies could be employed, Li-S batteries may certainly be the optical choice for specific missions [5,36].

## 14.2 Energy storage for (solar-) electric aircraft and high-altitude airships

In this section, we discuss some examples of aeronautical and suborbital applications of batteries such as high-altitude airships [45–47] and all-electric aircraft [48–50]. We discuss systems considerations and analyses for both types of high-altitude platforms, including alternative energy storage technology options, comparison to hydrocarbon-powered platforms, and a combined performance analysis of high-altitude platforms.

### 14.2.1 Batteries for solar-electric aircraft

In contrast to traditional airplanes, solar-powered aircraft harvest solar radiation and convert it into electrical energy via solar cell arrays; batteries are used to store excess energy for periods of diminished solar fluence. Table 14.1, adapted from reviews [45,46] (and other sources [54]), summarizes important solar-electric aircraft with power system details included [54–59]. Commercial battery technologies are typically employed for prototype solar-powered airplanes (Fig. 14.1); examples of U.S.-based solar-powered airplanes described in Table 14.1 are pictured in Fig. 14.2.

The SoLong airplane used Li-ion cells with an energy density of 220 Wh/kg [45]. Zephyr 6 and beyond utilize Li-S batteries, with an energy density that reached 350 Wh/kg [45,46]. Meanwhile, the Helios HP03, built for endurance and not maximum altitude, used hydrogen- and oxygen-based regenerative fuel cells, thus becoming the first solar-powered airplane that used fuel cells [54]. An alternative approach of primary or exclusive use of battery power for all-electric aircraft would require addressing inherent safety and endurance issues.

### 14.2.2 All-electric battery-powered aircraft

Chuck Yeager, flying the X-1, is credited with being the first person to break the sound barrier during level flight on October 14, 1947. The X-planes, joint NASA/military experimental aircraft, have been pushing through invisible barriers ever since. The latest X-plane (X-57 Mx) has a goal of producing a commercially viable, all-electric passenger aircraft [50]. To achieve this goal, many technological hurdles had to be surmounted, including unprecedented battery reliability and power; this was developed through an industry partnership. Two electric cruise motors on the very tips of the wings and six smaller motors to add to the flow over the wings, increasing lift during takeoff and landing, replaced gas-powered

**Table 14.1** Representative and important solar-electric aircraft with key flight and power details.

| Name | Year | Wingspan (m) | Mass (kg) | Power system details | Comments | Ref. |
|------|------|------|------|------|------|------|
| Sunrise I | 1974 | 9.8 | 12.3 | Commercial 10% Si solar cells | Wing area 8.4 m$^2$ | [45,55] |
| Gossamer Penguin (Piloted) | 1980 | 21.6 | 30[a] | 28 NiCd batteries[b] | 8/7/80—flew 1.95 mi 14 m 21 s | [54] |
| Solar Challenger (Piloted) | 1981 | 14.2 | 90[a] | 16,128 solar cells $P > 4.4$ kW | Flew 262 km France to United Kingdom | [45] |
| Pathfinder[c] | 1993 1997 | 29.5 | 250 (45)[d] | 7.5 kW PV + batteries | 15.4 km[e] 21.5 km[e] | [54] |
| Pathfinder-Plus[c] | 1998 | 36.9 | 315 (68)[d] | 12.5 kW PV + batteries | 24.4 km[e] | [54] |
| Centurion[c] | | 63.1 | 860 (270)[d] | 31 kW PV + batteries | | [54] |
| Helios HP01[c] | 2001 | 75.3 | 930 (325)[d] | 35 kW PV + LIBs | 29.5 km[e] (current world record) | [54] |
| SoLong | 2005 | 4.75 | | 5.6 kg—LIB 220 Wh/kg 12.6 kg—22% c-Si (76 cells total) | Wing area 1.5 m$^2$ April 2005 24 h[f] June 2005 48 h[f] | [45] |
| Zephyr 5-1 | 2006 | 12 | 31 | Solar cells and batteries | 18 h[f]—11.0 km[d] | [45,46] |
| Zephyr 5-2 | 2005 | 12 | 25 | Primary batteries | 6 h[f] | [45,46] |
| Zephyr 6 | 2007 2008 | 18 | 30[g] | PV + Li-S (3 kW h) batteries | 54 h[f]—17.7 km[d] 82 h[f]—18.3 km[d] | [45,46] |

*Continued*

**Table 14.1** Representative and important solar-electric aircraft with key flight and power details—cont'd

| Name | Year | Wingspan (m) | Mass (kg) | Power system details | Comments | Ref. |
|---|---|---|---|---|---|---|
| Zephyr 7 | 2010 | 22.6 | 53.1 | | 336 h[f] | [45,46] |
| Zephyr S[h] | 2018 | 25 | 75 (12)[d] | | 600 h[f] | [45,56] |
| Sky-Sailor | 2008 | 3.2 | 2.4 | 1.06 kg LIB | 27 h[f] | [57] |
| Solar Impulse HB-SIA (Piloted) | 2012 | 63.4 | 1600 | 11,628 PV panels 400 kg LB | Switzerland to Morocco Flight[i] | [45,58] |
| EAV-3 | 2016 | 19.5 | 53 | 21% c-Si: 1.5 kWh LIB: 3 kWh (230 Wh/kg) Mass—13 kg[j] | 6–10 m/s speed 14 km[e] | [59] |

[a]Mass without a pilot.
[b]First flight (4/7/80) replaced for the solar-powered flights (starting 5/18/80) by a panel of 3920 solar cells capability of producing 541 $W_p$.
[c]Environmental Research Aircraft and Sensor Technology (ERAST) project aircraft, a joint NASA-industry initiative.
[d]Maximum weight and (payload).
[e]Highest altitude above sea level achieved.
[f]Flight time.
[g]Ultra-lightweight carbon fiber body.
[h]Airbus—commercial product.
[i]First intercontinental flight.
[j]Three modules of solar cells, four battery packs.
Data from X. Zhu, Z. Guo, Z. Hou, Solar-powered airplanes: a historical perspective and future challenges, Prog. Aerosp. Sci. 71 (2014) 36–53; J. Gonzalo, D. López, D. Domínguez, A. García, A. Escapa, On the capabilities and limitations of high altitude pseudo-satellites, Prog. Aerosp. Sci. 98 (2018) 37–56; https://www.nasa.gov/centers/armstrong/news/FactSheets/FS-054-DFRC.html, Accessed 3 August 2020.

**Fig. 14.1** Typical energy storage technologies used in solar-powered airplanes. *(Reproduced with permission from X. Zhu, Z. Guo, Z. Hou, Solar-powered airplanes: a historical perspective and future challenges, Prog. Aerosp. Sci. 71 (2014) 36–53. Copyright (November 2014) Elsevier.)*

engines. A major challenge is safely delivering enough power to the engines. During takeoff, all of these motors combined can use more than 200 kW, enough energy to momentarily power more than 100 average American homes.

Under Small Business Innovation Research contracts, Electric Power (EP) Systems, a California-based company producing reliable batteries and power systems for aerospace, as well as ground transportation, medical, and military applications, built an 850-pound lithium-ion battery pack that could eventually safely do the job. Some modifications were required to address thermal runaway, a significant safety issue for LIBs that needed to be resolved. NASA had recently developed a lithium-ion battery casing technique that filled an aluminum block with holes to hold batteries a millimeter or less apart, isolating them and dissipating heat from any runaway event throughout the block. This approach had been successfully utilized for batteries on the International Space Station, the Orion space capsule, and tools used by astronauts on spacewalks; the technology was adapted and modified by EP Systems and tested successfully in late 2017 [49].

In March 2018, Bye chose the EP Systems' battery technology to power the Sun Flyer, its own all-electric Sun Flyer trainer certified under new

**Fig. 14.2** (A) Gossamer Penguin in flight above Rogers Dry Lakebed at Edwards, CA, showing the solar panel perpendicular to the wing and facing the sun; (B) pathfinder, a first-generation NASA ERAST solar-powered, remotely piloted aircraft, is shown, while it was conducting a series of science flights to highlight the aircraft's science capabilities while collecting imagery of forest and coastal zone ecosystems on Kauai, Hawaii; (C) the slow-flying Centurion solar-electric flying wing, one of several remotely piloted aircraft developed under NASA's ERAST project, glides in for a landing on Rogers Dry Lake following a test flight at NASA's Dryden Flight Research Center; (D) the solar-electric Helios Prototype (HP01) flying wing near the Hawaiian islands of Niihau and Lehua during its first test flight on solar power from the U.S. Navy's Pacific Missile Range Facility on Kauai, HI, July 14, 2001; this same aircraft achieved the current world record altitude on August 13, 2001; see entry in Table 14.1. *(Source: https://www.nasa.gov/centers/armstrong/news/FactSheets/FS-054-DFRC.html. Courtesy NASA.)*

Federal Aviation Administration (FAA) rules for electric aircraft [50]. Its first successful flight was made the following month. Bye Aerospace will produce two- and four-person versions of the plane, which is intended for training and can fly for about an hour on a full charge. The batteries used to power the Sun Flyer are similar to the X-57 batteries [49]. Electric motors are far more efficient than internal combustion engines and require much less maintenance, replacing the high heat and friction of combustion chambers and pistons with wound wires and magnets. They are 90% efficient or better, meaning almost all the power is turned into useful work; gas-powered engines are less than 30% efficient. Electric motors also generate lower noise and no emissions. Bye Aerospace calculates that the Sun Flyer will operate at a cost of $3/h [50]. Fig. 14.3 includes pictures of both aircraft and the EP Systems battery pack; the caption includes websites with further details.

**Fig. 14.3** Clockwise from upper-left: NASA's X-57 Maxwell, an all-electric aircraft with an industry-created battery pack that can power all of the plane's motors, weighs as little as possible, and meets stringent safety guidelines; right: Electric Power Systems (EP Systems) battery pack to power NASA's all-electric X-57 Maxwell airplane. The package houses thousands of off-the-shelf lithium-ion batteries and ensures that if one of them overheats, the problem will not spread; Lower-left: Bye Aerospace's Sun Flyer, an all-electric training airplane that is being flight tested. It is powered by a battery pack developed by EP Systems based on the one it built, with NASA funding and research partnering, as described in the text, for the X-57 aircraft. *(For further information, see https://www.nasa.gov/centers/armstrong/news/FactSheets/FS-109.html, and https://spinoff.nasa.gov/Spinoff2019/t_1.html, Accessed 2 August 2020. Courtesy NASA.)*

## 14.2.3 Batteries for high-altitude airships

A second suborbital application of electrochemical energy storage technologies is for high-altitude airships [47]. The traditional airship aims for a station-keeping strategy at a fixed point (FP) with constant (minimal) airspeed during day and night [60]. Unmanned high-altitude vehicles take advantage of weak stratospheric winds and (primarily) solar energy to operate without interfering with current commercial aviation and with enough endurance to provide long-term services as essentially suborbital satellites [46]. High-altitude (quasi-)stationary platforms are utilized for three main classes of missions: Earth-observing, telecommunications, and navigation, science, and special projects. Table 14.2 includes representative examples of actual high-altitude airships [46,61–67] and virtual airships [68–72] generated to study specific issues, operating conditions, and/or missions.

Several examples of high-altitude airships (HAAs) as well as a nominal design are shown in Fig. 14.4. Thales Alenia Space started developing the Stratobus in 2015, aiming at a stratospheric airship with 250-kg payload able to keep its position at 20-km altitude for 1 year (Fig. 14.4A). NASA is considering a 20-20-20 Airship challenge as part of the Centennial Challenges Program that uses financial incentives for the achievement of previously defined metrics as an alternative to negotiated contract awards. The goals for each Tier are as follows: a minimum altitude of 20 km; maintain the altitude for 20 h (200 h for Tier 2 competition); remain in a 20 km diameter station-keeping area (and navigate a designated path for Tier 2); successfully return the 20 kg payload (200 kg for Tier 2 competition) and payload data. The goal of a next phase is Airship scalability for longer duration flights carrying larger payloads after an in-depth NASA scalability review. The HiSentinel80 unmanned high-altitude airship completed its first successful flight demonstration, launching November 10, 2010, from Page, Ariz., tracking northeast toward Utah and Colorado. The airship payload, part of a U.S. Army Space and Missile Defense Command/Army Forces Strategic Command program, was recovered on November 11, north of Monticello, Utah. The HiSentinel80 airship is designed to launch like a weather balloon (Fig. 14.4D); it assumes the familiar airship shape at mission altitude (Fig. 14.4C).

One significant difference between HAAs and solar-electric aircraft is the significant impact of high winds; the FP strategy cannot fulfill the requirements of long-endurance missions. Similar to the gravity potential flight strategy for solar-powered aircraft [73], a position energy storage strategy

**Table 14.2** Parameters of airships and power systems.

| Name (Study) | Year(s) | Airship size (m) | | | Photovoltaics[a] | | Energy storage type[b] | Ref. |
| | | Length | Diam. | L/D | PV type | η (%) | | |
|---|---|---|---|---|---|---|---|---|
| HASPA | 1975 | 101.5 | 20.4 | 5.0 | Si N/P | 9.5–10 | AgO/Zn | [61] |
| HALROP | 1991 | 223 | 67 | 3.3 | c-Si | 14–16 | $H_2/O_2$ FC | [62] |
| Lotte | 1991–95 | 15.6 | 4 | 3.9 | a-Si | 10–12 | NiCd | [63] |
| SPF airship | 1998 | 245 | 61 | 4.0 | a-Si | 12 | FC/LIB[c] | [64] |
| SOUNDER | 2000 | 37.8 | 7.2 | 5.3 | ASTM[d] | N/A[e] | Li cells | [65] |
| HiSentinel | 2005–10 | 44.5 | 13 | 3.4 | a-Si | 16.8 | Li cells | [66] |
| HALE-D | 2009 | 73 | 21 | 3.5 | a-Si | N/S[f] | LIB | [67] |
| Thales Alenia Stratobus | 2015-Present | 100 | 33 | 3.3 | Not specified[g] | 24 | Fuel cells or Li cells[h] | [46] |
| HALE Airship (Study) | 2005 | 185 | 46 | 4.0 | a-Si | 8 | $H_2/O_2$ FC | [68] |
| MAAT (Study) | 2011 | 70 m height, 350 m diameter | | 0.2[i] | a-Si | 8.5 | $H_2/O_2$ FC | [69] |
| Stratospheric Airship (Study)[j] | 2017 | 72 | N/S[f] | N/A | N/S[f] | 8, 8.5 | LB | [70] |
| Stratospheric Airship (Study)[k] | 2018 | 220 | 54 | 4.1 | N/S[f] | N/S[f] | LB/FC[c] | [71] |
| Stratospheric Airship (Study)[l] | 2020 | 220 | 54 | 4.1 | Si-based | 20 | LB | [72] |

[a]Suborbital photovoltaic technologies are typically, silicon-based, either crystalline (c-Si) or thin-film amorphous (a-Si); the typical efficiency of crystalline is 15%–20% and amorphous is 8%–14%.
[b]Energy storage is unless designated a battery, typically a Li or Li-ion battery (LB/LIB) or fuel cell (FC).
[c]Employs both battery and fuel cell technology.
[d]Auto sun-tracking mode.
[e]Not applicable.
[f]Not specified for proprietary reasons or not important for the application or study.
[g]Given the 24% efficiency could be advanced c-Si or thin GaAs.
[h]Both fuels cells and high-performance battery technologies are being considered.
[i]Saucer-shaped.
[j]Endurance study.
[k]Energy system study.
[l]Position energy storage study.
Data from J. Gonzalo, D. López, D. Domínguez, A. García, A. Escapa, On the capabilities and limitations of high altitude pseudo-satellites, Prog. Aerosp. Sci. 98 (2018) 37–56; Y. Xu, W. Zhu, J. Li, L. Zhang, Improvement of endurance performance for high-altitude solar-powered airships: a review, Acta Astronaut. 167 (2020) 245–259.

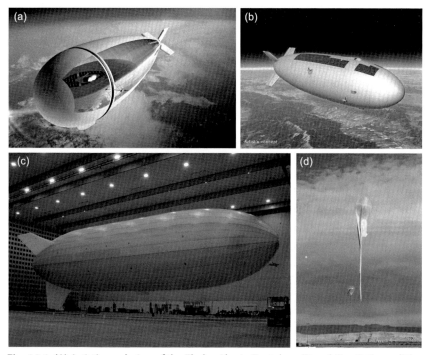

**Fig. 14.4** (A) Artist's rendering of the Thales Alenia Stratobus; (B and C) artist's rendition of an airship that achieves goals of the NASA 20-20-20 Airship Challenge; (C) picture of HiSentinel80 unmanned high-altitude airship moored indoors; (D) HiSentinel80 beginning a successful test November 10, 2010, launch from Page, Arizona. ((A) Creative Commons 4.0 (BY and CC) https://creativecommons.org/licenses/by-sa/4.0/legalcode; (B) Courtesy NASA; (C and D) Courtesy United States Army.)

(PES) can store extra energy during the daytime in the wind field potential energy to improve the wind resistance and endurance [72]. As shown in Fig. 14.5, an airship can be directed to an upper wind area at high speed by supplying surplus solar energy to the propulsion system during daytime while reducing propulsion power to float back to the original position at an airspeed below the wind speed during nighttime. In the top row, the airship sits at its dusk starting point; in the second row, it flies backward at low airspeed (nighttime); in the third row, solar energy is used to accelerate the airship, the airship flies to the upper wind area at airspeed, the excess energy charges the energy storage battery (daytime); in the fourth row: As the intensity of solar radiation decreases, the airship decelerates to its starting point (dusk). Adopting a PES strategy is one approach to enhance the utilization of solar energy by adjusting the altitude and the average airspeed of the

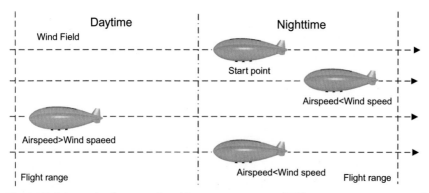

**Fig. 14.5** Schematic diagram of position energy storage (PES) strategy, noting time of day and relative wind and airship speeds. *(Reproduced with permission from C. Shan, M. Lv, K. Sun, J. Gao, Analysis of energy system configuration and energy balance for stratospheric airship based on position energy storage strategy, Aerosp. Sci. Technol. 101 (2020) 105844. Copyright (June 2020) Elsevier.)*

airship. As discussed in the study by Sun et al., this energy storage strategy is impacted by wind speed and irradiation intensity; a flight strategy and an energy management model are necessary to cope with variable and changing environmental conditions [72].

A comparison of the energy utilization of several different approaches in parallel studies is involving either combined energy storage technologies [71] or a PES approach [72]. Fig. 14.6 shows the total stored energy during a typical day of a lithium baseline battery (BB) or fuel cell (FC) for the energy storage study that relies mainly on battery usage at night, while the FC subsystem is mainly employed as an emergency backup power supply in case of battery failure [71]. For the PES case outlined above and shown in Fig. 14.5, Fig. 14.7 demonstrates the power condition (required and solar-generated) and state of charge (SOC) of the batteries. Morning ($t_1$ to $t_2$) involves speeding up of the airship, moving to higher altitude, charging of batteries; daytime ($t_2$ to $t_3$) charges batteries to maximum SOC at the higher altitude; late afternoon to evening ($t_3 \sim t_4$ to $t_5$) is when the airship drifts back to lower altitude and slows down; nighttime ($t_5$ to $t_1$), the study airship runs off battery power. Besides utilizing strategies such as the PES [72] or combining batteries and fuel cells [71], there are other approaches to enhancing solar energy harvesting. A summary of various methods [74–82] is given in Table 14.3. The following subsection addresses power issues, mainly related to the choice of energy storage technology, performance enhancement, and matching mission with appropriate choice of platform.

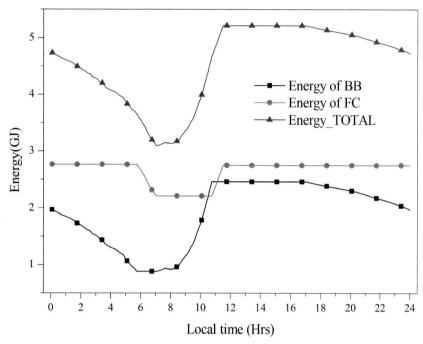

**Fig. 14.6** Daily time history of the energy of lithium baseline battery (BB) cells, fuel cell (FC) system, and total energy of a theoretical (study) airship. *(Reproduced by permission from J. Liao, Y. Jiang, J. Li, Y. Liao, H. Du, W. Zhu, L. Zhang, An improved energy management strategy of hybrid photovoltaic/battery/fuel cell system for stratospheric airship, Acta Astronaut. 152 (2018) 727–739. Copyright (November 2018) Elsevier.)*

### 14.2.4 High-altitude platforms: Power considerations and alternative technologies

A generic schematic for a solar-powered high-altitude aircraft or airship (alternatively known as high-altitude pseudo-satellites (HAPS) [46]) power system is shown in Fig. 14.8. Batteries are inherently simpler than fuel cell technology; thus, batteries are an attractive option for use in airships [47,83] and certainly advantageous for an all-electric aircraft, as discussed above. With few exceptions (i.e., Helios HP03), energy storage for SEAs relies on batteries, typically LIBs or high-performance Li-S batteries for commercially available Zephyr S [45,46]. In fact, the current world record for high altitude at 29.5 km is held by the Helios HP01, relying on LIB energy storage [45]. However, HAA energy storage can be supplied by batteries, fuel cells, or a combination of the two. A schematic of a representative HAA is shown in Fig. 14.9; obviously, fuel cell usage requires space for storage of constituent gases; typically, a HAA length exceeds 150 m; see Table 14.2.

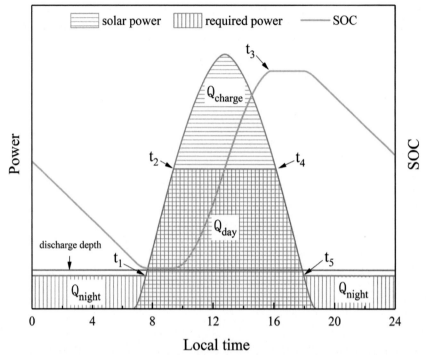

**Fig. 14.7** Energy balance (including battery state of charge (SOC)) based on PES strategy study of theoretical study. *(Reproduced with permission from C. Shan, M. Lv, K. Sun, J. Gao, Analysis of energy system configuration and energy balance for stratospheric airship based on position energy storage strategy, Aerosp. Sci. Technol. 101 (2020) 105844. Copyright (June 2020) Elsevier.)*

### *14.2.4.1 Energy storage technology options*

As shown in Fig. 14.10, because the cycle efficiency of RFCs (40%–60%) [84] is lower than that of LIBs (~90%) [85], fuel cells require a larger area photovoltaic array or more solar cells to charge, assuming a constant output energy; this results in an increased weight of the photovoltaic system. However, as the theoretical specific energy of RFCs is higher than LIBs [86], the energy storage system weight can be greatly reduced; utilizing weight-saving gas storage technologies further improves weight reduction [87]. The offsetting advantages of the two technologies are demonstrated by comparing the trends for PV array area and energy storage system weight indicated in the four corners of Fig. 14.10; a more in-depth analysis of battery choices for solar-powered HAAs can be found in a study by Pande and Verstraete [88]. Although outside of the scope of this chapter, other energy storage

Batteries for aeronautics and space exploration    545

**Table 14.3** Comparison of some methods to increase solar array output of high-altitude airships.

| Method | Energy output increase (%) | Further details/feature description | Ref. |
|---|---|---|---|
| MPPT[a] | 18 | Experimental validation | [74] |
| | 17 | MATLAB simulation[b] | [75] |
| Temperature control | 8 | Thermal protection structure | [76] |
| Solar array layout optimization | 50 | Installation angle optimization | [77] |
| | 12 | Layout optimization | [78] |
| | 8 | Layout optimization with thermal effect[c] | [79] |
| | 8 | Delimit the layout effective surface area | [80] |
| Airship attitude optimization | 100 | Rotatable solar array system | [81] |
| | 6 | Yaw angle optimization at quasi-zero wind layer[d] | [82] |

[a]Maximum power point tracking.
[b]Theoretical study.
[c]Increased temperature of arrays reducing power output.
[d]Control of rotation to optimize solar radiation on fixed solar arrays.
Reproduced with permission from Y. Xu, W. Zhu, J. Li, L. Zhang, Improvement of endurance performance for high-altitude solar-powered airships: a review, Acta Astronaut. 167 (2020) 245–259. Copyright (February 2020) Elsevier.

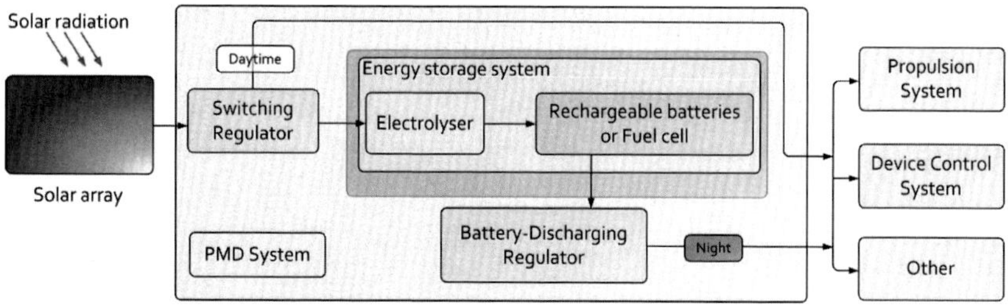

**Fig. 14.8** Components of energy system for high-altitude airship but also generally applicable to solar-powered aircraft. *(Reproduced by permission from Y. Xu, W. Zhu, J. Li, L. Zhang, Improvement of endurance performance for high-altitude solar-powered airships: a review, Acta Astronaut. 167 (2020) 245–259. Copyright (February 2020) Elsevier).*

options such as thermal energy storage (TES) and superconducting magnetic energy storage (SMES) for a variety of applications are discussed in greater detail in a recent review article [89]. The impact of energy storage technologies on the performance of HASPs has been studied and recently reported by Gonzalo et al. [46]. A comparison of operating metrics for representative energy storage options is shown in Table 14.4.

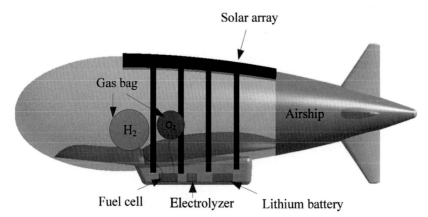

**Fig. 14.9** Schematic of a representative high-altitude airship that emphasizes the energy system using either batteries and/or fuel cells; gas storage infrastructure is included. *(Reproduced by permission from J. Liao, Y. Jiang, J. Li, Y. Liao, H. Du, W. Zhu, L. Zhang, An improved energy management strategy of hybrid photovoltaic/battery/fuel cell system for stratospheric airship, Acta Astronaut. 152 (2018) 727–739. Copyright (November 2018) Elsevier).*

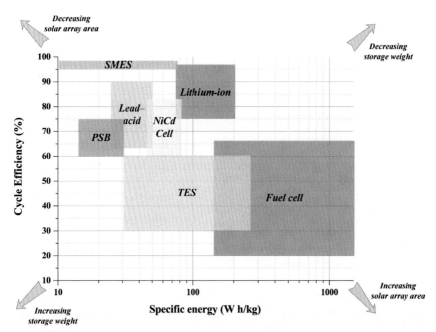

**Fig. 14.10** Comparison of specific energy and cycle efficiency of fuel cell, battery (lead-acid, polysulfide bromide (PSB), Li-ion, NiCd), thermal energy storage (TES), and superconducting magnetic energy storage (SMES). *(Reproduced with permission from Y. Xu, W. Zhu, J. Li, L. Zhang, Improvement of endurance performance for high-altitude solar-powered airships: a review, Acta Astronaut. 167 (2020) 245–259. Copyright (February 2020) Elsevier.*

**Table 14.4** Key parameters of energy storage solutions for high-altitude pseudo-satellites.

| Performance metric | Gasoline engine | Rechargeable batteries[a] | | | | | Regenerative fuel cells[b] | |
|---|---|---|---|---|---|---|---|---|
| | | NiCd | Ni-MH | Li-Po | Li-ion | Li-S | PEMFC | SOFC |
| Energy density (Wh/kg) | 12,800 | 40–60 | 30–80 | 130–200 | 160 | 250–350 | 300–2000 | 300–2000 |
| Operating temperature range (°C) | −50/60 | −20/60 | −20/60 | −20/60 | −40/60 | −20/60 | 50/100 | 700/1000 |
| Efficiency (%)[c] | 15–25 | 80 | 70 | >99 | >99 | >99 | 40–45 | 60 |
| Life cycles | N/A | 500 | 500–1000 | >1000 | 1200 | >100 | N/A | N/A |

[a]NiCd, nickel cadmium, Ni-MH, nickel metal hydride, Li-Po, lithium polymer.
[b]PEMFC, proton exchange membrane fuel cells, SOFC, solid oxide fuel cell.
[c]From energy input to electrical output.
Reproduced by permission from J. Gonzalo, D. López, D. Domínguez, A. García, A. Escapa, On the capabilities and limitations of high altitude pseudo-satellites, Prog. Aerosp. Sci. 98 (2018) 37–56. Copyright (April 2018) Elsevier.

### 14.2.4.2 Comparison to hydrocarbon-powered platforms

Obviously, given the much higher energy density of a hydrocarbon-fuel-powered HASP, such as the Global Hawk (RQ-4) [90], superior performance is expected. During the 1990s, the U.S. DOD DARPA-funded Advanced Concept Technology Demonstration program funded seven demonstration aircraft. The RQ-4 (6800 kg empty, 1360 kg payload, 14,630 kg gross) with a 40 m wingspan, a 14.5 m length, a cruising speed of 360 mph (160 m/s), and an 18.3 km ceiling has operated for over 20 years for a variety of nations' militaries and NASA (Fig. 14.11). In April 2001, a Global Hawk flew nonstop from a U.S. Air Force base in the California desert to a RAAF base in Australia, the first pilotless aircraft to cross the Pacific Ocean. The flight took 22 h, setting a world record for distance flown by a UAV of 13,220 km.

**Fig. 14.11** The NASA Global Hawk Unmanned Aircraft System (UAS) banks for a landing over Rogers Dry Lake at Edwards Air Force Base in California at the end of a test flight on October 23, 2009. The NASA Global Hawk is capable of flight altitudes greater than 18 km and flight durations of up to 30 h and thus is an ideal platform for investigations of hurricanes. The Hurricane and Severe Storm Sentinel (HS3) mission will utilize two Global Hawks, one with an instrument suite geared toward measurement of the environment and the other with instruments suited to inner-core structure and processes. *(Courtesy NASA.)*

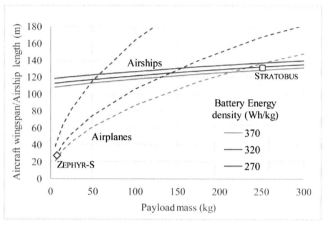

**Fig. 14.12** Aircraft size required for different battery energy densities (*dashed. lines*: airplanes; *solid lines*: airships). *(Reproduced with permission from J. Gonzalo, D. López, D. Domínguez, A. García, A. Escapa, On the capabilities and limitations of high altitude pseudo-satellites, Prog. Aerosp. Sci. 98 (2018) 37–56. Copyright (April 2018) Elsevier).*

### *14.2.4.3 Performance analysis of high-altitude platforms*

A plot of battery energy density and wingspan (or length) on payload mass capacity is given in Fig. 14.12 for two representative platforms: a HAA and a SEA [46]. Given the numerous assumptions made during the study, the trends are far more important than the absolute numbers generated by the methodology. A useful aspect of the study is adopting a generic approach as it seeks to compare the two major HAPS designs (HAA and SEA) using the same metrics. As expected, the payload capacity of a SEA is much more dependent on both wingspan and battery energy density.

Finally, Table 14.5 gives a summary of key results of a recent study that seeks to determine the dependence of the endurance of a HAA on different flight approaches, similar to a more recent study [72], and power system performance [70]. The study examines a virtual HAA of 72 m that begins flight at an altitude of 18.3 km starting in a mid-Summer evening at 120°E, 40°N. Some assumptions made during the simulation are as follows: (1) The LB storage system begins at a fully charged state and stops discharging when it reaches discharge depth; (2) when power is supplied by the solar array and lithium battery, the solar array power is consumed first, and any insufficiency is supplied by the LB; (3) a fixed-point wind resistance strategy is the baseline case, with an average wind speed of 10 m/s.

## 550  Lithium-sulfur batteries

**Table 14.5** Main parameters of power system with endurance times from HAA endurance study.

| Study case ⇒ <br><br><br> Parameter ⇓ | Fixed-point wind resistance strategy | Maneuverable wind resistance strategy | + Increased PV cell efficiency | + LB mass-specific energy increase |
|---|---|---|---|---|
| Endurance (h) | 32.3 | 56.5 | 130 | 177.5 |
| Solar cell η (%) | 8 | | 8.5 | 8.5 |
| LB-specific mass (Wh/kg) | 200 | | 200 | 212.5 |
| Length (m) | 72 | | | |
| Solar array area (m$^2$) | 65 | | | |
| LB mass (kg) | 130 | | | |
| LB charge η | 95 | | | |
| LB discharge η | 98 | | | |
| LB discharge depth | 95 | | | |
| Propulsion system power (W) | 1320 | | | |
| Payload power (W) | 500 | | | |
| Other devices power (W) | 500 | | | |
| Average wind speed (m/s) | 10 | | | |

Reproduced with permission from X. Yang, D. Liu, Renewable power system simulation and endurance analysis for stratospheric airships, Renew. Energy 113 (2017) 1070–1076. Copyright (December 2017) Elsevier.

Results of the optimized case are shown in Fig. 14.13. Given assumption (1) above, the virtual flight was ended when a 95% discharge (Table 14.5) was reached; thus, the depth of discharge, plotted in Fig. 14.13, is the determining factor. The important take-home lesson of this study is a more realistic assessment of component technologies and the overall system as it points out numerous considerations required in the design and performance of an actual power system including the utilization of power from the solar array and energy storage subsystems. Simply increasing the solar cell efficiency and battery energy density by approximately 5% each resulted in a tripling of the endurance time from an optimized flight strategy of 56.5–177.5 h. The study points to the need for further improvement(s) in power components, structures, and flight strategies to enable long-endurance missions (including station keeping) for HAPS of any design.

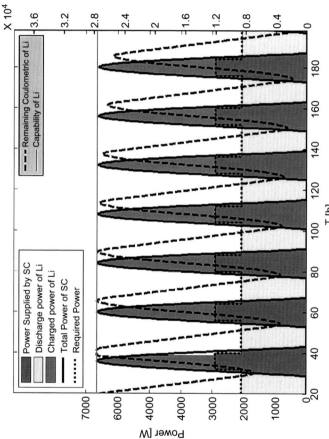

**Fig. 14.13** Simulation results for optimized case Table 14.5 with lithium battery specific energy of 212.5 Wh/kg. (Reproduced with permission from X. Yang, D. Liu, Renewable power system simulation and endurance analysis for stratospheric airships, Renew. Energy 113 (2017) 1070–1076. Copyright (December 2017) Elsevier.)

552 Lithium-sulfur batteries

## 14.3 Overview of energy storage for space exploration

In this section, we address the recent past and future energy storage technology needs for the exploration of the solar system. Despite the greater efficiency and power generated by multijunction III–V solar cells, with efficiencies in the range of 25%–30% [40,44,91], the total power required for spacecraft operations is not always available (or sufficient) from solar cells or arrays depending upon the orbit, mission duration, distance from the Sun, or peak loads. This will therefore necessitate stored, onboard energy, mainly from batteries.

Primary batteries are not rechargeable and thus are used only for short mission durations (typically a day, up to 1 week) [36,40]. As discussed above, there are numerous rechargeable or secondary energy storage options with various performance metrics; see Tables 14.1, 14.2 and 14.4, and Fig. 14.10 [46,83,88,89]. Secondary-type batteries include chemistries such as nickel-cadmium (NiCd), nickel-hydrogen (NiH$_2$), nickel metal hydride (Ni-MH), lithium-ion battery (LIB), lithium sulfide (Li-S), and lithium polymer (LiPo), and many have been used extensively in the past on small spacecraft [36,40] or HAPS [45,46]. Each battery type is associated with certain applications that depend on performance parameters, including energy density, cycle life, and reliability [40,83,88]. A comparison of state-of-the-art battery energy densities for key commercial manufacturers currently utilized in space missions is shown in Fig. 14.14 [92–95].

A recent NASA Jet Propulsion Laboratory report on Energy Storage Technologies for Future Planetary Science Missions [36] is highly recommended to the reader for an in-depth treatment of this intriguing topic with details of a wide variety of potential primary and rechargeable batteries for numerous current and future proposed missions. Table 14.6 gives a convenient overview of optimal battery type to be utilized for planetary science solar system exploration missions. Given the broad technology space of battery types and materials, the final section of this chapter will focus on a discussion of several interesting representative missions and practical aspects of (rechargeable) batteries for exploration. The instructive sampling of the recent past and upcoming exploration missions will be followed by a survey of future mission concepts, a more in-depth consideration of appropriate battery technologies for specific destinations, and end with some comments on the long-term prospects for humanity in our solar system and beyond.

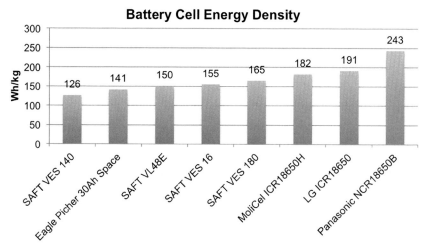

**Fig. 14.14** A comparison of secondary battery energy densities for key commercial manufacturers. (Reproduced from E. Agasid, R. Burton, R. Carlino, G. Defouw, A.D. Perez, A.G. Karacalıoğlu, B. Klamm, A. Rademacher, J. Schalkwyk, R. Shimmin, J. Tilles, S. Weston, State of the Art Small Spacecraft Technology, NASA/TP—2018–220,027, December 2018, 207 pp. Courtesy NASA.)

## 14.4 Recent NASA missions to Mercury, Mars, and small bodies

Table 14.7 presents a summary of rechargeable battery technologies for 2001–22 NASA missions [96–118]; note that after 2007, Li-ion battery technologies have displaced Ni-H$_2$ batteries due to superior performance. For readers interested in further information on deep space exploration missions, a recent NASA publication provides an excellent historical overview [51]. Several interesting and representative (inner planetary, Mars, outer planets (and their moons), and small bodies) missions with challenging environments or instructive battery technology lessons learned are discussed in further detail in this and the next subsections. Battery technologies for small satellites (i.e., CubeSats [40]), mainly used for LEO missions (but also Mars [44]), are discussed in more detail below, and local (Earth-Moon system) missions are cursorily addressed in the JPL report [36] and included in Table 14.7; the referenced online mission pages, several recent publications [119–121], as well as our own recent book chapter [122] are all recommended to the reader.

**Table 14.6** Energy storage technology needs for future planetary science mission concepts.

| Mission | | Battery type (Primary or rechargeable) | | Desirable battery performance parameters | | | Challenging destination environmental parameters and issues | | | |
|---|---|---|---|---|---|---|---|---|---|---|
| Destination | Type | P | R | Specific energy (Wh/kg) | Life (years) | Cycle life | High temp. (°C) | Low temp. (°C) | R[a] | PP[b] |
| Inner planets/ Venus | Orbital | | X | >250 | >10 | >50,000 | | | | |
| | Aerial | | X | >100 | >4 | >500 | 25–350 | | | |
| | Surface | X | | >200 | 0.5–1 | N/A | ~460 | | | |
| Mars | Orbital | | X | >250 | >15 | >50,000 | | | | |
| | Aerial | | X | >250[c] | >5 | >1000 | | −40 | | X |
| | Surface | | X | >250 | >5 | >1000 | | −40 | | X |
| | Sample return missions | | X | >250 | >10 | | | −40 | | X |
| | Human precursor missions | | X | >250 | >15 | >1000 | | | | X |
| Outer planets | Orbital | | X | >250 | >15 | 1000 | | | J | OW |
| | Surface | X | | >500 | >15 | N/A | | −180 | J | OW |
| | Probes | X | | >500 | >15 | N/A | | −180 | J | OW |
| Small bodies | Orbital | | X | >250 | >15 | >50,000 | | | | |
| | Surface | | X | >250 | >5 | >1000 | | −40 to 40 | | |
| | Sample return | X | | >500 | >15 | NA | | −40 to 40 | | |

[a]Radiation tolerance for Jupiter system missions (J)—see footnote.[d]
[b]Planetary protection protocols required on Mars (X) or Ocean Worlds (OW)—see footnote.[d]
[c]Specific power of 3 kW/kg.
[d]Two categories of outer planetary missions being considered for 2020–29: (a) missions to Ocean Worlds (OW) destinations include Callisto (J), Enceladus, Europa (J), Ganymede (J), and Titan and (b) missions to the Ice Giants (IG) of Neptune and Uranus.
Data from P.M. Beauchamp, Energy Storage Technologies for Future Planetary Science Missions, Jet Propulsion Laboratory, prepared for Planetary Science Division, Science Mission Directorate, NASA HQ, December 2017, JPL D-101146. Courtesy NASA.

**Table 14.7** Rechargeable batteries used in NASA missions.

| Mission | Launch date | Destination (Type)[a] | Battery system[b] | Ref.[c] |
|---|---|---|---|---|
| 2001 Mars Odyssey | April 2001 | Mars | $Ni-H_2$ | [96] |
| MER–Spirit[d] | June 2003 | Mars | Li-ion (NCO) | [97] |
| MER–Opportunity[d] | July 2003 | Mars | Li-ion (NCO) | [97] |
| Messenger | August 2004 | Mercury (IP) | $Ni-H_2$ | [98] |
| Deep Impact | January 2005 | Comet (SB) | $Ni-H_2$ | [99] |
| MRO[e] | August 2005 | Mars | $Ni-H_2$ | [100] |
| Phoenix | August 2007 | Mars | Li-ion (NCO) | [101] |
| Dawn | September 2007 | Vesta and Ceres (SB) | $Ni-H_2$ | [102] |
| Kepler | March 2009 | Earth Orbit (EM) | Li-ion (LCO) | [103] |
| LRO[f] | June 2009 | Moon (EM) | Li-ion (LCO) | [104] |
| LCROSS[g] | June 2009 | Moon (EM) | Li-ion (LCO) | [105] |
| Juno | August 2011 | Jupiter (OW) | Li-ion (NCO) | [106] |
| GRAIL[h] | September 2011 | Moon (EM) | Li-ion (NCO) | [107] |
| MSL-Curiosity[i] | November 2011 | Mars | Li-ion (NCO) | [108] |
| LADEE[j] | September 2013 | Moon (EM) | Li-ion (LCO) | [109] |
| MAVEN[k] | November 2013 | Mars | Li-ion (NCO) | [110] |
| OSIRIS-REx[l] | September 2016 | Asteroid Bennu (SB) | Li-ion | [111] |
| InSight | May 2018 | Mars | Li-ion (NCA) | [112] |
| Mars 2020-Perserverence | July 2020 (landing 2/2021) | Mars | Li-ion (NCA) | [113] |

*Continued*

**Table 14.7** Rechargeable batteries used in NASA missions—cont'd

| Mission | Launch date | Destination (Type) | Battery system | Ref. |
|---|---|---|---|---|
| DSCOVR[m] | February 2015 | L-1[n] (EM) | Li-ion (LCO) | [114] |
| MMS[o] | March 2015 | Various orbits (EM) | Li-ion (LCO) | [115] |
| TESS[p] | April 2018 | HEO Orbit[q] (EM) | Li-ion (NCO) | [116] |
| JWST[r] | December 2021[u] | L-2 (EM)[s] | Li-ion (LCO) | [117] |
| JPSS-2[t] | September 2022[u] | LEO (EM) | Li-ion (LCO | [118] |

[a]Mission type: Inner planets (IP), Earth-Moon (EM), small bodies (SB), Outer World (OW).
[b]NCO = LiNiCoO$_2$, LCO = LiCoO$_2$, NCA = LiNi$_{0.83}$Co$_{0.14}$Al$_{0.03}$O$_2$-based systems.
[c]Mission website.
[d]Mars Explorer Rover.
[e]Mars Reconnaissance Orbiter.
[f]Lunar Reconnaissance Orbiter.
[g]Lunar Crater Observation and Sensing Satellite.
[h]Gravity Recovery and Interior Laboratory.
[i]Mars Science Laboratory.
[j]Lunar Atmosphere and Dust Environment Explorer.
[k]Mars Atmosphere and Volatile EvolutioN.
[l]Origins, Spectral Interpretation, Resource Identification, and Security-Regolith Explorer.
[m]Deep Space Climate Observatory.
[n]L1 Lagrangian point is about 1.5 million km (0.01 au) from Earth toward the Sun.
[o]Magnetospheric Multiscale Satellites.
[p]Transiting Exoplanet Survey Satellite.
[q]Also known as a lunar-resonant orbit known as P/2.
[r]James Webb Space Telescope.
[s]L2 Lagrangian point is about 1.5 million km (0.01 au) from Earth away the Sun.
[t]Joint Polar Satellite System.
[u]Current projected launch date.
Modified from P.M. Beauchamp, Energy Storage Technologies for Future Planetary Science Missions, Jet Propulsion Laboratory, prepared for Planetary Science Division, Science Mission Directorate, NASA HQ, December 2017, JPL D-101146. Courtesy NASA.

## 14.4.1 Mercury Surface, Space Environment, Geochemistry and Ranging (MESSENGER) mission

Several useful insights into environmental issues (high temperature and variable pressure) to be faced by power systems intended for inner planetary missions [123] can be gained by considering the Mercury Surface, Space Environment, Geochemistry and Ranging (MESSENGER) mission (Fig. 14.15), the seventh Discovery-class mission, and the first spacecraft to orbit Mercury [98]. Its primary goal was to study Mercury's geology, magnetic field, and chemical composition, the first mission to the planet after Mariner 10, more than 30 years before. MESSENGER was launched at 2:15 a.m. EDT August 3, 2004, into an initial parking orbit around Earth. The six-and-a-half-year path to Mercury was facilitated by a number of gravity-assist maneuvers through the inner solar system, including one flyby of Earth, two flybys of Venus, and three flybys of Mercury. On March 18, 2011, MESSENGER made history by becoming the first spacecraft ever to orbit Mercury; 11 days later, the spacecraft captured the first image ever obtained from Mercury orbit (Fig. 14.15D). Originally planned as a 1-year orbital mission, the spacecraft orbited Mercury for 4 years, demonstrating several new spaceflight technologies and making new scientific discoveries about the origin and evolution of Mercury, our solar system's innermost planet.

A preflight paper by Santo et al. details a number of technological advances developed for the MESSENGER spacecraft to address the environmental challenges presented by a mission to Mercury (extreme temperature(s), intense solar radiation (up to 11 Earth-equivalent suns), and vacuum) [123]. We address two adaptations required for power systems to function under these extreme conditions: sealing of batteries and a sunshade (Fig. 14.15A), 3M Nextel 312 ceramic cloth covering both sides of a conventional all-Kapton multilayer blanket core, to protect the spacecraft from the direct solar illumination to the extent that it allowed for the use of standard space-grade electronics. The $Ni-H_2$ batteries were protected with COTS pressure-vessel technology, with 11 vessels and two cells per vessel. It was designed to function with the failure of one vessel. The power system maintained in excess of 20% power margin over the entire mission with a maximum battery discharge of 46% [123]. The only other major element exposed to the Sun is the solar array, which is composed of 30% cells (GaAs/Ge) and 70% mirrors. The large area of mirrors reduces the panel operating temperatures to that experienced in Earth-orbiting concentrator

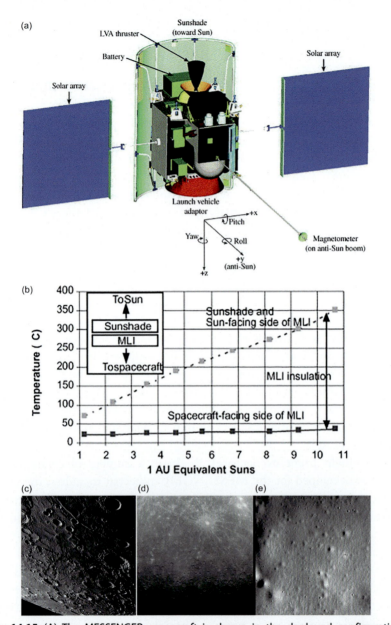

**Fig. 14.15** (A) The MESSENGER spacecraft is shown in the deployed configuration; (B) test results at NASA's Glenn Research Center show the insulating capability of the sunshade; (C) MESSENGER image looking toward Mercury's southern horizon, the rim of Rembrandt basin extends across the middle of the image; (D) first image of Mercury ever obtained from orbit by MESSENGER spacecraft on March 29, 2011; (E) the last of 277,000 images MESSENGER acquired and returned to Earth from Mercury orbit, on April 30, 2015; MESSENGER again made history, becoming the first spacecraft to impact the planet. ((A and B) Reproduced with permission from A.G. Santo, R.E. Gold, R.L. McNutt, Jr., et al., The MESSENGER mission to Mercury: spacecraft and mission design, Planet. Space Sci. 49 (14–15) (2001) 1481–1500. Copyright (December 2001) Elsevier; Photographs (C–E) Courtesy NASA/JHU Applied Physics Laboratory/Carnegie Institution of Washington.)

solar arrays. The sunshade ceramic cloth is rated in excess of 1000°C; under worst-case conditions, a 400°C temperature difference is predicted between the sunshade and the spacecraft [123]. A sample sunshade blanket was fabricated and tested under solar simulation in vacuum at NASA's Glenn Research Center. At 11 times the solar flux received at Earth, the sample shade surface measured 350°C and the spacecraft-facing side of the multilayer insulation measured 35°C (Fig. 14.15B); the temperature of the spacecraft-facing side of the shade was predicted to be flat over the varying solar distances. From this test, it was deduced that the shade's insulting properties are quite insensitive to the shade's optical surface qualities; if the shade surface degrades (effectively increasing the temperature of the Sun-facing surface), the temperature of the back side of the shade remained unchanged. The highly successful MESSENGER mission to Mercury ended on April 30, 2015. MESSENGER again made history, becoming the first spacecraft to impact the planet, and Fig. 14.15E shows the last of 277,000 images. MESSENGER, the first spacecraft ever to orbit Mercury, revolutionized our understanding of the solar system's innermost planet [98].

## 14.4.2 Battery technologies for Mars surface rovers

The technological advance(s) of LIBs is readily apparent by comparing the performance of energy storage systems over the past 15 years as documented by Smart and co-workers of the NASA/Caltech Jet Propulsion Laboratory [124]. As can be seen from Table 14.7, the Li-ion heritage NCO-based chemistry with an all carbonate-based electrolyte has provided excellent performance for a number of missions, including the 2003 Mars Explorer Rover (MER-Spirit and Opportunity) and 2011 Mars Science Laboratory (MSL) Curiosity Rovers. For both MER and MSL, heritage lithium-ion batteries far exceeded primary mission objectives; this has been attributed to a number of factors, including the durability of the chemistry, robustness of the cell and battery design, and the validity of cell balancing and battery operational strategies [124].

A brief review of the history of the three 21st-century Mars rovers is in order (see Fig. 14.16). MER-Spirit (MER-A) landed on January 4, 2004. Six years after the original mission limit, Spirit had covered a total distance of 7.73 km (4.80 mi); further progress ceased after it became trapped in sand. The last communication received from Spirit was on March 22, 2010; NASA ceased attempts to re-establish communication on May 25, 2011. MER-Opportunity (MER-B) landed on January 25, 2004. Opportunity

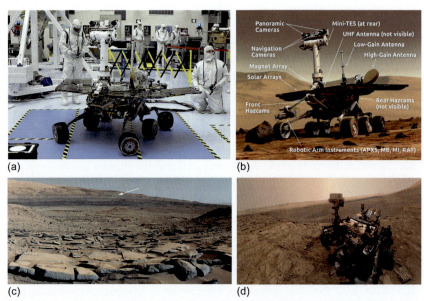

**Fig. 14.16** (A) Mars Exploration Rover-Spirit is tested for mobility and maneuverability in the Payload Hazardous Servicing Facility at NASA Kennedy Space Center, and the MER Mission consisted of two identical rovers designed to cover roughly 110 yards each Martian day; each rover will carry five scientific instruments that will allow it to search for evidence of liquid water that may have been present in the planet's past; (B) Mars Exploration Rover instruments diagram; (C) image taken on March 24, 2014 (Sol 580), when Curiosity was at the base of Mount Sharp, arrow indicates the rover's anticipated location as of July 30, 2020, about 5.5 km (3.5 mi) away; (D) self-portrait of NASA's MSL-Curiosity rover taken on October 11, 2019 (2553rd Martian day, or sol, of its mission), self-portraits are created using images taken by Curiosity's Mars Hand Lens Imager. *(Photographs courtesy of NASA (A and B) or NASA/JPL-Caltech (C and D).)*

surpassed the previous records for longevity after ~15 years (5498 days) from landing to mission end, after traversing 45.16 km (28.06 mi) of the Martian surface. A global 2018 Mars dust storm blocked the sunlight needed to recharge its batteries; Opportunity sent its last status update in June 2018; after repeated attempts to reactivate the rover, NASA declared the mission complete on February 13, 2019. The MSL mission by NASA was launched November 26, 2011, and landed at the Aeolis Palus plain near Aeolis Mons in Gale Crater on August 6, 2012. Curiosity is still operational as of the writing of this chapter.

Beginning with the Mars Insight [112] (including MarCO twin CubeSats [44]) and Mars-2020 missions [113], batteries with NCA-based cathode materials and advanced (i.e., ethylene carbonate/ethyl methyl carbonate/

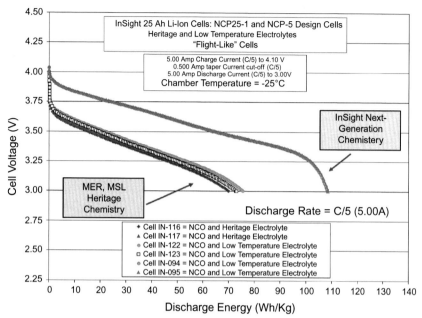

**Fig. 14.17** Discharge energy (Wh/kg) of heritage NCO and next-generation NCA 25 Ah nameplate Li-ion cells delivered at −25°C using 5.0 A charge to 4.10 V (with a C/50 taper current cutoff) and a 5.0 A discharge to 3.0 V. Cells were charged and discharged at −25°C. *(Reproduced by permission from M.C. Smart, B.V. Ratnakumar, R.C. Ewell, S. Surampudi, F.J. Puglia, R. Gitzendanner, The use of lithium-ion batteries for JPL's Mars missions, Electrochim. Acta 268 (2018) 27–40. Copyright (April 2018) Elsevier.)*

methyl propionate [125]) electrolytes have been employed [124]; see Fig. 14.17 for mission pictures. As seen clearly in Fig. 14.18, improved cells with NCA-based cathodes and low-temperature electrolyte delivered 31% more capacity at −25°C (>28.7 Ah corresponding to >108 Wh/kg) relative to NCO-based cells (~19.6 Ah corresponding to ~75 Wh/kg), displaying a much higher operating voltage throughout discharge. Furthermore, cells that employ next-generation chemistries have demonstrated improved retention of low-temperature capability after exposure to high-temperature cycling. As illustrated in Fig. 14.19, cells with the next-generation chemistry suffer less than a 1% loss in capacity at −25°C after being exposed to 35°C cycling (delivering over 29.3 Ah); this is three times better than the greater than 3% loss in the capacity for the heritage (NCO-based) cells. In summary, new cell technology based on improved materials, developed, ground tested, and patented by JPL [124,125] has produced next-generation batteries that have been demonstrated to provide improved performance, especially at

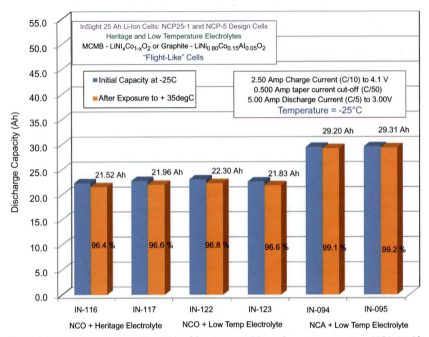

**Fig. 14.18** Discharge capacity (Ah) of heritage NCO and next-generation NCA 25 Ah nameplate Li-ion cells delivered at −25°C, before and after cycling at +35°C, using 5.00 A charge to 4.10 V (with a C/50 taper current cutoff) and a 5.00 A discharge to 3.00 V. Cells were both charged and discharged at −25°C. *(Reproduced by permission from M.C. Smart, B.V. Ratnakumar, R.C. Ewell, S. Surampudi, F.J. Puglia, R. Gitzendanner, The use of lithium-ion batteries for JPL's Mars missions, Electrochim. Acta 268 (2018) 27–40. Copyright (April 2018) Elsevier).*

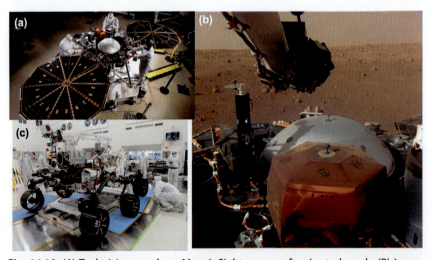

**Fig. 14.19** (A) Technicians work on Mars InSight spacecraft prior to launch; (B) image acquired on December 4, 2018 (Sol 8) from InSight's robotic-arm mounted Instrument Deployment Camera shows the instruments on the spacecraft's deck, with the Martian surface of Elysium Planitia in the background; (C) in a clean room at NASA's Jet Propulsion Laboratory, engineers observed the first driving test for Mars 2020 rover (Perseverance) on December 17, 2019. *(Photographs courtesy of NASA/JPL-Caltech.)*

Batteries for aeronautics and space exploration    **563**

very low temperatures [124] for Mars InSight [112], Mars-2020 [113], and future missions. In the next subsection, we will examine a preview of future exploration missions for later in this decade as well as new directions in lithium battery chemistries and structures to enable new missions and more aggressive planning for future solar system exploration in challenging environments under quite difficult circumstances with intriguing mission objectives.

### 14.4.3 Notable NASA exploration missions to comets and asteroids

We highlight three representative small body exploration missions (see Table 14.7) from the past 15 years. The primary mission of NASA's Deep Impact, launched January 12, 2005, was to probe beneath the surface of a comet (Fig. 14.20A) [99]. The dual-mission spacecraft (flyby and impactor) delivered its specialized impactor into the path of Tempel 1 on July 4, 2005, to reveal never-before-seen materials and provide clues about the internal composition and structure of a comet (Fig. 14.20B). The mission continued with a flyby of comet Tempel 1, a subsequent comet flyby of Hartley 2 (November 2010), and observation of two more comets (Garradd (C/2009 P1) in early 2012 and ISON (C/2012 S1) in early 2013); Deep Impact returned approximately 500,000 images of celestial objects before its mission was ended in September 2013.

Dawn (Fig. 14.20E), launched September 27, 2007, visited the two largest bodies (Vesta and Ceres) in the main asteroid belt and became the first spacecraft to orbit a body in the region between Mars and Jupiter [102]. Studying these planet-like intact survivors from the earliest days of the solar system's formation should provide insights into the original building blocks of our solar system. After a February 2009 gravity assist from Mars, Dawn arrived at Vesta (Fig. 14.20D) on July 16, 2011, and departed September 5, 2012, arriving at Ceres (Fig. 14.20F) on March 5, 2015, and the mission ended on November 1, 2018, when Dawn ran out of fuel orbiting Ceres. The Origins, Spectral Interpretation, Resource Identification, Security-Regolith Explorer (OSIRIS-REx) spacecraft (Fig. 14.20C) lifted off on September 8, 2016, and OSIRIS-REx [111] will be the first U.S. mission to sample an asteroid, retrieving at least two ounces of surface material and returning it to Earth for study. On October 20, 2020, after several sample collection practices, OSIRIS-REx used the Touch-and-Go-Sample-Acquisition-Mechanism (TAGSAM) instrument to collect a sample of regolith from Bennu (Fig. 14.20G). In May2021, several months after the window for departure from the asteroid opened, OSIRIS-REx began its return

**564** Lithium-sulfur batteries

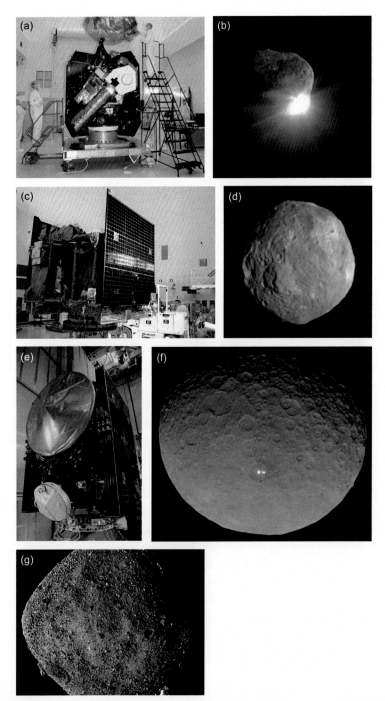

**Fig. 14.20** (A) Photograph of Deep Impact spacecraft with cover removed in a cleanroom at NASA Kennedy Space Center; (B) image of comet Tempel 1 taken 67 s after it obliterated Deep Impact's impactor spacecraft. The image was taken by the
*(continued)*

journey to Earth; it is scheduled to arrive on September 24, 2023 in the Utah (USA) desert. The asteroid, Bennu, may hold clues to the origin of the solar system and the source of water and organic molecules found on Earth.

## 14.5 Radiation issues and exploration missions to the Jupiter region

Beginning in 1972, NASA has conducted seven successful missions (Pioneer 10 and 11, Voyager 1 and 2, Galileo, Cassini-Huygens, and Juno) to the gas giants (Jupiter, Saturn, Uranus, and Neptune) and a number of their moons [51,126]. The latest mission, Juno (Fig. 14.21, Left), the second of the New Frontiers Program missions [127], was launched August 2011, arrived at Jupiter in July 2016, and has recently had its mission extended to September 2025 or until the spacecraft ceases to function. Its principal mission was to orbit Jupiter's poles (minimizing exposure to Jupiter's intense radiation [128]) to investigate the planet's origins, structures, atmosphere, and magnetosphere and finally to delve into the possible existence of a solid planetary core. The extended mission tasks Juno with including multiple rendezvous of Jupiter's most intriguing Galilean moons: Ganymede, Europa, and Io (Fig. 14.21, Right).

**Fig. 14.20—cont'd** high-resolution camera on the mission's flyby craft, scattered light from the collision saturated the camera's detector, creating the bright splash seen here, image reveals topographic features, including ridges, scalloped edges and possibly impact craters formed long ago; (C) inside the Payload Hazardous Servicing Facility at NASA Kennedy Space Center illumination testing is underway on the power-producing solar arrays for Origins, Spectral Interpretation, Resource Identification, Security-Regolith Explorer (OSIRIS-Rex) spacecraft; (D) Dawn obtained this image of the protoplanet Vesta with its framing camera on July 9, 2011, from a distance of about 26,000 miles (41,000 km) away; (E) Dawn spacecraft at the NASA Kennedy Space Center prior to encapsulation at its launch pad on July 1, 2007; (F) view of Ceres, the brightest spots within a crater in the northern hemisphere are revealed to be composed of many smaller spots, from an animation of sequences taken by Dawn spacecraft on May 4, 2015, the image resolution is 0.8 mile (1.3 km) per pixel, the brightest spots within a crater in the northern hemisphere are revealed to be composed of many smaller spots, their exact nature remains unknown; (G) preliminary shape model of asteroid Bennu was created from a compilation of images taken by OSIRIS-REx's PolyCam camera during the spacecraft's approach toward Bennu during the month of November 2018. This 3D shape model shows features on Bennu as small as 6 m. *(Photographs (A, C, E) courtesy of NASA, images (B and D) courtesy of NASA/Jet Propulsion Laboratory-Caltech, image (G) courtesy of NASA/University of Arizona.)*

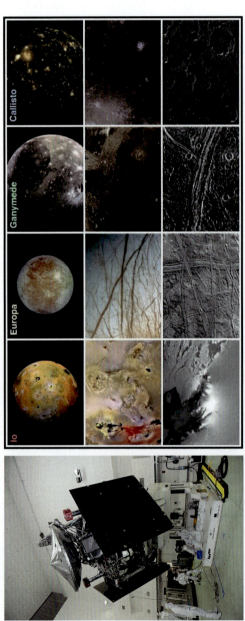

**Fig. 14.21** *(Left)* Juno spacecraft is tested for center of gravity, weighing and balancing on the rotation stand at Astrotech's payload processing facility in Titusville, Florida; *(Right)* Galilean moons mosaic, including images taken by NASA's spacecraft (i.e., Galileo and Voyager's mission). From left to right, Io, Europa, Ganymede, and Callisto moons. The top row displays the relative sizes of the moons in global views at relatively low resolution (10 km per picture element—pixel). Surfaces are affected by tectonic or volcanic changes in the moons' interiors or by exterior deposition. Middle row images (1000 × 750 km$^2$). Regional features include volcanic caldera fields on Io, tidally induced cracks thousands of kilometers long on Europa, bright grooved regions on Ganymede, and enormous impact basins on Callisto caused by impacts with primitive comets or asteroids. Bottom row views (100 × 75 km$^2$) show Io's volcanic plume vents, Europa's abundant ridges, Ganymede's fractured, grooved terrain, and Callisto's heavily eroded and mantled craters. (Photograph (Left) courtesy of NASA, images (Right) courtesy of NASA/JPL-Caltech, *https://photojournal.jpl.nasa.gov/catalog/PIA00743*.)

Every mission to the Jovian region of the solar system must deal with the intense radiation environment near the planet. A combination of Jupiter's strong magnetic field, radial diffusion of electrons, $SO_2$ (and SO and S fragments) from Io's prodigious volcanic activity, and resonant interactions between waves and electrons drive the planet's intense radiation [128]. In fact, the Juno mission successfully demonstrated a critical feature for enabling sustained exploration in such a heavy radiation environment, and the first radiation shielded electronics vault [106]. In this section, we discuss examples of exploration missions that are in progress or recently announced (Table 14.8) [52]. This overview of several missions to outer worlds provides a brief introduction to the next generation(s) of batteries that will need to function in a variety of challenging conditions and environments during exploration missions throughout the solar system.

## 14.5.1 Radiation in space and impact on rechargeable batteries

As discussed above, the radiation environment around Jupiter is a consequence of Jupiter's massive size and innermost moon [128]. However, though much smaller, Earth is surrounded by (much weaker) radiation belts [39], discovered in 1958 after the launch of the first American satellite in 1958, named after their discoverer, James Van Allen [141]. In fact, radiation encountered in space is actually made up of three sources of radiation: particles trapped in a planet's magnetic field, particles shot into space during solar flares, and galactic cosmic rays, high-energy protons, and heavy ions from outside our solar system [142].

As is clear from Table 14.7, there has been extensive use of commercial LIBs for planetary missions; thus, it was imperative to investigate the degradation of the various battery chemistries under (harsh) radiation environments, possibly leading to performance loss and premature cell failure. Several studies at JPL subjected various commercial lithium–ion cells to high levels of $\gamma$-radiation using a $^{60}Co$ source, including large-format NCO-based, large-format NCA-based [143], and LCO-based [144]. Both large-format 7 Ah prismatic cells and 9 Ah cylindrical cells demonstrated tolerance to cumulative levels as high as 25 Mrad, with less than 10% permanent capacity loss observed; some losses could be attributed to the cycling and storage associated with testing. In a similar study, commercial 18,650-size LIBs cells demonstrated tolerance to radiation levels as high as 18 Mrad, displaying 4%–6% capacity loss upon exposure.

**Table 14.8** Recently launched or confirmed future space missions.

| Mission | Date | Destination | Type | Comments | Ref. |
|---|---|---|---|---|---|
| Mars 2020-Perserverence | July 2020 (February 2021) | Mars | Rover Rotorcraft | Jezero crater landing | [113] |
| Hope | July 2020 (February 2021) | Mars | Orbiter | First Arab Mars Mission | [129] |
| Tianwen-1 | July 2020 (February 2021) | Mars | Orbiter, Lander, Rover | First Mars Surface China | [130] |
| Chang'e 5[a] | November 23, 2020 | EMS | Sample Return | Chinese Mission | [131] |
| Double Asteroid Redirection Test (DART) | November 24, 2021 | Small Body | Asteroid Impactor | NASA impactor to asteroids Didymos and Dimorphos | [132] |
| Lucy | October 16, 2021 | Small Body | Asteroid Flyby | NASA Multiple Trojan Asteroids | [133] |
| JWST | December 25, 2021 | L-2 (EM) | Astrophysics | Successor to Hubble Space Telescope | [117] |
| JPSS-2 | September 2022 | LEO (EM) | Earth Science | Environmental Mission | [118] |
| JUpiter ICy moons Explorer (JUICE) | April 2023 | Outer World | Jupiter System Orbiter | ESA mission to orbit both Jupiter and Ganymede | [134] |
| Korea Pathfinder Lunar Orbiter | August 2022 | EMS | Orbiter | First mission by Rep. of Korea | [135] |
| Psyche | August 2022 | Small Body | Orbiter | Main belt asteroid 16 Psyche | [136] |

*Continued*

**Table 14.8** Recently launched or confirmed future space missions—cont'd

| Mission | Date | Destination | Type | Comments | Ref. |
| --- | --- | --- | --- | --- | --- |
| ExoMars 2022 | September 2022 | Mars | Lander, Rover | ESA Mars rover, Russian surface platform | [137] |
| Chang'e 6 | 2024 | EMS | Sample Return | Second Chinese Lunar Sample Return | [131] |
| Europa Clipper | October 2024 | Outer World | Orbiter | NASA Jupiter Orbiter to study Europa | [138] |
| Hera | October 2024 | Small Body | Orbiter | ESA mission to asteroids Didymos and Dimorphos | [139] |
| Dragonfly[b] | 2027 | Outer World | Rotorcraft | NASA Titan explorer (Saturn's largest moon) | [140] |

[a]Chang'e 5's primary mission was completed in late 2020 but the mission has been extended; the orbiter was successfully captured by Sun-Earth Lagrange point 1 (March 15, 2021) and made a lunar flyby on September 9, 2021. Its current destination is unknown.
[b]Dragonfly is the fourth mission in the New Frontiers Program [127]; 1 = New Horizons (Pluto), 2 = Juno (Jupiter), 3 = OSIRIS REx (Asteroid Bennu Sample Return). Modified from: https://nssdc.gsfc.nasa.gov/planetary/upcoming.html, Accessed 19 May 2022. Courtesy NASA.

570 Lithium-sulfur batteries

A comparable study on recently developed commercially available high specific energy 18,650-size Li-ion cells was conducted on Panasonic NCR cells [44]. Minimal capacity fade and impedance growth were observed with the cells after being subjected to cumulative $\gamma$-radiation levels of 20 Mrad from a $^{60}$Co source (Fig. 14.22). After completing radiation testing, these cells also displayed good performance when subsequently cycled at full DOD at ambient temperatures using C/2 charge and discharge rates over the voltage range of 3.0–4.10 V, with over 700 cycles completed for each cell type [44]. Though not strictly relevant for use of commercial batteries in space exploration, irradiation of battery components followed by assembly resulted in cells with significant capacity fade, impedance growth, and premature battery failure [145]. However, a more relevant theoretical study of defects introduced into $Li_4Ti_5O_{12}$ (an advanced anode material) produced mixed results [146].

## 14.5.2 Upcoming missions to Jupiter and several of its icy moons

There are two currently scheduled missions (see Table 14.8) in the middle of this decade with a Jovian destination—JUICE and Europa Clipper; these spacecraft will also investigate two of its three ocean world moons [147]. Callisto, Europa, focus of the Europa Clipper, and Ganymede, to be orbited by JUICE, are three of the four so-called Galilean or largest moons of Jupiter, and the fourth is the innermost and volcanically active moon, Io. Given the level of interest over the next decade in exploration of the Jovian system, see Table 14.8 (and relevant mission websites); it is germane to this topic to include several recent studies.

Because Europa is bathed in radiation trapped in Jupiter's magnetic field, Europa Clipper's payload and other electronics will be enclosed in a thick-walled vault; this was successfully used for the first time by NASA's Juno spacecraft. The vault walls (made up of titanium and aluminum) will act as a radiation shield against most of the high-energy atomic particles, dramatically slowing down the aging effect that radiation has on the spacecraft's electronics [138]. Furthermore, the power subsystem architecture design approach involves implementing a robust single-fault tolerant design with small fault containment regions [148], based on a similar approach used for the Cassini mission to Saturn [149]. JUICE, the European Space Agency mission to Jupiter and Ganymede, will likely include a similar shielding technology approach [134,150]. While not the main topic of this chapter or book, it is necessary to point out that the power-generating component

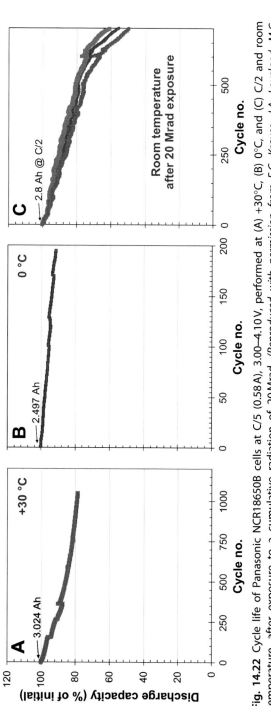

**Fig. 14.22** Cycle life of Panasonic NCR18650B cells at C/5 (0.58 A), 3.00–4.10V, performed at (A) +30°C, (B) 0°C, and (C) C/2 and room temperature, after exposure to a cumulative radiation of 20 Mrad. *(Reproduced with permission from F.C. Krause, J.A. Loveland, M.C. Smart, E.J. Brandon, R.V. Bugga, Implementation of commercial Li-ion cells on the MarCO deep space CubeSats, J. Power Sources 449 (2020) 227544. Copyright (February 2020) Elsevier.)*

## 572  Lithium-sulfur batteries

of the overall power system, solar cells, and arrays are exposed to a much greater degree to damage during missions into high radiation environments; there is an extensive literature in this area, and a cross section of several example publications delving into different types of solar cell technologies and radiation is included for the interested reader [151–156].

Finally, radiation effects have been examined for several primary cells: $Li/CF_x$-$MnO_2$ and $Li/CF_x$ D-sized cells, as well as prototype 3-electrode cylindrical $Li/CF_x$ cells. With up to 10 Mrad of total ionizing radiation from a $^{60}Co$ source, relatively small effects on cell capacity and energy were observed for both cell types when discharged within days of radiation dosing. These results suggest that high doses of radiation on the surface of an icy moon such as Europa should not be a significant issue for a primary $Li/CF_x$ battery [157].

## 14.6 Next generation(s) of battery technologies for space exploration

Table 14.9 provides a summary of battery technologies that could be employed for a variety of future space exploration missions [36]; this is a battery-focused version of Tables 14.7 and 14.8. For six of the seven types of enhanced battery types covered, references for review articles or novel concepts are included below. Additionally, a number of references included in this chapter discuss advanced battery technologies for future space exploration to overcome environmental and durability challenges due to lengthy missions [15–17,22,36,38,39,83,89,124,125,128,141–144].

As we conclude this chapter, we will consider the next steps in continued exploration and possible settlement [53] of the accessible and habitable regions of the solar system during the remainder of the 21st century and beyond. We will revisit several topics that we have addressed previously. Our perspective will be forward-looking, toward future exploration missions, feasibility and concept studies, locales for essential resources, and intriguing enabling technologies to facilitate the next steps in human exploration of the solar system.

### 14.6.1 Future space exploration: Battery technology options and considerations

Aside from continued progress to develop batteries with longer endurance, in general, discussed in this chapter and book, other approaches include conversion of nuclear energy to electricity [158,159] or novel use of materials

**Table 14.9** Advanced rechargeable batteries for future NASA space exploration missions.

| Area(s) of improvement | Potential benefits | Applications | Issues | Solutions | Potential capabilities[a] |
|---|---|---|---|---|---|
| Long(er)-life LIBs | Extend life of mission or reduce replacement cost | Planetary missions, satellite, or ISS | − Electrolyte and cathode degradation <br> − SEI formation <br> − Interfacial impedance increase | − Surface-modified and high-capacity cathodes <br> − Improved electrolytes (additives/co-solvents) | 150–200 Wh/kg <br> 300–400 Wh/L <br> $10^5$ cycles @ 30% DoD <br> CL = 20 y <br> −10°C to 25°C |
| Low(er)-temperature LIBs | Mission enabling for outer worlds and surface missions | Ocean worlds and icy moons <br> Small rovers and landers | − Li diffusivity <br> − Interfacial stability <br> − Electrolyte ionic mobility | − Additives to modify SEI <br> − New electrode materials with enhanced kinetics (Li intercalation; Li + diffusion) | 150–200 Wh/kg <br> 300–400 Wh/L <br> 500 cycles @ 100% DoD <br> CL = 5 y <br> −60°C to 30°C |
| Lithium solid-state inorganic electrolyte batteries | Long calendar life and radiation tolerance (>20 Mrad) | Deep space missions <br> Micro-/nano-spacecraft <br> Advanced instruments | − Low TRL[b] <br> − Low power density and capacity <br> − New oxides are needed | − Inorganic electrolytes <br> − Fabrication methods are needed | 250–350 Wh/kg <br> 300–400 Wh/L <br> $>10^4$ cycles @100% DoD <br> CL > 20 y <br> 10–80°C |
| High-energy LIBs | Potential for lightweight and compact batteries | − Planetary rovers, landers, and probes <br> − CubeSats <br> − EVA | − Structures <br> − Electrolyte/metal ion solubility <br> − Cathode and Hi-V performance | − Surface passivation or composition gradients of cathodes <br> − Advanced electrolytes via additives | 150–200 Wh/kg <br> 300–400 Wh/L <br> >500 cycles @ 100% DoD <br> CL = 20 y <br> −20°C to 40°C |

*Continued*

**Table 14.9** Advanced rechargeable batteries for future NASA space exploration missions—cont'd

| Area(s) of improvement | Potential benefits | Applications | Issues | Solutions | Potential capabilities |
|---|---|---|---|---|---|
| Advanced Li-S batteries | Higher specific energy and energy density Smaller format payloads | – CubeSats<br>– Rovers<br>– Small aerial vehicles and submersibles | – Low TRL[b]<br>– Identify compatible electrolyte<br>– Dendrites<br>– Sulfur | – S-control cathode and electrolyte<br>– New solid electrolytes | 250–300 Wh/kg<br>300–350 Wh/L<br>100–500 cycles @ 100% DoD<br>CL = 5 y<br>−10°C to 25°C |
| Rechargeable LIBs with alternate anodes | Higher specific energy and energy density good low-T performance | – Planetary rovers, landers, and probes<br>– CubeSats<br>– Aerial vehicles | – Si anode issues<br>– SEI<br>– Li dendrites<br>– Safety | – Nanomaterials<br>– Prelithiation<br>– Polymer/SS electrolytes | 250–350 Wh/kg<br>300–400 Wh/L<br>∼500 cycles @ 100% DoD<br>CL = 5 y<br>−40°C to 30°C |
| High-temperature LIBs | Mission enabling:<br>Long duration@25–350°C<br>Short duration@460°C | Venus aerial and surface operations | – Corrosion<br>– Thermal stability<br>– High pressure<br>– Space test | LiAl-FeS$_2$, Na-S, and Na-NiCl$_2$<br>β-alumina<br>Solid electrolyte | 100 Wh/kg<br>150 Wh/L<br>2000 cycles @ 100% DoD<br>5–10 y<br>190–460°C |

[a]Specific energy; energy density; cycle life @ target depth of discharge (DOD); calendar life (y); operating temperature.
[b]Technology readiness level (TRL)—https://www.nasa.gov/topics/aeronautics/features/trl_demystified.html, Accessed August 17, 2020.
Data from P.M. Beauchamp, Energy Storage Technologies for Future Planetary Science Missions, Jet Propulsion Laboratory, prepared for Planetary Science Division, Science Mission Directorate, NASA HQ, December 2017, JPL D-101146. Courtesy NASA.

Batteries for aeronautics and space exploration **575**

available locally or via recycling [160,161]; we will address this topic in more detail in the next sections of this chapter. For many years, LIBs have been subjected to in-depth studies [162] and continually improved, to increase their functionality at lower temperatures, particularly for Martian surface rovers and landers, as discussed above [44,124,125,163]. Self-heating [164] or novel materials [165,166] technologies are more recent advances for enabling the use of LIBs at lower temperatures. Other approaches include simply keeping the battery warm via spacecraft electronics [36,51] or advanced primary batteries [36,157,167]. Lithium solid-state inorganic electrolyte batteries have many potential applications, including those for future space exploration missions, and are the subject of numerous recent reviews [168–171] as a consequence of intense worldwide interest and research activity.

High-energy rechargeable batteries have numerous potential applications; as future space exploration missions are of relevance for this chapter, reviews that include references to aerospace are emphasized. Studies and review articles range from theoretical approaches [146,172,173] to materials studies [174,175], to practical considerations [176,177]. Advanced Li-S batteries are being intensely studied worldwide for numerous applications; we include references to recent reviews that focus on improved performance [24,178–181]. As discussed above for the MESSENGER mission to Mercury, there are alternative methods involving spacecraft design [98,123] to enable high-temperature operations. However, given the level of interest in safe battery operations at high(er) temperatures, there is an abundant literature with numerous reviews of batteries for high-temperature applications [182–186]; finally, we include several reports of novel battery concepts and materials technologies relevant for exploration of Venus [187–189].

## 14.6.2 Upcoming missions to three major classes of asteroids

From 2021 to 2024, four missions will eventually be launched by NASA and the ESA to various asteroids (the total number is approaching $1 \times 10^6$ [190–192]) in our solar system. The destination for these missions will explore the three major classes of asteroids: near-Earth objects (Double Asteroid Redirection Test (DART) [132] and Hera [139]) [193], a main belt asteroid (i.e., Psyche [136]) [191,192], and Jupiter Trojan (i.e., Lucy [133]) asteroids [194]. The DART and Hera missions to Didymos and Dimorphos (Fig. 14.23), a binary synchronous system, are notable as the

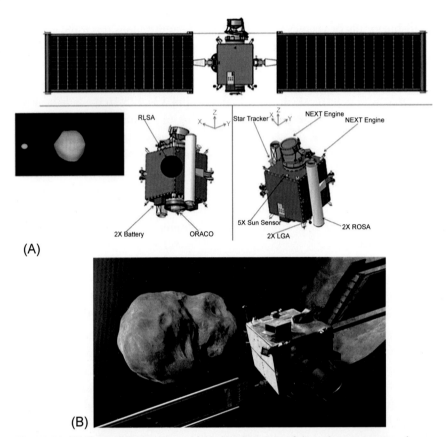

**Fig. 14.23** (A) Three different views of the DART spacecraft bus, the top portion shows an overview of the DART spacecraft with the Roll Out Solar Arrays (ROSA) extended (full extension, including the spacecraft bus, measures ∼12.5 m across), the bottom isometric views show the key aspects of the DART spacecraft bus, the DRACO (a nested acronym for the Didymos Reconnaissance and Asteroid Camera for OpNav) is based on the LOng-Range Reconnaissance Imager (LORRI) high-resolution imaging instrument from the New Horizons spacecraft. The left view also shows the radial line slot array (RLSA) antenna with (solar arrays rolled up). The isometric view on the right shows a clearer view of the (NASA Evolutionary Xenon Thruster—Commercial) NEXT-C ion engine, an electric propulsion xenon thruster developed at the NASA Glenn Research Center; (B) artist's conception of the DART spacecraft approaching the (65803) Didymos system, the primary body is about 800 m in diameter and the moonlet (Dimorphos) is approximately 150 m across, they are separated by just over a kilometer (*inset* to (A) is a telescopic image). The primary body rotates once every 2.26 h, while the tidally locked moonlet revolves about the primary once every 11.9 h, almost one-sixth of the known near-Earth asteroid (NEA) population are binary or multiple-body systems. *((A) Reproduced with permission from R. Landis, L. Johnson, Advances in planetary defense in the United States, Acta Astronaut. 156 (2019) 394–408. Copyright (August 2019) Elsevier; (B) (and inset to (A)) Courtesy: NASA/Johns Hopkins APL.)*

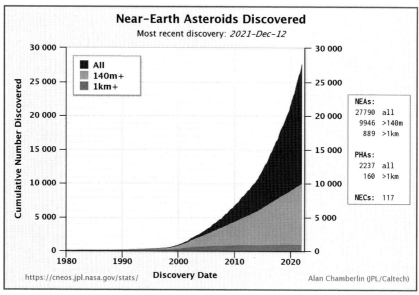

**Fig. 14.24** Known near-Earth asteroid (NEA) population as of December 12, 2021; the total number of NEAs as of this writing stands at 27,790. Of this total, there are 2237 potentially hazardous asteroids (PHAs) with 160 > 1 km in diameter, also, there are 117 "near-Earth comets" (NECs); "PHA-km" and "NEA-km" are PHAs and NEAs with diameters roughly 1 km and larger, "NEA-140 m" are NEAs with diameters roughly 140 m and larger. *(Source: https://cneos.jpl.nasa.gov/stats/totals.html, courtesy NASA.)*

first example of a "planetary defense" [195] or "applied planetary science" mission.

The DART/Hera International Asteroid Impact and Deflection Assessment (AIDA) collaboration has the primary objective to assess kinetic impact as a method for redirection of any future asteroids found to be on a trajectory to impact Earth, and this is the source of the term "potentially hazardous asteroid (PHA)" (Fig. 14.24) [193]. The collaborative mission is also notable for the intended use of a 6U CubeSat (Light Italian CubeSat for Imaging Asteroids (LICIACube) [196]) to image an impact into Dimorphos, the smaller member of the binary PHA. The DART mission is the first managed by the Planetary Defense Coordination Office [195]; the primary mission of the PDCO is to address a potential near-Earth object NEO (includes asteroids and comets) impact hazard [197]. One example of a future NEO close approach will occur on April 13, 2029, when asteroid Apophis passes Earth at an altitude of 32,000 km [52]. While the prospect of a collision between any NEA and the Earth is certainly a potentially catastrophic event, as

**578** Lithium-sulfur batteries

discussed below, asteroids hold great potential for resource harvesting in the future [198–203]. Although outside the purview of this chapter, there are clear limits to missions to asteroids related to spacecraft flight dynamics (or $\Delta v$ (measured in m/s)) [91,198,204–207].

### 14.6.3 Aerial exploration of other planetary bodies: Mars, Titan, and Venus

In Section 14.2, we addressed the myriad technologies, demonstrations, and analyses of terrestrial HAPS, both high-altitude aircraft and airships. However, it is certainly feasible to develop aerial exploration vehicles above the surface of other bodies (with atmospheres) in the solar system. In fact, two missions in this decade, the successful and currently on-going Mars 2020 Ingenuity Helicopter (Fig. 14.25A–E) [113,208] and the upcoming Dragonfly mission (see Fig. 14.25F for an artist's conception of the mission concept) [140] to Saturn's largest moon Titan, an ocean world [147], are intended to demonstrate the operation of nonterrestrial aerial exploration technology.

NASA's Ingenuity Mars Helicopter received a checkout and recharge of its power system on August 7, 2020, 1 week into its near 7-month journey to Mars with the Perseverance rover. The 2-kg ($\sim$4.5 lb) helicopter (Fig. 14.25A) was stowed on Perseverance's belly (Fig. 14.25B) during the flight to Mars; and received its charge from the rover's power supply. This is the first time that Ingenuity has been powered up with its batteries being charged in the space environment. During this (8 h) operation, the performance of the rotorcraft's 6 lithium–ion batteries was analyzed as the team brought their charge level up to 35%. Mars 2020 flight controllers have determined that a lower charge state is optimal for battery health during the trip to Mars. Once Ingenuity is deployed on the Martian surface after touchdown of Perseverance, its batteries will only be charged by the helicopter's own solar panel. Assuming that during its preflight checkout, Ingenuity survives the cold Martian nights, the team will proceed with testing (Fig. 14.25C and D). Fig. 14.25E is a still frame of a video from Perseverance that captures ingenuity in flight. The interested reader is encouraged to visit the Perseverance and Ingenuity websites for updated information, images, and videos [113,208].

Our discussion of future space exploration missions has almost entirely addressed past, present, or near-future missions [33,36,51]. However, an examination of the selection process for the fourth New Frontiers Program (NFP-4) [127] mission that resulted in the selection of Dragonfly

**Fig. 14.25** (A) The flight model of the Mars 2020 Ingenuity Helicopter; (B) the Ingenuity Mars Helicopter can be seen between the left and center wheels of the Mars 2020 Perseverance rover, the image was taken in the vacuum chamber at JPL on October 1, 2019; (C) in this illustration, the Ingenuity Mars Helicopter stands on the Red Planet's surface as the Perseverance rover (partially visible on the left) rolls away; (D) an artist's conception of the Ingenuity Mars Helicopter in flight, while the Perseverance rover rolls over the Martian surface; (E) NASA's Mars Perseverance rover acquired this image using its left Mastcam-Z camera, located high on the rover's mast. This is a still frame from a sequence captured by the camera while taking video. This image was acquired on April 22, 2021. (F) Dual-quadcopter lander Dragonfly mission concept of entry, descent, landing, surface operations, and flight at Titan. *(Courtesy NASA/JPL-Caltech.)*

(Fig. 14.25F) had six mission themes: (1) Comet Surface Sample Return; (2) Lunar South Pole–Aitken Basin Sample Return; (3) Saturn Atmospheric Probe; (4) Trojan Asteroid Tour and Rendezvous; (5) Venus In Situ Explorer; (6) Ocean Worlds (Titan and/or Enceladus). A mission addressing theme (1) named Comet Astrobiology Exploration SAmple Return

(CAESAR) was the runner-up to Dragonfly (addressed Theme 6) in the selection process for NFP-4, and instrument development continues for potential future selection or inclusion of hardware for other missions [209]. Theme 4 will be addressed by Lucy [133], a mission launched in 2021 to Jupiter Trojan [194] asteroids (see Table 14.8).

The very challenging environments of the atmosphere and surface of Venus (Theme 5) remain a significant impediment to the exploration of the second planet from the Sun [210]. However, as with many destinations in the solar system [36], there are certainly numerous proposed missions and studies developed in academia, industry, and various government laboratories, including NASA research centers; one example is the COMPASS team at the NASA Glenn Research Center [211]. A relevant concept study from this team is a Venus high-temperature analysis with proposals for a descent instrument package and a relay orbiter to demonstrate the viability of an integrated atmospheric and in situ geologic exploratory mission, and the proposed mission, Venus high-temperature atmospheric dropsonde and extreme-environment seismometer, is designated with a fitting acronym (HADES) [212]. An example of a Venus exploration study from NASA Langley involves several airship concepts, both manned and unmanned, called a high-altitude Venus operational concept (HAVOC) [213]; Fig. 14.26 provides an interesting size comparison of baseline HAVOC vehicle concepts with other (well-known)

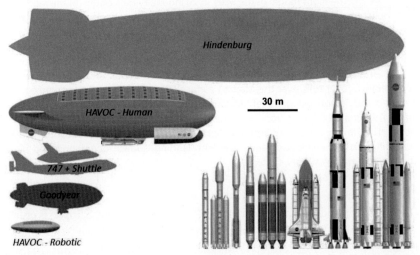

**Fig. 14.26** Size comparison of baseline HAVOC vehicle concepts with other aerospace vehicles. *(Courtesy: Space Mission Analysis Branch, NASA Langley Research Center.)*

aerospace vehicles. As discussed frequently during this chapter, the JPL report *Energy Storage Technologies for Future Planetary Science Missions* [36] is a valuable resource to gain insights into future planetary exploration missions; reviews [214] and straightforward literature searches can also generate a multitude of concept studies or proposed space exploration missions for spacecraft of all sizes, including CubeSats [40,43,44,91,206] throughout the solar system [36,37,51,52] and beyond.

### 14.6.4 Off-world utilization of local resources for inhabited settlements

Our final topic concerns the off-world utilization of local resources; this exploration paradigm or collection of technologies is known as in situ resource utilization (ISRU) [53,214]. The ISRU approach to exploration is to both enable human habitation on the Moon [53,215–218], Mars [219–223], and beyond [200,201,222] and facilitate far-flung exploration of the solar system [53,214,224,225]. The basic goal of ISRU is to vastly reduce the cost and logistical complexity of developing the infrastructure of a base and/or human settlement by using local resources to produce essential supplies and consumables such as water, oxygen, propellants, structural parts, habitats, and power from regolith and extent atmosphere using imported materials and available energy sources. The extensive energy input required will be provided by nuclear and/or solar energy; thermal energy storage, batteries, and fuel cells will certainly be important ancillary power (system) considerations [226]. As can be seen in Figs. 14.27 and 14.28, there are numerous aspects to the manifold considerations of this research topic (including imagination): labor required by personnel (Fig. 14.27), robotic and additive manufacturing processes (Fig. 14.28), and mining and basic processing (Fig. 14.27). Given the significant amount of water, silicon, metals, and oxygen readily available in the regolith (and where applicable) atmosphere of the Moon, Mars, and numerous smaller bodies in the solar system, it is not unreasonable to envision the manufacture of Si solar cells, solid-state batteries, and fuel cells in an off-world production facility in the future (Fig. 14.27) [53,214,226].

## 14.7 Conclusions

Energy storage for aerospace power applications presents unique challenges such as temperature fluctuations, rapid gravitational fluctuations, high-energy particles and radiation environments, atomic oxygen, hard-ultraviolet light,

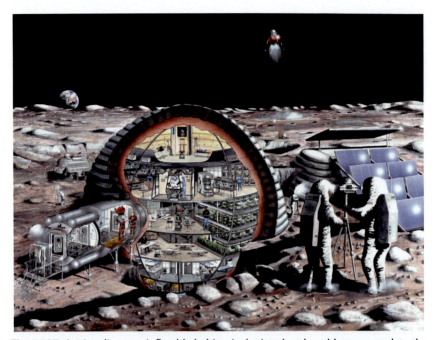

**Fig. 14.27** A 16 m diameter inflatable habitat is depicted and could accommodate the needs of a dozen astronauts living and working on the surface of the Moon. Illustrated are power and mining infrastructure (outside), astronauts exercising, a base operations center, a pressurized lunar rover, a small clean room, a fully equipped life sciences laboratory, a lunar lander, selenological work, hydroponic gardens, a wardroom, private crew quarters, dust-removing devices for lunar surface work, and an airlock. The top level shows joggers required to run with their bodies almost parallel to the floor as a result of the low gravity. *(Source: This artist's concept reflects the evaluation and study at NASA Johnson Space Center by the Man Systems Division and Johnson Engineering personnel. S89-20084, July 1989. Courtesy G. Kitmacher, NASA.)*

thermal management, and the necessity or weight- and space savings. We reviewed a variety of battery technologies for current aeronautics and Earth-Moon system applications, including high-altitude pseudo-satellite (both aircraft and airships), electric aircraft, and small satellites. A summary of energy storage options and issues for space exploration missions introduced the follow-on discussion. Given the broad technology space of battery types and materials, the final sections of this chapter focused on a consideration of several interesting representative missions and practical aspects of energy storage (with a focus on rechargeable (Li-ion or Li-S) batteries, when appropriate) for space exploration.

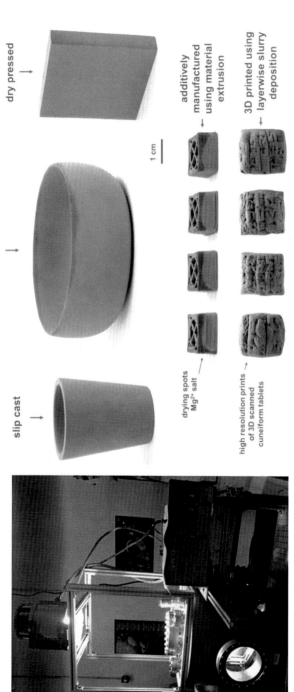

**Fig. 14.28** Left—concentrated solar-regolith additive manufacturing equipment at the Colorado School of Mines, Right—shaping of green bodies with the versatile Mars global simulant (~70% by weight of silicon, aluminum, and iron oxides) slurry system using classic pottery (potter's wheel), slip casting, material extrusion (robocasting/direct ink writing), 3D printing (layerwise slurry deposition with binder jetting) and as a reference dry pressing. (Left: Reproduced with permission from H. Williams, E. Butler-Jones, Additive manufacturing standards for space resource utilization, Addit. Manuf. 28 (2019) 676–681. Copyright (August 2019) Elsevier; Right: Reproduced with permission from D. Karl, T. Duminy, P. Lima, F. Kamutzki, A. Gili, A. Zocca, J. Günster, A. Gurlo, Clay in situ resource utilization with Mars global simulant slurries for additive manufacturing and traditional shaping of unfired green bodies, Acta Astronaut. 174 (2020) 241–253. Copyright (2020) Elsevier.)

The instructive discussion of recent successful and upcoming exploration missions was followed by a brief survey of appropriate battery technologies for specific destinations, a consideration of unique aspects of future missions to asteroids, and ocean worlds, including off-world atmospheric flight exploration, and Venus mission concepts, and ended with some comments on the long-term prospects for humanity to explore and possibly settle in habitable regions of our solar system. The use of local resources (water, regolith, solar energy, etc.) may one day enable off-world production of consumables, power components, and structural materials to facilitate the construction of settlements and the exploration of the farthest regions of the solar system.

An important take-home lesson is the need to develop energy storage technologies and power systems that can withstand the radiation fluxes and temperature extremes encountered in the solar system (even in Earth orbit); this will be critical for electronic devices, advanced instrumentation, and small off-world exploration vehicles such as rovers, helicopters, drones, and airships. While a variety of battery technologies could be employed for numerous planned and proposed aerospace applications and missions, further development of Li-S batteries should definitely be actively pursued.

## References

[1] J.N. Mrgudich, P.J. Bramhall, J.J. Finnegan, Thin-film rechargeable solid-electrolyte batteries, IEEE Trans. Aerosp. Electron. Syst. AES-1 (3) (1965) 290–296.

[2] D.K. Coates, C.L. Fox, L.E. Miller, Hydrogen-based rechargeable battery systems: military, aerospace, and terrestrial applications—I. Nickel-hydrogen batteries, Int. J. Hydrogen Energy 19 (9) (1994) 743–750.

[3] R.A. Marsh, S. Vukson, S. Surampudi, B.V. Ratnakumar, M.C. Smart, M. Manzo, P.J. Dalton, Li ion batteries for aerospace applications, J. Power Sources 97–98 (2001) 25–27.

[4] M.D. Kankam, V.J. Lyons, M.A. Hoberecht, R.R. Tacina, A.F. Hepp, Recent GRC aerospace technologies applicable to terrestrial energy systems, J. Propuls. Power 18 (2) (2002) 481–488.

[5] B.V. Ratnakumar, M.C. Smart, Aerospace applications: II. Planetary exploration missions (orbiters, landers, rovers and probes), in: M. Broussely, G. Pistoia (Eds.), Industrial Applications of Batteries: From Cars to Aerospace and Energy Storage, Elsevier, 2007, pp. 327–393.

[6] F. Cheng, J. Liang, Z. Tao, J. Chen, Functional materials for rechargeable batteries, Adv. Mater. 23 (15) (2011) 1695–1715.

[7] S. Megahed, B. Scrosati, Lithium-ion rechargeable batteries, J. Power Sources 51 (1–2) (1994) 79–104.

[8] R.A. Huggins, Lithium alloy negative electrodes, J. Power Sources 81-82 (1999) 13–19.

[9] P. Poizot, S. Laruelle, S. Grugeon, L. Dupont, J.-M. Tarascon, Nano-sized transition-metal oxides as negative-electrode materials for lithium-ion batteries, Nature 407 (6803) (2000) 496–499.

[10] J.-M. Tarascon, M. Armand, Issues and challenges facing rechargeable lithium batteries, Nature 414 (2001) 359–367.

[11] M.W. Verbrugge, P. Liu, Electrochemical characterization of high-power lithium ion batteries using triangular voltage and current excitation sources, J. Power Sources 174 (1) (2007) 2–8.

[12] W.Q. Walker, Rechargeable lithium batteries for aerospace applications, in: A.A. Franco (Ed.), Rechargeable Lithium Batteries: From Fundamentals to Applications, Woodhead Publishing Series in Energy, vol. 81, Elsevier, 2015, pp. 369–383.

[13] J.B. Goodenough, Y. Kim, Challenges for rechargeable Li batteries, Chem. Mater. 22 (3) (2010) 587–603.

[14] C. Mikolajczak, M. Kahn, K. White, R.T. Long, Lithium-Ion Batteries Hazard and Use Assessment, Exponent Failure Analysis Associates, Inc., Menlo Park, CA, July 2011. 94025, prepared for Fire Protection Research Foundation, Quincy, MA 02169–7471. Doc. no. 1100034.000 AOFO 0711 CMOl.

[15] X.-B. Cheng, R. Zhang, C.-Z. Zhao, Q. Zhang, Toward safe lithium metal anode in rechargeable batteries: a review, Chem. Rev. 117 (15) (2017) 10403–10473.

[16] B. Xu, D. Qian, Z. Wang, Y.S. Meng, Recent progress in cathode materials research for advanced lithium ion batteries, Mater. Sci. Eng. R 73 (5–6) (2012) 51–65.

[17] N. Nitta, F. Wu, J.T. Lee, G. Yushin, Li-ion battery materials: present and future, Mater. Today 18 (5) (2015) 252–264.

[18] H.-P. Feng, L. Tang, G.-M. Zeng, Y. Zhou, Y.-C. Deng, X. Ren, B. Song, C. Liang, M.-Y. Wei, J.-F. Yu, Core-shell nanomaterials: applications in energy storage and conversion, Adv. Colloid Interface Sci. 267 (2019) 26–46.

[19] P.J. Hanumantha, B. Gattu, P.M. Shanthi, S.S. Damle, Z. Basson, R. Bandi, M.K. Datta, S. Park, P.N. Kumta, Flexible sulfur wires (Flex-SWs)—a new versatile platform for lithium-sulfur batteries, Electrochim. Acta 212 (2016) 286–293.

[20] B. Yan, X. Li, Z. Bai, X. Song, D. Xiong, M. Zhao, D. Li, S. Lu, A review of atomic layer deposition providing high performance lithium sulfur batteries, J. Power Sources 338 (2017) 34–48.

[21] P.M. Shanthi, P.J. Hanumantha, B. Gattu, M. Sweeney, M.K. Datta, P.N. Kumta, Understanding the origin of irreversible capacity loss in non-carbonized carbonate—based metal organic framework (MOF) sulfur hosts for lithium—sulfur battery, Electrochim. Acta 229 (2017) 208–218.

[22] M. Liu, X. Qin, Y.-B. He, B. Li, F. Kang, Recent innovative configurations in high-energy lithium-sulfur batteries, J. Mater. Chem. A 5 (11) (2017) 5222–5234.

[23] X. Zhang, H. Xie, C.-S. Kim, K. Zaghib, A. Mauger, C.M. Julien, Advances in lithium—sulfur batteries, Mater. Sci. Eng. R. Rep. 12 (2017) 1–29.

[24] H.-J. Peng, J.-Q. Huang, X.-B. Cheng, Q. Zhang, Review on high-loading and high-energy lithium–sulfur batteries, Adv. Energy Mater. 7 (24) (2017), 1700260.

[25] Y. He, Z. Chang, S. Wu, H. Zhou, Effective strategies for long-cycle life lithium-sulfur batteries, J. Mater. Chem. A 6 (15) (2018) 6155–6182.

[26] X. Chen, T. Hou, K. Persson, Q. Zhang, Combining theory and experiment in lithium—sulfur batteries: current progress and future perspectives, Mater. Today 22 (2019) 142–158.

[27] A. Fotouhi, D.J. Auger, K. Propp, S. Longo, R. Purkayastha, L. O'Neill, S. Walus, Lithium–sulfur cell equivalent circuit network model parameterization and sensitivity analysis, IEEE Trans. Veh. Technol. 66 (9) (2017) 7711–7721.

[28] P.R. Shearing, L.R. Johnson, Toward practical demonstration of high-energy-density batteries, Joule 4 (7) (2020) 1359–1361.

[29] J.P. Maranchi, A.F. Hepp, P.N. Kumta, High-capacity, reversible silicon thin-film anodes for lithium ion batteries, Electrochem. Solid St. 6 (9) (2003) A198–A201.

[30] R.P. Raffaelle, J.D. Harris, D. Hehemann, D. Scheiman, G. Rybicki, A.F. Hepp, A facile route to thin-film solid state lithium microelectronic batteries, J. Power Sources 89 (1) (2000) 52–55.

[31] R.P. Raffaelle, A.F. Hepp, G.A. Landis, D.J. Hoffman, Mission applicability assessment of integrated power components and systems, Prog. Photovolt. Res. Appl. 10 (6) (2002) 391–397.

[32] J.P. Maranchi, I.S. Kim, A.F. Hepp, P. Kumta, Nanostructured electrochemically active materials: opportunities for aerospace power applications, in: AIAA Proceedings of the 1st International Energy Conversion Engineering Conference, Portsmouth, VA, USA, 2003, 14 pp. AIAA-2003-5952.

[33] R.P. Raffaelle, B.J. Landi, J.D. Harris, S.G. Bailey, A.F. Hepp, Carbon nanotubes for power applications, Mater. Sci. Eng. B 116 (3) (2005) 233–243.

[34] E.J. Simburger, J.H. Matsumoto, P.A. Gierow, A.F. Hepp, Integrated Thin Film Battery and Circuit Module, U.S. Patent 7,045,246, May 16, 2006.

[35] A.F. Hepp, J.S. McNatt, S.G. Bailey, R.P. Raffaelle, B.J. Landi, S.-S. Sun, C.E. Bonner, K.K. Banger, D. Rauh, Ultra-lightweight space power from hybrid thin-film solar cells, IEEE Aerosp. Electron. Syst. Mag. 23 (9) (2008) 31–41.

[36] P.M. Beauchamp, Energy Storage Technologies for Future Planetary Science Missions, Jet Propulsion Laboratory, prepared for Planetary Science Division, Science Mission Directorate, NASA HQ, December 2017. JPL D-101146.

[37] A.C. Angleman, NASA Space Technology Roadmaps and Priorities: Restoring NASA's Technological Edge and Paving the Way for a New Era in Space, Steering Committee for NASA Technology Roadmaps, Aeronautics and Space Engineering Board, Division on Engineering and Physical Sciences, National Research Council, National Academy Press, Washington, DC, 2012. 376 pp.

[38] J.A. Caffrey, D.M., Hamby, a review of instruments and methods for dosimetry in space, Adv. Space Res. 47 (4) (2011) 563–574.

[39] J. Gonzalo, D. Domínguez, D. López, On the challenge of a century lifespan satellite, Prog. Aerosp. Sci. 70 (2014) 28–41.

[40] E. Agasid, R. Burton, R. Carlino, G. Defouw, A.D. Perez, A.G. Karacalıoğlu, B. Klamm, A. Rademacher, J. Schalkwyk, R. Shimmin, J. Tilles, S. Weston, State of the Art Small Spacecraft Technology, NASA/TP—2018–220027, December 2018, 207 pp. Available at https://www.nasa.gov/sites/default/files/atoms/files/soa2018_final_doc-6.pdf. to be updated Fall 2020, Accessed July 30, 2020.

[41] A. Probst, R. Förstner, Spacecraft design of a multiple asteroid orbiter with re-docking lander, Adv. Space Res. 62 (8) (2018) 2125–2140.

[42] D. Lau, N. Song, C. Hall, Y. Jiang, S. Lim, I. Perez-Wurfl, Z. Ouyang, A. Lennon, Hybrid solar energy harvesting and storage devices: the promises and challenges, Mater. Today Energy 13 (2019) 22–44.

[43] K. Moyer, C. Meng, B. Marshall, O. Assal, J. Eaves, D. Perez, R. Karkkainen, L. Roberson, C.L. Pint, Carbon fiber reinforced structural lithium-ion battery composite: multifunctional power integration for CubeSats, Energy Storage Mater. 24 (2020) 676–681.

[44] F.C. Krause, J.A. Loveland, M.C. Smart, E.J. Brandon, R.V. Bugga, Implementation of commercial Li-ion cells on the MarCO deep space CubeSats, J. Power Sources 449 (2020), 227544.

[45] X. Zhu, Z. Guo, Z. Hou, Solar-powered airplanes: a historical perspective and future challenges, Prog. Aerosp. Sci. 71 (2014) 36–53.

[46] J. Gonzalo, D. López, D. Domínguez, A. García, A. Escapa, On the capabilities and limitations of high altitude pseudo-satellites, Prog. Aerosp. Sci. 98 (2018) 37–56.

[47] Y. Xu, W. Zhu, J. Li, L. Zhang, Improvement of endurance performance for high-altitude solar-powered airships: a review, Acta Astronaut. 167 (2020) 245–259.

[48] https://spinoff.nasa.gov/Spinoff2019/t_1.html. (Accessed 31 July 2020).
[49] https://www.nasa.gov/specials/X57/. (Accessed 3 August 2020).
[50] https://byeaerospace.com/sun-flyer/. (Accessed 3 August 2020).
[51] A.A. Siddiqi, Beyond Earth: A Chronicle of Deep Space Exploration, 1958–2016, second ed., The NASA History Series, National Aeronautics and Space Administration, Office of Communications, NASA History Division, Washington, DC, 2018. NASA SP2018-4041 https://www.nasa.gov/connect/ebooks/beyond_earth_detail.html. (Accessed 27 August 2020).
[52] https://nssdc.gsfc.nasa.gov/planetary/upcoming.html. (Accessed 31 August 2020).
[53] N. Kalapodis, G. Kampas, O.-J. Ktenidou, A review towards the design of extraterrestrial structures: from regolith to human outposts, Acta Astronaut. 175 (2020) 540–569.
[54] https://www.nasa.gov/centers/armstrong/news/FactSheets/FS-054-DFRC.html. (Accessed 3 August 2020).
[55] http://www.projectsunrise.info/First_Solar_Powered_Aircraft.html. (Accessed 3 August 2020).
[56] https://www.airbus.com/defence/uav/zephyr.html. (Accessed 3 August 2020).
[57] A. Noth, M.W. Engel, R. Siegwart, Flying solo and solar to Mars, IEEE Robot. Autom. Mag. 13 (3) (2006) 44–52, https://doi.org/10.1109/MRA.2006.1678138.
[58] https://solarimpulse.com. (Accessed 3 August 2020).
[59] S.-J. Hwang, S.-G. Kim, C.-W. Kim, Y.-G. Lee, Aerodynamic design of the solar-powered high altitude long endurance (HALE) unmanned aerial vehicle (UAV), Int. J. Aeronaut. Space Sci. 17 (1) (2016) 132–138.
[60] J. Li, J. Liao, Y. Liao, H. Du, S. Luo, W. Zhu, M. Lv, An approach for estimating perpetual endurance of the stratospheric solar-powered platform, Aerosp. Sci. Technol. 79 (2018) 118–130.
[61] F. Petrone, P. Wessel, HASPA design and flight test objectives, in: Lighter than Air Technology Conference, American Institute of Aeronautics and Astronautics, Snowmass, CO, July 1975, https://doi.org/10.2514/6.1975-924. paper 1975-924.
[62] M. Onda, Y. Morikawa, High-altitude lighter-than-air powered platform, in: Proceedings 29th International Pacific Air and Space Technology Conference and Aircraft Symposium, Gifu, Japan, Oct. 1991, 8 pp. https://doi.org/10.4271/912054. Paper 912054.
[63] B. Kroeplin, I. Schaefer, Experiences by design and operation of the solar powered airships, 'Lotte 1–3', in: 11th Lighter-Than-Air Systems Technology Conference, Clearwater Beach, FL, USA, May 1995, https://doi.org/10.2514/6.1995-1613. AIAA 1995-1613.
[64] H. Naito, K. Eguchi, T. Hoshino, S. Okaya, T. Fujiwara, S. Miwa, Y. Nomura, Design and analysis of solar power system for SPF airship operations, in: 13th Lighter-Than-Air Systems Technology Conference, Norfolk, VA, USA, June–July 1999, https://doi.org/10.2514/6.1999–3913. AIAA 1999–3913.
[65] L.G. Seely, M.S. Smith, A realistic view of the future of scientific ballooning, Adv. Space Res. 30 (5) (2002) 1125–1133.
[66] I. Smith, M. Lee, M. Fortneberry, R. Judy, HiSentinel80: flight of a high altitude airship, in: 11th AIAA Aviation Technology, Integration, and Operations (ATIO) Conference, Virginia Beach, VA, USA, September 2011, 14 pp. https://doi.org/10.2514/6.2011–6973. AIAA 2011–6973.
[67] M. Lee, I. Smith, S. Androulakakis, High-altitude LTA airship efforts at the, U.S. Army SMDC/ARSTRAT, in: 18th Lighter-Than-Air Systems Technology Conference, Seattle, WA, USA, May 2009, https://doi.org/10.2514/6.2009–2852. AIAA 2009–2852.

[68] A. Colozza, J.L. Dolce, High-Altitude, Long-Endurance Airships for Coastal Surveillance, NASA/TM—2005-213427, February 2005, 21 pp. E–14961.

[69] T. Smith, C. Bingham, P. Stewart, R. Allarton, J. Stewart, MAAT high altitude cruiser feeder airship concept, in: Electrical Systems for Aircraft, Railway and Ship Propulsion (ESARS) 2012, October 2012, pp. 1–6, https://doi.org/10.1109/ESARS.2012.6387386. Bologna, Italy.

[70] X. Yang, D. Liu, Renewable power system simulation and endurance analysis for stratospheric airships, Renew. Energy 113 (2017) 1070–1076.

[71] J. Liao, Y. Jiang, J. Li, Y. Liao, H. Du, W. Zhu, L. Zhang, An improved energy management strategy of hybrid photovoltaic/battery/fuel cell system for stratospheric airship, Acta Astronaut. 152 (2018) 727–739.

[72] C. Shan, M. Lv, K. Sun, J. Gao, Analysis of energy system configuration and energy balance for stratospheric airship based on position energy storage strategy, Aerosp. Sci. Technol. 101 (2020), 105844.

[73] Y. Huang, J. Chen, H. Wang, G. Su, A method of 3D path planning for solar-powered UAV with fixed target and solar tracking, Aerosp. Sci. Technol. 92 (2019) 831–838.

[74] M. Qiang, A Solar Power System for High Altitude Airships, The University of Toledo, 2011, p. 28, ISBN: 9781124693132. ProQuest Dissertations and Theses; Thesis (PEG), Publication Number: AAT 3458075. Source: Dissertation Abstracts International, vols. 72–10, Section: B.

[75] K.W. Sun, M. Zhu, G.M. Liang, D.D. Xu, Design and simulation to composite MPPT controller on the stratospheric airship, Appl. Mech. Mater. 672–674 (2014) 1765–1769.

[76] J. Meng, S. Liu, Z. Yao, M. Lv, Optimization design of a thermal protection structure for the solar array of stratospheric airships, Renew. Energy 133 (2019) 593–605.

[77] C. Lee, Y.S. Seok, A study on solar panel installation angles for stratospheric platform airship power systems, J. Korean Soc. Aeronaut. Space Sci. 30 (2002) 148–155.

[78] M. Lv, J. Li, H. Du, W. Zhu, J. Meng, Solar array layout optimization for stratospheric airships using numerical method, Energ. Conver. Manage. 135 (2017) 160–169.

[79] W. Zhu, Y. Xu, J. Li, H. Du, L. Zhang, Research on optimal solar array layout for near-space airship with thermal effect, Sol. Energy 170 (2018) 1–13.

[80] K. Ghosh, A. Guha, S.P. Duttagupta, Power generation on a solar photovoltaic array integrated with lighter-than-air platform at low altitudes, Energ. Conver. Manage. 154 (2017) 286–298.

[81] M. Lv, J. Li, W. Zhu, H. Du, J. Meng, K. Sun, A theoretical study of rotatable renewable energy system for stratospheric airship, Energ. Conver. Manage. 140 (2017) 51–61.

[82] W. Zhu, J. Li, Y. Xu, Optimum attitude planning of near-space solar powered airship, Aerosp. Sci. Technol. 84 (2019) 291–305.

[83] G. Zubi, R. Dufo-López, M. Carvalho, G. Pasaoglu, The lithium-ion battery: state of the art and future perspectives, Renew. Sustain. Energy Rev. 89 (2018) 292–308.

[84] S. Mekhilef, R. Saidur, A. Safari, Comparative study of different fuel cell technologies, Renew. Sustain. Energy Rev. 16 (2012) 981–989.

[85] L.O. Valøen, M.I. Shoesmith, The effect of PHEV and HEV duty cycles on battery and battery pack performance, in: Plug-in Hybrid Electrical Vehicle 2007 Conference, Winnipeg, MB, Canada, November 1-2, 2007, 9 pp.

[86] C.K. Dyer, Fuel cells for portable applications, J. Power Sources 106 (1–2) (2002) 31–34.

[87] G. Li, D. Ma, M. Yang, Research of near space hybrid power airship with a novel method of energy storage, Int. J. Hydrogen Energy 40 (30) (2015) 9555–9562.

Batteries for aeronautics and space exploration **589**

[88] D. Pande, D. Verstraete, Impact of solar cell characteristics and operating conditions on the sizing of a solar powered nonrigid airship, Aerosp. Sci. Technol. 72 (2018) 353–363.

[89] S. Koohi-Fayegh, M.A. Rosen, A review of energy storage types, applications and recent developments, J. Energy Storage 27 (2020), 101047.

[90] https://www.northropgrumman.com/what-we-do/air/globalhawk/. (Accessed 15 August 2010).

[91] A. Poghosyan, A. Golkar, CubeSat evolution: analyzing CubeSat capabilities for conducting science missions, Prog. Aerosp. Sci. 88 (2017) 59–83.

[92] https://www.saftbatteries.com/products-solutions/products. (Accessed 21 August 2020).

[93] https://www.eaglepicher.com/markets/space/. (Accessed 21 August 2020).

[94] http://www.molicel.com/products-applications/. (Accessed 21 August 2020).

[95] https://www.batteryspace.com/18650seriesli-ioncells.aspx. (Accessed 21 August 2020).

[96] https://mars.nasa.gov/mars-exploration/missions/odyssey/. (Accessed 28 August 2020).

[97] https://www.nasa.gov/mission_pages/mer/index.html. (Accessed 28 August 2020).

[98] https://www.nasa.gov/mission_pages/messenger/main/index.html. (Accessed 28 August 2020).

[99] https://www.jpl.nasa.gov/missions/deep-impact/. (Accessed 28 August 2020).

[100] https://www.nasa.gov/mission_pages/MRO/mission/index.html. (Accessed 28 August 2020).

[101] https://www.nasa.gov/mission_pages/phoenix/main/index.html. (Accessed 28 August 2020).

[102] https://www.nasa.gov/mission_pages/dawn/main/index.html. (Accessed 28 August 2020).

[103] https://www.nasa.gov/mission_pages/kepler/overview/index.html. (Accessed 28 August 2020).

[104] https://www.nasa.gov/mission_pages/LRO/overview/index.html. (Accessed 28 August 2020).

[105] https://www.nasa.gov/mission_pages/LCROSS/overview/index.html. (Accessed 28 August 2020).

[106] https://www.nasa.gov/mission_pages/juno/main/index.html. (Accessed 28 August 2020).

[107] https://solarsystem.nasa.gov/missions/grail/in-depth/. (Accessed 28 August 2020).

[108] https://www.jpl.nasa.gov/missions/mars-science-laboratory-curiosity-rover-msl/. (Accessed 28 August 2020).

[109] https://www.nasa.gov/mission_pages/ladee/main/index.html. (Accessed 28 August 2020).

[110] https://www.nasa.gov/mission_pages/maven/overview/index.html. (Accessed 28 August 2020).

[111] https://www.nasa.gov/content/osiris-rex-overview. (Accessed 28 August 2020).

[112] https://mars.nasa.gov/insight/. (Accessed 28 August 2020).

[113] https://mars.nasa.gov/mars2020/. (Accessed 28 August 2020).

[114] https://www.nesdis.noaa.gov/content/dscovr-deep-space-climate-observatory. (Accessed 28 August 2020).

[115] https://www.nasa.gov/mission_pages/mms/index.html. (Accessed 28 August 2020).

[116] https://www.nasa.gov/tess-transiting-exoplanet-survey-satellite. (Accessed 28 August 2020).

[117] https://www.nasa.gov/mission_pages/webb/main/index.html. (Accessed 28 August 2020).

[118] https://eospso.nasa.gov/missions/joint-polar-satellite-system-2. (Accessed 28 August 2020).

[119] B.H. Foing, G. Racca, A. Marini, D. Koschny, D. Frew, B. Grieger, O. Camino-Ramos, J.L. Josset, M. Grande, SMART-1 science and technology working team, SMART-1 technology, scientific results and heritage for future space missions, Planet. Space Sci. 151 (2018) 141–148.

[120] G. Johnson, Memories and safety lessons learned of an Apollo electrical engineer, J. Space Saf. Eng. 7 (1) (2020) 18–26.

[121] T. Hoshino, S. Wakabayashi, M. Ohtake, Y. Karouji, T. Hayashi, H. Morimoto, H. Shiraishi, T. Shimada, T. Hashimoto, H. Inoue, R. Hirasawa, Y. Shirasawa, H. Mizuno, H. Kanamori, Lunar polar exploration mission for water prospection—JAXA's current status of joint study with ISRO, Acta Astronaut. 176 (2020) 52–58.

[122] A.F. Hepp, P.N. Kumta, O. Velikokhatnyi, R.P. Raffaelle, Batteries for integrated power and CubeSats: recent developments and future prospects, in: A.F. Hepp, P.N. Kumta, O. Velikokhatnyi, M.K. Datta (Eds.), Silicon Anode Systems for Lithium-Ion Batteries, Elsevier, New York, 2021.

[123] A.G. Santo, R.E. Gold, R.L. McNutt Jr., S.C. Solomon, C.J. Ercol, R.W. Farquhar, T.J. Hartka, J.E. Jenkins, J.V. McAdams, L.E. Mosher, D.F. Persons, D.A. Artis, R.S. Bokulic, R.F. Conde, G. Dakermanji, M.E. Goss Jr., D.R. Haley, K.J. Heeres, R.H. Maurer, R.C. Moore, E.H. Rodberg, T.G. Stern, S.R. Wiley, B.G. Williams, C.L. Yen, M.R. Peterson, The MESSENGER mission to mercury: spacecraft and mission design, Planet. Space Sci. 49 (14–15) (2001) 1481–1500.

[124] M.C. Smart, B.V. Ratnakumar, R.C. Ewell, S. Surampudi, F.J. Puglia, R. Gitzendanner, The use of lithium-ion batteries for JPL's Mars missions, Electrochim. Acta 268 (2018) 27–40.

[125] M.C. Smart, B.V. Ratnakumar, Electrolytes for Wide Operating Temperature Lithium-Ion Cells, United States Patent 9,293,773 B2, March 22, 2016.

[126] http://solarsystem.nasa.gov/galileo/. (Accessed 4 September 2020).

[127] https://www.nasa.gov/planetarymissions/newfrontiers.html. (Accessed 4 September 2020).

[128] I. Yoshikawa, F. Suzuki, R. Hikida, K. Yoshioka, G. Murakami, F. Tsuchiya, C. Tao, A. Yamazaki, T. Kimura, H. Kita, H. Nozawa, M. Fujimoto, Volcanic activity on Io and its influence on the dynamics of the Jovian magnetosphere observed by EXCEED/Hisaki in 2015, Earth Planets Space 69 (2017) 110, https://doi.org/10.1186/s40623-017-0700-9.

[129] https://www.emiratesmarsmission.ae/. (Accessed 4 September 2020).

[130] https://www.planetary.org/space-missions/tianwen-1. (Accessed 4 September 2020).

[131] https://nssdc.gsfc.nasa.gov/planetary/lunar/cnsa_moon_future.html. (Accessed 4 September 2020).

[132] https://www.nasa.gov/planetarydefense/dart. (Accessed 4 September 2020).

[133] https://www.nasa.gov/content/goddard/lucy-overview. (Accessed 4 September 2020).

[134] https://sci.esa.int/web/juice. (Accessed 4 September 2020).

[135] https://nssdc.gsfc.nasa.gov/nmc/spacecraft/display.action?id=KPLO. (Accessed 4 September 2020).

[136] https://www.jpl.nasa.gov/missions/psyche/. (Accessed 4 September 2020).

[137] https://exploration.esa.int/web/mars/-/48088-mission-overview. (Accessed 4 September 2020).

[138] https://europa.nasa.gov/mission/about/. (Accessed 4 September 2020).

[139] https://www.esa.int/Safety_Security/Hera/Hera. (Accessed 4 September 2020).

[140] https://www.nasa.gov/planetarymissions/newfrontiers.html#dragonfly. (Accessed 4 September 2020).

[141] J.A. Van Allen, L.A. Frank, Radiation around the earth to a radial distance of 107,400 km, Nature 183 (1959) 430–434. 4659.

[142] https://www.nasa.gov/analogs/nsrl/why-space-radiation-matters. (Accessed 6 September 2020).

[143] B.V. Ratnakumar, M.C. Smart, L.D. Whitcanack, E.D. Davies, K.B. Chin, F. Deligiannis, S. Surampudi, Behavior of Li-ion cells in high-intensity radiation environments, J. Electrochem. Soc. 151 (4) (2004) A652–A659.

[144] B.V. Ratnakumar, M.C. Smart, L.D. Whitcanack, E.D. Davies, K.B. Chin, F. Deligiannis, S. Surampudi, Behaviour of Li-ion cells in high-intensity radiation environments: II. Sony/AEA/ComDEV Cells, J. Electrochem. Soc. 152 (2) (2005) A357–A363.

[145] C. Tan, D.J. Lyons, K. Pan, K.Y. Leung, W.C. Chuirazzi, M. Canova, A.C. Co, L.R. Cao, Radiation effects on the electrode and electrolyte of a lithium-ion battery, J. Power Sources 318 (2016) 242–250.

[146] A. Samin, M. Kurth, L. Cao, Ab initio study of radiation effects on the $Li_4Ti_5O_{12}$ electrode used in lithium-ion batteries, AIP Adv. 5 (2015), https://doi.org/10.1063/1.4917308, 047110.

[147] F. Nimmo, R.T. Pappalardo, Ocean worlds in the outer solar system, J. Geophys. Res. Planets 121 (8) (2016) 1378–1399, https://doi.org/10.1002/2016JE005081.

[148] A. Ulloa-Severino, G.A. Carr, D.J. Clark, S.M. Orellana, R. Arellano, M.C. Smart, R.V. Bugga, A. Boca, S.F. Dawson, Power subsystem approach for the Europa mission, in: 11th European Space Power Conference, 03–07 October 2016, Thessaloniki, Greece, vol. 16, 2017, 13004. 8 pp. E3S Web Conf. https://doi.org/10.1051/e3sconf/20171613004. Published online 23 May 2017.

[149] https://solarsystem.nasa.gov/missions/cassini/overview/. (Accessed 6 September 2020).

[150] M. Tulej, S. Meyer, M. Lüthi, D. Lasi, A. Galli, D. Piazza, L. Desorgher, D. Reggiani, W. Hajdas, S. Karlsson, L. Kalla, P. Wurz, Experimental investigation of the radiation shielding efficiency of a MCP detector in the radiation environment near Jupiter's moon Europa, Nucl. Instrum. Methods Phys. Res., Sect. B 383 (2016) 21–37.

[151] A. Khan, M. Yamaguchi, Y. Ohshita, N. Dharmaraso, K. Araki, V.T. Khanh, H. Itoh, M. Ohshima, M. Imaizumi, S. Matsuda, Strategies for improving radiation tolerance of Si space solar cells, Sol. Energy Mater. Sol. Cells 75 (1–2) (2003) 271–276.

[152] S. Hubbard, C. Bailey, S. Polly, C.D. Cress, J. Andersen, D.V. Forbes, R.P. Raffaelle, Nanostructured photovoltaics for space power, J. Nanophotonics 3 (1) (2009), https://doi.org/10.1117/1.3266502, 031880.

[153] M. Imaizumi, T. Nakamura, T. Takamoto, T. Ohshima, M. Tajima, Radiation degradation characteristics of component subcells in inverted metamorphic triple-junction solar cells irradiated with electrons and protons, Prog. Photovolt. Res. Appl. 25 (2) (2017) 161–174.

[154] P. Espinet-Gonzalez, E. Barrigón, G. Otnes, G. Vescovi, C. Mann, R.M. France, A.J. Welch, M.S. Hunt, D. Walker, M.D. Kelzenberg, I. Åberg, M.T. Borgström, L. Samuelson, H.A. Atwater, Radiation tolerant nanowire array solar cells, ACS Nano 13 (11) (2019) 12860–12869.

[155] N. Gruginskie, F. Cappelluti, G.J. Bauhuis, P. Mulder, E.J. Haverkamp, E. Vlieg, J.J. Schermer, Electron radiation–induced degradation of GaAs solar cells with different architectures, Prog. Photovolt. Res. Appl. 28 (4) (2020) 266–278.

[156] H. Karadeniz, A study on triple-junction $GaInP_2/InGaAs/Ge$ space grade solar cells irradiated by 24.5 MeV high-energy protons, Nucl. Instrum. Methods Phys. Res., Sect. B 471 (2020) 1–6.

[157] J.-P. Jones, S.C. Jones, K.J. Billings, J. Pasalic, R.V. Bugga, F.C. Krause, M.C. Smart, E.J. Brandon, Radiation effects on lithium $CF_x$ batteries for future spacecraft and landers, J. Power Sources 471 (2020), 228464.

[158] S. Wang, C. He, Design and analysis of nuclear battery driven by the external neutron source, Ann. Nucl. Energy 72 (2014) 455–460.

[159] V. Elkina, M. Kurushkin, Promethium: to strive, to seek, to find and not to yield, Front. Chem. 8 (2020) 588, https://doi.org/10.3389/fchem.2020.00588.

[160] Y. Nie, C. Kacica, M.E. Meyer, R.D. Green, P. Biswas, Graphene synthesized as by-product of gas purification in long-term space missions and its lithium-ion battery application, Adv. Space Res. 62 (5) (2018) 1015–1024.

[161] M. Mammarella, P.M. Vernicari, C.A. Paissoni, N. Viola, How the lunar space tug can support the cislunar station, Acta Astronaut. 154 (2019) 181–194.

[162] G. Gave, Y. Borthomieu, B. Lagattu, J.-P. Planchat, Evaluation of a low temperature Li-ion cell for space, Acta Astronaut. 54 (8) (2004) 559–563.

[163] B.V. Ratnakumar, M.C. Smart, C.K. Huang, D. Perrone, S. Surampudi, S.G. Greenbaum, Lithium ion batteries for Mars exploration missions, Electrochim. Acta 45 (8–9) (2000) 1513–1517, https://doi.org/10.1016/S0013-4686(99)00367-9.

[164] C.-Y. Wang, G. Zhang, S. Ge, T. Xu, J. Yan, X.-G. Yang, Y. Leng, Lithium-ion battery structure that self-heats at low temperatures, Nature 529 (2016) 515–518.

[165] M.C. Smart, B.V. Ratnakumar, S. Surampudi, Use of organic esters as cosolvents in electrolytes for lithium-ion batteries with improved low temperature performance, J. Electrochem. Soc. 149 (4) (2002) A361–A370.

[166] S. Kim, V.I. Hegde, Z. Yao, Z. Lu, M. Amsler, J. He, S. Hao, J.R. Croy, E. Lee, M.M. Thackeray, C. Wolverton, First-principles study of lithium cobalt spinel oxides: correlating structure and electrochemistry, ACS Appl. Mater. Interfaces 10 (16) (2018) 13479–13490.

[167] F.C. Krause, J.-P. Jones, S.C. Jones, J. Pasalic, K.J. Billings, W.C. West, M.C. Smart, R.V. Bugga, E.J. Brandon, M. Destephen, High specific energy Lithium primary batteries as power sources for deep space exploration, J. Electrochem. Soc. 165 (10) (2018) A2312–A2320.

[168] J.C. Bachman, S. Muy, A. Grimaud, H.-H. Chang, N. Pour, S.F. Lux, O. Paschos, F. Maglia, S. Lupart, P. Lamp, L. Giordano, Y. Shao-Horn, Inorganic solid-state electrolytes for lithium batteries: mechanisms and properties governing ion conduction, Chem. Rev. 116 (1) (2016) 140–162.

[169] C. Sun, J. Liu, Y. Gong, D.P. Wilkinson, J. Zhang, Recent advances in all-solid-state rechargeable lithium batteries, Nano Energy 33 (2017) 363–386.

[170] S. Ramakumar, C. Deviannapoorani, L. Divya, L.S. Shankar, R. Murugan, Lithium garnets: synthesis, structure, $Li^+$ conductivity, $Li^+$ dynamics and applications, Prog. Mater. Sci. 88 (2017) 325–411.

[171] R.C. Xu, X.H. Xia, S.Z. Zhang, D. Xie, X.L. Wang, J.P. Tu, Interfacial challenges and progress for inorganic all-solid-state lithium batteries, Electrochim. Acta 284 (2018) 177–187.

[172] B. Aoun, C. Yu, L. Fan, Z. Chen, K. Amine, Y. Ren, A generalized method for high throughput in-situ experiment data analysis: an example of battery materials exploration, J. Power Sources 279 (2015) 246–251.

[173] A. Urban, I. Matts, A. Abdellahi, G. Ceder, Computational design and preparation of cation-disordered oxides for high-energy-density Li-ion batteries, Adv. Energy Mater. 6 (15) (2016) 1600488.

[174] Y. Lu, L. Yu, X.W. Lou, Nanostructured conversion-type anode materials for advanced lithium-ion batteries, Chem 4 (5) (2018) 972–996.

[175] X. Shen, H. Liu, X.-B. Cheng, C. Yan, J.-Q. Huang, Beyond lithium ion batteries: higher energy density battery systems based on lithium metal anodes, Energy Storage Mater. 12 (2018) 161–175.

[176] M. Akbulut, H. Erol, Determination of the topology of lithium-ion battery packs for space equipment and validation through experimental investigation, J. Energy Storage 30 (2020), 101417.

[177] C. Sun, X. Zhang, C. Li, K. Wang, X. Sun, Y. Ma, Recent advances in prelithiation materials and approaches for lithium-ion batteries and capacitors, Energy Storage Mater. 32 (2020) 497–516.

[178] J.-Q. Huang, Q. Zhang, F. Wei, Multi-functional separator/interlayer system for high-stable lithium-sulfur batteries: progress and prospects, Energy Storage Mater. 1 (2015) 127–145.

[179] X. Yang, J. Luo, X. Sun, Towards high-performance solid-state Li-S batteries: from fundamental understanding to engineering design, Chem. Soc. Rev. 49 (7) (2020) 2140–2195.

[180] J. Wu, F. Ciucci, J.-K. Kim, Molybdenum disulfide based nanomaterials for rechargeable batteries, Chem. A Eur. J. 26 (29) (2020) 6296–6319.

[181] Y. Feng, G. Wang, J. Ju, Y. Zhao, W. Kang, N. Deng, B. Cheng, Towards high energy density Li–S batteries with high sulfur loading: from key issues to advanced strategies, Energy Storage Mater. 32 (2020) 320–355.

[182] A. Manthiram, X. Yu, S. Wang, Lithium battery chemistries enabled by solid-state electrolytes, Nat. Rev. Mater. 2 (4) (2017) 16103.

[183] R.J. Gummow, G. Vamvounis, M.B. Kannan, Y. He, Calcium-ion batteries: current state-of-the-art and future perspectives, Adv. Mater. 30 (39) (2018) 1801702.

[184] S. Ma, M. Jiang, P. Tao, C. Song, J. Wu, J. Wang, T. Deng, W. Shang, Temperature effect and thermal impact in lithium-ion batteries: a review, Prog. Nat. Sci.: Mater. Int. 28 (6) (2018) 653–666.

[185] X. Hong, J. Mei, L. Wen, Y. Tong, A.J. Vasileff, L. Wang, J. Liang, Z. Sun, S.X. Dou, Nonlithium metal–sulfur batteries: steps toward a leap, Adv. Mater. 31 (5) (2019) 1802822.

[186] W. Wu, S. Wang, W. Wu, K. Chen, S. Hong, Y. Lai, A critical review of battery thermal performance and liquid based battery thermal management, Energ. Conver. Manage. 182 (2019) 262–281.

[187] G.A. Landis, R. Harrison, Batteries for Venus surface operation, J. Propuls. Power 26 (4) (2010) 649–654.

[188] D.E. Glass, J.-P. Jones, A.V. Shevade, D. Bhakta, E. Raub, R. Sim, R.V. Bugga, High temperature primary battery for Venus surface missions, J. Power Sources 449 (2020), 227492.

[189] D.E. Glass, J.-P. Jones, A.V. Shevade, R.V. Bugga, Transition metal phosphorous Tri-sulfides as cathode materials in high temperatures batteries, J. Electrochem. Soc. 167 (11) (2020), 110512.

[190] https://solarsystem.nasa.gov/asteroids-comets-and-meteors/overview/. (Accessed 8 September 2020).

[191] https://solarsystem.nasa.gov/asteroids-comets-and-meteors/asteroids/in-depth/#asteroid_classifications. (Accessed 8 September 2020).

[192] https://nssdc.gsfc.nasa.gov/planetary/planets/asteroidpage.html. (Accessed 8 September 2020).

[193] https://cneos.jpl.nasa.gov. (Accessed 8 September 2020).

[194] https://nineplanets.org/trojans/. (Accessed 8 September 2020).

[195] https://www.nasa.gov/planetarydefense. (Accessed 8 September 2020).

[196] P. Tortora, V. Di Tana, LICIACube, the Italian witness of DART impact on Didymos, in: Proceedings of 2019 IEEE 5th International Workshop on Metrology for AeroSpace (MetroAeroSpace), Torino, Italy, June 2019, pp. 314–317, https://doi.org/10.1109/MetroAeroSpace.2019.8869672.

[197] R. Landis, L. Johnson, Advances in planetary defense in the United States, Acta Astronaut. 156 (2019) 394–408.

[198] Q. Dong, C. Pingyuan, W. Yamin, Opportunity options for rendezvous, flyby and sample return mission to different spectral-type asteroids for the 2015–2025, Acta Astronaut. 72 (2012) 143–155.

[199] M. Elvis, T. Esty, How many assay probes to find one ore-bearing asteroid? Acta Astronaut. 96 (2014) 227–231.

[200] M. Klas, N. Tsafnat, J. Dennerley, S. Beckmann, B. Osborne, A.G. Dempster, M. Manefield, Biomining and methanogenesis for resource extraction from asteroids, Space Policy 34 (2015) 18–22.

[201] D.D. Mazanek, R.G. Merrill, J.R. Brophy, R.P. Mueller, Asteroid redirect mission concept: a bold approach for utilizing space resources, Acta Astronaut. 117 (2015) 163–171.

[202] N. Anthony, M.R. Emamia, Asteroid engineering: the state-of-the-art of near-earth asteroids science and technology, Prog. Aerosp. Sci. 100 (2018) 1–17.

[203] A.M. Hein, R. Matheson, D. Fries, A techno-economic analysis of asteroid mining, Acta Astronaut. 168 (2020) 104–115.

[204] R.G. Stacey, M. Connors, Delta-v requirements for earth co-orbital rendezvous missions, Planet. Space Sci. 57 (7) (2009) 822–829.

[205] A. Taylor, J.C. McDowell, M. Elvis, A Delta-V map of the known main belt asteroids, Acta Astronaut. 146 (2018) 73–82.

[206] R. Jedicke, J. Sercel, J. Gillis-Davis, K.J. Morenz, L. Gertsch, Availability and delta-v requirements for delivering water extracted from near-earth objects to cis-lunar space, Planet. Space Sci. 159 (2018) 28–42.

[207] K. Lemmer, Propulsion for CubeSats, Acta Astronaut. 134 (2017) 231–243.

[208] https://mars.nasa.gov/technology/helicopter/. (Accessed 8 September 2020).

[209] http://caesar.cornell.edu/. (Accessed 8 September 2020).

[210] L.S. Glaze, C.F. Wilson, L.V. Zasova, M. Nakamura, S. Limaye, Future of Venus research and exploration, Space Sci. Rev. 214 (2018) 89.

[211] https://re.grc.nasa.gov/compass/. (Accessed 8 September 2020).

[212] N.J. Boll, D. Salazar, C.J. Stelter, G.A. Landis, A.J. Colozza, Venus high-temperature atmospheric dropsonde and extreme-environment seismometer (HADES), Acta Astronaut. 111 (2015) 146–159.

[213] D.C. Arney, C.A. Jones, High altitude venus operational concept (HAVOC): an exploration strategy for Venus, in: AIAA SPACE 2015 Conference and Exposition 31 Aug-2 Sep, Pasadena, California, AIAA 2015-4612, 16 pp, 2015, https://doi.org/10.2514/6.2015-4612. Published Online: Aug. 28, 2015.

[214] P. Ehrenfreund, C. McKay, J.D. Rummel, B.H. Foing, C.R. Neal, T. Masson-Zwaan, M. Ansdell, N. Peter, J. Zarnecki, S. Mackwell, M.A. Perino, L. Billings, J. Mankins, M. Race, Toward a global space exploration program: a stepping stone approach, Adv. Space Res. 49 (1) (2012) 2–48.

[215] G.H. Heiken, D.T. Vaniman, B.M. French (Eds.), Lunar Sourcebook, Cambridge University Press, Cambridge, 1991. https://www.lpi.usra.edu/publications/books/lunar_sourcebook/. (Accessed 10 September 2020).

[216] A.F. Hepp, D.L. Linne, G.A. Landis, M.F. Wade, J.E. Colvin, Production and use of metals and oxygen for lunar propulsion, J. Propuls. Power 10 (6) (1994) 834–840.

[217] H. Williams, E. Butler-Jones, Additive manufacturing standards for space resource utilization, Addit. Manuf. 28 (2019) 676–681.

[218] M.F. Palos, P. Serra, S. Fereres, K. Stephenson, R. González-Cinca, Lunar ISRU energy storage and electricity generation, Acta Astronaut. 170 (2020) 412–420.

[219] R.L. Ash, W.L. Dowler, G. Varsi, Feasibility of rocket propellant production on Mars, Acta Astronaut. 5 (9) (1978) 705–724.

[220] B. Kading, J. Straub, Utilizing in-situ resources and 3D printing structures for a manned Mars mission, Acta Astronaut. 107 (2015) 317–326.

[221] H. Chen, T.S. du Jonchay, L. Hou, K. Ho, Integrated in-situ resource utilization system design and logistics for Mars exploration, Acta Astronaut. 170 (2020) 80–92.

[222] D. Karl, T. Duminy, P. Lima, F. Kamutzki, A. Gili, A. Zocca, J. Günster, A. Gurlo, Clay in situ resource utilization with Mars global simulant slurries for additive manufacturing and traditional shaping of unfired green bodies, Acta Astronaut. 174 (2020) 241–253.

[223] S.O. Starr, A.C. Muscatello, Mars in situ resource utilization: a review, Planet. Space Sci. 182 (2020), 104824.

[224] S. Lim, V.L. Prahbu, M. Anand, L.A. Taylor, Extra-terrestrial construction processes—advancements, opportunities and challenges, Adv. Space Res. 60 (7) (2017) 1413–1429.

[225] M. Hassanalian, D. Rice, A. Abdelkefi, Evolution of space drones for planetary exploration: a review, Progress in Aerospace Sciences 97 (2018) 61–105.

[226] P. Fleith, A. Cowley, A.C. Pou, A.V. Lozano, R. Frank, P.L. Córdoba, R. González-Cinca, In-situ approach for thermal energy storage and thermoelectricity generation on the Moon: modelling and simulation, Planet. Space Sci. 181 (2020), 104789.

# Index

Note: Page numbers followed by *f* indicate figures and *t* indicate tables.

## A

Activated carbon nanofibers (ACNF), 307–309

Adaptive neuro-fuzzy inference systems (ANFIS), 525–527

All-solid-state Li-Se$_x$S$_y$ batteries, 253–256, 254–255*f*

Asteroid Impact and Deflection Assessment (AIDA), 577–578

Asteroids, 532

NASA exploration missions, 563–565

upcoming missions, 575–578

## B

Battery

anode, electrolyte, 9–10

cathodes, 8–9

commercialized, 12

early lithium-sulfur, 5

electric vehicle, 4

lithium-ion, 5–6

lithium-sulfur, 3–8, 8*f*

low cell voltage, 10–11, 11*f*

sulfur (S$_8$), 6–7

Battery management system (BMS), 492

Binders, 314–321

multifunctional polar, 314–317, 315–317*f*

PAA/PEDOT, PSS as functional, 319–321, 320–321*f*

polyamidoamine dendrimer based, 317–319, 318*f*

## C

Capacity loss, 421

Carbon nanotubes (CNT), 307–309

Carboxymethyl cellulose (CMC), 317–318

Cathode-electrolyte interphase (CEI), 251

Cathode/oxide-based electrolyte interface, 468, 469–472*t*

Cathode/solid polymer electrolytes interface, 474, 475–476*t*

Cathode/sulfide-based electrolyte interface, 468–474

Cell voltage, 10

Chemical vapor deposition (CVD), 309–310

CNT. *See* Carbon nanotubes (CNT)

Coin cells, hybrid solid electrolyte-coated battery separators, 373–376

Comet Astrobiology Exploration SAmple Return (CAESAR), 578–580

Complex energy storage challenges, solution, 298–300, 300*f*

Complex framework materials (CFM), 435

Constitutional underpotential plating (CUP), 263–264

Coulombic efficiency (CE), 332–333, 382

Cracking, 330–331

Crystallographic structures, 21–22

## D

Dendrite formation, 357–358

Density functional theory (DFT), 124, 310–311

Directly deposited sulfur architectures, 376–392

advanced materials and processing approaches, 377–380

pouch cell fabrication, 380–382

practical battery systems, 382–383

Directly derived sulfur assembled (DDSA) electrodes, 382

Directly doped sulfur architectures, 385–392

characterization, 386

electrochemical performance, 387–389

sulfur-infiltrated sulfur-copper-bipyridine, 392

synthesis, 385

X-ray photoelectron and electrochemical impedance, spectroscopy, 387–389

Dissolved polysulfide ions, nonaqueous electrolytes, 164–166

597

598   Index

Double Asteroid Redirection Test (DART), 568–569t, 575–577
Dry solid polymer electrolytes (dry-SPEs), 39

# E

Electric vehicles (EVs), 123–124
Electrocatalysts, 393–420
  bifunctional catalyst cathode material synthesis, 402–403
  bifunctional catalyst chemical, electrochemical characterization, 403–405
  computational results, 395–399
  hybrid active material (HBA) synthesis and characterization, 413–415
  inorganic framework materials, 416–420
    boron nitride-S materials, characterization, 416–417
    synthesis of zeolite (ZSM-5)-S materials, 417–420
  Li-ion-coated bifunctional, chemical, electrochemical characterization, 406–407
  theoretical methodology, 393–395
  titanium oxide-based catalyst material preparation, 400–402
  novel complex framework material processing, 407–411
    EC (electronic conductor)-CFMs, chemical, electrochemical, 410–411
    printing complex framework material-graphene architecture, 407–408
    synthesis of electrical conductor coated, complex framework materials, 409–410
    3D-printed materials, chemical, electrochemical, 408–409
  synthetic polymer binder, carbon framework materials, 411–413
  synthesis of lithium-ion conductor coated on bifunctional catalyst, 405
Electrochemical impedance spectroscopy (EIS), 390
Electrode, electrolyte modeling, interfaces lithium-sulfur batteries
  cathode porous electrode, 213–218

concentrated electrolyte transport, 218–220
Li-S electrochemistry, 201–202t, 226–227
polysulfide shuttle, 221–225
porous electrode performance, 206–213
variability sources, 225–226
Electrolytes, 205–206, 236
Energy density, 3, 338
Energy dispersion spectrometer (EDS), 386
Energy sources, rechargeable batteries, 298, 299–300f
Energy storage, (solar-) electric aircraft, 533–551
  all-electric battery-powered aircraft, 533–538, 538f
  batteries, 533, 534–535t, 536–537f
  high-altitude airship batteries, 539–542, 540t, 541–544f, 545t
  high-altitude platforms, 543–551, 545–546f, 549f, 550t, 551f
    energy storage technology options, 544–547, 546f, 547t
    hydrocarbon-powered platforms, 548, 548f
  space exploration, 552, 553f, 554t
Energy storage systems, 491–492
Engineering dendrite-free anodes, 421–434
  electrochemical cycling, Li-SIA (structurally isomorphous alloys) alloys, 427–430
  multicomponent alloys (MCAs)/dendrite-free anodes, 432–434
  theoretical strategies, diffusion barrier SIA, 427–430
Equivalent circuit network (ECN), 496
Ethyl methyl carbonate (EMC), 238
Extended Kalman filter (EKF), 504, 525–527

# F

Federal Aviation Administration (FAA), 536–538
First-principles approach, 125
Forgetting factor recursive least squares (FFRLS), 506
Functional catalyst (FC), 376

Functional modules, BMS, lithium-sulfur EV, 492–493, 492*f*
Future design strategies, Se$_x$S$_y$-based battery systems, 256–257

**G**

Galvanostatic intermittent titration (GITT), 247–248
Gelatin PEI composite (GPC), 315
Gibbs-Thomson parameter (GTP), 435
Gravimetric energy density, 328–330
Growth mechanism, dendrites, 262

**H**

High-altitude airships (HAAs), 539
High-altitude pseudo-satellites (HAPS), 543
High altitude Venus operational concept (HAVOC), 580–581
High-energy-density batteries, 253–254
High-energy X-ray diffraction (HEXRD), 237
Highly concentrated electrolytes (HCEs), 247
Hollow graphene spheres (HGs), 301–302
Host materials, 300–314
  carbon nanotube cathodes, 305–309, 305–308*f*
  hierarchical network macrostructure, 309–313, 309–311*f*, 313*f*
  in situ wrapping process, 313–314, 314*f*
  micro-mesoporous graphitic carbon spheres, 304, 304*f*
  reduced graphene oxide nanocomposite/ nitrogen-doped carbon framework, 302, 302*f*
  three-dimensional graphene hollow spheres, 301–302, 301*f*
  three-dimensional porous carbon composites, 303, 303*f*
HPLC-MS method, 159–164
  liquid chromatograph, 159–162
  mass spectroscopic, detectors, 162–164
  *vs.* other analytical techniques, 166–170, 168–169*t*
Hybrid electrolyte (HE), 267
Hydroxypropyl cellulose (HPC), 318–319

**I**

Inorganic solid electrolytes (ISEs), 20–38
  garnet type, 29–32, 30–31*f*
    argyrodite, 35–36, 36*f*
    glass, glass-ceramics, 36–38, 38*t*
    thiophosphates, 32–35, 35*f*
  sodium superionic conductor (NASICON) type electrolytes, 22–29, 23–24*f*, 26–27*f*, 29*f*
  sulfide type electrolytes, 32–38
Inorganic solid-state electrolytes, 469–472*t*
In situ resource utilization (ISRU), 581
Interfacial modification, 460–462

**L**

LABs. *See* Lithium-air batteries (LABs)
Layer-by-layer (LbL), 305–306
Li anode/oxide-based solid-state electrolytes, 464–466
Li anode/solid polymer electrolytes, 466–467
Li anode/sulfide-based solid-state electrolyte, 460–463
Light Italian CubeSat for Imaging Asteroids (LICIACube), 577–578
Li-ion conductivity, 444–445
Li-ion conductor (LIC), 358–359
LiOPAN (Li-ion conducting polymer), 359–360
Li-S BMS, state estimation methods, 497–525, 497*f*
  computer science techniques, 511–525
    ANFIS, discharge current pulses, 511–516, 512–515*f*, 513*t*, 517*f*, 518*t*
    ANFIS, UDDS (urban dynamometer driving schedule) cycle current profile, 516–522, 519–521*f*, 520*t*, 522*t*
    SVM classifier, Millbrook London transport bus (MLTB) test, 522–525, 523*f*, 524*t*, 526*t*
  direct methods, Li-S SoC estimation, 497–499
    coulomb counting method, 498
    open-circuit voltage method, 498–499, 499*f*

**600** Index

Li-S BMS, state estimation methods *(Continued)*
indirect methods, Li-S SoC estimation, 500–507
Li-S state estimation methods, control theory, 500–505, 501–503*f*
nonmodel Li-S state computer science-based estimation techniques, 505–507, 506*f*
Li-S battery testing, 507–508, 507*f*
recursive Bayesian filters, 508–511, 509*f*, 510–511*t*
Li-Se$_x$S$_y$ batteries, 237
Lithium-air batteries (LABs), 236
Lithium bis(trifluoromethanesulfonyl)imide (LiTFSI), 247–248
Lithium-conducting glass ceramic (LICGC), 40
Lithium dendrite growth, suppression, lithium-sulfur batteries (Li-S), 261–262
anode, 269–278
Li powder anode, 277–278
surface treatment, 272–277
3D anode, 269–272
dendritic growth, 263–266
crystallography, 264–265
kinetics, 264
thermodynamics, 263–264
electrolyte, 278–287
additive, 281–285
ionic liquid electrolyte, 278–281
novel electrolyte, 285–286
solid polymer electrolyte (SPE), 285–286
separator, 267–269
suppression method, 266–287
Lithium hexafluorophosphate (LiPF$_6$), 238
Lithium-ion batteries (LIBs), 5–6, 123–124, 235–236, 532
Lithium metal anodes (LMA), 382
Lithium nitrate (LiNO$_3$), 9–10
Lithium polysulfides (LiPs), 236, 311, 332–333
Lithium sulfide (Li$_2$S), 6, 383–384
Lithium-sulfur batteries (LSBs), 236, 399, 532

advantages, 327–331
lithium metal electrodes, 339–345
path forward, 345–346
electrochemical models, 493–495, 494–495*f*
equivalent circuit network models, 496–497, 496*f*
in situ optical spectroscopic, 100–104
in situ infrared spectroscopy, 102–103
in situ Raman spectroscopy, 103–104
in situ UV-Vis spectroscopy, 100–102
in situ X-ray techniques, 93–99
in situ powder X-ray diffraction, 93–97
in situ XAS, 98–99
in situ X-ray microscope, 97–98
Liquid electrolytes, 238–253
carbonate-based, 238–242, 239–240*f*
ether-based, 242–247, 243–244*f*, 246*f*
fluorinated, 251–253, 252*f*
highly concentrated, 247–251, 249–250*f*
Lowest unoccupied molecular orbital (LUMO), 247

# M

Mars Science Laboratory (MSL), 559
Metallic Li anodes, 356
Metal-organic frameworks (MOFs)
anode, 88–92
carbon materials, 72–73
composites, 62–68
carbon-based, 63–67
MOF/conductive polymer composites, 67–68
electrolytes, 85–88
metal compounds/carbon, 74–82
separator/interlayer, 82–84
sulfur hosts, 54–82
MOF composites, 62–68
MOF-derived materials, 68–82
pristine, 57–61
Millbrook London Transport Bus (MLTB), 507
MOF/carbon nanotubes composites, carbon- based composites, 65–67
MOF-derived metal/C composites, 74–75, 78–79

MOF-derived metal nitride/C composites, 79–82
MOF-derived metal oxide/C composites, 75–77
MOF-derived metal sulfide/C composites, 77–78
MOF/graphene composites, carbon- based composites, 63–65
MSL. *See* Mars Science Laboratory (MSL)
Multicomponent alloys (MCA), 435
Multiscale modeling, 124–125
  parasitic reactions in anode, 144–152
    mesoscale model, self discharge, 147–152
    metallic Li anode, passivation, 144–147
  precipitation growth, 125–144
    atomistic insights, growth process, 127–129
    exposed surface of solid $Li_2S$ film, 125–126
    formation/graphite interface, 129–134
    interfacial model, 134–144
Multiwalled carbon nanotubes (MWCNTs), 238, 305–306

## N

Nano-sulfur/multiwall carbon nanotube (MWCNT) composites, 36
Nernst glower, glowing rod, 19
New European Driving Cycle (NEDC), 505, 508
Next generation(s), battery technologies, 572–581, 573–574t
  aerial exploration, 578–581, 579–580f
  future space exploration, 572–575
  off-world utilization, 581, 582–583f
  upcoming missions, 575–578, 576–577f
Nitrogen-doped carbon framework (NCF), 302
Nucleation aid/silver nanoparticles, 343–344

## O

One-step derivatization, separation, determination, 171–174, 174t
Open-circuit voltage (OCV), 124, 497–498
Organic liquid electrolytes, 442–443

## P

Pair distribution function (PDF), 238–240
Particle filter (PF), 505, 525–527
Passivation metrics, 213–214
Polyamidoamine (PAMAM), 317–318
Poly(ethylene oxide) (PEO), 20
Polymer electrolytes (PEs), 20, 39–43, 338–339
  gel polymer electrolytes (GPEs), 42–43, 43f
  solid polymer electrolytes (SPEs), 39–41, 41–42f
Polypropylene (PP), 25
Polysulfide (PS), 8, 356–357, 382
Polyvinylidene fluoride (PVDF), 318–319
Porous cathode, 206
Position energy storage strategy (PES), 539–542
Pouch cells, 360–361
Pristine MOFs, 57–61
  metal-containing units, 59–60
  organic ligands, 61
  pore structure, 57–59

## R

Radiation issues, exploration missions, 565–572, 566f, 568–569t
  radiation in space, impact, rechargeable batteries, 567–570, 571f
  upcoming missions, Jupiter, 570–572
Recent NASA missions, 553–565, 555–556t
  Mars surface rovers, 559–563, 560–562f
  Mercury Surface, Space Environment, Geochemistry, Ranging mission, 557–559, 558f
  notable NASA exploration missions, comets, asteroids, 563–565, 564–565f
Reduced graphene oxide (rGO), 34

## S

Safety hazards, 357–358
Se doping, 241–242
Side reaction, 213
Solid electrolyte interphase (SEI), 241–242, 263–264, 341
Solid electrolytes, 19–20, 20f

**602** Index

Solid polymer (SPEs), 39
Solid-state approach, 441–443
  anode/solid-state electrolyte interface, 460–463
  cathode/solid-state electrolyte interface, 468
  ceramic hybrid composite electrolytes, 456–457
  oxide-based ceramics electrolytes, 448–450
  PEO-based solid polymer electrolytes, 444–445
  solid electrolytes, 443–456
  solid polymer electrolytes, 443–446
  single-ion-conducting SPEs, 445–446
  solid-state thin-film batteries, 474–477
  stable Li metal anodes for all-solid-state Li-S batteries, 458–467
  sulfide-based ceramics electrolytes, 450–456
  sulfur-based cathode composites, 467–474
Solid state electrolyte (SSE) separator, 267
Solid-state Li-S batteries (SS-LiSBs), 442–443
Spherical particles, 214
Sport utility vehicles (SUVs), 4
State-of-charge (SOC), 5–6
State-of-health (SoH), 496
Structurally isomorphous alloys (SIA), 435
Styrene butadiene rubber (SBR), 317–318
Sulfide-impermeable solid electrolytes, 23
Sulfur cathodes novel confinement architectures, 359–366
  assembly and testing of pouch cells, 366–373
    pouch cell fabrication process, overview, 366
    super P-containing pouch cell cycling capacity studies, 367–371
    YP-80F-containing pouch cell cycling capacity studies, 371–373
  chemical, electrochemical characterizations, 362–363

complex framework materials, 363–366
  LiOPAN-coated super P/YP-80F-sulfur composites, 365–366
  polyacrylonitrile coating on super P and YP-80F, 365
  polyacrylonitrile polymer processing, 364–365
Li-ion conductors, novel carbon framework, 361–362
Sulfur electrodes conventional, liquid electrolytes, 331–339
Sulfur redox reaction, 175–193
  investigation, 178–193
    chemical equilibrium, 191–193, 194$t$
    dissolved polysulfide, 188–191
    first reduction wave of sulfur, sulfur polysulfide, 184–187, 187$t$
  mechanism studies, 175–178
Superconducting magnetic energy storage (SMES), 544–545
Support vector machine (SVM), 525–527
Surface passivation, 125

**T**

Thermal energy storage (TES), 544–545
3D current collector, 343

**U**

Unscented Kalman filter (UKF), 504–505, 525–527

**V**

Volumetric energy density, 328–330

**X**

X-ray absorption near-edge structure (XANES), 238–240
X-ray diffraction (XRD), 264–265, 386
X-ray photoelectron spectroscopy (XPS), 389

**Z**

Zinc-air batteries (ZABs), 236

Printed in the United States
by Baker & Taylor Publisher Services